Advances in Comparative and Environmental Physiology 18

Advances in
Comparative and Environmental Physiology   18

Biomechanics of Feeding in Vertebrates

Guest Editors:
V.L. Bels · M. Chardon · P. Vandewalle

With Contributions by
P. Aerts · V.L. Bels · T.W. Beneke · H. Berkhoudt · M. Chardon
A.W. Crompton · F. De Vree · P. Dullemeijer · J.-P. Ewert
T.H. Frazzetta · C. Gans · S.W. Herring · A. Huyssene · K.V. Kardong
G.V. Lauder · S.M. Reilly · E. Schürg-Pfeiffer · W.W. Schwippert
D.N. Stern · J.C. Vanden Berge · P. Vandewalle · W. Verraes
A. Weerasuriya · W.A. Weijs · C.B. Wood · G.A. Zweers

With 94 Figures

Springer-Verlag Berlin Heidelberg GmbH

Guest Editors:

Prof. Vincent L. Bels

Institut Agricole
Centre Agronomique de Recherches
Appliquées du Hainaut
rue Paul Pastur, 11
B-7800 Ath, Belgique

Prof. Michel Chardon

Morphologie Fonctionnelle
Institut Ed. Van Beneden
Université de Liège
Quai Van Beneden, 22
B-4020 Liège, Belgique

Dr. Pierre Vandewalle

Morphologie Fonctionnelle
Institute de Zoologie
Université de Liège
Quai Van Beneden, 22
B-4020 Liège, Belgique

ISBN 978-3-642-63399-7   ISBN 978-3-642-57906-6 (eBook)
DOI 10.1007/978-3-642-57906-6

Typesetting: Macmillan India Ltd., Bangalore-25
31/3145/SPS – 5 4 3 2 1 0 – Printed on acid-free paper

Foreword to the Series

The aim of the series is to provide comprehensive, integrated reviews giving sound, critical and provocative summaries of our present knowledge in environmental and comparative physiology, from the molecular to the organismic level.

Living organisms have evolved a widespread range of basic solutions to cope with the different problems, both organismal and environmental, with which they are faced. A clear understanding of these solutions is of course of fundamental interest for all biologists, zoologically or medically oriented. They can be best comprehended in the framework of the environmental and/or comparative approaches. These approaches demand either wide surveys of animal forms or a knowledge of the specific adaptive features of the species considered. This diversity of requirements, both at the conceptual and technological level, together with the fact that physiology and biochemistry have long been mainly devoted to the service of medicine, can account for the fact that these approaches emerged only slowly amongst the other new, more rapidly growing disciplines of the biological sciences.

The field has now gained the international status it deserves and the organization of a series devoted to it appeared timely to me in view of its actual rapid development and of the interest it arouses for a growing number of biologists, physiologists, and biochemists, independently of their basic, major orientation.

Liège, Belgium, Spring 1988   *Raymond Gilles*

List of Editors

Preface

Studies on feeding in vertebrates have developed enormously over the past 20 years and have seen great advances in techniques and theories. We feel that this is an opportune time to produce an extensive review of the knowledge in biomechanics interconnected with functional morphology and sensori-motor control of this behaviour in aquatic and terrestrial environments.

Feeding is currently one of the functions that have interested biologists from many disciplines. There is a great deal of diversity in studies of feeding behaviour. We have attempted to present a book reflecting a broad series of exciting questions in fundamental biology. Each chapter helps to elucidate some of the problems and to provide a comprehensive discussion of general concepts regarding feeding mechanisms and designs of vertebrates. In addition, it offers extensive literature on all the subjects treated in each chapter. We hope that such a review of feeding biology will be useful for fundamental research and undergraduate or graduate students.

Many people and organizations have assisted directly and indirectly during the production of this book. We would like to thank all the contributors who accepted our invitation to participate in the writing of this book, in particular, Carl Gans and Piet Dullemeijer who provided the Introduction and Conclusion for this volume. Their contributions were of primary importance, emphasizing the continuity of the chapters. We are also very grateful to Raymond Gilles for his support as Series Editor.

Ath, Belgium   VINCENT L. BELS
Liège, Belgium   MICHEL CHARDON
Winter 1993   PIERRE VANDEWALLE

Contents

Chapter 4

Feeding in Tetrapods
F. De Vree and *C. Gans*

Chapter 5

Sensorimotor Processes That Underlie Feeding Behavior
in Tetrapods
J.-P. Ewert, T.W. Beneke, E. Schürg-Pfeiffer, W.W. Schwippert
and *A. Weerasuriya*

Chapter 6

Amphibian Feeding Behavior: Comparative Biomechanics
and Evolution
G.V. Lauder and *S.M. Reilly*

Chapter 7

Biomechanics of the Hyolingual System in Squamata
V.L. Bels, M. Chardon and *K.V. Kardong*

Chapter 8

Behavioral Mechanisms of Avian Feeding
G.A. Zweers, H. Berkhoudt and *J.C. Vanden Berge*

Chapter 9

Evolutionary Approach of Masticatory Motor Patterns
in Mammals
W.A. Weijs

Chapter 10

Differential Wear of Enamel: A Mechanism
for Maintaining Sharp Cutting Edges
A.W. Crompton, C.B. Wood and *D.N. Stern*

Conclusion: A General Theory for Feeding Mechanics?
P. Dullemeijer

Introduction

C. Gans

The biomechanics of feeding may be introduced by considering some aspects of the ecology and behavior of this process. Beyond this it seems appropriate to note the benefits of placing results in a comparative framework. The field has recently encountered many new techniques and it seems pertinent to raise the issue of some cost-benefit considerations applicable to instrumentation. Finally, this seems a good moment to review the matter of role phenotype matching, its implication, its achievement, and its possible merits.

Animals harvest diverse, patchy, and continuously changing resources, the nature of which commonly responds to the harvesting patterns. The occurrence of a great diversity of predator and prey species assures a level of unpredictability. This unpredictability determines that unusual prey is likely to be encountered by predators at any particular time, but also that unusual predators are likely to be encountered by any particular prey object.

Hence, prey must be prepared to avoid the attention of and, if discovered, to deter a spectrum of potential predators. Avoidance may imply cryptic behavior, selection of inaccessible habitats, speed, and endurance; deterence may be an aposematic warning of noxious characteristics or their use in active defenses. Recognition of the predator may be a useful strategy. Commonly, the magnitude of the defense reflects the success of past predation pressure and incorporates the capacity for variation of escape and avoidance responses.

In parallel, predators can afford only limited specialization of their feeding patterns. The systems for control and coordination of harvesting and the biomechanics of the prey acquisition and handling systems must retain some capacity for change. They can only be matched to a narrowly defined prey type if this represents a significant resource that is likely to be available over the long term, both through the life cycle of the individual and over the historical time seen by the species. Even then the matching is likely to incorporate enough uncertainty to incur advantage to behavioral and phenotypic variants which permit the predator to shift to alternate prey should this become common.

The study of feeding types and ingestion patterns, topics of this volume, consequently offers great opportunities for understanding evolutionary patterns. The enormous diversity of patterns disclosed includes numerous specializations, but also many variants of suboptimum structural utilization. One

Department of Biology, The University of Michigan, Ann Arbor, Michigan 48109-1048, USA learns that there are numerous circumstances in which the capacity to utilize a resource is much more important to the predator than is the effectiveness of the utilization. Historical factors, in part reflecting a recent shift of food type, commonly provide traces from the past. The unraveling of such adaptive patterns consequently involves more than the matching of mechanism to the most common or even the most extreme task.

The title of this volume refers to the biomechanics of feeding; however, most of the chapters actually address topics in the functional morphology of feeding, and rightly so. Feeding involves the development of hunger, the identification and positioning of the predator relative to the prey, and the acquisition of the entire or part of the prey object. Much of this initial phase involves the central and peripheral nervous systems and only indirectly the mechanical portions of the predator. However, sensory and motor coordination also serves to limit potential failure by controlling the distribution of mechanical stress. Critical here are the materials out of which the animals are formed, the physical properties of the structures and the forces and movements generating potentials of the motor apparatus.

Thereafter comes a series of decisions about the kinds of food to be acquired and the way this is to be accomplished. Whereas animals can acquire prey by unspecialized mechanisms, this second stage of substrate utilization is that at which major biomechanical specializations are commonly encountered. Decisions about prey selection transcend such categories as herbivory and carnivory. Here, one generally sees the major specializations, i.e., for nondiscriminating filter feeding in contrast to selection of particular prey, such as fruits or seeds; or predators opt for ingestion of whole prey in contrast to removing portions thereof, as in grazing and browsing. Various pickup devices, such as multiple projectile tongues, are among the most spectacular specializations.

The next stage of feeding involves reduction, and this again is commonly associated with major biomechanical specializations. Crushing, grinding, and repeated cutting are basic mechanical modifications. Each appears multiple times, in the buccal cavity and the posterior pharynx, as do the various gut grinding patterns that become associated with chemical reduction. Also important is the pattern of muscle utilization; the application of tetanic pulses during the crushing sequence is a version that has until recently been ignored.

Whereas the biomechanical rules remain common, the various kinds of acquisition and loss of specializations occur repeatedly. However, study of the details of each pattern makes it clear that the commonalities generally apply to subunits, whereas the overall acquisition and reduction patterns are formed of disparate variants. This makes it critical to establish a clear basis for the comparison of such systems, as an aid to understanding both adaptation and the evolutionary changes that have occurred in historical time.

It has become common to start functional and biomechanic comparisons by mapping their states on phylogenetic diagrams. Naturally, one attempts to obtain the most up-to-date classification, as errors in the phylogenetic framework will obviously induce faults in the decisions about homology and homoplasy. Also, one notes that characters must be considered from ancestral to recent species, rather than by starting, for instance, with some mammals and then "descending" from here.

By mapping on phylogenetic diagrams one can see how often particular phenotypic states are likely to have arisen and which of them represent synapomorphies, rather than convergences. Moreover, one can determine the number of occasions in which equivalent behaviors have allowed the development of parallel systems of food acquisition and handling that are based on non-homologous phenotypes. The analyses clearly test whether the unpredictability mentioned in the ecological discussion represents reality. Furthermore, they test whether this differs among groups, or rather whether it peaks during various ecological or biogeographical circumstances.

Mapping of functional or better role-associated aspects on cladograms demands separation of analogical and homologous states. Traditionally, this separation has occurred for morphological states and we have a prolonged series of tests for evaluating structural homology. Unfortunately, comparison of behaviors commonly proceeds much less carefully and without explicit standards. This applies to overall behavior and to its phases. For instance, approximately equivalent activation times of jaw opening muscles need to be evaluated against possible alternatives that would produce equivalent mechanical effects. Furthermore, one would like to know how the details of the activation sequences compare. Also, the neuronal input to motoneurons deserves analysis and should not be assumed as equivalent.

Phylogenetic mapping has other advantages also, a major one is that it facilitates a self-test for functional study. For any state, it shows the magnitude of the data base and with this the sufficiency of the evidence (or lack thereof). This permits the student to avoid variants of the "two-species fallacy" which assumes that the phenotypic or functional difference between any two (or more) species now being compared is due to the most obvious ecological difference between them. Each such comparison pair also incorporates other ecological differences as well as phylogenetic differences which are not obviously functionally associated. Only the simultaneous comparison of multiple pairs is likely to partition the multiple effects due to phylogeny and function. Whenever the phylogenetic diagrams disclose a single or a most limited number of comparison pairs, they represent an admission of inadequacy of the study.

The kind of biomechanics reported here represents one of the pure sciences, for which we inevitably encounter only limited research support, thus implying that cost-benefit analysis remains significant at least in order to determine where scarce funds should be invested. The utility of biomechanical approaches to sports medicine, kinesiology, and orthopedics, as well as to robotics, has provided a rapid expansion of new research approaches and many new tools. It is now possible to measure movements, forces, and muscular actions in a much more detailed and easily comparable fashion. These methods of measurement have generated a new industry and, whereas the new off-the-shelf approaches may be cheap relative to the homemade versions of past generations, they indeed proceed at a substantially increased cost, a cost customary to the applied fields of medicine and technology in which the benefits yielded are highly desired. Unfortunately, these methodological applications to organismic bio-mechanics do not always evaluate the cost of increasing the number and rate of measurements.

One should not take such arguments as representing an antitechnological bias. However, the availability of an instrument only provides justification for its use after it has been demonstrated that the information could not have been generated by other (and perhaps more cost-effective) approaches. Similarly, we should stop noting that our ancestors did not use tools not yet available in their day. The task before us remains the elucidation of the fascinating and complex ways in which organisms solve questions. The tools and techniques by which this is done are those demanded by the question and must be justified by the answers they provide.

Finally, it seems useful to remember why the magnitude of adaptation is of interest. The role is that functional aspect of the phenotype that enhances the survival or evolutionary fitness of the individuals. Aspects such as inherent variability and excessive construction, not even to speak of environmental variability, insure that the role–phenotype match will be imperfect. Indeed, there is an overconstruction, including what is commonly referred to as a factor of safety, thus each phenotypic characteristic is able to match more than the mean circumstances expected to be encountered. Naturally, the factor of safety is a statistical abstraction that is established for whatever conditions are assumed to be limiting at the time. Its magnitude will depend not only on the mean demands on the species or population, but also on the biotopes these currently occupy. Furthermore, the factor of safety is commonly defined as pertaining to one or another role; yet each phenotypic aspect will support multiple roles. Hence, environmental matching is not simple to analyze.

From the viewpoint of our field, analysis of adaptation always depends on a detailed characterization of suites of interactions, mechanical in the present case. These, rather than some global statement of physiological performance, com-monly assumed to represent an optimum and to reflect the best possible state for multiple interlocked processes, need to be understood. Individual resolution of the relative magnitude of many such processes is required in order to establish the possible reality of such optima. This can be a major task and the major benefit of our field. The papers that follow should document how far we have come in this direction and how far we still must go.

Chapter 1

Functional Properties of the Feeding Musculature

S.W. Herring

Contents

1 Introduction

1.1 Overview

The focus of this review is on the characteristics of the individual fibers that compose the feeding muscles. The literature on muscle biochemistry and physiology is extensive and has been reviewed many times. Among the most useful recent contributions are Hoyle's (1983) comparison of muscle cells across animal phyla, the summaries of vertebrate muscle structure and molecular diversity by Ogata (1988) and Pette and Staron (1990), respectively, and the discussion of slow fibers by Morgan and Proske (1984). With few exceptions (notably the excellent review of Rowlerson 1990), these compendia reflect the

Department of Orthodontics, SM-46, University of Washington, Seattle, Washington 98195, USA

Advances in Comparative and Environmental Physiology, Vol. 18
© Springer-Verlag Berlin Heidelberg 1994

primary literature in that they deal almost exclusively with muscles of the locomotor system. It is common to hear feeding apparatus researchers complain that the field of muscle physiology concentrates too much on locomotor muscles (especially cat hindlimb muscles), and that the unique feeding system does not receive adequate attention. Actually, these reviews make it clear that cat hindlimb muscles are not even representative of postcranial musculature in general. Skeletal muscles are extremely diverse, not only between groups of muscles but also among species. As will become clear below, a dichotomous view of locomotor vs feeding muscles is probably just as misleading as an extrapolation from cat hindlimbs to cranial muscles. Muscle properties vary and depend on many factors, ranging from metabolism to behavior; at this stage in our understanding it seems prudent to avoid categorization.

Even though feeding muscles seldom receive much emphasis in general reviews, they have been much studied, particularly with regard to their enzymatic makeup. However, it is surprising to note how randomly the information seems to be distributed among vertebrate groups. It is expected that there should be more interest in mammalian, especially human, than in, say, crocodilian feeding muscles. But other discrepancies are not expected. For example, even though chicken (*Gallus domesticus*) pectoralis and latissimus muscles have been used extensively for studies in muscle adaptation, almost nothing is known about the contractile properties of chicken jaw muscles. Other areas of interest, such as the contractile properties of whole feeding muscles, have received very little attention in any species, probably due to technical difficulties related to the anatomy of the muscles. This review necessarily reflects the available literature, and must give uneven coverage in many areas.

One area that is not covered herein is muscle architecture and its biomechanical consequences. There are two reasons for omitting this important subject, the first being space and the second being the existence of many previous treatments. The interested reader is referred to Loeb and Gans (1986) for an introduction and to Gans and De Vree (1987), Otten (1988), and Herring (1992) for features particularly relevant to jaw muscles.

1.2 Definitions

The first problem in dealing with functional properties of the feeding musculature is, of course, deciding what constitutes that musculature. In some organisms such as ram-feeding whales, every skeletal muscle in the body could be considered involved in feeding, a definition clearly too broad for the scope of this review. For the present purpose, the feeding muscles will be defined as those which (1) move the jaws or branchial skeleton, (2) manipulate oral soft tissues such as the tongue or (in mammals) the lips, and (3) are involved in bolus transport. This definition thus includes somitic muscles (tongue and hyoid) as well as branchiomeric (or somitomeric; Noden 1983) muscles innervated by the vagus, glossopharyngeal, facial, and trigeminal nerves. The bulk of published information pertains to the trigeminal system, particularly the jaw-closing muscles of mammals.

2 Functional Properties of Individual Muscle Fibers

2.1 Sarcomere Construction and the Length-Tension Relationship

It is often assumed that sarcomere lengths are the same in all vertebrate muscles. Although relatively uniform compared to the situation in invertebrates, myofilament lengths do vary in vertebrates. Thick filaments, while varying little in relaxed length, may shorten in contracted muscle (reviewed by Pollack 1983). Thin filament lengths, on the other hand, are species-specific and seem to increase with increasing body size (Walker and Schrodt 1974). The data compiled by Hoyle (1983, p. 270) suggest intraindividual variation as well, with slower muscles having longer actin filament lengths than faster muscles in all tetrapod groups. The same phenomenon occurs in cranial muscles; Akster (1981) found that in perch (*Perca fluviatilis*) m. levator operculi, "red" fiber types had longer actin filaments as well as thicker Z-lines than did "white" types. Similar results were reported by Gradwell and Walcott (1971) for differing fiber types in m. interhyoideus of bullfrog tadpoles (*Rana catesbeiana*). The functional significance of longer myofilaments is that such muscles should have slower twitches (Hoyle 1983) and a more even tension production at various sarcomere lengths (Akster 1981). Unfortunately, there are few measurements of filament lengths in feeding muscles, the only examples being Akster's work and that of Muhl et al. (1978) on the m. digastricus of rabbits (*Oryctolagus cuniculus*). However, both studies found lengths to be similar (within a species) in feeding muscles and in postcranial muscles.

Although myofilament lengths are little known in feeding muscles, sarcomere lengths have been studied in a variety of taxa. If a reasonable estimate of myofilament lengths can be made, data on sarcomere length when the muscle is stretched or shortened can be used to estimate the isometric length-tension curve. This method is based on the classic work of Gordon et al. (1966) and subsequent authors, showing that active tension production is maximal at optimal overlap of the filaments and falls off at both shorter and longer sarcomere lengths. Busbey (1989) has used these considerations to analyze the functional roles of the jaw adductor muscles of *Alligator mississippiensis*, as have several investigators working on mammalian adductors (rat, *Rattus norvegicus*: Nordstrom and Yemm 1972; Rayne and Crawford 1972; Nordstrom et al. 1974; rabbit: Hertzberg et al. 1980; Weijs and van der Wielen-Drent 1982, 1983; pig, *Sus scrofa*: Herring et al. 1984). In a few cases the general method has been used even in the absence of sarcomere length data, as exemplified by Thexton et al.'s (1977) suggestion that simultaneous activation of the tongue protractor and retractor in the salamander *Bolitoglossa occidentalis* can result in protraction followed by retraction because the muscles are at different points on their length-tension curves when activated. The general validity of this method has been established by comparison with actual length-tension measurements (Kiliaridis and Shyu 1988 for rats; Muhl et al. 1978 for rabbits; Anapol and Herring 1989 for pigs); however, Muhl and colleagues have pointed out that the measured optimal sarcomere length may not be identical to the length at which myofilaments are maximally overlapped.

The major interest in isometric length-tension relationships has not been the correlation with sarcomere construction, but rather the question of whether jaw-closing force is largest at jaw-opened positions (where the teeth might first encounter a prey item or a large bolus) or at occlusal positions (where, in mammals, shearing contacts occur between the teeth). Although earlier workers assumed that the jaw-closed positions would have most occlusal force, experimental work – unfortunately dealing only with mammals – has indicated otherwise. Measurements on jaw-closing muscles of opossum (*Didelphis virginiana*, Thexton and Hiiemae 1975), fruit bat (*Pteropus giganteus*, De Gueldre and De Vree 1991), rat (Nordstrom and Yemm 1974; Kiliaridis and Shyu 1988), cat (*Felis catus*, Mackenna and Türker 1978), pig (Anapol and Herring 1989), and even *Homo sapiens* (Manns et al. 1979; Fields et al. 1986) have shown clearly that maximum forces are produced when the jaws are opened, indeed, sometimes at openings approaching maximal. Muscles vary as to the gape at which maximal tension is produced. Passive tension also rises with gape because of the stretching of connective tissues, so the total tension can probably be considered to increase monotonically with gape. As might be expected, the digastric muscle produces most force when the jaws are closed (or superclosed) (Muhl et al. 1978; Mackenna and Türker 1978; Anapol and Herring 1989; De Gueldre and De Vree 1991; but not Thexton and Hiiemae 1975).

The dynamic aspects of sarcomeres are far more difficult to examine than isometric properties. Faulkner et al. (1982) found fiber bundles from mm. masseter and temporalis of rhesus monkeys (*Macaca mulatta*) to have speeds of 7.4 and 9 muscle lengths per second, respectively. Force velocity in whole muscles and in vivo changes in muscle length have been studied only by Muhl and his colleagues (Muhl and Newton 1982; Anapol et al. 1987), again using the digastric muscle of the rabbit. Maximum sarcomere shortening velocity was estimated at 26 μm/s, slightly slower than in fast hindlimb muscles in the cat but much faster than slow hindlimb muscles. In contrast, typical rates of shortening during mastication were estimated at 7 μm/s.

2.2 Metabolism

Muscle fibers metabolize glycogen using anaerobic pathways and lipids using aerobic pathways associated with mitochondria. The latter mechanism is associated with resistance to fatigue. The relative importance of these two energy sources can in theory be assessed at the light microscope level by staining for glycogen (periodic acid-Schiff reaction) or for lipids (Sudan black B). In practice, studies using these techniques (e.g., Taylor et al. 1973; Suzuki 1977) have found it difficult to make distinctions between fiber types, reflecting an underlying, continuous variation and lability in these features. Ultrastructural quantification of mitochondria and glycogen granules would probably be more helpful, but ultrastructure has been little studied in feeding muscles, although there are some exceptions (e.g., Akster 1981).

More commonly, metabolism has been examined by using histochemical reactions for various enzymes. Phosphorylase has been used to demonstrate the anaerobic glycolytic pathway (Masuda et al. 1974; Suzuki 1977), but far more popular has been the use of mitochondrial enzymes such as succinic dehydrogenase and NADH diaphorase. By labeling the mitochondria, these reactions provide a measure of oxidative capacity. Oxidative enzymes have been used alone (Katsura et al. 1981, 1982) or in combination with other enzymes to classify fibers. As a rough generalization, slow fibers (whether twitch or tonic, see Morgan and Proske 1984) are well supplied with mitochondria and may or may not have extensive glycogen stores as well. Fast twitch fibers may be mitochondria-poor, in which case they are easily fatigued, or mitochondria-rich, in which case they are fatigue-resistant. In mammals, jaw muscles are typically richer in mitochondria than are limb muscles, making fiber classification on this basis difficult if not dubious (Rowlerson 1990).

Few studies have addressed the question of fatigability directly. In the carp (*Cyprinus carpio*) m. hyohyoideus Granzier et al. (1983) found that force produced by mitochondria-poor "white" fibers declined by 50% after only 45 s, whereas mitochondria-rich "red" fibers reached that point after 10 min. In strips of cat jaw-closing muscles, Taylor et al. (1973) showed a 50% dropoff after about 40 s of tetanus, but mitochondrial content is known to vary in cat muscles (see Sect. 3.4.3) and was not investigated in the strips used for fatigue testing. Cat hyoid complex muscles (mm. genioglossus, geniohyoideus, sternohyoideus, and sternothyroideus) were comparable to the diaphragm in resisting fatigue (van Lunteren et al. 1990; van Lunteren and Manubay 1992). In human jaw muscles (where recording a stimulated tetanus is not feasible) fatigability is somewhat controversial. Van Boxtel et al. (1983) asked human subjects to contract specific muscles at 50% maximum electromyographic (EMG) activity levels (presumably not tetanus) until "unbearable" discomfort or pain. Using this criterion, the average time to fatigue was 1.7 min for m. masseter, 6 min for m. temporalis, about 10 min for two facial muscles associated with the mouth, and 26–34 min for three facial muscles associated with the eyes. These figures receive some support from Clark et al. (1988), who found that 50% maximal force levels could be maintained for about 2 min in masseter and temporalis muscles. In their study of single motor unit fatigue in the masseter, Nordstrom and Miles (1990) reported a broad range of variation that was generally comparable to fatigability of units in limb muscles. The implication that human feeding muscles are relatively fatigable is surprising, given their generally mitochondria-rich composition (see Sect. 3.4.3).

2.3 Contraction Speed

2.3.1 General Considerations

Although contraction speed is often deduced from the reaction properties of myosin adenosine triphosphatase (ATPase), it is important to remember that the splitting of ATP may not be the rate-limiting step. Other relevant factors include the length of myofilaments (Sect. 2.1), the quantity and organization of the T-tubules and sarcoplasmic reticulum (Akster 1981), and the kinetics of enzymes responsible for sequestering and transporting calcium (Pette and Staron 1990). All these factors are known to vary in vertebrate locomotor muscles, but have not been investigated in feeding muscles. In contrast, myosins and their ability to cleave ATP have been extensively studied.

2.3.2 Myofibrillar ATPase and Myosin Isoforms

Myofibillar ATPase is associated with the heavy chain component of myosin, and varying activity levels are presumably associated with the particular mixture of heavy chain isoforms in individual fibers (reviewed by Grove 1989). Based on early biochemical correlations of ATPase activity and contraction speed (Bárány 1967) and later demonstration of contractile properties in identified motor units (Burke et al. 1971), the classical treatment has been to classify fibers which react strongly under alkaline conditions as "fast" and those which react strongly under acid conditions as "slow" (Brooke and Kaiser 1974). Apart from the empirical evidence, one possible explanation for the correlation is the necessity for the frequently contracting slow fibers to maintain their activity even under lactic acid buildup. In addition to the difference in contraction speed, slow fibers are more energy-efficient and tend to be recruited first and used most often, while fast fibers are required to produce maximal movement (Rome et al. 1988). Although the ATPase classification has provided a useful basis for comparing muscles, numerous problems with functional interpretations have surfaced. First, the several histochemical techniques employed are not fully compatible (Pette and Staron 1990). Second, reaction differences between fibers give information only about relative contraction speeds; absolute values are not only unknown but clearly differ among muscles and species (Hoyle 1983). Third, as a greater variety of muscles has been sampled, it has become obvious that many fibers do not follow the expected patterns; many exhibit variable staining or react strongly regardless of pH. This third problem is the worst, because a fiber that reacts under all conditions gives few hints of its contractile properties. The problem arises from the recently discovered and hitherto unimagined diversity in myosin isoforms that has been revealed using immunohistochemical, biochemical, and molecular techniques. Not only are there variants of heavy chains (presumably responsible for the ATPase reactions), but also variants of light chains and a large number of possible combinations of four light chains and two heavy chains to form the hexameric myosin protein. Many studies are based on electrophoresis of native (whole) myosin or antibodies raised to native myosin, and hence the differences found may be due to chain variants or to unusual combinations of chains. Further, isoform hybrids are common. In the rabbit m. masseter, Mabuchi et al. (1984) found at least 30% of histochemically identified fast fibers to be hybrids. A definitive discussion of the correlation between myosin isoforms and conventional histochemical fiber typing is therefore not yet possible, and may never be so, as attention turns away from the indirect evidence of histochemistry and toward the more specific assessments of biochemistry.

The functional significance of myosin diversity is not clear, since the proliferation of molecular information has not been matched by an advance in understanding the functional attributes of myosin. Nevertheless, the subject is of considerable interest to feeding researchers, because the myosin composition of the branchial musculature is particularly unusual. Many of the early reports of anomalous ATPase reactions dealt with mammalian jaw-elevator muscles (Taylor et al. 1973, for cat; Schiaffino 1974, for rabbit; Maxwell et al. 1980, for rhesus monkeys), leading to a new focus on the jaw musculature. In general, mammalian jaw muscles have a large number of "transitional" (usually type IIC) fibers which express both fast and slow myosins and hence react for ATPase at almost all pH levels routinely used (Rowlerson 1990). In addition to mixed myosins, there are special myosins. Rowlerson et al. (1981) reported a novel myosin from cat jaw-elevating muscles; both heavy and light chains were found to differ from limb muscle myosins. In a later study that surveyed the trigeminal muscles of several carnivorans and several primates using myosin antisera and conventional histochemistry, Rowlerson et al. (1983) found the unusual fibers (designated type IIM) in the jaw-closing (but not the jaw-opening) muscles of all carnivorans sampled save the lesser panda (*Ailurus fulgens*) and all primates sampled save man. Mm. tensor veli palatini and tensor tympani usually had IIM fibers as well. IIM fibers react strongly to ATPase regardless of pH (Rowlerson 1990).

Subsequent work on cat muscles had indicated that the unique myosin is specific to the myogenic lineage (Hoh and Hughes 1988, 1989) and appears relatively late in ontogeny (Hoh et al. 1988), although it is not expressed in culture (although the collagen substrate did pick up the antibody, raising some doubts about specificity; Hill et al. 1989). On the basis of the distribution of the IIM fibers and because cat jaw muscles are known to contract rapidly (Taylor et al. 1973; but see Sect. 3.4.1), Rowlerson et al. (1983) suggested that the novel myosin was associated with rapid and powerful biting and termed it "superfast," a term also used for a heavy chain in extraocular muscles (reviewed by Pette and Staron 1990). Direct experimental proof that masticatory myosins are actually superfast is scanty and relies primarily on studies by Hoh and Hughes (1988, 1991), correlating isometric contraction times of regenerating cat posterior temporalis with the replacement of fetal and slow myosins by the superfast isoform. Temporalis regenerates stabilized at contraction times (presumably time-to-peak tension, TPT) of 11.5 ms in the 1988 study and 16 ms in the 1991 study, compared to 21 ms for extensor digitorum longus regenerates (1988). Thus, the magnitude of the difference between superfast and fast isoforms is not great; furthermore, statistical information on the significance of the difference was not presented. Therefore, it is probably premature to use the term "superfast" to describe this myosin.

An additional problem with the functional interpretation of these isoforms as fast is that cat temporalis may not be representative of other species. For example, the heavy chain of opossum masticatory muscles does not comigrate with that of cat temporalis (Sciote 1991), and even the more closely related dog (*Canis familiaris*) is thought to have a different temporalis IIM variant (Shelton et al. 1988). Variants of myosin isoforms are found in other mammalian taxa as well. d'Albis et al. (1986) compared the myosins of m. masseter in three rodents, mouse (*Mus musculus*), rat, and guinea pig (*Cavia porcellus*). All had different components, even the closely related mouse and rat. Rat and mouse jaw muscles also contain many IIX fibers, which have a fast heavy chain also found in oxidative postcranial muscles (Pette and Staron 1990; Rowlerson 1990).

In addition to hybrid fibers, hybrid myosins, and unique isoforms, jaw muscles contain myosins found in developing muscles and in cardiac (atrial) muscle. Mouse masseter muscle was distinguished by a high proportion of neonatal myosins even in adults (d'Albis et al. 1986). Similarly, Butler-Browne et al. (1988) found developmental isoforms of myosin in adult human m. masseter, and Soussi-Yanicostas et al. (1990a) have reported that during postnatal development in man, embryonic and fetal heavy chains and an embryonic light chain are expressed much longer in m. masseter than in m. quadriceps femoris. However, other species, including pig, macaque (species?) and "pygmy lemur" (*Cheirogaleus*?), did not express the embryonic or fetal isoforms (Soussi-Yanicostas et al. 1990b). However, it is not clear whether embryonic isoforms from these species would cross-react with the human-derived probes. In a more recent study based on antibody developed against bovine (*Bos taurus*) myosin, neonatal myosin was found in the m. tensor tympani of juvenile and adult pigs and cattle (Scapolo et al. 1991). Interestingly, the embryonic light chain is also found in adult atria (reviewed by Collins 1991), and Bredman et al. (1990, 1991) have identified atrial heavy chain, which is intermediate in contraction speed, in all rabbit cranial muscles, regardless of function or development origin. Bredman et al. (1990) have suggested a relationship between the distribution of atrial heavy chains and that of connective tissues derived from neural crest.

Outside of jaw muscles, there is little information on myosin isoforms in the mammalian feeding apparatus. The additional rabbit cranial muscles in which Bredman et al. (1990) identified atrial heavy chains were the mm. buccinator and stylohyoideus (but not m. platysma), m. cricothyroideus (but not the pharynx or the esophagus), tongue and hyoid muscles (but not mm. trapezius or sternocleidomastoideus). A characteristic myosin heavy chain is found in the esophagus of cat and dog, but not other species (Mascarello et al. 1984). Outside of mammals, the only information on feeding muscle myosins is the report by Alley and Reiser (1991) showing that isoforms in larval frog jaw adductor are distinct from those of the adult muscle and those from limb muscles; further, adult fibers contracted 50% faster than larval fibers.

Although biochemical verification is lacking, ATPase results suggest that unusual myosins are present in other mammalian muscles and in the jaw muscles of other vertebrates. Peculiar histochemical staining, possibly due to the neonatal myosin mentioned above, has been reported in m. tensor tympani (a former feeding muscle!) of several mammals, whereas m. stapedius is not unusual (Veggetti et al. 1982). Akster (1983) noted anomalous histochemical staining in carp cranial muscles, although she was able to classify fibers using immuno-histochemistry. Similarly, Condon (1987) found a variety of fiber types in lizard (*Varanus niloticus*) jaw adductors, including several in which altering pH did not affect ATPase activity.

3 Fiber Types and Contractile Properties in Vertebrate Feeding Muscles

Because of the preliminary and confused status of the literature on myosin isoforms, functional insights must rely for now on conventional histochemical fiber typing. Fortunately, several important assumptions of the method remain valid. With regard to metabolism, a wide range of mitochondrial enzyme activity is found in feeding muscles, and that activity is generally correlated with usage patterns. With regard to myofibrillar ATPase, fibers that stain strongly under alkaline conditions and do not stain under acidic conditions can probably be considered faster than other fibers in the same muscle with the reverse response. Fibers which do not meet these conditions should not be characterized as to relative speed, as they may be either slower (atrial myosin) or faster ("superfast" myosin) than typical fast fibers.

3.1 Multiply Innervated Fibers

In vertebrates other than mammals, skeletal muscles (including feeding muscles) contain fibers with multiple motor end plates. In mammals, multiply innervated fibers are common in extraocular muscles and muscles of the middle ear but are rarely if ever found in mature feeding muscles. Until recently, such fibers were referred to as "tonic," as opposed to focally innervated "twitch" fibers. They were thought not to conduct action potentials and not to relax in depolarizing solutions. This picture is certainly true for some fibers, but not all. In birds and lizards, multiply innervated fibers can conduct action potentials and hence twitch (Gleeson and Johnston 1987; earlier literature reviewed by Morgan and Proske 1984). In many teleosts, fibers of all types appear to be multiply innervated (reviewed by Bone 1964). Indeed, some polyneuronally innervated fibers in sculpin (*Myoxocephalus scorpius*) myotome showed similar contractile properties to focally innervated fibers in the eel *Anguilla anguilla* (Altringham and Johnston 1988). Bewick et al. (1991) have pointed out that polyneuronal innervation has advantages in muscles that grow continuously by adding muscle fibers, a situation common in all vertebrate groups except mammals. Therefore, a priori classification of multiply innervated muscles as tonic (or even slow) is not justified.

Within a given muscle, however, multiply innervated fibers are usually slower than focally innervated fibers (e.g., dogfish, *Scyliorhinus canicula* myotome; Bone et al. 1986), and it seems reasonable to assume that such would also be the case in feeding muscles. The correlation of ATPase activity with relative contraction speed appears to be valid for multiply innervated fibers. True tonic fibers have low ATPase at any pH and often low mitochondrial content as well. Other multiply innervated fibers have variable histochemistry (reviewed by Morgan and Proske 1984).

Few studies have specifically investigated whether multiple innervation occurs in feeding muscles. In some teleosts, their occurrence is universal in jaw as well as trunk muscles. Flitney and Johnston (1979) examined two different fiber types, both multiply innervated, from m. adductor operculi in the cichlid *Tilapia mossambica*. One type was found to be tonic; the other resembled typical twitch fibers except that it produced graded, fused tetani. In carp Akster (1983) found virtually all muscles to be multiply innervated, although m. hyohyoideus was more so than the axial musculature. Within m. hyohyoideus, all fibers tested produced twitches, but the slower fibers had a greater number of nerve terminals per unit fiber length (Akster 1983; Granzier et al. 1983). Thus, in these fish multiple innervation is a quantitative property correlated with degree of "tonicity," not a qualitative differentiating factor.

The same considerations may hold for amphibians. Although frog limb muscles provided the original example of tonic fibers, a study of m. intermandibularis posterior ("submaxillaris") in *Rana temporaria*, *Xenopus* sp., and "newt" showed the fibers to have twitch properties despite being multiply innervated (Miledi and Uchitel 1984). The reconstruction of jaw muscles at metamorphosis also affects their innervation patterns. In *Rana pipiens* the larval adductor muscles are diffusely supplied by multiple end plates, each multiply innervated, whereas the adult muscle fibers have fewer (but still multiple) end plates, each typically served by just one axon (Alley et al. 1992).

In the lizard *Varanus niloticus* Condon (1987) found that only one jaw muscle, the deep part of m. adductor mandibulae externus, was composed exclusively of focally innervated fibers. The other adductors were mixed, often showing a gradient or localized distribution of multiply innervated fibers. One, m. adductor mandibulae posterior, had a majority of multiply innervated fibers.

3.2 Fishes, Amphibians, Reptiles, and Birds

ATPase and metabolic enzyme activity have been assessed in the feeding muscles of a few teleosts, a few reptiles, and many mammals. A smaller number of reports deal with the direct measurement of contraction speed and fatigability. A search of the literature reveals no information on other groups of fish and single reports for amphibians and birds.

Fish adductor muscles function for respiration as well as for feeding, with the former activity presumably requiring less vigorous but more frequent contractions. Akster and Osse (1978) compared perch muscles which are active only during feeding (e.g., m. sternohyoideus) with muscles active during respiration as well as feeding (e.g., m. adductor operculi). The first group had only fast, low-mitochondria fibers, while the second group had high-mitochondria fibers of variable ATPase activity. Barends (1979) and Barends et al. (1983) compared various parts of the m. adductor mandibulae in the perch and in the rosy barb, *Barbus conchonius*. In both taxa the adductor had two parts, one with small, low-ATPase, oxidative fibers active during respiration and the other with larger, less oxidative fibers active only during feeding. Barends et al. (1983) also reviewed earlier work on teleost jaw adductors, mostly showing a similar division between aerobic and anaerobic portions. It should be noted that all of the above authors distinguished multiple fiber types, usually four to five, making the reduction to two groups an oversimplification. Nevertheless, the emergent pattern seems to be one of anaerobic fast fibers being primarily used for sporadic feeding movements and aerobic fibers of unknown but variable speed used additionally for respiration.

A direct assessment of contractile properties has been performed on the two parts of m. hyohyoideus in the carp, as mentioned above (Granzier et al. 1983). The more aerobic (and more multiply innervated) portion had a time-to-peak tension of 59 ms, as opposed to 32 ms in the anaerobic portion. The aerobic portion also had lower frequencies for tetanic fusion. In general, the aerobic fibers were similar to slow twitch fibers in cat limb muscles, while the anaerobic fibers were similar to fast twitch fatigable fibers. No fibers were strictly tonic, even though all were multiply innervated. Because of the enormous diversity in fish feeding mechanisms (including the ultrafast prey engulfment in antennariid anglerfish; Grobecker and Pietsch 1979), it seems prudent not to generalize about the feeding musculature from such a small sampling of cases.

Analogously, m. interhyoideus of bullfrog tadpoles demonstrated two portions of differing structure and function (Gradwell and Walcott 1971). The "pink" portion was rich in mitochondria, fatigue-resistant, showed fused contractions at low stimulation frequencies, and was rhythmically active during continuous buccal pumping. The "white" portion was low in mitochondria, fatigable, could not be tetanized, and active only during sporadic expulsion of detritus from the mouth.

The histochemistry of *Alligator mississippiensis* jaw muscles indicates a variety of different types, including tonic fibers, as well as differences among muscles in fiber type composition (Sato et al. 1992). These authors also noted a general correspondence between fiber type composition and the functional roles of the various muscles.

In lizards interest in jaw muscle histochemistry has focused on explaining irregular patterns of electromyographic (EMG) activity (e.g., Smith 1982). Fiber type analyses have been performed by Throckmorton and Saubert (1982) on the teiid *Tupinambis nigropunctatus* and by Condon (1987) on the varanid *Varanus niloticus*. The former study identified three fiber types using alkaline ATPase and metabolic enzymes only; the latter used acid ATPase as well and identified six types. Thus, not only are the fiber types not directly comparable to mammalian limb muscle types, they are not directly comparable in the two studies. Nevertheless, there are areas of agreement. Both groups of authors identified one type as a probable tonic fiber and found regionalization of fiber types within some of the muscles. In *Tupinambis*, the only muscles that were completely mixed were mm. depressor mandibulae and levator anguli oris. The adductor muscles were either purely fast fatigable (as deduced from histochemistry) or compartmentalized, with the fast fatigable fibers usually peripheral to a more deeply placed region of tonic and fast fatigue-resistant fibers (Throckmorton and Saubert 1982). In *Varanus* all fiber types had moderate levels of mitochondrial enzyme activity, so no metabolic distinctions were possible. The predominant type showed strong ATPase activity at acidic pH and moderate activity at alkaline pH, similar to the unusual staining properties of many mammalian feeding muscle fibers (Condon 1987). The conclusion that this is a "superfast" myosin is tempting but clearly unwarranted. Some, but not all of the multiply innervated fibers in Condon's study had tonic staining characteristics. The diversity in lizard feeding muscles was thought by the authors of these studies to be related to the wide range of feeding behaviors. In particular, the irregular EMG recordings could be caused by the presence of tonic fibers, which may not conduct potentials (in which case no signal would be detectable) and which may have a very long relaxation time (in which case force production would outlast the signal). If these speculations are correct, then multiply innervated fibers behave differently in lizards and in fish, where all those tested have produced twitches.

The single investigation on avian feeding muscle histochemistry is a developmental study on the chicken by Dubale and Muralidharan (1970). Only fiber size and metabolic features were assessed. In animals at least 10 days post-hatching, the fibers could be classified as either glycolytic or oxidative. The oxidative fibers were smaller. Although there was individual and possibly ontogenetic variability, four adductor muscles and m. depressor mandibulae all averaged about 50% of each fiber type.

3.3 Mammals: Hyoid, Tongue, Facial, and Pharyngeal Muscles

Most of the studies on the mammalian feeding apparatus involve jaw muscles, but there have been a few exceptions. The infrahyoid muscles of the rat were examined by Müntener et al. (1980) using alkaline ATPase reactions only. No type I (presumably slow twitch) fibers were found in m. omohyoideus and very few in m. sternohyoideus; about 15% in m. sternothyroideus was type I. A similar finding was reported by Rokx et al. (1984), who looked at mitochondrial enzymes as well and found both (presumed) fatigable and fatigue-resistant fibers in the muscles; they also analyzed m. geniohyoideus and found that it resembled the majority of infrahyoid muscles in having only sparse slow fibers. Müntener and colleagues attributed the presence of slow fibers in m. sternothyroideus to the fact that it (but not the other muscles) is used during inhalation as well as feeding (but see Roberts et al. 1984, who show that m. sternohyoideus is also active during breathing in rabbits, although less effective than m. sternothyroideus in opening the airway). In cats mm. geniohyoideus and sternohyoideus were very similar, both having about 50% oxidative fibers, including 15% slow twitch (Dick and van Lunteren 1990).

An ultrastructural examination of tongue muscle in the bat *Myotis grisescens* revealed types resembling "twitch" and "tonus" fibers, but many intermediate forms were present (Reger and Holbrook 1974). Histochemistry of cat tongue musculature resulted in the identification of four fiber types, one a slow-twitch oxidative type and three fast-twitch types of differing oxidative capacity (Hellstrand 1980). The fast-twitch fibers were larger than the slow-twitch fibers, and within the fast-twitch category, the least oxidative fibers were largest. In the extrinsic muscles, the four types were roughly equally represented. In the intrinsic muscles, slow-twitch fibers were seldom seen; of the fast-twitch types, the more oxidative were most common. An interesting aspect of this study is that the author considered the longitudinal intrinsic muscles to be continuations of the extrinsic m. hyoglossus (superiorly) and m. styloglossus (inferiorly), yet the internal and external parts of these muscles were histochemically quite different.

Facial muscle histochemistry has been examined in man (Schwarting et al. 1982), various rodents (Wineski et al. 1991), dog (Braund et al. 1991), and cat (Edström and Lindquist 1973). Atypically staining fibers have been reported for guinea pig and rat (Wineski et al. 1991; pers. comm.) and in small numbers for man (Schwarting et al. 1982). Of the conventional types, slow fibers compose only a small proportion (0–16%) of rodent vibrissal muscles and cat mm. orbicularis oris and orbicularis oculi, but approximately one-third of human facial muscles and dog m. buccinator. This differential may be an effect of body size. With regard to metabolic activity, Wineski and colleagues found greater oxidative enzyme activity in the vibrissal muscles of the golden hamster, *Mesocricetus auratus*, than in homologous muscles of the guinea pig; the difference was attributed to the nonmotile behavior of the vibrissae in guinea pigs. In addition to species differences, there are differences among facial muscles. In man, mitochondria-poor fibers comprise 44% of m. platysma but are negligible in the muscles around the mouth (Schwarting et al. 1982), presumably reflecting the use of the oral muscles in speech and feeding (the latter usage is documented by Blanton et al. 1970; Schieppati et al. 1989) as well as in facial expression. [However, the data on fatigability of these muscles (Sect. 2.2) are in conflict with this interpretation.] Mitochondria-rich fibers also predominate in the m. orbicularis oris of cat, and contrast this muscle with the more fatigable (but much faster contracting, despite a similar proportion of fast-staining fibers) m. orbicularis oculi (Edström and Lindquist 1973). The latter muscle, which does not function in feeding but presumably is important in the blink reflex, has a contraction time of 8.5 ms, similar to the very fast contracting extraocular muscles, in contrast to 33 ms recorded for the m. orbicularis oris, a typical value for a mixed limb muscle (Lindquist 1973).

Histochemical fiber typing of the esophagus has been summarized by Mascarello et al. (1984). Results vary from pure fast fibers in rat, guinea pig, and rabbit, to pure slow in the marmoset *Callithrix* (although Mascarello and colleagues suggest that these fibers were actually the same unusual myosin they found in esophageal muscles of dog and cat). Most species, including various primates and ungulates, have a mixture of fast and slow fibers. The two ruminants studied, sheep (*Ovis aries*) and cow (*Bos taurus*) had an increasing proportion of slow fibers caudally, a feature possibly related to sphincteric action. There are several reports on contractile properties of esophageal striated muscle. Floyd and Morrison (1975) studied strips of striated esophageal muscle from cat and sheep. The two species were quite similar. Both showed a time-to-peak contraction time of about 80 ms and a tetanic fusion frequency of about 30 Hz, typical slow-twitch values. An investigation of the m. levator veli palatini in dogs (Kogo et al. 1991) indicates that it is faster than the esophagus, as might be expected from the rapid closure of the palatopharyngeal sphincter during swallowing. Contraction times averaged 43 ms, an intermediate value corresponding to a fiber type composition reported as 39% "red" and 61% "intermediate" (Nakaji and Oota 1978, cited in Kogo et al. 1991).

3.4 Mammals: Jaw Muscles

3.4.1 Caveats

Histochemical fiber typing has been carried out on rodents (rats, mice, and guinea pigs), rabbits, a bat, a variety of carnivores and primates (including man) and several domestic ungulates (sheep, pig, and cow). Information on contractile properties is available for many of these species as well. However, the reader should be cautioned about the state of the art.

As histochemistry presents at best indirect evidence on functional properties, and at worst is uninterpretable in terms of function, contractile properties would seem to be the most reliable guide and will be stressed. Nevertheless, there are problems with measuring contractile properties. Aside from the usual caveats about isometric muscle physiology [for example, sensitivity of the techniques to fatigue (Dobose et al. 1987) and possible irrelevance to normal kinematics (Marsh 1988)], there are difficulties with jaw muscles. Most jaw muscles have spread attachments which resist connection to force transducers. Thus, many studies have used strips of muscle rather than whole muscles or have substituted occlusal force for muscle force. Further, the trigeminal innervation is inaccessible, so that most studies have stimulated the muscles directly rather than using the motor nerve. Except in a few studies which have studied limb muscles using the same techniques as jaw muscles (notably Masuda et al. 1974), these differences in technique make comparison questionable.

These technical variations may be responsible for a particularly alarming disagreement among research groups that have examined cat jaw muscles. Three groups have measured time-to-peak tension (TPT) in cat masseter, and four in cat temporalis. The figures given for masseter range from 13 ms (Taylor et al. 1973) to 30 ms (Mackenna and Türker 1978); an intermediate value of 18 ms was reported by Tamari et al. (1973). For temporalis, 11.5 ms was found by Taylor and colleagues (and Hoh et al. 1988), rising to 16 ms (Hoh and Hughes with the same techniques in 1991), 28 ms (Mackenna and Türker 1978), and 34 ms (Tamari et al. 1973 using the same procedure that measured masseter at 18 ms). This inconsistency is unexplained, and is worrisome not only because of the technical problems suggested, but also because of the doubt created in attempts to correlate ATPase or myosin isoforms with contraction speed (see Sect. 2.3.2 above).

3.4.2 Digastric Muscle

The histochemical composition of m. digastricus (and the related m. mylohyoideus, where examined) is similar to that of the infrahyoid muscles; that is, unusual myosins are not found and hybrid fibers (such as IICs) are rare. Slow twitch fibers are found in frequencies ranging from 0% in mouse (Scapolo et al. 1981, cited by Rowlerson 1990) to 60% in *Macaca fascicularis* (Clark and Luschei 1981). Unlike the jaw-closing musculature, mitochondria-poor fibers are also common; they have been reported in *Pteropus* (De Gueldre and De Vree 1991), rat (Rokx et al. 1984), guinea pig (Masuda et al. 1974), cat and "panther" (Rowlerson et al. 1983), pig (Anapol and Herring in prep.), and man (Eriksson et al. 1982).

The twitch contractile properties of the digastric muscle have been studied in opossum (Thexton and Hiiemae 1975), rat (Matthews and Smith 1972), guinea pig (Masuda et al. 1974), rabbit (Anapol et al. 1987), cat (Mackenna and Türker 1978), and pig (Anapol and Herring unpubl. results). The results fall into two categories. In opossum and rat, TPT was 10–11 ms; in both cases this was slightly faster than the jaw-closing muscles (16–18 ms in opossum, 14–18 ms in rat; Nordstrom and Yemm 1974; Hiraiwa 1977; Easton and Carlson 1990). Both of these species have masseters that are relatively uniform and intermediate in fiber composition. Because the digastric muscle has both faster (IIB) and slower (I) fibers, its faster contraction suggests that TPT is dominated by the fastest fibers in the muscle. The other category consists of guinea pig, rabbit, and cat, the digastric muscles of which all had TPTs of about 31 ms, and pig, in which an average of 47 ms was recorded. These values were similar to those measured for jaw-closing muscles in the same studies [24–33 ms for various guinea pig adductors, 26 ms for rabbit temporalis (Guelinckx et al. 1986), 27–30 ms for cat adductors, and 56 ms for pig masseter].

3.4.3 Masseter, Temporal, and Pterygoid Muscles

As mentioned earlier, mitochondria-poor fibers are uncommon in mammalian jaw-closing muscles. Most often, the (presumed) fatigable fibers have been found in m. temporalis but not mm. masseter or pterygoideus medialis (rat, Rokx et al. 1984; guinea pig, Masuda et al. 1974; rabbit, Schiaffino 1974 and Guelinckx et al. 1986). In man (Eriksson and Thornell 1983) and in *Macaca* (Clark and Luschei 1981; Maxwell et al. 1981; Miller and Farias 1988), the masseter also has fatigable fibers, but they are less frequent than in the temporalis. In cat, however, large numbers of fatigable fibers are found in all jaw-closing muscles (Taylor et al. 1973; Gorniak 1986), and this is also true for *Pteropus*, except for the lateral pterygoid muscle (De Gueldre and De Vree 1991).

It is generally accepted that oxidative activity is correlated with how often a muscle is used. Gorniak's (1986) EMG data from cats supported this premise, although Miller and Farias' (1988) study on macaques did not. Nevertheless, there is considerable circumstantial evidence in favor of the premise.

One of the most interesting aspects of the correlation between oxidative status of muscle fibers and usage is whether the correlation is established "genetically" or whether it is a phenotypic response to behavior. While the former factor cannot be eliminated, there is much better documentation of the latter. Studies on normal ontogeny of jaw muscles include those of Katsura et al. (1981, 1982), demonstrating a postnatal increase in oxidative capacity of rat masseter and medial pterygoid muscles, and Nakata (1981), showing an increasing percentage of oxidative fibers with age in mouse masseter and temporalis, but a decreasing percentage in digastric muscle. These changes correspond to the increasing importance of chewing and decreasing importance of suckling with age. In addition, a number of investigations have altered oxidative capacity in jaw muscles by experimentally interfering with function. Maeda et al. (1981) were able to retard the change to a more oxidative metabolism in rat masseter by postponing weaning, and Kiliaridis et al. (1988) were able to decrease the percentage of oxidative fibers in the same muscle (but not the digastric) by feeding the rats a soft diet. Maxwell et al. (1981) found reduced oxidative capacity in masseter and temporalis (but not digastric) muscles of rhesus macaques after procedures which altered muscle length. In the other direction, Guelinckx et al. (1986) increased the fatigue resistance of rabbit temporalis by removing the masseter. Because of the plastic nature of oxidative activity, it is necessary to be cautious about generalizing from domestic and laboratory animals.

The correlation between oxidative capacity and usage can also be established on a more comparative level. First, the only animal with a majority of nonoxidative fibers in all jaw muscles is a specialized carnivore (cat). A carnivorous diet is usually thought of as requiring relatively few masticatory movements, and especially considering that the animals examined were lab-bred and fed a soft diet, there were probably minimal demands on the jaw muscles. *Pteropus* runs a close second to the cat; these bats subsist on soft fruit pulp and juices (De Gueldre and De Vree 1991). Second, Rowlerson (1990) has noted the preponderance of nonoxidative fibers in jaw muscles which are sexually dimorphic (usually temporalis). Her interpretation is that the dimorphism is related to the need of the male (larger and less oxidative) muscle for a rapid and powerful bite, but a simpler interpretation is that the "extra" fibers of dimorphic muscles, as they are not necessary for mastication, are only seldom used and hence do not develop much oxidative capacity. Third, many studies have shown variation within jaw muscles in oxidative activity. Two trends are evident: oxidative capacity is greater anteriorly than posteriorly, and in deeper than in more superficial parts of muscles. Where there are myosin differences, these trends are correlated with the percentage of slow twitch (type I) fibers (Katsura et al. 1981, 1982 for rat; Schiaffino 1974 for guinea pig; Herring et al. 1979 and Anapol 1985 for pig; Rowlerson et al. 1988 for rabbit and macaque; Eriksson et al. 1981 and Eriksson and Thornell 1983 for man). The anterior and/or deep fibers of these muscles tend to be the most vertically oriented, and hence are typically the first recruited for jaw closing (e.g., Herring et al. 1979). Further, the vertical fibers, which provide occlusal force but little tendency to move the jaw, can be used during a greater variety of movements than more inclined fibers (Gorniak 1985).

In summary, it seems reasonable to interpret the high oxidative level of most jaw muscles in terms of the functional demands placed on them, and even intramuscular regional variations can be related to differential usage patterns.

The situation with regard to myosins and contraction speed is less clear than that for oxidative capacity, in large part because of the confused state of the literature. Rowlerson (1990) has summarized the distribution of fiber types in various mammals, and that information is superimposed on a cladogram of mammals in Fig. 1. Keeping in mind that conclusions must be provisional at this stage, the following generalizations can be suggested.

1. Several orders of placental mammals and the opossum, representing the outgroup marsupials, have neither standard slow (type I) nor standard fast (type IIA or B, acid-labile) fibers. The absence of slow fibers in the rabbit, rodents, and hedgehog can therefore be considered primitive, and their presence in carnivorans (with secondary loss from the lesser panda) and the two ungulate groups can be considered new acquisitions. In the primates, type I fibers appear at the base of the Catarrhini and therefore must have been lost from the chimpanzee.

2. The development of type IIM myosin appears to have occurred twice, convergently, in the carnivorans and the anthropoid primates, as the prosimian pygmy lemur does not have it and as the orders Carnivora and Primates are not considered sister groups. Further, the absence of IIM myosin in one carnivoran (the lesser panda) and one primate (man) must be secondary (a loss), because the sister taxa of both have IIM (mustelid carnivores and chimpanzee, respectively).

3. The loss of type II fibers must have occurred independently in the perissodactyls and the selenodont artiodactyls (sheep, cow), because the pig retains the full complement of fiber types; alternatively, type II fibers might have

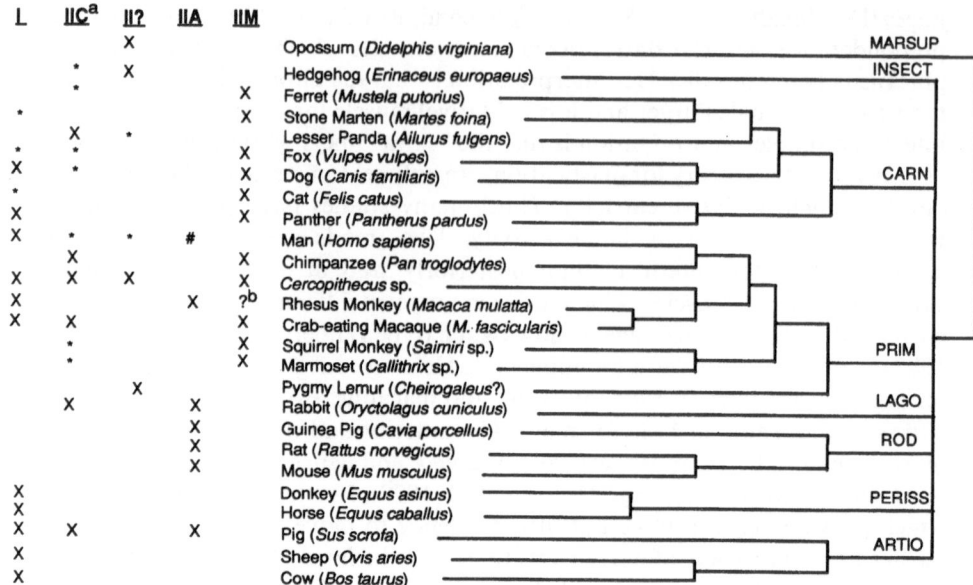

Fig. 1. Summary of myosins found in mammalian jaw-closing muscles and related to phylogeny. Myosin data are from Rowlerson (1990, Table 1.2) and the original sources listed below; the phylogeny is based on Novacek et al. (1988) for the orders of mammals and Flynn et al. (1988) for carnivorans. Fiber types for the hedgehog, rhesus macaque, mouse, and two of the artiodactyls (sheep, cow) are derived from histochemistry only. Type *I* fibers are acid-stable and alkaline-labile; they are presumed slow. *IIC* fibers ("mixed") are considered slow by Rowlerson (1990). *II?* fibers include those that are unique (neonatal myosin in humans) and fibers which have not been fully characterized (opossum, hedgehog, etc.). The *IIA* category is for fibers with myosins resembling limb muscle myosins. Most are relatively oxidative, but as mentioned in the text, there are some exceptions, and this category includes some IIB fibers as well. *IIM* fibers are identified biochemically only. Orders of mammals: *MARSUP* Marsupialia (or Metatheria); *INSECT* Insectivora; *CARN* Carnivora; *PRIM* Primates; *LAGO* - Lagomorpha; *ROD* Rodentia; *PERISS* Perissodactyla; *ARTIO* Artiodactyla. Myosins: *X* major component;* minor component; # IIB fibers only; *a* the category IIC includes IM fibers as well (these are both thought to be "mixed" fast and slow myosins); *b* fibers identified as IIA and IIB by routine histochemistry might have been IIM. Original sources: opossum (Sciote 1991); hedgehog (Lindman et al. 1986); all carnivorans and all primates except rhesus monkey and pygmy lemur (Rowlerson et al. 1983); dog (Suzuki 1977; Orvis and Cardinet 1981; Tuxen and Kirkeby 1990); cat (Taylor et al. 1973; Gorniak 1986; Rowlerson et al. 1988; Tuxen and Kirkeby 1990); man (Ringqvist 1974; Vignon et al. 1980; Eriksson et al. 1981; Ringqvist et al. 1982; Eriksson and Thornell 1983); pygmy lemur (Soussi-Yanicostas et al. 1990b); rhesus macaque (Maxwell et al. 1979; Miller and Farias 1988); crab-eating macaque (Clark and Luschei 1981; Rowlerson et al. 1988); rabbit and various rodents (Schiaffino 1974; Suzuki 1977; Rowlerson et al. 1988; Tuxen and Kirkeby 1990); rat (Rokx et al. 1984); guinea pig (Masuda et al. 1974); perissodactyls (donkey and horse) (Mascarello et al. 1979, cited by Rowlerson 1990); pig (Suzuki 1977; Herring et al. 1979; Horák 1988; Soussi-Yanicostas et al. 1990b; Tuxen and Kirkeby 1990); sheep and cow (Suzuki 1977)

been redeveloped in the lineage leading to the pig after a loss at the base of the ungulates.

Altogether, there are many instances of convergence or parallelism, suggesting that fiber types are malleable in evolution. The evidence that myosins are malleable in ontogeny as well is not as convincing as for metabolic enzymes, but it has been suggested (e.g., by Ulrici et al. 1985) that experimental interference with function can alter myosin type in jaw muscles. An increased percentage of type I fibers was induced in the rat lateral pterygoid muscle by a protrusive appliance (Easton and Carlson 1990), while a decreased percentage resulted from edentulousness in rhesus monkey temporalis and posterior masseter (Carlson and Poznanski 1982). It is not clear, however, whether these changes are the product of conversion or selective atrophy of fiber types.

As in the case of facial muscles, there is a clear trend for type I fibers to be associated with large body size. This is true not only among orders (ungulates, primates and carnivorans are on average larger than rodents and insectivores), but also within groups. For example, man (with I) is larger than the chimpanzee (without), just as the stone marten (with I) is larger than the ferret (without); the catarrhine primates (with I) are all larger than the two platyrrhines and the prosimian primate (without). The relatively large dog and panther have relatively more type I fibers than their smaller relatives, fox and cat. The correlation of type I fibers with body size may also account, at least in part, for the association with rate of chewing noted by Rowlerson (1990). Differential distribution of type I fibers within muscles (correlated, of course, with oxidative activity) points again to their energy efficiency for these most used regions.

In contrast to the appearance of type I fibers, the distribution of IIM myosin suggests some phylogenetic constraint. If IIM fibers are indeed associated with the need for a fast, strong bite (Rowlerson et al. 1983), then one might expect them in many other taxa for which prey capture is important (for example, the opossum or hedgehog) or in which agonistic biting encounters are prevalent (for example, horses). Considering the IIM fibers as characteristic of a lineage also removes the functional problem of explaining their occurrence in a variety of primates whose requirement for a fast, strong bite seems less than fully documented. Indeed, since the presence of IIM myosin is primitive in the carnivorans and higher primates, one need only explain why it is lost in the lesser panda and man, and not why it is present in the other taxa. Rowlerson et al. (1983) hypothesize for the lesser panda that loss of IIM fibers is correlated with a herbivorous diet, a suggestion that could be tested by examining other herbivorous carnivorans (other procyonids and bears). The situation in humans is less clear, since we are not more herbivorous than our close relatives, and even if we no longer require a strong bite, that does not constitute a likely selective basis for loss of IIM myosin. Actually, humans are even more unique in their masticatory muscle fiber types than is indicated in Fig. 1, because human muscles are dominated, both in size and in frequency, by type I fibers (Eriksson and Thornell 1983). In contrast, nothing about human mastication is unique. If a functional explanation is relevant at all to the human case, perhaps the use of jaw muscles in speech should be investigated as a possible correlate of fiber type composition.

The correlation of myosin fiber type with twitch contraction time is not overwhelming, possibly due to problems with experimental determination of contractile properties (see Sect. 3.4.1). Some good correlations of TPT and percentage of type I fibers have been reported, for example, for the adductors of *Pteropus* (De Gueldre and De Vree 1991). On the other hand, the masseters of rats and guinea pigs have similar, uniform fiber compositions, and yet the former has a TPT of 14–18 ms (Nordstrom and Yemm 1974; Hiraiwa 1977; Easton and Carlson 1990) and the latter 24 ms (Masuda et al. 1974). A similar discrepancy has been observed when alterations in TPT were produced by experimental manipulation. Ellis et al. (1988) documented a postsurgical shortening from 43 to 39 ms in a combined contraction of jaw muscles in rhesus monkeys, which they attributed to a soft diet. Lengthened TPTs were produced in rabbit temporalis (25.5 to 30.1 ms) after masseter removal (Guelinckx et al. 1986) and in rat superficial masseter and lateral pterygoid (18.6 to 21.2 ms and 18.2 to 22.4 ms, respectively) after protrusive treatment (Easton and Carlson 1990). Of all these examples, only the change in rat lateral pterygoid was accompanied by a correlated change in the proportion of type I fibers. Because most of these muscles are mixed, however, it is probable that the association would be better if individual fibers or motor units were treated. For example, Clark et al. (1978) examined motor units of rhesus monkey temporalis and found TPTs ranging from 17 to 31 ms, which compares very favorably with the TPT of 22 ms reported for (presumably mixed) strips by Faulkner et al. (1982). In human masseter, which lacks fast IIM fibers but has many slow I fibers, Yemm (1976) found motor units with TPTs from 24–91 ms, Goldberg and Derfler (1977) an average of 49 ms, and McMillan et al. (1990) a range from 25–67 ms.

Acknowledgments. I am very grateful to Keith Alley, Dave Carlson, Keith Condon, Dean Dessem, Bill LaFramboise, Kiisa Nishikawa, Wim Weijs, Larry Wineski, and especially Jim Sciote for help with the literature. I thank Dave Wright for advice on the cladogram and computer use and Jim Sciote for helpful suggestions on the manuscript. The writing of this review was supported in part by PHS grant DE 08513.

References

Akster HA (1981) Ultrastructure of muscle fibres in head and axial muscles of the perch (*Perca fluviatilis* L.). A quantitative study. Cell Tissue Res 219: 111–131

Akster HA (1983) A comparative study of fibre type characteristics and terminal innervation in head and axial muscle of the carp (*Cyprinus carpio* L.): a histochemical and electron-microscopical study. Neth J Zool 33: 164–188

Akster HA, Osse JWM (1978) Muscle fibre types in head muscles of the perch *Perca fluviatilis* (L), Teleostei. A histochemical and electromyographical study. Neth J Zool 28: 94–110

Alley KE, Reiser PS (1991) Molecular and contractile features of frog jaw myofibers. J Dent Res 70: 420

Alley KE, Omerza FF, Reiser PS (1992) Cellular aspects of neuromuscular accommodation during rapid craniofacial morphogenesis. In: Davidovitch Z (ed) The biological mechanisms of tooth movement and craniofacial adaptation. Ohio State Univ, Columbus, pp 531–540

Altringham JD, Johnston IA (1988) The mechanical properties of polyneuronally innervated, myotomal muscle fibres isolated from a teleost fish (*Myoxocephalus scorpius*). Pflügers Arch Eur J Physiol 412: 524–529

Anapol FC (1985) Electromyographic and histochemical diversity within pig masseter muscle. Am Zool 25: 121A

Anapol FC, Herring SW (1989) Length-tension relationships of masseter and digastric muscles of miniature swine during ontogeny. J Exp Biol 143: 1–16

Anapol FC, Muhl ZF, Fuller JH (1987) The force-velocity relation of the rabbit digastric muscle. Arch Oral Biol 32: 93–99

Bárány M (1967) ATPase activity of myosin correlated with speed of muscle shortening. J Gen Physiol 50: 197–218

Barends PMG (1979) The relation between fiber type composition and function in the jaw adductor muscle of the perch (*Perca fluviatilis*, L.). A histochemical study. Proc K Ned Akad Wet Ser C 82: 147–164

Barends PMG, van Leeuwen JL, Taverne-Thiele AJ (1983) Differentiation of the jaw adductor muscle of the rosy barb, *Barbus conchonius* (Teleostei, Cyprinidae, L.), during development. Neth J Zool 33: 1–20

Bewick GS, Rowlerson A, Tonge DA, Holder N (1991) Organization of motor units in the axolotl: a continuously growing animal. J Comp Neurol 303: 551–562

Blanton PL, Biggs NL, Perkins RC (1970) Electromyographic analysis of the buccinator muscle. J Dent Res 49: 389–394

Bone Q (1964) Patterns of muscular innervation in the lower chordates. Int Rev Neurobiol 6: 99–147

Bone Q, Johnston IA, Pulsford A, Ryan KP (1986) Contractile properties and ultrastructure of three types of muscle fibre in the dogfish myotome. J Muscle Res Cell Motil 7: 47–56

Braund KG, Mehta JR, Amling KA (1991) Fibre type proportions of the buccinator muscle in clinically normal adult dogs. Res Vet Sci 50: 371–373

Bredman JJ, Weijs WA, Moorman AFM (1990) Expression of 'cardiac-specific' myosin heavy chain in rabbit cranial muscles. In: Maréchal G, Carraro U (eds) Muscle and motility, vol 2 Proc XIXth Eur Conf, Brussels. Intercept, Hampshire, pp 329–335

Bredman JJ, Wessels A, Weijs WA, Korfage JAM, Soffers CAS (1991) Demonstration of 'cardiac-specific' myosin heavy chain in masticatory muscles of human and rabbit. Histochem J 23: 160–170

Brooke MH, Kaiser KK (1974) The use and abuse of muscle histochemistry. Ann N Y Acad Sci 228: 121–144

Burke RE, Levine DN, Zajac FE III, Tsairis P, Engel WK (1971) Mammalian motor units: physiological-histochemical correlation in three types in cat gastrocnemius. Science 174: 709–712

Busbey AB III (1989) Form and function of the feeding apparatus of *Alligator mississippiensis*. J Morphol 202: 99–127

Butler-Browne GS, Eriksson P-O, Laurent C, Thornell L-E (1988) Adult human masseter muscle fibres express myosin isozymes characteristic of development. Muscle Nerve 11: 610–620

Carlson DS, Poznanski A (1982) Experimental models of surgical intervention in the growing face: histochemical analysis of neuromuscular adaptation to altered muscle length. In: McNamara JA Jr, Carlson DS, Ribbens KA (eds) The effect of surgical intervention on craniofacial growth. Univ of Mich Craniofacial Growth Series Monogr 12, Ann Arbor, pp 73–98

Clark GT, Carter MC, Beemsterboer PL (1988) Analysis of electromyographic signals in human jaw closing muscles at various isometric force levels. Arch Oral Biol 33: 833–837

Clark RW, Luschei ES (1981) Histochemical characteristics of mandibular muscles of monkeys. Exp Neurol 74: 654–672

Clark RW, Luschei ES, Hoffman DS (1978) Recruitment order, contractile characteristics, and firing patterns of motor units in the temporalis muscle of monkeys. Exp Neurol 61: 31–52

Collins JH (1991) Myosin light chains and troponin C: structural and evolutionary relationships revealed by amino acid sequence comparisons. J Muscle Res Cell Motil 12: 3–25

Condon KW (1987) A study of cranial kinesis in the Nile Monitor, *Varanus niloticus*. PhD Thesis, Univ of Illinois, Chicago

d'Albis A, Janmot C, Bechet J-J (1986) Comparison of myosins from the masseter muscle of adult rat, mouse and guinea-pig. Eur J Biochem 156: 291–296

De Gueldre G, De Vree F (1991) Fibre composition of the masticatory muscles of *Pteropus giganteus* (Brunnich, 1782) (Megachiroptera). Belg J Zool 121: 279–294

Dick TE, van Lunteren E (1990) Fiber subtype distribution of pharyngeal dilator muscles and diaphragm in the cat. J Appl Physiol 68: 2237 2240

Dubale MS, Muralidharan P (1970) Histochemical studies on the fiber types in the developing jaw-muscles of domestic fowl (*Gallus domesticus*). Life Sci 9: 949–959

Dubose L, Schelhorn TB, Clamann HP (1987) Changes in contractile speed of cat motor units during activity. Muscle Nerve 10: 744–752

Easton JW, Carlson DS (1990) Adaptation of the lateral pterygoid and superfical masseter muscles to mandibular protrusion in the rat. Am J Orthod Dentofac Orthop 97: 149–158

Edström L, Lindquist C (1973) Histochemical fiber composition of some facial muscles in the cat in relation to their contraction properties. Acta Physiol Scand 89: 491–503

Ellis E III, Dechow PC, Carlson DS (1988) A comparison of stimulated bite force after mandibular advancement using rigid and non-rigid fixation. J Oral Maxillofac Surg 46: 26–32

Eriksson P-O, Thornell L-E (1983) Histochemical and morphological muscle-fibre characteristics of the human masseter, the medial pterygoid and the temporal muscles. Arch Oral Biol 28: 781–795

Eriksson P-O, Eriksson A, Ringqvist M, Thornell L-E (1981) Special histochemical muscle-fibre characteristics of the human lateral pterygoid muscle. Arch Oral Biol 26: 495–507

Eriksson P-O, Eriksson A, Ringqvist M, Thornell L-E (1982) Histochemical fibre composition of the human digastric muscle. Arch Oral Biol 27: 207–215

Faulkner JA, McCully KK, Carlson DS, McNamara JA Jr (1982) Contractile properties of the muscles of mastication of rhesus monkeys (*Macaca mulatta*) following increase in muscle length. Arch Oral Biol 27: 841–845

Fields HW, Proffit WR, Case JC, Vig KWL (1986) Variables affecting measurements of vertical occlusal force. J Dent Res 65: 135–138

Flitney FW, Johnston IA (1979) Mechanical properties of isolated fish red and white muscle fibres. J Physiol 295: 49P–50P

Floyd K, Morrison JFB (1975) The mechanical properties of oesophageal striated muscle in the cat and sheep. J Physiol 248: 717–724

Flynn JJ, Neff NA, Tedford RH (1988) Phylogeny of the Carnivora. In: Benton MJ (ed) The phylogeny and classification of the tetrapods, vol 2. Clarendon Press, Oxford, pp 73–115

Gans C, De Vree F (1987) Functional bases of fiber length and angulation in muscle. J Morphol 192: 63–85

Gleeson TT, Johnston IA (1987) Reptilian skeletal muscle: contractile properties of identified, single-twitch and slow fibers from the lizard *Dipsosaurus dorsalis*. J Exp Zool 242: 283–290

Goldberg LJ, Derfler B (1977) Relationship among recruitment order, spike amplitude, and twitch tension of single motor units in human masseter muscle. J Neurophysiol 40: 879–890

Gordon AM, Huxley AF, Julian FJ (1966) The variation in isometric tension with sarcomere length in vertebrate muscle fibres. J Physiol 184: 170–192

Gorniak GC (1985) Trends in the actions of mammalian masticatory muscles. Am Zool 25: 331–337

Gorniak GC (1986) Correlation between histochemistry and muscle activity of jaw muscles in cats. J Appl Physiol 60: 1393–1400

Gradwell N, Walcott B (1971) Dual functional and structural properties of the interhyoideus muscle of the bullfrog tadpole (*Rana catesbeiana*). J Exp Zool 176: 193–218

Granzier HLM, Wiersma J, Akster HA, Osse JWM (1983) Contractile properties of a white and a red-fibre type of the m. hyohyoideus of the carp (*Cyprinus carpio* L.). J Comp Physiol 149: 441–449

Grobecker DB, Pietsch TW (1979) High-speed cinematographic evidence for ultrafast feeding in antennariid anglerfishes. Science 205: 1161–1162

Grove BK (1989) Muscle differentiation and the origin of muscle fiber diversity. CRC Crit Rev Neurobiol 4: 201–234

Guelinckx P, Dechow PC, Vanrusselt R, Carlson DS (1986) Adaptations in the temporalis muscles of rabbits after masseter muscle removal. J Dent Res 65: 1294–1299

Hellstrand E (1980) Morphological and histochemical properties of tongue muscles in cat. Acta Physiol Scand 110: 187–198

Herring SW (1992) Muscles of mastication: architecture and functional organization. In: Davidovitch Z (ed) The biological mechanisms of tooth movement and craniofacial adaptation. Ohio State Univ, Columbus, pp 541–548

Herring SW, Grimm AF, Grimm BR (1979) Functional heterogeneity in a multipinnate muscle. Am J Anat 154: 563–576

Herring SW, Grimm AF, Grimm BR (1984) Regulation of sarcomere number in skeletal muscle: a comparison of hypotheses. Muscle Nerve 7: 161–173

Hertzberg SR, Muhl ZF, Begole EA (1980) Muscle sarcomere length following passive jaw opening in the rabbit. Anat Rec 197: 435–440

Hill MA, Ecob-Prince MS, Hoh JFY (1989) Regeneration of cat posterior temporalis muscle in culture. Cell Diff Dev 28: 145–152

Hiraiwa T (1977) The effects of motortrigeminal denucleation on rat masticatory muscles. Jpn J Physiol 27: 617–641

Hoh JFY, Hughes S (1988) Myogenic and neurogenic regulation of myosin gene expression in cat jaw-closing muscles regenerating in fast and slow limb muscle beds. J Muscle Res Cell Motil 9: 59–72

Hoh JFY, Hughes S (1989) Immunocytochemical analysis of the perinatal development of cat masseter muscle using anti-myosin antibodies. J Muscle Res Cell Motil 10: 312–325

Hoh JFY, Hughes S (1991) Expression of superfast myosin in aneural regenerates of cat jaw muscle. Muscle Nerve 14: 316–325

Hoh JFY, Hughes S, Chow C, Hale PT, Fitzsimons RB (1988) Immunocytochemical and electrophoretic analyses of changes in myosin gene expression in cat posterior temporalis muscle during postnatal development. J Muscle Res Cell Motil 9: 48–58

Horák V (1988) Histochemical fiber type composition in 12 skeletal muscles of miniature pigs. Anat Anz 167: 231–238

Hoyle G (1983) Muscles and their neural control. Wiley, New York

Katsura S, Ishizuka H, Matsumoto H, Nakae Y (1981) Histochemical studies on the histogenesis of rat masseter muscle. Jpn J Oral Biol 23: 677–684

Katsura S, Ishizuka H, Matsumoto H, Nakae Y (1982) Histochemical studies on the histogenesis of rat medial pterygoid muscle. Acta Histochem Cytochem 15: 701–709

Kiliaridis S, Shyu BC (1988) Isometric muscle tension generated by masseter stimulation after prolonged alteration of the consistency of the diet fed to growing rats. Arch Oral Biol 33: 467–472

Kiliaridis S, Engström C, Thilander B (1988) Histochemical analysis of masticatory muscle in the growing rat after prolonged alteration in the consistency of the diet. Arch Oral Biol 33: 187–193

Kogo M, Inoue K, Matsuya T, Nishio J, Hamamura Y, Yasui Y, Miyazaki T (1991) Mechanical contraction property of the levator veli palatini muscle. Cleft Pal J 28: 221–225

Lindman R, Eriksson P-O, Thornell L-E (1986) Histochemical enzyme profile of the masseter, temporal and lateral pterygoid muscles of the European hedgehog (Erinaceus europaeus). Arch Oral Biol 31: 51–55

Lindquist C (1973) Contraction properties of cat facial muscles. Acta Physiol Scand 89: 482–490

Loeb GE, Gans C (1986) Electromyography for experimentalists. Univ Chicago Press, Chicago

Mabuchi K, Pinter K, Mabuchi Y, Sreter F, Gergely J (1984) Characterization of rabbit masseter muscle fibers. Muscle Nerve 7: 431–438

Mackenna BR, Türker KS (1978) Twitch tension in the jaw muscles of the cat at various degrees of mouth opening. Arch Oral Biol 23: 917–920

Maeda N, Hanai H, Kumegawa M (1981) Postnatal development of masticatory organs in rats. III. Effect of mastication on the postnatal development of the M. masseter superficialis. Anat Anz 150: 424–427

Manns A, Miralles R, Palazzi C (1979) EMG, bite force, and elongation of the masseter muscle under isometric voluntary contractions and variations of vertical dimension. J Prosthet Dent 42: 674–682

Marsh RL (1988) Ontogenesis of contractile properties of skeletal muscle and sprint performance in the lizard *Dipsosaurus dorsalis*. J Exp Biol 137: 119–139

Mascarello F, Aureli G, Veggetti A (1979) Muscoli masticatori. Determinazione istochimica dei tipi di fibre muscolari in mammiferi. Quad Anat Prat 35: 193–211

Mascarello F, Rowlerson A, Scapolo PA (1984) The fibre type composition of the striated muscle of the oesophagus in ruminants and carnivores. Histochemistry 80: 277–288

Masuda K, Takahashi S, Kuriyama H (1974) Studies on the fibre types of the guinea pig masticatory muscles. Comp Biochem Physiol 47A: 1171–1184

Matthews B, Smith BH (1972) An investigation into the presence of slow-graded fibres in the anterior belly of the rat digastric muscle. Arch Oral Biol 17: 473–478

Maxwell LC, Carlson DS, McNamara JA Jr, Faulkner JA (1979) Histochemical characteristics of the masseter and temporalis muscles of the rhesus monkey (*Macaca mulatta*). Anat Rec 193: 389–402

Maxwell LC, Carlson DS, Brangwyn CE (1980) Lack of 'acid reversal' of myofibrillar adenosine triphosphatase in masticatory muscle fibres of rhesus monkeys. Histochem J 12: 209–219

Maxwell LC, Carlson DC, McNamara JA Jr, Faulkner JA (1981) Adaptation of the masseter and temporalis muscles following alteration in length, with or without surgical detachment. Anat Rec 200: 127–137

McMillan AS, Sasaki K, Hannam AG (1990) The estimation of motor unit twitch tensions in the human masseter muscle by spike-triggered averaging. Muscle Nerve 13: 697–703

Miledi R, Uchitel OD (1984) A study of the submaxillaris muscle of the frog. J Physiol 350: 279–291

Miller AJ, Farias M (1988) Histochemical and electromyographic analysis of craniomandibular muscles in the rhesus monkey, *Macaca mulatta*. J Oral Maxillofac Surg 46: 767–776

Morgan DL, Proske U (1984) Vertebrate slow muscle: its structure, pattern of innervation, and mechanical properties. Physiol Rev 64: 103–169

Muhl ZF, Newton JH (1982) Change of digastric muscle length in feeding rabbits. J Morphol 171: 151–157

Muhl ZF, Grimm AF, Glick PL (1978) Physiologic and histologic measurements of the rabbit digastric muscle. Arch Oral Biol 23: 1051–1059

Müntener M, Gottschall J, Neuhuber W, Mysicka A, Zenker W (1980) The ansa cervicalis and the infrahyoid muscles of the rat. I. Anatomy; distribution, number and diameter of fiber types; motor units. Anat Embryol 159: 49–57

Nakaji S, Oota I (1978) Histochemical observation on soft palate muscle of cat. Sapporo Med J 47: 525–533

Nakata S (1981) Relationship between the development and growth of cranial bones and masticatory muscles in postnatal mice. J Dent Res 60: 1440–1450

Noden DM (1983) The embryonic origins of avian cephalic and cervical muscles and associated connective tissues. Am J Anat 168: 257–276

Nordstrom MA, Miles TS (1990) Fatigue of single motor units in human masseter. J Appl Physiol 68: 26–34

Nordstrom SH, Yemm R (1972) Sarcomere length in the masseter muscle of the rat. Arch Oral Biol 19: 895–902

Nordstrom SH, Yemm R (1974) The relationship between jaw position and isometric active tension produced by direct stimulation of the rat masseter muscle. Arch Oral Biol 19: 353–359

Nordstrom SH, Bishop M, Yemm R (1974) The effect of jaw opening on the sarcomere length of the masseter and temporal muscles of the rat. Arch Oral Biol 19: 151–155

Novacek MJ, Wyss AR, McKenna MC (1988) The major groups of eutherian mammals. In: Benton MJ (ed) The phylogeny and classification of the tetrapods, vol 2. Clarendon Press, Oxford, pp 31–71

Ogata T (1988) Morphological and cytochemical features of fiber types in vertebrate skeletal muscle. CRC Crit Rev Anat Cell Biol 1: 229–275

Orvis JS, Cardinet GH III (1981) Canine muscle fiber types and susceptibility of masticatory muscles to myositis. Muscle Nerve 4: 354–359

Otten E (1988) Concepts and models of functional architecture in skeletal muscle. In: Pandolf KB (ed) Exercise and sport sciences reviews, vol 16. MacMillan, New York, pp 89–137

Pette D, Staron RS (1990) Cellular and molecular diversities of mammalian skeletal muscle fibers. Rev Physiol Biochem Pharmacol 116: 1–76

Pollack GH (1983) The cross-bridge theory. Physiol Rev 63: 1049–1113

Rayne J, Crawford GNC (1972) The relationship between fibre length, muscle excursion and jaw movements in the rat. Arch Oral Biol 17: 859–872

Reger JF, Holbrook JR (1974) The fine structure of tongue muscle in the bat, *Myotis grisescens*, with particular reference to twitch and slow muscle fiber morphology. J Submicrosc Cytol 6: 1–13

Ringqvist M (1974) A histochemical study of temporal muscle fibers in denture wearers and subjects with natural dentition. Scand J Dent Res 82: 28–39

Ringqvist M, Ringqvist I, Eriksson PO, Thornell L-E (1982) Histochemical fibre-type profile in the human masseter muscle. J Neurol Sci 53: 273–282

Roberts JL, Reed WR, Thach BT (1984) Pharyngeal airway-stabilizing function of sternohyoid and sternothyroid muscles in the rabbit. J Appl Physiol 57: 1790–1795

Rokx JTM, van Willigen JD, Jansen HWB (1984) Muscle fibre types and muscle spindles in the jaw musculature of the rat. Arch Oral Biol 29: 25–31

Rome LC, Funke RP, Alexander RM, Lutz G, Aldridge H, Scott F, Freadman M (1988) Why animals have different muscle fibre types. Nature 335: 824–827

Rowlerson AM (1990) Specialization of mammalian jaw muscles: fibre type compositions and the distribution of muscle spindles. In: Taylor A (ed) Neurophysiology of the jaws and teeth. Macmillan, London, pp 1–51

Rowlerson A, Pope B, Murray J, Whalen B, Weeds AG (1981) A novel myosin present in cat jaw-closing muscles. J Muscle Res Cell Motil 2: 415–438

Rowlerson A, Mascarello F, Veggetti A, Carpene E (1983) The fiber-type composition of the first branchial arch muscles in Carnivora and Primates. J Muscle Res Cell Motil 4: 443–472

Rowlerson A, Mascarello F, Barker D, Saed H (1988) Muscle-spindle distribution in relation to the fibre-type composition of masseter in mammals. J Anat 161: 37–60

Sato I, Shimada K, Sato T, Kitagawa T (1992) Histochemical study of jaw muscle fibers in the American alligator (*Alligator mississippiensis*). J Morphol 211: 187–199

Scapolo PA, Mascarello F, Veggetti A, Carpene E (1981) Caratterizzazione istochimica ed immunoistochimica del muscola digastrica. Atti Soc It Sci Vet 35: 331–332

Scapolo PA, Rowlerson A, Mascarello F, Veggetti A (1991) Neonatal myosin in bovine and pig tensor tympani muscle fibres. J Anat 178: 255–263

Schiaffino S (1974) Histochemical enzyme profile of the masseter muscle in different mammalian species. Anat Rec 180: 53–62

Schieppati M, DiFrancesco G, Nardone A (1989) Patterns of activity of perioral facial muscles during mastication in man. Exp Brain Res 77: 103–112

Schwarting S, Schröder M, Stennert E, Goebel HH (1982) Enzyme histochemical and histographic data on normal human facial muscles. J Otorhinolaryngol (Basel) 44: 51–59

Sciote JJ (1991) Myosin heavy chain isoform and histochemical fiber type characteristics of masticatory and selective limb muscles in the American opossum. Eur Muscle Club, 1991, Proc

Shelton GD, Cardinet GH III, Bandman E (1988) Expression of fiber type specific proteins during ontogeny of canine temporalis muscle. Muscle Nerve 11: 124–132

Smith KK (1982) An electromyographic study of the function of the jaw adducting muscles in *Varanus exanthematicus* (Varanidae). J Morphol 173: 137–158

Soussi-Yanicostas N, Barbet JP, Laurent-Winter C, Barton P, Butler-Browne GS (1990a)

Transition of myosin isozymes during development of human masseter muscle. Development 108: 239–249

Soussi-Yanicostas N, Breuer EM, Dang DC, Butler-Browne GS (1990b) The masseter, a very specialized muscle. In: Maréchal G, Carraro U (eds) Muscle and motility, vol 2. Proc XIX Eur Conf Brussels. Intercept, Hampshire, pp 63–70

Suzuki A (1977) A comparative histochemical study of the masseter muscle of the cattle, sheep, swine, dog, guinea pig, and rat. Histochemistry 51: 121–131

Tamari JW, Tomey GF, Ibrahim MZM, Baraka A, Jabbur SJ, Bahuth N (1973) Correlative study of the physiologic and morphologic characteristics of the temporal and masseter muscles of the cat. J Dent Res 52: 538–543

Taylor A, Cody FWJ, Bosley MA (1973) Histochemical and mechanical properties of the jaw muscles of the cat. Exp Neurol 38: 99–109

Thexton AJ, Hiiemae KM (1975) The twitch-contraction characteristics of opossum jaw musculature. Arch Oral Biol 20: 743–748

Thexton AJ, Wake DB, Wake MH (1977) Tongue function in the salamander Bolitoglossa occidentalis. Arch Oral Biol 22: 361–366

Throckmorton GS, Saubert CW IV (1982) Histochemical properties of some jaw muscles of the lizard Tupinambis nigropunctatus (Teiidae). Anat Rec 203: 345–352

Tuxen A, Kirkeby S (1990) An animal model for human masseter muscle. J Oral Maxillofac Surg 48: 1063–1067

Ulrici V, Vogel A, Pieper K-S, Scharschmidt F, Schumacher G-H (1985) Veränderungen im feinstrukturellen Aufbau des M. masseter durch unilaterale Okklusionsstörungen. Anat Anz 160: 9–15

van Boxtel A, Goudswaard P, van der Molen GM, van den Bosch WEJ (1983) Changes in electromyogram power spectra of facial and jaw-elevator muscles during fatigue. J Appl Physiol 54: 51–58

van Lunteren E, Manubay P (1992) Contractile properties of feline genioglossus, sternohyoid, and sternothyroid muscles. J Appl Physiol 72: 1010–1015

van Lunteren E, Salomone RJ, Manubay P, Supinski GS, Dick TE (1990) Contractile and endurance properties of geniohyoid and diaphragm muscles. J Appl Physiol 69: 1992–1997

Veggetti A, Mascarello F, Carpenè E (1982) A comparative histochemical study of fibre types in middle ear muscles. J Anat 135: 333–352

Vignon C, Pellissier JF, Serratrice G (1980) Further histochemical studies on masticatory muscles. J Neurol Sci 45: 157–176

Walker SM, Schrodt GR (1974) I segment lengths and thin filament periods in skeletal muscle fibers of the rhesus monkey and the human. Anat Rec 178: 63–82

Weijs WA, van der Wielen-Drent TK (1982) Sarcomere length and EMG activity in some jaw muscles of the rabbit. Acta Anat 113: 178–188

Weijs WA, van der Wielen-Drent TK (1983) The relationship between sarcomere length and activation pattern in the rabbit masseter muscle. Arch Oral Biol 28: 307–315

Wineski LE, Pitts SA, Weeks OI (1991) Histochemical profiles of the vibrissae-operating facial muscles in the golden hamster and guinea pig. Anat Rec 229: 93A

Yemm R (1976) The properties of their motor units, and length-tension relationships of the muscles. In: Anderson DJ, Matthews B (eds) Mastication. Wright, Bristol, pp 25–30

Chapter 2

Feeding Mechanisms in Sharks and Other Elasmobranchs

T.H. Frazzetta

Contents

1 Chondrichthyian Fishes

Earliest vertebrates were fish-like forms, but lacked definitive jaws and paired appendages (of the sort possessed by gnathostomes), and were represented in the fossil record as long ago as the later Cambrian (Carroll 1988). Gnathostomes – fishes with a fully formed jaw apparatus, and an appendicular skeleton comparable to that of all other jawed groups – appear suddenly in the Silurian. These first gnathostomes, the Acanthodii (Moy Thomas and Miles 1971; Carroll 1988) are interpreted variously, as being related to either, both, or neither of the two major classes of fish, the Chondrichthyes and the Osteichthyes (Romer 1966; Moy Thomas and Miles 1971; Schaeffer and Williams 1977; Jarvik 1980; Carroll 1988). Another gnathostome group, the Placodermi, is known at or near the

Department of Ecology, Ethology, and Evolution, University of Illinois, 515 Morrill Hall, 505 South Goodwin Avenue, Urbana, Illinois, USA

Advances in Comparative and Environmental Physiology, Vol. 18
© Springer-Verlag Berlin Heidelberg 1994

Silurian-Devonian boundary. The placoderms include many shark-shaped forms, but have a skeleton containing extensive dermal bone (see Stensiö 1963; Romer 1966; Jarvik 1980; Carroll 1988) as well as endochondrial ossification. The relationships of placoderms to acanthodians and or to chondrichthyians (if to either) are still debated.

For many years it had been thought that chondrichthyians (possessing no internal bone, but instead a skeleton of cartilage) were ancestral to jawed, bony fishes, including the osteichthyians. This view was based upon an uncritical application of the "biogenetic law" ("ontogeny recapitulates phylogeny"), in which cartilage was seen as an embryonic precursor of bone tissue. More recently, thanks to a more careful consideration of biogenesis, and a more informative (and better interpreted) fossil record, this idea of a chondrichthyian ancestry to bony vertebrates has been rejected. Instead, chondrichthyians and osteichthyians are usually conjectured as having arisen independently from some earlier, bony, jawed ancestor. Until the last few years it seemed that the Chondrichthyes and Osteichthyes arose sometime in the Devonian, but more recent evidence (see Zangerl 1981; Carroll 1988) reveals that the Chondrichthyes and the Osteichthyes are older, both groups being represented by individual integumentary dermal scales recovered from the Late Silurian.

This more recent evidence moves both the chondrichthyian and osteichthyian fishes back, closer to the known advent of jaws. The relatively primitive body form of early sharks, and the retention in modern sharks of many characteristics of their ancient forebears, opens the intriguing possibility that studies of the jaw mechanics of living sharks may provide insights into the evolutionary acquisition of jaws in the first gnathostomes. The origin of jaws was a major transition that involved remodeling the anterior branchial (visceral) skeleton from gill supports and musculature to become jaws and their morphological accessories. The transition is further fascinating in that it may have resulted in additions of branchial-skeletal (pharyngobranchial) elements to the cartilaginous braincase (chondrocranium) as suggested by Jarvik (1954; for opposing views, see e.g. Nelson 1969; Jollie 1971). As the branchial skeleton is derived from neural-crest ectoderm, not from embryonic mesoderm, any possibility that branchial tissue may have contributed to the chondrocranium could bear upon recent, broader discoveries of neural-crest contributions to cranial structures in vertebrates generally (Le Douarin 1982; Gans and Northcutt 1983; Northcutt and Gans 1983; Noden 1984; Hall 1987; Thomson 1987).

2 Elasmobranchs

The class Chondrichthyes is composed of two identifiable subclasses, the Elasmobranchii and Holocephali. The latter, a once successful group, exists today as the modern chimaeras. While sharing several features with elasmobranchs (e.g. cartilaginous endoskeleton), they are distinct in many others.

Modern elasmobranchs are divisible as selachians, the sharks, and batoids, the skates and rays. The fossil record strongly suggests that the first chondrichthyians were elasmobranch "sharks," streamlined, active predators.

Modern sharks arose in the Mesozoic (Schaeffer 1967; Maisey 1977, 1984; Carroll 1988) and are represented today by several descendent groups of fast-swimming, powerful predators. These latter differ from the most ancient sharks in their evolution away from a terminally placed mouth to one below the snout, alterations in jaw shape, tooth form, and other features. Nevertheless, there are significant similarities between modern and ancient sharks, and an analysis of extant forms should certainly illuminate many aspects of the most primitive sharks. Skates and rays (batoids) also first appear as Mesozoic fossils, and are clearly derived from a shark ancestor.

Knowledge of the elasmobranch feeding mechanism is supported by a surprising paucity of studies. In stark contrast is the much richer literature on jaw mechanics in osteichthyian fishes (e.g. Tchernavin 1953; Schaeffer and Rosen 1961; Alexander 1967a, b, 1969, 1970; Liem 1967, 1970, 1986; Thomson 1967; Nyberg 1971; Lauder 1980a, b, 1983a, b; Lauder and Liem 1980). Most detailed work on elasmobranch jaws has focused upon active sharks such as lamniforms and carcharhiniforms (except for an important paper on heterodontiforms, Nobiling 1977). Because our knowledge of jaw mechanisms in the active, predatory sharks is so much better than in other elasmobranchs, perforce the present chapter will emphasize the fast, predaceous members of the carchariniforms and lamniforms.

3 The Shark Jaw Apparatus

3.1 Structure

3.1.1 Skeleton

While the (squaliform) "classroom-laboratory dogfish" is anatomically well known (e.g. through college-level dissection manuals, e.g. Gans and Parsons 1964; Gilbert 1973; Bohensky 1981; Ashley and Chiasson 1988), morphological accounts of lamnids and carcharhinids are few. The descriptions below are based upon Gohar and Mazhar (1964), Compagno (1977, 1988), and the present author's work.

The head skeleton consists of three entities, the chondrocranium, jaws, and hyoid arch (Fig. 1), and can be defined to include the postchondrocranial branchial arches (not included here). The chondrocranium, a mostly closed braincase of cartilage, is mounted upon a slightly flexible vertebral column, and can be considered in terms of basic regions: the rostrum, nasal capsules, orbits, and otic capsules. Below the orbits the chondrocranium is developed as a suborbital shelf which juts outwardly beneath each eye; the shelf is present in

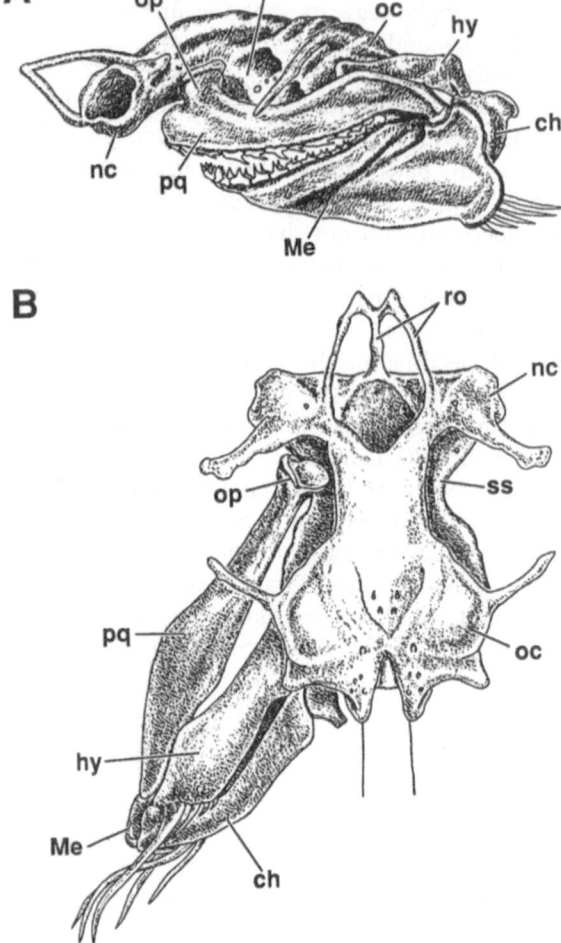

Fig. 1. A Lateral; **B** dorsal views of the head skeleton of a carcharhinid shark (*Negaprion brevirostris*) in a nonfeeding mode; jaw and hyoid elements are removed from the right side of the chondrocranium in **B**. *ch* Ceratohyal; *hy* hyomandibula; *Me* Meckel's cartilage (lower jaw); *nc* nasal capsule; *oc* otic capsule; *op* orbital process; *or* orbit; *pq* palatoquadrate (upper jaw); *ro* rostrum; *ss* suborbital shelf

carcharhinids but not in all sharks. Anteriorly, the margin of the shelf bends laterally, its contours flowing into the base of the ipsilateral nasal capsule. The chondrocranium supports the jaws and hyoid skeleton.

Three articulated cartilages compose the hyoid arch. The most dorsal, the hyomandibula, is a flattened, sculptured bar, whose upper border enters a depression in the posterior otic region of the chondrocranium. Here, it is movably joined, and the body of the element reaches downward and backward toward the jaw articulation. Its rotational movement about its articulation with the chondrocranium is largely constrained to permit swinging in an oblique plane – in a pendulum-like fashion – anteriorly, ventrally, and laterally. Distally, the hyomandibula extends along the medial surface of the jaw articulation, to which it is flexibly fastened by fibrous tissue.

The hyomandibula joins the ceratohyoid near the level of the jaw joint. This latter cartilage is firmly bound to the medial side of the lower jaw (Meckel's cartilage), and provides the most substantial tie between hyoid arch and jaws (in contrast to several accounts). Ventrally, the ceratohyoid joins the basihoid, a flat, median cartilage that braces the floor of the mouth and forms the skeletal support of the stiff "tongue."

Each upper jaw (palatoquadrate) piece curves medially near its anterior end, and becomes a smoothly bent cylinder (Fig. 1). Thus, the anterior portions of the upper jaws lie beneath the ethmoid region of the chondrocranium, where they meet in the midline, entering into a movable symphysis formed by fibrous tissue. Typically, there is a cartilaginous prominence – the orbital process – arising from the dorsal surface of the upper jaw (it is lacking in *Eugomphodus* = *Odontaspis*). Usually, this process occupies the anterior orbital region of the chondrocranium, and lies against the suborbital shelf, in the location of the shelf's lateral expansion (Fig. 1). Here, the edge and underside of the shelf bear a soft, cartilaginous pad, and give rise to the ethmopalatine ligament, a short, stout but flexible cord that joins a pliable, cartilaginous crown atop the orbital process of the upper jaw. Thus, the ethmopalatine ligament supplies the anterior suspension to the jaws, and when the shark is not feeding, the ligament is folded.

Near its posterior end, the upper jaw is flattened vertically, producing an expanded surface for attachment of portions of the adductor muscle. Similarly, the posterior part of the lower jaw is flattened and expanded vertically, but the expansion is much greater than in the case of the upper, its outer surface serving for the attachment of the lower half of the jaw-adductor musculature. More anteriorly, the lower jaw becomes more cylindrical, and curves inward to meet its fellow in the midline in a movable joint, formed by flexible, fibrous connective tissue.

At the mandibular joint, upper and lower jaws enter a virtually double articulation with one another. Laterally, the upper jaw bears a ventrally directed condyle that enters a dorsal concavity within a short, subcylindrical prominence sculptured from the lateral face of the lower jaw. Medially, and slightly anterior to the first articulation, a second joint is found. Here, the lower jaw has a small, flattened condyle that extends dorsally to enter a shallow receptacle in the medial surface of the upper jaw. Movement of the jaws occurs by rotation at the lateroposterior joint, and sliding between upper and lower elements at the medioanterior contact.

This skeletal morphology found in active predatory sharks, such as the fast, predaceous lamniforms and carcharhiniforms, is roughly comparable to that of the more familiar dogfish shark (*Squalus*). The major differences include the relative sizes of the jaws and their musculature, size and angulation (from the chondrocranium) of the hyomandibula, the more completely chondrified rostrum in *Squalus* and the absence of a well-developed suborbital shelf in the latter, and in shape differences between comparable skeletal parts. Suspension of the jaws from the chondrocranium differs among shark groups, in that several have a postorbital suspension, or other connections, a subject treated by Compagno (1977) and Maisey (1980).

3.1.2 Teeth

Typically, in modern sharks, the lingual surfaces of both upper and lower jaws bear organized ranks of nonerect, developing teeth, which continuously shift forward toward the edges of the jaw cartilages as the teeth complete development. The outermost tooth of each rank is moved into position on the jaw margin, where it becomes erect. The fully developed teeth, thus positioned to bite prey, are never implanted in sockets, but are attached to the jaws by fibrous connective tissues. Their tenantship in this position is relatively short, perhaps between 1 and 2 weeks before the tooth is dropped, as its vacated site becomes available to receive a new tooth (Moss 1967, 1984; Reif et al. 1978; Luer et al. 1990); this high rate of replacement, insuring that an individual shark will not be long encumbered by a worn or broken tooth, may slow as the shark ages (but see Luer et al. 1990). In carcharhiniforms the bases of the teeth tend to overlap (Compagno 1984, 1988). Often there is a disparity in both shape and arrangement between upper and lower teeth (Fig. 2). In many active predators, especially those feeding upon large prey, the upper teeth may have nearly triangular, broad crowns, which have serrations along their edges (Bigelow and Schroeder 1948; Applegate 1965; Peyer 1968; Garrick 1982; Compagno 1984, 1988; Frazzetta 1988). Their front faces are flat, although in many forms the tip may be angled slightly forward.

The lower teeth crowns in many species are narrow, smooth-edged, and are often curved backward except for their tips, which are reversed to curve

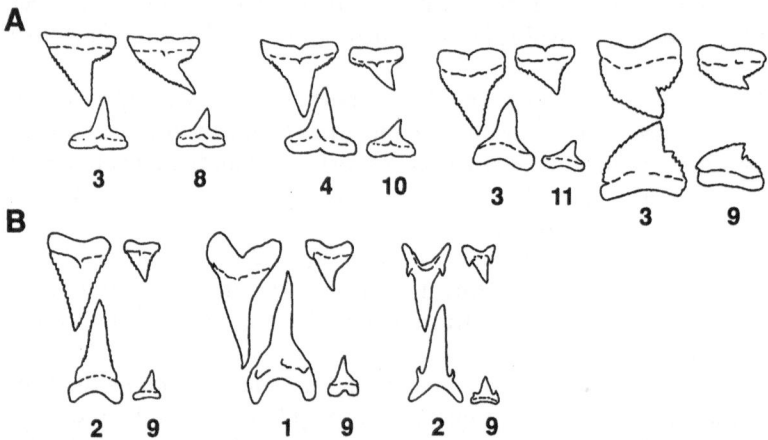

Fig. 2. Representative teeth of **A** carcharhiniforms; **B** lamniforms. Each group of teeth shows, on the *left*, an upper and lower tooth from a relatively central locus of the tooth row; to the *right* of each group is a relatively posterior upper and lower tooth. Positions of the teeth from the jaw's center are shown by the *number* below each. The species in row **A** are, from *left to right*: blacknose (*Carcharhinus arcronotus*), lemon (*Negaprion brevirostris*), oceanic whitetip (*C. longimanus*), tiger (*Galeocerdo cuvier*). In row **B**, the species are: great white (*Carcharodon carcharias*), short-fin mako (*Isurus oxyrhinchus*), sand tiger (*Eugomphodus* [= *Odontaspis*] *taurus*). (Redrawn from Bigelow and Schroeder 1948)

anteriorly (shown in Fig. 9). The disparity of form is considered to reflect the differing roles of upper and lower teeth. The lower teeth readily enter prey, holding or capturing it, with little resistance to penetration. However, the upper teeth produce greater damage in penetration, and are important in sawing pieces from a large prey.

In carcharhinids, the anteriormost teeth in the upper jaw form a single row with alternatively anteriorly and posteriorly overlapping bases (Fig. 3). Thus, in this anterior tooth row, one tooth will be slightly external to its adjacent neighbors on each side, its base anteriorly overlapping a portion of each neighbor's base; each neighbor, therefore, lies slightly internal to the tooth between them, and the base of each *posteriorly* overlaps those of its immediate neighbors. This system alternates along the entire tooth row. Behind each of the externally positioned teeth in the anterior row, there is often a semi-erect tooth that is just completing development. Usually, however, behind the internal teeth of the front row, the developing tooth has not yet achieved semi-erectness. This produces a second, more posterior tooth row, made up of alternating semi-erect and nonerect teeth (see Fig. 3).

A roughly similar arrangement is found for the lower teeth, except frequently the rows are not as close together as the upper tooth rows. Also, the second row usually contains a full set of semi- or fully erect teeth. And a third row, with alternately semi-erect and nonerect teeth is commonly present. The greater spacing between rows, and the greater number of erect teeth that are capable of engaging prey, produce a battery of spaced spikes to increase the likelihood of penetration.

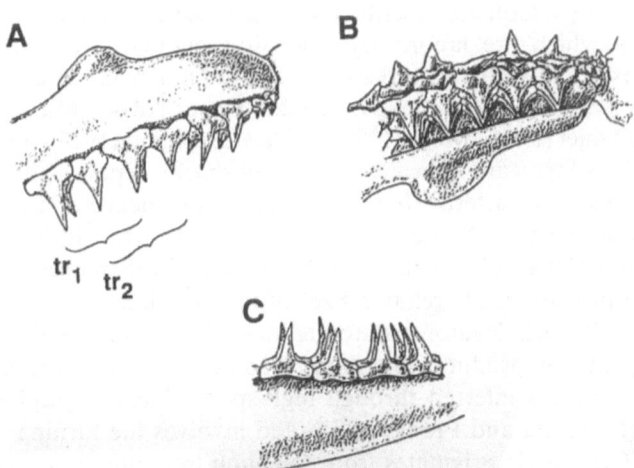

Fig. 3. A Anterior view of several upper teeth attached to the right side of the jaw, a short distance from jaw symphysis, in the lemon shark (*Negaprion brevirostris*). **B** The same teeth as in **A**, in ventral view to show second tooth row and developing teeth. **C** As in **A**, for lower teeth. The *brackets* labeled tr_1 and tr_2 correspond to two "overlapping triads" in the front upper tooth row (see text)

Tooth form is not constant along the extent of the jaws, and the more posterior teeth tend to become smaller, their crowns being virtually "bent" sideways to produce a marked curvature (cf. Fig. 2, and illustrations in Bigelow and Schroeder 1948; Garrick 1982).

In many sharks, including all carcharhinids examined, and the great white shark (*Carcharodon*), the erect teeth of both jaws are not rigidly fixed to the jaw margins. While the tooth-jaw attachment is strong (produced by binding of the tooth bases to the jaw through fibrous tissue), the teeth in fresh (and in some tests, live) shark specimens are easily rotated about their bases in both forward and backward directions. At the limit in either direction (from 20° to more than 40° from the undisturbed position), the tooth reaches a limiting rotational displacement (controlled by its fibrous connections to the jaw) and can be rotated no further. To a lesser extent the teeth can also be shifted from side to side. In carcharhinids, especially in the upper tooth rows, it can be seen that the overlapping bases can transmit movement from one tooth to another.

In the swimming mode, the upper teeth are positioned nearly horizontally, their tips pointing posteriorly and/or medially (depending upon whether the tooth in question is an anterior or posterior tooth). In contrast, the lower teeth are erect, and may even tilt slightly forward from their jaw attachments. In some sharks (e.g. *Eugomphodus* = *Odontaspis*) this is very pronounced and the lower teeth seem to "bristle" from the jaw.

3.1.3 Musculature

Many references describe the head muscles in the dogfish shark (*Squalus*), most of which are laboratory dissection manuals (such as those given above as examples). Other sharks have received much less attention; important literature on varied shark groups includes Vetter (1874), Marion (1905), Luther (1909), Daniel (1934), Moss (1972, 1977), Compagno (1977, 1988), and Nobiling (1977). This literature indicates that many sharks depart significantly from the dogfish muscular pattern. However, published evidence and the author's studies show that the actively predaceous lamnids and carcharhinids are myologically roughly comparable to the dogfish, although differences exist in the pattern, and especially in the relative sizes of the muscles.

The M. levator palatoquadrati (Fig. 4) lifts and protracts the jaws, and probably produces longitudinal rotation of the upper jaw rami. This rotational motion is inferred through high-speed cinematographic studies of live sharks (Frazzetta and Prange 1987), and involves the turning outward of each ramus. The muscle originates from the chondrocranium, above the anterior extent of the orbit, and sweeps downward and backward to insert upon the dorsal surface of the upper jaw behind the eye. Another muscle, the M. levator hyomandibulae, also lifts the upper jaws, but probably functions as a retractor, restoring the previously protracted jaws to their posterior position (but note that this view contradicts Moss 1972). The M. levator hyomandibulae originates from fascia

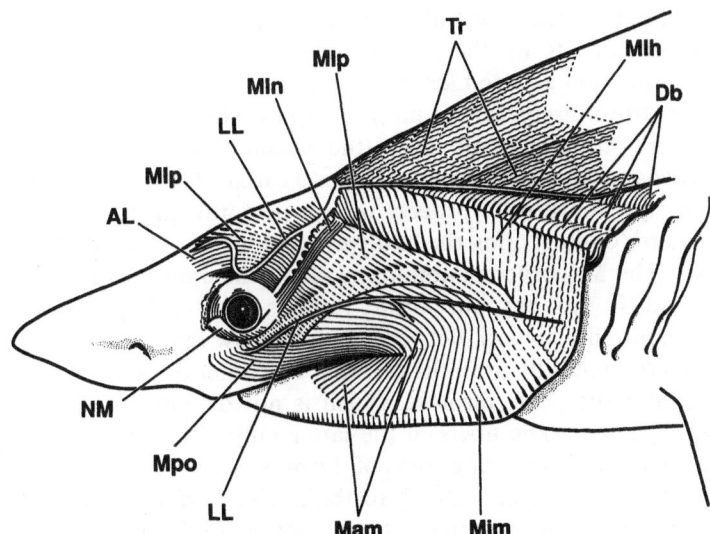

Fig. 4. Lateral view of head muscles in a representative carcharhinid shark (based on the blacktip, *Carcharhinus limbatus*). *AL* Bundle of Lorenzini ampullae; *Db* dorsal branchial muscles; *LL* lateral line; *Mam* M. adductor mandibulae; *Mim* M. intermandibularis; *Mlh* M. levator hyomandibulae; *Mln* M. levator nictitans; *Mlp* M. levator palatoquadrati; *Mpo* M. preorbitalis; *NM* nictitating membrane; *Tr* trunk musculature. During life, much of the dorsal portion of the M. adductor mandibulae is covered by a thick fascial sheet

covering the anterior trunk muscles, and the connective tissue associated with the lateral-line canal, and extends ventrally, its fibers focusing to an apex that inserts upon the hyomandibula and the connective tissue that ties it to the region of the jaw joint. A small slip of muscle, the M. levator nictitans, is found in carcharhinids (Compagno 1988, pers. obser.) extending from above and behind the eye, and coursing downward and forward to insert upon the eye's nictitating membrane.

A prominent, lateral, and lateroventral muscle is the M. adductor mandibulae (= M. quadratomaxillaris of many authors) that is subdivided into two major portions. The anteriormost part, called the M. preorbitalis, arises from the ethmoid region of the chondrocranium, medial to the bulbous expansion of the ipsilateral nasal capsule. From this anchorage the cylindrical muscle passes laterally, around the anterior and lateral surfaces of the orbital process of the upper jaw, and runs backward along the outer face of the upper jaw. It is partially divisible into two (in some cases three) portions. The anterior part inserts upon the lateral surface of the upper jaw, at roughly the level of the eye. The posterior portion attaches to the outer jaw surface near the corner of the mouth, and some fibers continue to insert upon the connective tissue associated with the jointed labial cartilage, which defines the rictal angle beneath the skin.

The major portion of the adductor muscle possesses fibers tightly bound to the outer surfaces of the upper and lower jaws by a tough connective tissue sheet.

The fibers extend ventrally, curving around the corner of the mouth, to an attachment site on the lateral surface of the lower jaw. This musculature is powerful, and acts to close the mouth.

Ventral musculature is more complex, and includes the M. coracohyoideus and M. coracomandibularis (the author has not found defined, common coracoarcual muscles in any of several carcharhinid species). These muscles arise from the "coracoid" process of the pectoral girdle, and respectively extend forward to insert upon the basihyoid and the anteriomedial surfaces of the lower jaws. The M. coracomandibularis opens the mouth, while the M. coracohyoideus aids in this, but more firmly retracts the stiff tongue and draws the floor of the mouth downward. Between the lower jaw rami are two muscle sheets, whose fibers run laterally from the midline. The anteriormost is the M. intermandibularis, and the posteriormost, which is not distinct from the former, is the M. interhyoideus. The fibers of the latter muscle run up behind the jaw angle to attach to fascia covering some of the musculature behind the head. The fibers of the intermandibularis attach to the medial surface of the lower jaw, and to a connective tissue sheet that extends over the tissues that bind the teeth to the jaw margin (pers. obser.; Nobiling 1977 describes the connective tissue sheets covering the jaws in *Heterodontus*).

3.1.4 Overlying Skin

The integument of the head is frequently neglected in functional studies of vertebrate jaws. Yet, the integument probably contributes constraints to jaw movement, acting as an enveloping or overlying ligament. Regionalization of the integument, varying between relative rigidity, flexibility, or extensibility, may have important controlling influences upon the cranial movements. Where the integument has been considered in functional research, its effects are found to be important (e.g. in snakes, Gans 1952; Frazzetta 1966). The present author and associates, G.R. Lawson and T.E. Audo, have demonstrated (in prep.) that the mechanical properties of the skin in the blacktip shark (*Carcharhinus limbatus*) are regionalized in a functionally sensible manner.

4 Feeding Behavior

4.1 Approaches to Bite Objects

Several patterns of engagement by a shark and its victim are reported in the existing literature. At the outset it is necessary to discriminate between approaches that are associated with feeding behavior, and those that seem to be involved with defensive or other motivations. For example, some defensive attacks (see Baldridge 1975) are less involved with actual bites than with a raking of the teeth to slash the attack object.

Other patterns of engagement are less clear in their intent. Adult lemon sharks (*Negaprion brevirostris*) have been observed by this author to take a large fish or other object in their jaws, forcefully enough to pull against a resistance, but leave little evidence of tooth penetration. It is uncertain whether this is a "test" bite to determine if the object is suitable as food, or related to some defensive action.

Engagements directly involved with feeding in active predaceous sharks have been described by Springer (1961) for several shark groups, by Tricas and McCosker (1984) and Tricas (1985) for the white shark (Lamnidae: *Carcharodon carcharias*), and by Moss (1972) and Frazzetta and Prange (1987) for several species of Carcharhinidae (in the case of the last paper, *Negaprion brevirostris*, *Carcharhinus acronotus*, *C. limbatus*). With minor exceptions, these accounts of shark feeding are in agreement.

Several feeding modes were documented in Frazzetta and Prange (1987), and were correlated with prey size relative to the feeding shark. A "small" prey is readily engulfed and does not require an initial grasp. "Medium" prey necessitates a grasp between upper and lower jaws, to hold or capture it, before swallowing. And "large" prey must have pieces excised from it by the shark as the entire prey object is too large to be ingested whole. The modes of feeding include the following:

1. Through suction a small prey is drawn toward the shark when it is within a few centimeters of the shark's mouth. Presumably, the suction is produced by reduction of pressure in the buccal cavity, mediated by hyoid musculature and the synchronized operation of gill-slit openings and closings, analogous to the mode reported for osteichthyian fishes (e.g. Alexander 1967a); perhaps this feeding operation occurs through a modification of respiratory movements (reported for a squaliform by Hughes and Ballintijn 1965).

2. The shark swims toward a small prey, opens its mouth, and engulfs it by virtually swimming over it.

3. A medium-sized prey is approached by the shark (Fig. 5) as it opens its mouth; once in range, the mouth closes upon the prey to secure it (perhaps, in some cases, also to partially disable it), then it reopens its mouth to ingest the captured object.

4. A large prey is seized by the shark (Fig. 6), often "mouthed" many times, as if to find either an excisable portion, or a usable angle of bite. Once this region of the prey is located, the shark may bite many times, possibly to gain a more secure purchase for its jaws. Following this, a firm bite is taken, the head is shaken rapidly from side to side, and a piece from the large prey is thus removed. Very often, the kerf formed is smooth and neatly cut.

In the last two modes, where the jaws clamp the prey with their teeth, it is often the lower teeth that make the first, penetrating engagement (Springer 1961; Gilbert 1962; Moss 1972).

A sequence of jaw actions has been noted for both the (lamnid) white shark (Tricas and McCosker 1984; Tricas 1985), and the carcharhinids (Frazzetta and Prange 1987), when medium-sized prey is grasped. The shark swims toward the

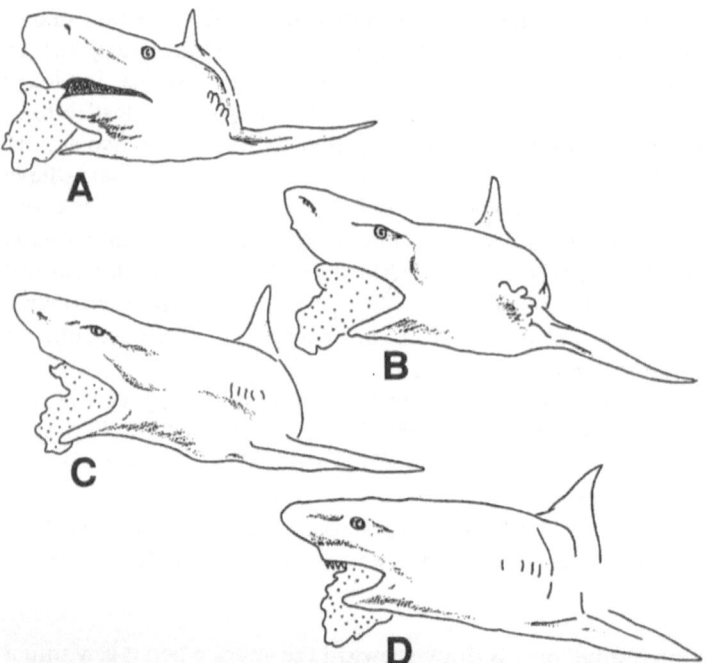

Fig. 5. Feeding in a lemon shark (*Negaprion brevirostris*) on a chunk of fish. **A** As the shark approaches, the mouth begins to open. **B** Mouth at greatest opening, fish is within grasp, impaled by lower teeth. **C, D** Mouth begins closing, protraction occurs (note bulge in upper jaw, exposed teeth), upper teeth are oriented vertically, probably a result of longitudinal twisting of upper jaw rami (see text). (After Frazzetta and Prange 1987)

subject, and within a small distance begins to open its mouth. As this occurs, the entire head is lifted, tilted upward through flexion of the anterior vertebral column. When the prey is within the range of the jaws, the mouth opening is maximal, and within the next fraction of a second, the jaws begin to close.

Closure of the jaws is not simple. The lower jaws lift, but as they do, the entire head is tilted downward. As this occurs, usually both the upper and lower jaws are displaced forward (protraction), while the upper jaw descends, relative to the chondrocranium (see Fig. 5). In viewing high-speed motion pictures, the upper teeth often appear to become erect, just as the bite begins. Many films, both those made for study and for commercial entertainment, can mislead a casual viewer; the upper teeth appear to become erect *relative* to the jaw cartilages. In fact, I have found no mechanism for repositioning the teeth relative to their jaws. Frazzetta and Prange (1987) report that, through analyses of their motion picture films, the erectness of the upper teeth may occur by the outward rotation of the paired upper jaws around their longitudinal axes. This action turns each jaw ramus and its teeth *together* until the teeth are oriented vertically.

The result of the described jaw-element displacements is that, at the moment of capture, the prey is stabbed by the lower teeth, as the upper teeth descend

Fig. 6. Feeding of a lemon shark (*Negaprion brevirostris*) on a whole fish. **A, B** After seeking a bite site on the prey, the jaws begin to protract *before* closing (cf Fig. 5); lower teeth are engaged. **C** Jaws close, upper teeth are visible in a vertical orientation. **D** Upper teeth now impale prey. After prey is clamped, the head is shaken, and a piece of food neatly excised (see text). (After Frazzetta and Prange 1987)

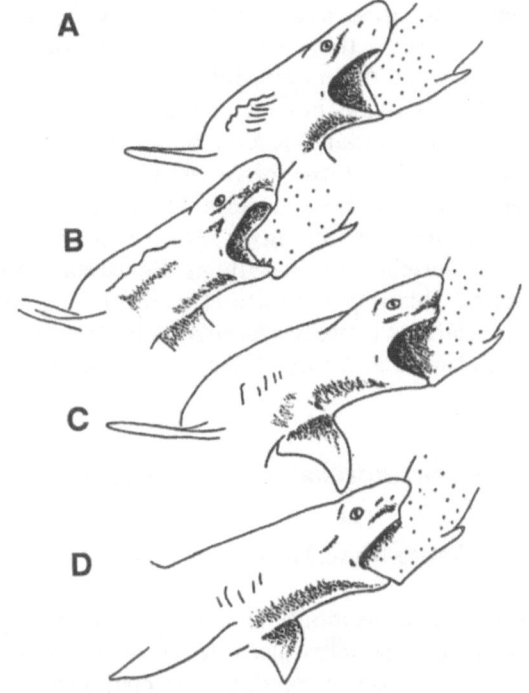

vertically upon it. These events occur simultaneously with the articulation between upper and lower jaws being thrust forward, downward, and outward, thus laterally expanding the head region posteriorly (Fig. 5).

Deviations from this pattern exist: in the white shark, Tricas and McCosker (1984) indicate that jaw protraction did not occur when feeding under water, but did so regularly above the water's surface. This was not the case in the carcharhinids studied by Frazzetta and Prange (1987), in which the jaws of most of their underwater sharks were protracted while feeding. These latter authors also note that while feeding upon large prey, whether above or below the water's surface, carcharhinids may drive the upper teeth forcefully into the food object well after the jaws have been protracted (Fig. 6). During such feeding episodes, the mouth will open and, before it begins to close, protraction will occur just *prior to* the forceful bite.

A recent paper (Motta et al. 1991), appearing after the present chapter had been submitted, is of considerable interest in that it documents an electromyographic study of jaw movements in an actively predaceous shark (the lemon shark, *Negaprion brevirostris*). While the techniques of electomyography in investigations of feeding in several vertebrates has produced promising insights, its application to active sharks is difficult at best. Such sharks require considerable swimming space, which they actively use, thus risking the tangling of connecting wires. There are other problems inherent in the reactions of these sharks to stress, and according to the experiences of several workers (including pers. obser.), they easily lose motivation for feeding. Motta et al. (1991) approached the subject differently: these researchers did not attempt to record feeding activity, but challenged their lemon sharks to bite defensively. The investigators found little or no jaw protraction during the bites, an observation at odds with those of other authors (see above). Possibly, then, defensive activities involve a pattern of jaw movements that differ from movements associated with feeding (as Motta et al. suggest; and which has been reported for other vertebrates, e.g. pythons, Frazzetta 1966). Although such questions can be raised, this paper (Motta et al. 1991) is a valuable contribution regarding its correlations between muscular activity and jaw movements, and has additional importance in raising those very questions.

5 Morphomechanics and Function

5.1 Skeleto-Muscular Aspects of Jaw Movements

Opening of the mouth probably occurs through the action of the coracomandibular and, perhaps also, the coracohyoid musculature. Contraction of these muscles places a caudally directed force upon the lower jaws and hyoid elements; this results in an initially small torque acting to depress the lower jaws. As depression continues, the magnitude of the torque increases.

Closure of the mouth is produced by the action of the jaw adductor musculature, whose active tension pulls upper and lower jaws together, thus elevating the lower jaw. Jaw protraction, which usually occurs as the mouth closes, is largely produced by action of the preorbitalis muscle. Because of its connection to rictal tissues, this muscle may also have a mild effect in elevating the lower jaw.

In several carcharhinids studied the forward protraction of the jaws involves several concomitant events (Fig. 7). The hyomandibula swings outward, forward, and downward as it is virtually pulled by the anterior shifting of the jaws (see Moss 1972). Thus, the posterior ends of both upper and lower jaws become splayed outward as they slip anteriorly. The hyomandibula seems strongly constrained to swing in an oblique plane that combines both horizontal and vertical components; thus, during protraction, the hyomandibula controls the displacement of the posterior jaw angle rather rigidly. Anteriorly, as the jaws are drawn forward, the orbital process of each upper jaw ramus is pressed against the expansion of the suborbital shelf (Fig. 7). This results in a sliding of the orbital processes downward against the shelf's expansion, as well as forward.

As the upper jaw descends during protraction, the ethmopalatine ligament becomes unfolded, and when taut it limits further protraction. Before it is completely extended, it may offer little in the way of constrainment to the moving mechanism. Hence, other factors of muscular tension, and resistances in

Fig. 7. Diagrams of head-skeleton displacements occurring during jaw closure and protraction. **A** Lateral; **B** dorsal (cf. with corresponding views in Fig. 1). *Wide, black arrows* show movements from rest to protraction. *el* Elevation of lower jaw; *lr* longitudinal rotation of upper jaw; *or* outward rotational swing of jaw joint region; *st* sliding translational movement of orbital process on the suborbital shelf (see text)

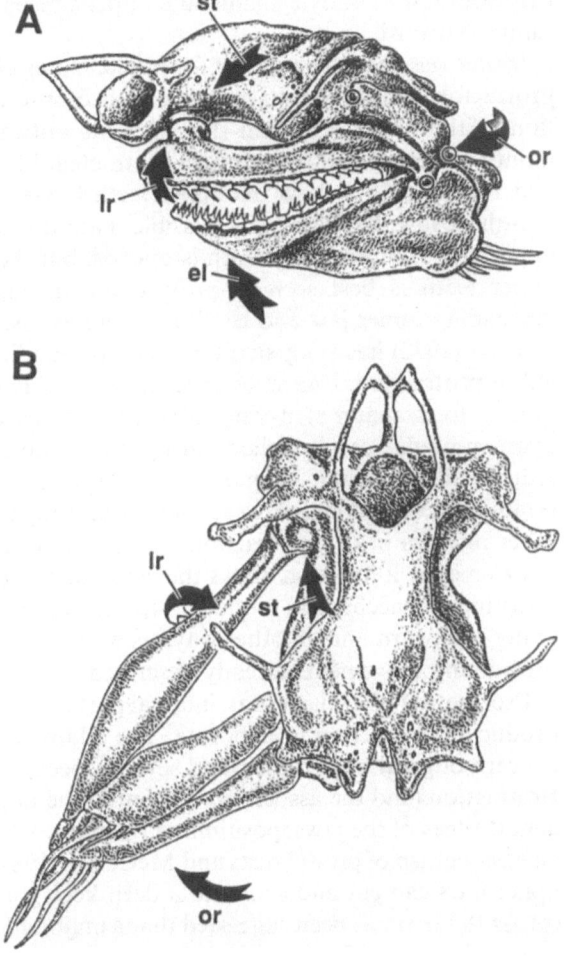

connective tissues elsewhere, presumably act to reduce the kinematic degrees of freedom to allow relatively precise control over the cephalic linkage (see Frazzetta 1966, for examples in snakes: Shigley 1961, for a general treatment of constrained mechanical linkages). It is likely that the protractor palatoquadrati musculature may aid the preorbitalis muscle in jaw protraction.

Longitudinal rotation of the jaws seems to be produced by at least two factors. In dissections of fresh sharks, mandibular depression results in an outward rotation of the upper jaws (thus bringing the attached teeth to a more vertical position), and a slightly inward longitudinal rotation of the lower jaws. The longitudinal movements are mediated by the medioanterior jaw joint which, as described above, is a sliding articulation. As the mouth opens, the joint surfaces glide upon each other, and are shaped to produce a lateral "wedging" of the lower jaw's condyle against the upper's receptacle, thus turning each upper ramus outward.

Moreover, manipulation of dissected material suggests that tension in the protractor palatoquadrati muscle can also produce outward longitudinal rotation of the upper jaw rami (Fig. 7). This outward rotation can occur as the mouth is closing and the jaws are protracted. Motion picture films of Frazzetta and Prange (1987) show the upper teeth to be vertically oriented only during mouth closure. However, it is possible that outward longitudinal rotation of the upper jaw begins as the mouth is opened, but the more vertical position of the upper teeth is best seen as protraction (during the bite) occurs, when the descending upper jaw and teeth are more exposed to view.

Moss (1972) has suggested that the hyomandibular levator muscle may also aid in protraction. I have no experimental evidence to support this; manipulation of fresh material during dissection shows that the fibers of the levator hyomandibulae are stretched during protraction, but that restoring *retraction* reduces fiber tension. Hence, I tentatively assume that the hyomandibular levator is involved mainly in retraction, although it may act synergistically with other muscles in the protraction phase to adjust longitudinal rotations of the jaw elements. Retraction draws the jaws back to their more posterior positions, restoring the mechanism to the non-feeding mode. It forces the hyomandibula to swing backward and, in other ways as well, retraction involves a reversal of the protracting movements already described.

Protraction of the jaws is an integral part of shark feeding, and the mechanism producing this action is elaborate. The adaptive basis of protraction is thus a concern of great importance, and several speculations have been brought forth. Protraction (and the associated descent of the upper jaws) expose the teeth and dental edges of the jaws, positioning them away from the snout, and permitting an easier grasp of prey (Tricas and McCosker 1984). Also, the jutting, protracted upper jaws can cut and sink into a deep kerf, gouging pieces from a large prey (Moss 1977). It has been suggested that a major adaptive role of protraction is its production of a fast, forward reach of the jaws to grasp prey. I find this last suggestion mostly unconvincing as a basic adaptive aspect in that the structural complexity of the protraction apparatus seems too intricate for so simple an effect; moreover, the direction of movement of the mouth axis is at least as much downward as forward, and thus not in line with a forwardly directed, fast-swimming attack upon prey.

Frazzetta and Prange (1987) have considered that protraction – involving the descent of the upper jaw – results in bringing the upper teeth down upon the prey, while the mandibular teeth are driven upward; this rough simultaneity of tooth contact may insure a better grasp of potentially elusive prey. By analogy, for certain reptiles having kinetic upper jaws (Frazzetta 1983), it has been proposed that the momentum of the lower jaw, as the mouth closes, may deflect a prey away from the mouth at the intended moment of capture. However a near simultaneous strike by *both* upper and lower jaws – moving to come together from opposite directions upon the prey – minimizes the risk of deflection. Clearly, much caution should accompany comparisons between such different animals as kinetic reptiles and sharks, especially when one captures prey terrestrially, the other under water. In sharks the lower teeth will usually pierce, hence transfix, prey reducing the tendency to deflect it. However, should penetration by the lower teeth fail, as might occur if a hard portion of the prey is engaged, descent of the upper teeth may have an important role in reducing the risk of losing the prey. Even if penetration by the lower teeth occurs, the prey may still not be secured, and deflection may yet occur after momentarily impaling the prey. It is most likely that protraction is adaptive in not merely one role, but in several.

These accounts of jaw movements through different phases of feeding activities omit the action of the gill skeleton and musculature. In fact, however, it has been noted that the gill slits are opened and closed in concerted fashion throughout the feeding cycle (Frazzetta and Prange 1987). Consideration of gill function, perhaps in altering intraoral pressure through different phases of a feeding bout, will be important in any broad assessment of feeding mechanics. It is clear that the combined actions of jaw and gill structures are highly complex (see Hughes and Ballintijn 1965 for an analysis of gill movements during respiration in a squaliform shark).

Presumably, buccal pressure reduction is the means by which sharks can suck prey into their mouths from some small distance away (see Sect. 4.1). Tanaka (1973) measured this reduction for nurse sharks (*Ginglymostoma cirratum*); he found that the intraoral pressure could be reduced by 1 atm.

Bite forces produced by large carcharhinids have been measured by Snodgrass and Gilbert (1967) who concluded that a large requiem shark can produce about 3 metric tons of force per cm^2. This large figure is based upon a measurement method that is very indirect in that it involves a rather subjective estimate of unit pressure, derived from marks made on a metallic device, and subsequent calculation of the total bite force per cm^2 from the assumed pressure and an estimate of tooth-tip areas.

The loosely-fixed teeth, already mentioned, raise questions about the nature of the bite. Are the teeth free to flex on the jaw cartilages? If so, how are they controlled – if at all – during penetration of prey? I have attempted some experimental approaches to these questions (Frazzetta unpubl.) based on analyses of tooth marks made during bites. Blocks of dental bite wax were molded with hard plastic or metal disks embedded within them. Each block was sewn into a large "prey" fish, and retrieved for examination after the bite. While the study has not yet been completed, interim results suggest that in some cases an individual tooth can deviate (out of line from the others) around a hard inclusion, whereas in other tests, the tooth is rigid and unaffected by engagement with the hard inclusions. These tentative results suggest that the shark may have some voluntary control in the ability to vary the strength of tooth fixation. The observations require fuller substantiation, but invite speculation.

It is relatively easy to conjecture about lower tooth fixation. Tension in the intermandibularis muscles, artificially produced during manipulation of dissected specimens, readily tightens the ligamentous sheath covering the dental surfaces of the lower jaw. The sheath is securely attached to the bases of the lower teeth and, when taut, holds the teeth securely, enough so that they resist movement in response to forces placed upon their crowns. Fixation of the upper teeth is not so straightforward. To date, no particular group of muscles has been recognized as having a direct effect upon holding the upper teeth rigidly upon their jaw cartilages. Moreover, tension per se, in the ethmopalatine ligament has but a minor effect on fixing the upper teeth (by increasing tension in a covering sheath, like that described for the lower jaw). An intriguing possibility is that the combination of tension in the ligament, and simultaneous contraction of the preorbitalis muscle, may produce tension in the sheath covering the soft tissues of the upper jaw. Direct manipulation of this sheath during dissection shows that tension in the sheath does, in fact, fix the upper teeth through its attachment to the tooth bases; but the question of what, if anything, produces sheath tension in normal feeding is not resolved.

Whether, during the bite, the attachment of the teeth to the jaw margins is loose, rigid, or can be varied under control by the shark, is of the greatest importance, and has implications for all aspects of jaw function analyses. As sufficient, clarifying data are incomplete, it is risky to speculate on the possible adaptive significance of loosely affixed teeth. There are, however, several obvious possibilities. One of them is that there may be *no* mechanical-adaptive role for loose teeth, the looseness being merely a consequence of the rapid dental replacement, making a rigid attachment structurally impossible if an old tooth is to be readily discarded shortly after its initial placement on the jaw (see above). Even if the shark is capable of fixing the teeth during the bite, the fixation process may be a momentary effect to *counter* the basic looseness; hence, the looseness itself has no direct, adaptive value.

A second possibility is that looseness – especially if its degree can be controlled by the shark – may mechanically favor unrestricted cutting when pieces are to be removed from large prey. As the tooth row must slice through the prey tissues in a saw-like fashion, a resisting bone or other hard part in the prey could impede one or a few teeth, and halt the slicing movement of the entire tooth row. However, if the impeded tooth were loose, it need not rigidly restrict the movement of the total dental battery. Another, and closely related possibility, is that while the shark tooth rows are roughly semicircular, they are not rotated about the circle's center during cutting. Unpublished observations on feeding lemon sharks suggest that the instantaneous center of rotation (the Instant Center; see e.g. Shigley 1961) of the jaws' cutting movements shifts rapidly during excision of pieces from prey; but while the Instant Center continuously changes its location during the bite, it usually lies well behind the jaws. Hence, at any given time, only one or a few adjacent teeth are in a position to cut, with their edges in line with the jaw movement (Fig. 8). The cutting edges of the remaining teeth, at that instant, will be out of the line of the jaw movement, and indeed, several of these teeth will have the flat surfaces of their crowns forced against the prey tissues into which they are embedded. If this view is correct, at each instant, many or most of the teeth will act as impediments to slicing jaw movement, *unless* they are loose enough to flex at their bases,

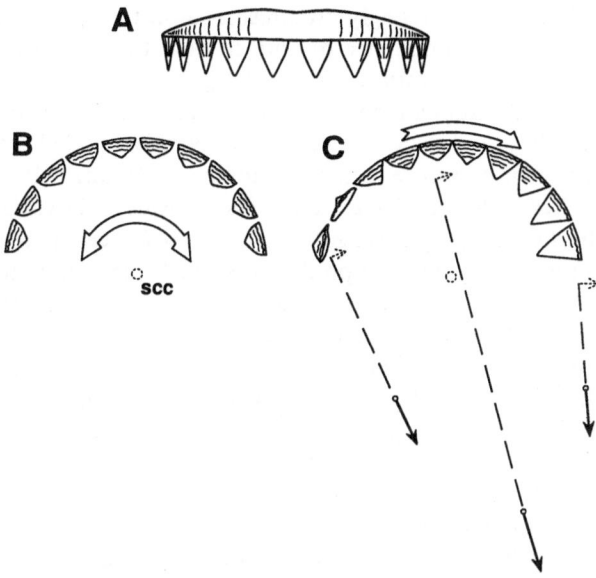

Fig. 8. Diagram of hypothetical cutting device with a semicircular base and corresponding row of tooth-like blades flexibly attached to it. **A** Front view of device; **B, C** Ventral views. **B** Orientation of "teeth" if cutting occurs by rotation about the semicircle's center, *scc*; the *double-headed arrow* shows that cutting can occur equally in either direction, and indicates the cutting direction: hence, the arrow is concentric with the tooth row. In this mode, all "teeth" share in the cutting operation. **C** The Instant Center of rotation (IC; see text) is *not* located at point *scc*, but for the instant of cutting shown, is located at the convergence of the three arrows (off the page). In this case, the semicircle of "teeth" is not rotated about the center. The cut occurs at right angles to any axis drawn from the IC (note *dotted arrows* from the three axes shown). If cutting occurs by jaw movement to the right (*wide arrow*) of the IC, only the teeth near the middle axis (of the three axes shown) apply their sharp edges in the direction of cutting. The more lateral teeth (left and right) are out of line to cut, their *broad* surfaces in fact are pressed upon the prey tissues in which they are embedded. If the left teeth can rotate outward upon the base, and the right inward, the cutting action will not be impeded. As the IC shifts in the next instants, another group of teeth will be poised to cut

permitting the few teeth in a cutting position to move edgewise – without restriction – against the prey tissues (see Fig. 8).

5.2 Cutting Mechanics of Shark Teeth

The material composition and physical hardness of shark teeth have been studied and related to tooth function (Reif 1973; Preuschoft et al. 1974). The variety of forms of shark teeth have been described and illustrated by a number of authors (Bigelow and Schroeder 1948; Garrick 1982; Compagno 1984, 1988) for extant sharks; fossil shark teeth have been treated by Case (1973) and Capetta (1987).

There is surprisingly little research on the actual cutting mechanics of shark teeth (or of similar teeth) and, in fact, there is not much in the mechanical-engineering literature on the design of cutting implements for piercing and severing compliant materials (such as animal tissues, which are compliant in that they tend to deform considerably before actual cutting occurs). The existing, pertinent literature includes Prindle and Walden (1976) and Frazzetta (1966 on snakes; 1988 on shark teeth). The considerable variety of tooth forms among sharks makes generalization risky. However, some recurrent themes prevail among many actively predaceous sharks. (See note following References)

Often the lower teeth of such sharks are smooth, or nearly so, and while they may be recurved over much of the crown, the tooth tip is frequently seen to possess a reversed curvature. This orients the tip forward permitting a closer alignment of the tip axis with the direction of the major force of contact between shark and prey, thus greatly increasing the liklihood of initial penetration (Frazzetta 1988; see Fig. 9). Once the tooth has entered the prey, even slightly, it tends to resist slipping along the prey's surface. Following the initial penetration of the tip, the rest of the tooth follows through the puncture. The major part of the tooth's crown is curved backward, adding security to the capture after initial penetration. The unserrated edges of so many lower shark teeth sacrifices some sideways slicing ability for greater ease in piercing food.

The upper teeth of actively predaceous sharks are more variable (see Fig. 2). It is common for some serrations to be present near the bases of the crowns, even if they occur nowhere else on an upper tooth. Frazzetta (1988) suggests that the row of teeth is analogous to a woodworking saw, and that the serrations near the bases act as a "clearance" device, common in most types of commercial saws. Substrate material that is compressed between tooth bases impedes movement of the saw – or of tooth rows – and must be either cut free or allowed to escape compression. Serrations cut tissues as they become compressed between adjacent teeth; these tissues, if left uncut, would produce a jamming of the teeth in the tissue substrate. Thus, because the potentially jamming tissue between teeth is cut as it is compressed, that tissue is in fact released from an impending jam.

In general, smooth, sharp blades (such as the unserrated tooth edges of certain sharks) cut by shearing the substrate material through the friction produced by drawing the edge against the object to be sliced (Frazzetta 1988). In the case of serrated blades, however, the cut is made by forcing the substrate to conform, even if but very slightly, to the points and intervening spaces of the scalloped edge. Drawing of a serrated blade against the substrate tears out of the material between the serration cusps. Hence, serrated teeth are not dependent upon friction, and their cutting force is not limited by the maximum possible friction value. Such teeth can make use of all the available bite force, and are potentially far more damaging than smooth teeth. But serrated teeth penetrate and withdraw with less ease than a smooth tooth, and are thus more likely to bind in the prey material. Other aspects of the mechanics of shark tooth form are discussed in Frazzetta (1988).

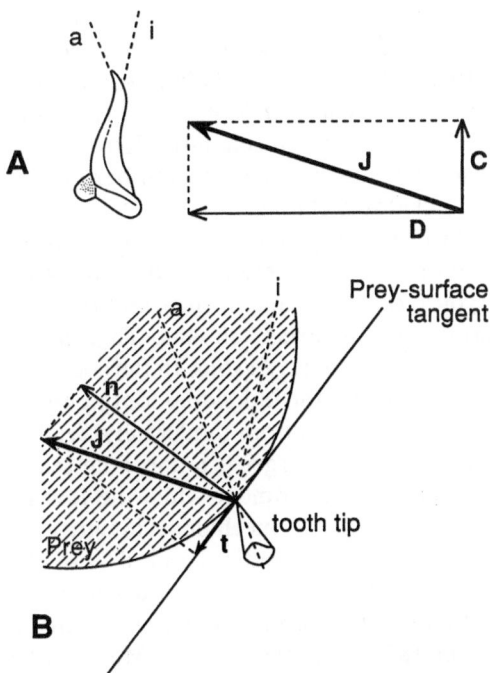

Fig. 9. A Side view of a relatively anterior lower tooth in the lemon shark (*Negaprion brevirostris*) showing its reversed curvature near tip. Line *a* is the actual central axis of the tip; line *i* is the imaginary axis if the tips's curvature were not reversed, but in line with the body of the tooth. To the *right* is a vector addition of forces of the tooth tip upon a prey surface. Vector *D* is the force generated from the forward swimming of the shark. Vector *C* is the force produced by elevation of the lower jaw during mouth closure. **B** Vector diagram of tooth tip as it is forced against the prey surface. The resultant vector *J* is resolved into two components relative to the prey's surface (note line showing tangent to that surface). Components are *n* and *t*, the former being perpendicular to the prey surface, the latter being tangential to it. If friction along the skin's surface is zero, the only applied force will be *n*; however, if friction is as great as − *t*, the applied force will be represented by *J*. And if the frictional value lies between these two extremes, the actual vector direction and magnitude, of the tooth-tip force against the prey, will be altered accordingly. Whatever the case, the actual tooth-tip axis *a* willl lie nearer to coincidence with the vector line than will *i*, thus increasing the relative likelihood of penetration when the tip axis is oriented anteriorly (as is *a*; see Frazzetta 1988, for further analysis)

6 Other Elasmobranchs

6.1 Sharks

The frilled shark, *Chlamydoselachus anguineus*, has many features (see Allis 1923; Schaeffer 1967) that could suggest some affinity with ancient forms such as the Devonian *Cladoselache*, and its jaw mechanics may be of interest in providing clues to the feeding function of early sharks. Luther (1909), Compagno (1977), and Maisey (1980) suggested that the jaws of *Chlamydoselachus* are movable. Moreover, Lauder's (1980b) reconstruction of a preorbital muscle for the Devonian *Cladoselache* implies a view that, even in this ancient shark, there was a jaw protraction mechanism. However, Compagno (1977) regards *Chlamydoselachus* as fitting well into the hexanchiforms, and diminishes its comparison with cladoselachidans.

Moss (1977) provides a classification of feeding types among elasmobranchs that distinguishes between cutters and gougers; the former cuts pieces from prey (typical of many squaliforms), and the latter digs into large prey, with jaws well protracted, to excise chunks. This distinction is not always clear, as in the cookiecutter shark, *Isistius*. In the cookiecutter, the lower tooth row is formed as a curved band, the individual tooth points acting as saw teeth. These sharks can readily cut pieces from moderately-sized prey (Compagno 1984), but frequently they will attack larger fish and even sea mammals, using a suctorial mouth to maintain an attachment while the lower tooth band slices and scoops out a chunk of flesh. The mechanical integrity of the lower tooth band is conserved by replacement of the entire band of teeth at each replacement period, rather than by loss and renewal at individual tooth loci (Strasburg 1963).

In some lamniforms and ctenacanths, as well as in hexanchids (see Gegenbauer 1872; Compagno 1977; Maisey 1980; Stahl 1988), there is a postorbital suspension of the jaw apparatus, in addition to the two suspensory connections already described. Such sharks possess a cartilaginous postorbital expansion of the chondrocranium which has a ligamentous connection with the palatoquadrate. It is quite likely that in sharks having this connection, the kinematics of their jaws differ in some respects from the jaw mechanism described above for carcharhinids.

Other structural differences in the feeding apparatus include the very large labial cartilages in certain sharks (e.g. hexanchids, orectolobiforms; Gegenbauer 1872; Daniel 1934; Moss 1977) which may have special mechanical significance, although it has not been investigated. The very peculiar lamniform shark, *Mitsukarina*, has a jaw configuration that seems permanently in a state of protraction (so described by Luther 1909).

As noted before, the diversity of tooth design among sharks is extentisive, too great to document here in any detail. Within that diversity are teeth that are serrated, or smooth-edged; having crowns whose axes are vertical, or angled to the side; teeth suited to the capture of elusive prey and/or to the removal of chunks of tissue from large food objects. Other configurations of teeth are designed to crush crustaceans and mollusks. Included in this last category are the small but numerous teeth, with small, sharp cusps, of the nurse shark, *Ginglymostoma cirratum*. Pavement-like teeth, appropriate for crushing, occur in several sharks (as the triakid *Mustelus*), and in myliobatid rays.

The largest forms, like the whale shark (*Rhiniodon*), basking shark (*Cetorhinus*), and the recently discovered megamouth (*Megachiasma*), are filter feeders on planktonic organisms, possessing large, filtering gill rakers. Saw sharks

(Pristiophoridae), like the batoid sawfish (see below) presumably use their tooth-studded rostral bill to disable small fish and squid, after which the prey is readily engulfed.

6.2 Batoids

Batoid sawfish, *Pristis*, have been observed by Breder (1952) to slash small food objects with the toothed bill, then recover them at leisure. More typical batoids are the skates and rays, which rely heavily upon jaw protraction in feeding. Luther (1909) noted that in these forms, the jaws extended ventrally, and flexed at the mandibular symphysis to open the mouth as a roughly diamond-shaped aperture. This permitted an extension of the mouth toward the sea bottom, enabling the grasping and swallowing of moderately sized prey. Moreover, the enlarged, subcircular mouth opening was also appropriate for sucking prey from the substrate. Moss (1977) noted that rajid batoids have jaws and teeth adapted for grasping small prey. The totally inferior mouth, even when the jaws are protruded, requires that prey taken from open water be overswum by the rajid (McCormick et al. 1963).

Myliobatids, with their crushing dentition, concentrate on sucking food from the substrate, and breaking small mollusks and crustaceans (Moss 1977). Stingrays (dasyatids) will burrow in soft bottom substrates seeking prey (Babel 1967). Batoids will often uncover prey by blowing a jet of water against the substrate upon which they lie (Gregory et al. 1979). The pair of spiracular gill openings is located dorsally, and functions in water intake. Water is presumably blown through the ventrally located gill slits, as a valvular arrangement impedes water exhalation through the mouth (Daniel 1934).

7 Conclusions

Fast, actively predaceous sharks have been the focus of this account, in part because more is known of their jaw mechanics, but also because it is likely that early sharks were fast predators. Many questions remain about the ancient sharks, and surely the operation of their jaws was different from today's carcharhinids. Whether early sharks had movable jaws, analogous to the mechanism so widespread in extant elasmobranchs, is a matter of importance in understanding the biology and evolution of chondrichthyians.

Structural diversity in elasmobranchs is limited in comparison with bony fishes. Members of this latter group have a number of dermal bones that can be fashioned as jaw-machine components in the numerous osteichthyian evolutionary radiations. Within this limitation, however, elasmobranchs seem to have tested a number of adaptive styles. The great variety of shark tooth shapes is indicative of numerous prey and feeding specializations. Different behaviors and habitat choices add to the possible diversity of feeding niches, as have varied jaw and muscle proportions and arrangements, and modifications of the entire body form.

Throughout the elasmobranchs, the protractable jaws appear to be a major motif in the design of the feeding mechanism. While the components of this apparatus are constructed of calcified cartilage, not bone, elasmobranchs have been variously adapted to subdue and reduce prey that may be active, or possess durable body coverings. However, the significance of certain aspects of the feeding mechanism, such as the loosely affixed teeth in modern sharks, is not understood, despite their importance in any analysis of jaw mechanics.

Compared to the relatively numerous studies of bony-fish jaw systems, investigations of elasmobranch cranial function are sparse. Sharks and their allies are an important vertebrate group, whose lineage is long and which may represent an early gnathostomous evolutionary line.

Acknowledgments. I am greatful to C.S. Fucciolo (Illinois State Geological Survey) and G.R. Lawson (Department of Ecology, Ethology, and Evolution, University of Illinois) who read the manuscript, and to A.A. Prickett (Artist Service, School of Life Sciences, University of Illinois) who aided in the preparation of the illustrations.

References

Alexander RMcN (1967a) Functional design in fishes. Hutchinson University Library, London

Alexander RMcN (1967b) Mechanisms of the jaw of some atheriniform fishes. J Zool (Lond) 151: 233–255

Alexander RMcN (1969) Mechanics of the feeding action of a cyprinid fish. J Zool (Lond) 159: 1–15

Alexander RMcN (1970) Mechanics of the feeding action of various teleost fishes. J Zool (Lond) 1902: 145–156

Allis EPJr (1923) The cranial anatomy of *Chlamydoselachus anguineus.* Acta Zool 4: 123–221

Applegate SP (1965) Tooth terminology and variation in sharks with special reference to the sand shark, *Carcharias taurus* Rafinesque. Los Ang Cty Mus Contrib Sci 86, 18 pp

Ashley LM, Chiasson RB (1988) Laboratory anatomy of the shark. WC Brown, Dubuque, Iowa

Babel JS (1967) Reproduction, life history, and ecology of the round stingray, *Urolophus halleri* Cooper. Calif Resour Agency Dep Fish Game: Fish Bull 137, pp 1–104

Baldridge HD (1975) Shark attack. Berkeley Medallion Books, New York

Bigelow HB, Schroeder JWC (1948) Fishes of the western North Atlantic. Mem Sears Found, Mar Res Part 1. Yale University Press, New Haven

Bohensky F (1981) Photo manual and dissection guide of the shark. Avery, Wayne, NJ

Breder CM (1952) On the utility of the saw in the sawfishes. Copeia 1952: 90–91

Capetta H (1987) Chondrichthyes II. Mesozoic and Cenozoic Elasmobranchii. In: Schultze HP (ed) Handbook of paleoichthyology, 3B. Fischer, Stuttgart, pp 1–193

Carroll RL (1988) Vertebrate paleontology and evolution. Freeman, New York

Case GR (1973) Fossil sharks. Pioneer Litho, New York

Compagno LJV (1977) Phyletic relationships of living sharks and rays. Am Zool 17: 303–322

Compagno LJV (1984) Sharks of the world. FAO species catalog, 2 vols. FAO Fish Synops, Rome

Compagno LJV (1988) Sharks of the order Carcharhiniformes. Princeton University. Princeton University Press, Princeton

Daniel JR (1934) The elasmobranch fishes. University of California, Berkeley

Frazzetta TH (1966) Studies on the morphology and function of the skull in the Boidae (Serpentes). Part II. Morphology and function of the jaw apparatus in *Python sebae* and *Python molurus*. J Morphol 118: 217–296

Frazzetta TH (1983) Adaptation and function of cranial kinesis in reptiles. Time-motion analysis of feeding in alligator lizards. In: Rhodin AGJ, Miyata K (eds) Advances in herpetology and evolutionary biology: essays in honor of Ernest E. Williams. Mus Comp Zool, Harvard University, Cambridge, pp 222–244

Frazzetta TH (1988) The mechanics of cutting and the form of shark teeth (Chondrichthyes, Elasmobranchii). Zoomorphology 108: 93–107

Frazzetta TH, Prange CD (1987) Movements of cephalic components during feeding in some requiem sharks (Carcharhiniformes: Carcharhinidae). Copeia 4: 979–993

Gans C (1952) The functional morphology of the egg-eating adaptations in the snake genus *Dasypeltis*. Zoologica 37: 209–244

Gans C, Northcutt RG (1983) Neural crest and the origin of vertebrates: a new head. Science 220: 268–274

Gans C, Parsons TS (1964) A photographic atlas of shark anatomy; the gross morphology of *Squalus acanthias*. Univ Chicago Press, Chicago

Garrick JAF (1982) Sharks of the genus *Carcharhinus*. Natl Ocean Atmos Admin (USA), Natl Mar Fish Serv Circ 445: i–viii, 1–194

Gegenbauer C (1872) Untersuchungen zur vergleichenden Anatomie der Wirbelthiere. Drittes Heft. Das Kopfskelet der Selachier, ein Beitrag zur Erkenntniss der Genese des Kopfskeletes der Wirbelthiere. Wilhelm Engelmann, Leipzig

Gilbert PW (1962) The behavior of sharks. Sci Am 207: 60–68

Gilbert SG (1973) Pictorial anatomy of the dogfish. Univ of Washington Press, Seattle

Gohar HAF, Mazhar FM (1964) The internal anatomy of Selachii from the north western Red Sea. Publ Mar Biol Stn Al-Ghardaqa 13: 145–240

Gregory MR, Ballance PF, Gibson GW, Ayling AM (1979) On how some rays (Elasmobranchia) excavate feeding depressions by jetting water. J Sediment Petrol 49: 1125–1130

Hall BK (1987) Tissue interactions in the development and evolution of the vertebrate head. In: Maderson PFA (ed) Development and evolutionary aspects of neural crest. Wiley Series in Neurobiol, Wiley, New York, pp 215–259

Hughes GM, Ballintijn CM (1965) The muscular basis of the respiratory pumps in the dogfish (*Scyliorhinus canicula*). J Exp Biol 43: 363–383

Jarvik E (1954) On the visceral skeleton in *Eusthenopteron* with a discussion of the parasphenoid and palatoquadrate in fishes. Sven Vetenskapsakad Handl 5: 1–104

Jarvik E (1980) Basic structure and evolution of vertebrates, 2 vols. Academic Press, London

Jollie M (1971) A theory concerning the early evolution of the visceral arches. Acta Zool 52: 85–96

Lauder GVJr (1980a) Evolution of the feeding mechanism in primitive actinopterygian fishes: a functional anatomical analysis of *Polypterus*, *Lepisosteus* and *Amia*. J Morphol 168: 283–317

Lauder GVJr (1980b) On the evolution of the jaw adductor musculature in primitive gnathostome fishes. Breviora 460: 1–10

Lauder GVJr (1983a) Functional design and the evolution of the pharyngeal jaw apparatus in euteleostean fishes. Zool J Linn Soc 77: 1–38

Lauder GVJr (1983b) Prey capture hydrodynamics in fishes: experimental tests of two models. J Exp Biol 104: 1–13

Lauder GVJr, Liem KF (1980) The feeding mechanism and cephalic myology of *Salvelinus fontinalis*: form, function, and evolutionary significance. In: Balon EK (ed) Charrs. Salmonid fishes of the genus *Salvelinus*. Perspectives in vertebrate science. W Junk, The Hague, pp 365–390

Le Douarin N (1982) The neural crest. Dev Cell Biol Ser 12: i–xi, 1–259 Dev Cell Biol Ser 12

Liem KF (1967) Functional morphology of the head of the anabantoid teleost fish *Helostoma temmincki*. J Morphol 121: 135–158

Liem KF (1970) Comparative functional anatomy of the Nandidae (Pisces: Teleostei). Fieldiana Zool 56: 1–166

Liem KF (1986) The pharyngeal jaw apparatus of the Embiotocidae (Teleostei): a functional and evolutionary perspective. Copeia 2: 311–323

Luer CA, Blum PC, Gilbert PW (1990) Rates of tooth replacement in the nurse shark, *Ginglymostoma cirratum.* Copeia 1990: 182–190

Luther AF (1909) Untersuchungen über die vom N. trigeminus innervierte Muskulatur der Selachier (Haie und Rochen) unter Berücksichtigung ihrer Beziehungen zu benachbarten Organen. Acta Soc Fenn 36: 1–176

Maisey JG (1977) The fossil selachian fishes *Paleospinax* Egerton, 1872 and *Nemascanthus* Agassiz, 1837. Zool J Linn Soc 60: 259–273

Maisey JG (1980) An evaluation of jaw suspension in sharks. Am Mus Novit 2706: 1–17

Maisey JG (1984) Higher elasmobranch phylogeny and biostratigraphy. Zool J Linn Soc 82: 33–54

Marion GE (1905) Mandibular and pharyngeal muscles of Acanthias and Raia. Am Nat 39: 891–924

McCormick HW, Allen T, Young W (1963) Shadows in the sea; the sharks, skates and rays. Chilton, Weathervane Books, New York

Moss SA (1967) Tooth replacement in the lemon shark, *Negaprion brevirostris.* In: Gilbert PW, Mathewson RF, Rall DP (eds) Sharks, skates and rays. Johns Hopkins Univ Press, Baltimore

Moss SA (1972) The feeding mechanism of sharks of the family Carcharhinidae. J Zool (Lond) 167: 423–436

Moss SA (1977) Feeding mechanisms in sharks. Am Zool 17: 355–364

Moss SA (1984) Sharks: an introduction for the amateur naturalist. Prentice Hall, Engelwood Cliffs

Motta PJ, Hueter RE, Tricas TC (1991) An electromyographic analysis of the biting mechanism of the lemon shark, *Negaprion brevirostris*: functional and evolutionary implications. J Morphol 210: 55–69

Moy-Thomas JA, Miles RS (1971) Paleozoic fishes. Saunders, Philadelphia

Nelson G (1969) Gill arches and the phylogeny of fishes, with notes on the classification of vertebrates. Bull Am Mus Nat Hist 141: 475–552

Nobiling G (1977) Die Biomechanik des Kieferapparates beim Stierkopfhai (*Heterodontus portusjacksoni = Heterodontus philippi*). Adv Anat Embryol Cell Biol 52: 1–52

Noden DM (1984) Cranio-facial development: new views on old problems. Anat Rec 208: 1–13

Northcutt RG, Gans C (1983) The genesis of neural crest and epidermal placodes: a reinterpretation of vertebrate origins. Q Rev Biol 58: 1–28

Nyberg DW (1971) Prey capture in the largemouth bass. Am Midl Nat 86: 128–144

Peyer B (1968) Comparative odontology. Univ Chicago Press, Chicago

Preuschoft H, Reif W-E, Muller WH (1974) Funktion-anpassungen in Form und Struktur an Haifischzähnen. Anat Entwickl Gesch 143: 315–344

Prindle B, Walden RG (1976) Deep-sea fishbite manual. Natl ocean Atmos Admin (USA); Natl Data Bouy Office, Bay St Louis, MS

Reif W-E (1973) Morphologie und Skulptur der Haifisch-Zahnkronen. Neues Jahrb Geol Palaeontol Abh 143: 29–44

Reif W-E, McGill D, Motta P (1978) Tooth replacement rates of the sharks *Triakis semifasciata* and *Ginglymostoma cirratum.* Zool Jahrb Anat 99: 151–156

Romer AS (1966) Vertebrate paleontology. Chicago University Press, Chicago

Schaeffer B (1967) Comments on elasmobranch evolution. In: Gilbert PW, Mathewson RF, Rall DP (eds) Sharks, skates and rays. Johns Hopkins Univ Press, Baltimore, pp 3–35

Schaeffer B, Rosen D (1961) Major adaptive levels in the evolution of the actinopterygian feeding mechanism. Am Zool 1: 187–204

Schaeffer B, Williams M (1977) Relationships of fossil and living elasmobranchs. Am Zool 17: 293–302

Shigley JE (1961) The theory of machines. McGraw-Hill, New York

Snodgrass JM, Gilbert PW (1967) A shark bite meter. In: Gilbert PW, Mathewson RF, Rall DP (eds). Sharks, skates and rays. Johns Hopkins Univ Press, Baltimore

Springer S (1961) Dynamics of the feeding mechanism of large galeoid sharks. Am Zool 1: 183–185

Stahl BJ (1988) Reconstruction of the head skeleton of the fossil elasmobranch, *Phoebodus heslerorum* (Pisces, Chondrichthyes). Copeia 1988: 858–865

Stensiö EA (1963) Anatomical studies of the arthrodiran head. K Sven Vetenskapsakad Hadl 9: 1–419

Strasburg DW (1963) The diet and dentition of *Isistius brasiliensis*, with remarks on tooth replacement in other sharks. Copeia 1963: 33–40

Tanaka SK (1973) Suction feeding by nurse sharks. Copeia 3: 605–608

Tchernavin VV (1953) The feeding mechanism of a deep sea fish, *Chauliodus sloani* Schneider. Br Mus Nat Hist (Lond): 1–101

Thomson KS (1967) Mechanisms of intracranial kinetics in fossil rhipidistian fishes (Cross-opterygii) and their relatives. J Linn Soc Lond Zool 46: 223–253

Thomson KS (1987) Speculations concerning the role of neural crest in morphogenesis and evolution of the vertebrate skeleton. In: Maderson PFA (ed) Development and evolutionary aspects of neural crest. Wiley Series in Neurobiol, Wiley, New York, pp 301–338

Tricas TC (1985) Feeding ethology of the white shark *Carcharodon carcharias*. Mem South Calif Acad Sci 9: 81–91

Tricas TC, McCosker JE (1984) Predatory behavior of the white shark (*Carcharodon carcharias*), with notes on its biology. Proc Calif Acad Sci 43: 221–238

Vetter B (1874) Untersuchungen zur vergleichenden Anatomie der Kiemen- und Kiefermus-kulatur der Fische. Jen Naturwiss 8: 405–458

Zangerl R (1981) Chondrichthyes 1. Paleozoic Elasmobranchii. In: Schultze HP (ed) Hand-book of paleoichthyology, 3A. Fischer, Stuttgart, pp 1–115

NOTE ADDED IN PROOF.

An important paper on the distribution of tooth types among vertebrates, and aspects of cutting mechanics, has appeared since the present chapter has gone to press. Interested readers are directed to:

Abler WL (1992) The serrated teeth in tyrannosaurid dinosaurs, and biting structures in other animals. Paleobiol 18: 161–183

Chapter 3

The Pharyngeal Apparatus in Teleost Feeding

P. Vandewalle[1], A. Huyssene[2], P. Aerts[3] and W. Verraes[4]

Contents

1 Introduction

Because of the high density and viscosity of water, interactions between food particles and fish, on the one hand, and water, on the other, complicate aquatic feeding considerably. Fish have to create a water flow through their buccopharyngeal cavity in order to take up their food. This can be achieved by simply swimming with both the mouth and the opercula widely opened (filter feeding), or by means of active suction (Lauder 1985). In the latter case the volume of the buccopharyngeal and opercular cavities is rapidly expanded so that water and food are drawn in (Muller and Osse 1984; Lauder 1985). Even when the food item or prey is first seized between the oral jaws (e.g. by biting pieces from a large prey), further intrabuccal transport and manipulation in most cases require subsequent suction activity (i.e. hydrodynamic tongue; cf. Liem 1990).

[1] Research Associate at the National Fund for Scientific Research, Zoological Institute, University of Liège, Belgium
[2] Senior Research Assistant at the National Fund for Scientific Research, Zoological Institute, University of Ghent, Belgium
[3] Research Associate at the National Fund for Scientific Research, Department of Biology, University of Antwerp, Belgium
[4] Zoological Institute, University of Ghent, Belgium

Advances in Comparative and Environmental Physiology, Vol. 18
© Springer-Verlag Berlin Heidelberg 1994

The mechanical conditions under which fish have to feed hinder the oral jaws from taking part in food processing as well. Indeed, movements of head parts might cause local flow patterns tending to carry the food particle away from the oral jaws. Therefore, if food processing takes place, it mostly occurs at the back of the buccopharyngeal cavity by means of (more or less) specialized pharyngeal elements.

The first, and major, part of this chapter deals with morphology, function and evolutionary differentiation of these pharyngeal elements. Parallel to the functional shift from a transporting towards a true processing device, the involved adaptations of the pharyngeal apparatus have changed from simple dentition of the branchial arches to the formation of an efficient, highly specialized pharyngeal mill.

The role of fish teeth as a causal agent in the morphogenesis of the skeletal parts of the pharyngeal apparatus, on the one hand, and the constraints which they impose upon the elements that bear them, on the other hand, are central in the second part of this chapter.

2 Morphology

2.1 Osteology

The branchial basket of teleost fish is built up of five skeletal arches (Fig. 1; for general information, see Nelson 1969). The first three are complete, with a medioventral basibranchial and a bilateral series of four elements: the hypo-, cerato-, epi- and pharyngobranchial elements. The fourth arch always includes a cerato- and an epibranchial element. A pharyngobranchial, basibranchial and rarely a hypobranchial element can be present as well (Nelson 1969; Rosen 1974). Mostly, the fifth arch is built up solely of the paired ceratobranchial elements. The fifth ceratobranchial and the pharyngobranchial elements of both body sides constitute the base for the lower (LPJs) and upper pharyngeal jaws (UPJs) respectively. The perichondral (endochondral) ossifications of these elements are covered and/or associated with dermal tooth plates.

In primitive teleosts such as elopomorphs, the branchial basket is suspended from the neurocranial base by means of the branchial levator muscles. The pharyngeal jaws are small (Fig. 1; Forey 1973; Taverne 1974). The LPJs are paired and carry a series of independent, small tooth plates. This tooth-bearing area is extended to the front by tooth plates supported by the basibranchial elements and the basihyal. Ceratobranchial and hypobranchial elements and ceratohyals also bear tooth plates. Facing the ventral dentition is an assemblage of dorsal tooth plates. Behind, the nearly independent pharyngobranchial elements (the dorsal jaws s.s.) are associated with tooth plates and are flanked by toothed epibranchial elements. Rostrally, tooth plates are associated with the parasphenoid, the vomer, the palatine and the ecto- and endopterygoid bones.

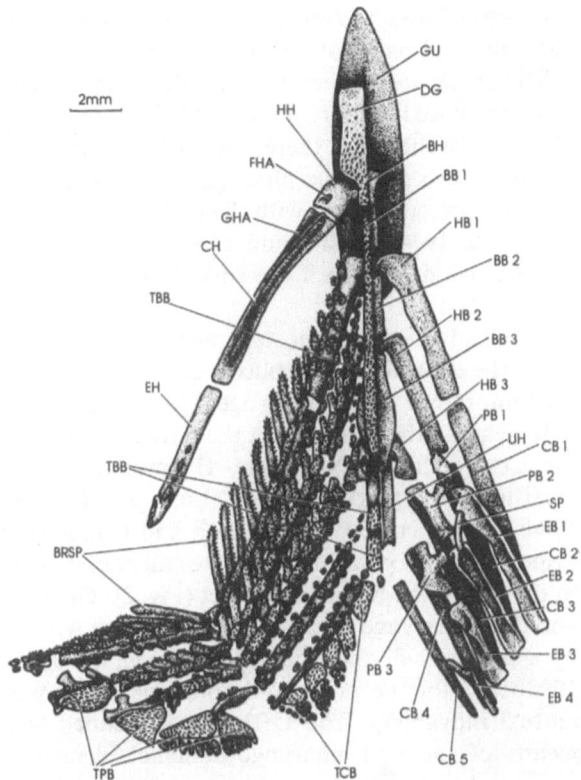

Fig. 1. Dorsal view of hyoid and branchial skeleton of *Elops lacerta*. The right branchiospines and tooth plates are removed. *BB 1,2,3* Basibranchial 1,2,3; *BH* basihyal; *BRSP* branchiospines; *CB 1,2,3,4,5* ceratobranchial *1,2,3,4,5*; *CH* ceratohyal; *DG* dermentoglosse; *EB 1,2,3,4* epibranchial *1,2,3,4*; *EH* epihyal; *FHA* foramen of the hypobranchial artery; *GHA* groove of the hypobranchial artery; *GU* gular plate; *HB 1,2,3* hypobranchial 1,2,3; *HH* hypohyal; *PB 1,2,3* pharyngobranchial 1,2,3; *TBB* tooth plates of the basibranchial elements; *TCB* tooth plates of the ceratobranchial elements; *TPB* tooth plates of the pharyngobranchial elements; *UH* urohyal. (After Taverne 1974, Fig. 22)

In this way, the buccal cavity of elopomorphs is toothed from the mouth aperture in the front to the entrance of the oesophagus at the back, and the pharyngeal jaws s.s. just form part of this continuous system. The teeth are simple and pointed.

During evolution, the extent of the intra-oral dentition is generally reduced and the pharyngeal jaws gradually become the most important food-processing apparatus. In part of the osteoglossomorphs the tooth plates on the epibranchial and ceratobranchial elements have disappeared but the dorsal and ventral median dentition is still well developed (Nelson 1968; Greenwood 1973; Taverne 1977; Lauder and Liem 1983). According to Sanford and Lauder (1989, 1990), osteoglossomorphs such as *Notopterus chilata* are thus characterized by three sets of jaws: the oral jaws in front, the pharyngeal jaws at the back and the parasphenoid-basihyal set in between.

The protacanthopterygians (i.e. the primitive euteleosts; cf. Lauder and Liem 1983) embrace families such as the Salmonidae and the Esocidae (Greenwood et al. 1966). In these fish there is a further tendency of reduction of the intra-oral dentition, except for the pharyngeal jaws s.s.. The dermal tooth plates of these elements are often fused with the underlying endoskeletal elements (Rosen 1974; Weitzman 1974; Lauder and Liem 1983). This reinforces the cohesion of the pharyngeal jaws, even if the endoskeletal parts remain largely independent of each other.

In the advanced euteleosts or acanthopterygians (i.e. fish with protrusive oral jaws), the dentition of the buccal cavity tends to disappear completely, whereas the importance of the pharyngeal jaws increases progressively.

In general, the dentition in the buccal cavity of primitive representatives like the Centrarchidae, Nandidae, Haemulidae, Serranidae and Girellidae only persists on the pterygoid bone (Liem and Greenwood 1981; Benmouna et al. 1984; Connes et al. 1988; Johnson and Fritzsche 1989). Despite fusion to their respective tooth plate, the LPJs are still paired. Rostrally, the jaws are connected to the rest of the branchial basket (Fig. 2). The UPJs are built up by the second and third pharyngobranchial elements each with an associated tooth plate, and by a posterior dermal tooth plate (Liem 1970; Rosen 1973; Vandewalle et al. 1992b). As for the protacanthopterygians, the elements of each UPJ retain a mutual movability. The UPJs are suspended from the neurocranial base by means of the first pharyngobranchial elements and the branchial levator muscles. In general, the teeth are simple and pointed.

In advanced acanthopterygians, such as the Cichlidae, Embiotocidae, Scaridae and Labridae, no dentition is found in the buccal cavity (Vandewalle 1972; Liem and Osse 1975; Liem 1978). One single, triangular lower pharyngeal jaw (LPJ) is formed by the fusion of the left and right fifth ceratobranchial elements (Fig. 2; Nelson 1967a; Thys van den Audenaerde 1970; Liem 1973; Barel et al. 1976, 1977; Hoogerhoud 1984, 1986; Witte and van Oijen 1990). The LPJ faces

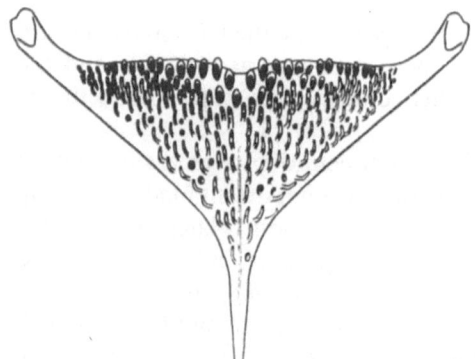

Fig. 2. Dorsal view of the LPJ of *Tilapia louka*. (After Thys van den Audenaerde 1970)

the paired UPJs, each of which is constituted in cichlids by the toothed, small second pharyngobranchial element and the assemblage of the third and fourth pharyngobranchial element (but see Liem 1978; Stiassny 1981; Ismail et al. 1982; Lauder and Liem 1983). In the Embiotocidae the UPJs are formed by the third and fourth pharyngobranchial elements (Nelson 1967a; Liem 1986). Both in cichlids and embiotocids the UPJs articulate with a posterior apophysis of the parasphenoid bone in the neurocranial base, via the third pharyngobranchial elements. The teeth are often specialized in relation to the specific trophic regime (uni-, bi-, tricuspid, molariform, etc.).

The LPJ of the Labridae and Scaridae articulates posteriorly with the pectoral girdle at the level of the cleithra (Fig. 3; Quignard 1962; Nelson 1967a; Yamaoka 1978; Gobalet 1989). The upper pharyngeal elements are built up by the voluminous third pharyngobranchial elements (Monod 1951; Nelson 1967a), which articulate with the pharyngeal process on the parasphenoid bone (Kaufman and Liem 1982; Liem and Sanderson 1986). The dentition can be specialized (Yamaoka 1978).

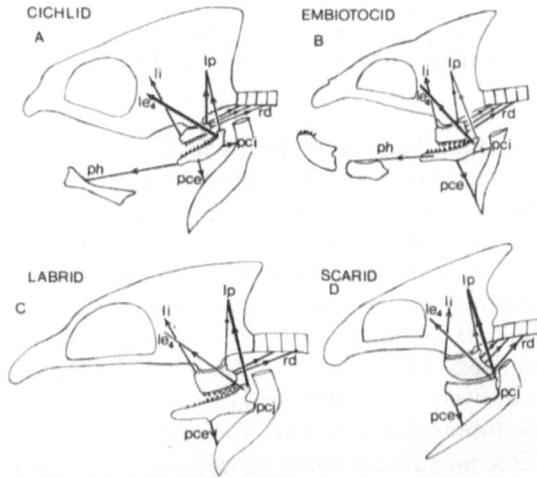

Fig. 3A–D. Diagrammatic representation of the principal components of the pharyngeal jaw apparatus of cichlids, embiotocids, labrids and scarids. *Shaded* elements are the UPJs and LPJs. The neurocranium and first vertebra are drawn as complete structures. In each diagram the dorsal half of the cleithrum has been eliminated. In the cichlid and embiotocid the hyoid is indicated in a simplified form, while in the labrid and scarid the hyoid has been omitted. The muscles are represented as *lines* and the principal direction of force has been indicated by *arrows*. Anatomical and functional dominance is indicated by a *heavier line*. In the cichlid-embiotocid the non-articulated LPJ is suspended in a muscular sling, of which the fourth levator externus (le_4) is dominant. In the labrid and scarid, the LPJ articulates with the cleithrum, and the dominant levator posterior (lp) in conjunction with the pharyngocleithralis externus (pce) form a force couple. le_4 Fourth levator externus; li levator internus; lp levator posterior; pce pharyngocleithralis externus; pci pharyngocleithralis internus; ph pharyngohyoideus; rd retractor dorsalis. (After Liem and Greenwood 1981, Fig. 9)

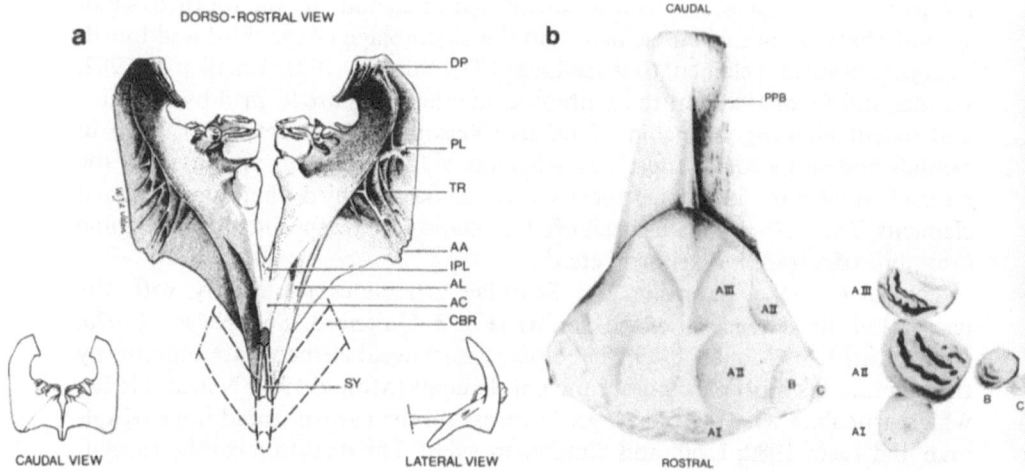

Fig. 4. a Pharyngeal bones and teeth; **b** ventral view of the chewing pad and the pharyngeal process of *Cyprinus carpio*. **b** A dorsal view of pharyngeal teeth for comparison at the same scale. Note the pitted anterior surface of the chewing pad, reflecting the crushing action of the A I teeth, and the grinding facets of the corresponding furrowed teeth. *AA* Anterior angle; *AC* anterior cartilage; *AL* anterior limb; *CBR* ceratobranchial; *DP* dorsal process; *IPL* interpharyngeal ligament; *PL* posterior limb; *PPB* pharyngeal process of the basioccipital; *SY* symphysis; *TR* trabecula (After Sibbing 1982, Fig. 2)

Within the Ostariophysi a similar evolution can be observed. In general, the Characiformes have non-protrusive, strong and heavily toothed oral jaws and this situation is comparable to the protacanthopterygian condition (Weitzman 1962; Roberts 1969; Miquelarena and Aramburu 1983).

In contrast to the characids, the cyprinids have developed a specific protrusive mouth, analogous to the acanthopterygian type (Ballintijn et al. 1972; Vandewalle 1978). The oral jaws and the walls of the buccal cavity are entirely toothless (Ramaswami 1955a, b; Vandewalle 1975) and food uptake relies exclusively upon suction. The only teeth are found on the well-developed fifth ceratobranchial elements, forming the left and right pharyngeal jaws (Fig. 4). These jaws face a horny chewing pad supported by the basioccipital process of the neurocranium (Ramaswami 1955a, b; Sibbing 1982). The pharyngobranchial elements thus do not participate in the pharyngeal jaw apparatus of cyprinids.

2.2 Myology

In a general way, the anatomical pattern of the muscles related to the pharyngeal jaws has been rather conserved within the teleosts. Nevertheless, some morphological shifts with a considerable functional impact urge one to consider two types of branchial musculature: the primitive teleost and the acanthopterygian architecture. Summarizing the literature (Vetter 1878; Dietz 1912;

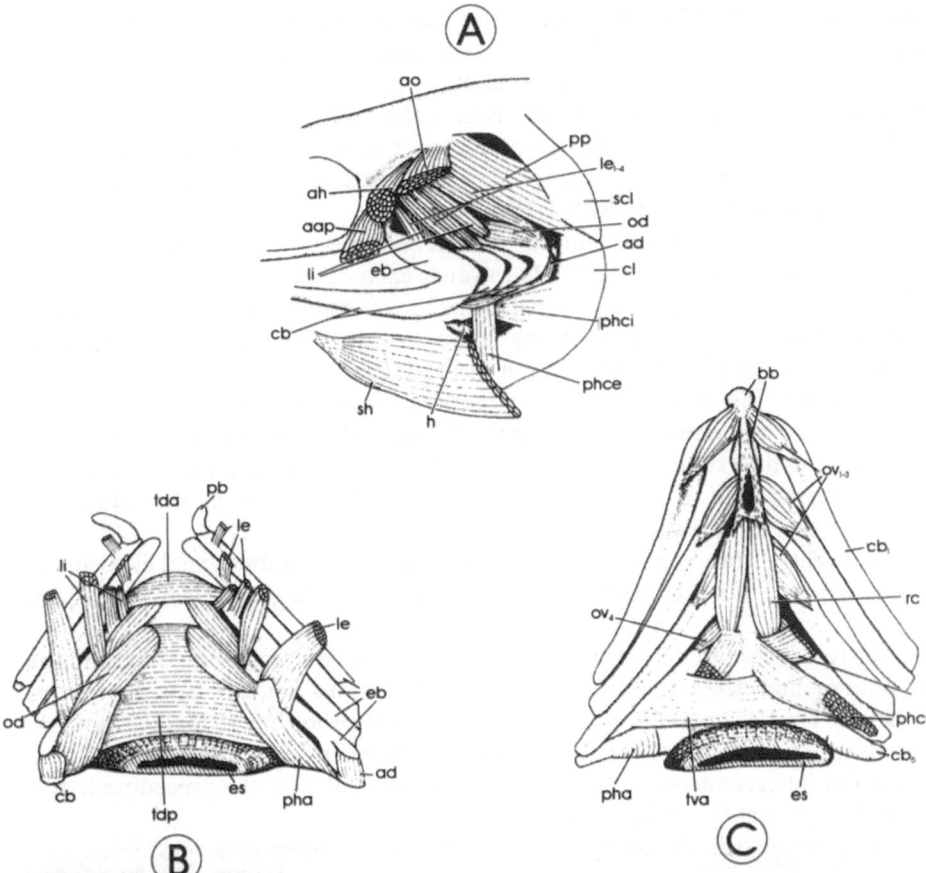

Fig. 5. A Lateral; **B** dorsal; **C** ventral views of the branchial basket and associated muscles of *Salvelinus fontinalis*. In **C**, the right pharyngocleithralis muscles are removed. *aap* Adductor arcus palatini; *ad* adductor branchialis 5; *ah* adductor hyomandibulae; *ao* adductor operculi; *bb* basibranchial; *cb* ceratobranchial; *cl* cleithrum; *eb* epibranchial; *es* oesophagus; *h* heart; *le* levator externus; *li* levator internus; *od* obliquus dorsalis; *ov* obliquus ventralis; *pb* pharyngobranchial; *pha* obliquus posterior; *phce* pharyngocleithralis externus; *phci* pharyngocleithralis internus; *pp* protractor pectoralis; *rc* rectus communis; *scl* supracleithrum; *sh* sternohyoideus; *tda* transversus dorsalis anterior; *tdp* transversus dorsalis posterior; *tva* transversus ventralis. (After Lauder and Liem 1980, Figs. 7–9)

Holstvoogd 1965; Bishai 1967; Nelson 1967b; Greenwood 1971; Winterbottom 1974; Lauder and Liem 1980; Lauder 1983b), the branchial musculature of the primitive teleosts (elopomorphs, osteoglossomorphs and protacanthopterygians) is composed of the muscles listed below (Fig. 5). Functions postulated primarily on the basis of the general orientation of the muscles are added.

– There are four external levator muscles interconnecting the four epibranchial elements with the neurocranium, and two or three internal levators (Holstvoogd

1965; Nelson 1967b) running from the second and third pharyngobranchial elements to the neurocranial base. In most cases it is assumed that these muscles lift and protract the UPJ. However, in many primitive euteleosts some of these levators are inclined posteriorly and retract the UPJ (Holstvoogd 1965; Lauder 1983b). As the insertion of the external levator muscles on the neurocranium is generally situated lateral to their pharyngeal attachments, contraction can also cause lateral spreading of both UPJs.
– Two dorsal transversal muscles, interconnecting the left and right epibranchial elements move the UPJs towards each other.
– The third and fourth epibranchial elements are connected to the corresponding pharyngobranchial elements by means of two obliquus dorsalis muscles. Contraction modifies the orientation of the UPJs.
– If present, the first four adductor muscles interconnect the epi- and ceratobranchial elements of the same arch. The fifth adductor, always present, runs between the fifth ceratobranchial and the fourth epibranchial element and lifts the posterior part of the corresponding LPJ. Contraction of the adductors reduces the gape between the UPJs and LPJs.
– Two ventral transversal muscles interconnect the ceratobranchial elements of the fourth and fifth arch respectively. Contraction moves the LPJs towards each other.
– The rectus communis generally runs between the fifth ceratobranchial and the hypobranchial element of the third arch, but insertions can vary between species (Dietz 1912; Nelson 1969; Greenwood 1971; Winterbottom 1974). This muscle depresses the LPJs caudally. In addition to the rectus communis, several recti ventrales (interconnecting a ceratobranchial element with the preceding hypobranchial element) can be present.
– The pharyngocleithralis externus and internus muscles connect the LPJs to the pectoral girdle. The internus muscle retracts the LPJs, whereas the externus moves them downwards. Activity of these muscles also makes the LPJs move apart.
– Most likely, the obliquus ventralis muscles (between the first hypobranchial and ceratobranchial elements) have little effect on the pharyngeal jaws.
– In addition to these branchial muscles, the sternohyoideus and the geniohyoideus (= protractor hyoidei) must be mentioned. They connect the hyoid to the pectoral girdle (+ hypaxial muscles) and to the mandibula respectively. The sternohyoideus moves the hyobranchial system caudoventrally. The geniohyoideus pulls it rostrodorsally.

With this muscular system the pharyngeal jaws can be lifted and depressed, protracted and retracted, moved apart and brought together.

The branchial musculature of the acanthopterygians essentially differs from that of the primitive teleosts in the following morphological characters (Fig. 6; Liem 1970, 1973, 1978, 1986; Vandewalle 1972; Hoogerhoud and Barel 1978; Yamaoka 1978, 1980; Lauder 1983a, b; Vandewalle et al. 1992b). Again, postulated functions are added.

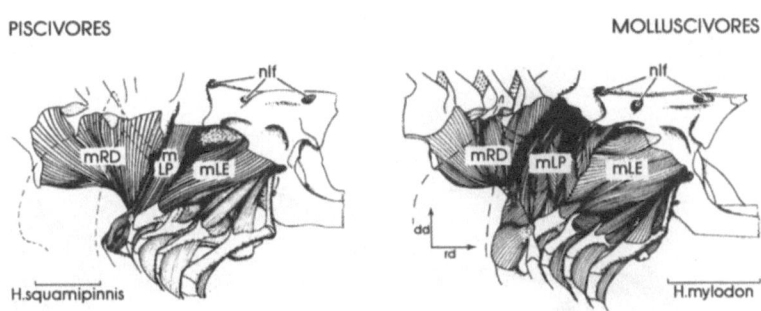

Fig. 6. Structure and relative size of the three larger extrinsic branchial muscles of *Haplochromis squamipinnis* and *H. mylodon*. *dd* dorsad; *mLE* levator externus; *mLP* levator posterior; *mRD* retractor dorsalis; *nlf* neurocranial lateral line foramina; *rd* rostrad. (After Hoogerhoud and Barel 1978, Fig. 3)

– The external and internal levator muscles all run in a rostrodorsal direction towards the neurocranium. They lift, protract and abduct the UPJs (but see further).
– The fourth levator externus muscle is strongly developed. Part of it can be "fused" with a bundle of the obliquus posterior muscle (a muscle interconnecting the fourth epibranchial and the fifth ceratobranchial element) (Aerts 1982; Liem 1986). Thus, the fourth levator externus becomes a levator of the LPJ as well.
– A vertically oriented levator posterior muscle runs from the neurocranium to the fourth epibranchial. As for the fourth levator externus, this muscle can be extended by a bundle of the obliquus posterior muscle. In this way the levator posterior becomes a prominent levator of the LPJ (Claeys and Aerts 1984, Liem 1986). Its importance is well illustrated in the Labridae, where the neurocranial insertion area of the levator posterior reaches dorsal aspects of the skull (Yamaoka 1978).
– The important retractor dorsalis muscle inserts at least on the third pharyngobranchial element and runs to the ventral side of the neurocranium and/or the ventral side of the anterior vertebrae. This muscle retracts and inclines the UPJs caudally (antagonistic to the levator complex).
– The pharyngohyoideus runs from the fifth ceratobranchial element to the urohyal (= parahyoid). According to Lauder (1983b), this muscle is homologous to the rectus communis. The presence of this long muscle increases the kinematic potential of the LPJ, especially when the high mobility of the urohyal is taken into account.

All these morphological specializations of the branchial musculature, proper to the acanthopterygians, are obviously related to an increased participation of the pharyngeal jaws in the processing of the food.

The peculiar evolution of the skeletal elements of the pharyngeal jaw apparatus of the Ostariophysi coincides with the development of a specific musculature (Takahasi 1925; Holstvoogd 1965; Winterbottom 1974; Vandewalle 1975;

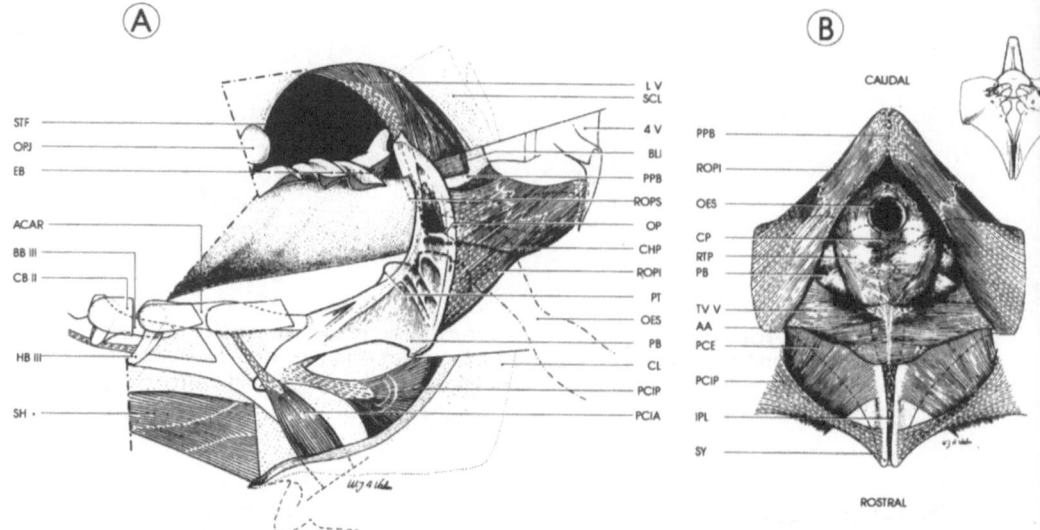

Fig. 7. A View of the wall of the posterior pharynx, between teeth and chewing pad **B** ventrocaudal view of the pharyngeal bones and related structures of *Cyprinus carpio*. In **B**, the *broken lines* indicate the pharyngeal bones as figured with the chewing pad at the *upper right*. *ACAR* Anterior cartilage; *BB III* basibranchial III; *BLI* Baudelot's ligament; *CB II* ceratobranchial II; *CHP* chewing pad; *CL* cleithrum; *CP* constrictor pharyngis; *EB* epibranchial; *HB III* hypobranchial III; *LV* levator arcus branchialis V; *OES* oesophagus; *OP* obliquus posterior; *OPJ* opercular joint; *PB* pharyngeal bone; *PCE* pharyngocleithralis externus; *PCIA* pharyngocleithralis internus anterior; *PCIP* pharyngocleithralis internus posterior; *PPB* pharyngeal process of the basioccipital; *PT* pharyngeal teeth; *ROPI* retractor os pharyngeus inferior; *ROPS* retractor os pharyngeus superior; *RTP* replace-tooth patch; *SCL* supracleithrum; *STF* subtemporal fossa; *TV V* transversus ventralis V; *4 V* fourth vertebra. (After Sibbing 1982, Fig. 5)

Sibbing 1982). The branchial musculature of the characids strongly resembles that of the primitive teleosts (Gijsen and Chardon 1976). In cyprinids the robust LPJs (fifth ceratobranchial elements) are operated by a large retractor pharyngeus (cf. Winterbottom 1974; Sibbing 1982) running to the posterior process of the basioccipital, and a voluminous levator arcus branchialis 5 (Fig. 7). Ventrally, there are the pharyngocleithralis muscles and a stout transversus muscle. The latter moves the jaws towards each other and rotates them in such a way that the teeth face the neurocranial chewing pad (Sibbing 1982).

3 Role of the Pharyngeal Apparatus in Food Processing

When compared to the process of food uptake (suction, biting etc.), the kinematics and muscle activity patterns of food processing by the pharyngeal apparatus have been the subject of only a limited number of studies. This is undoubtedly coupled to the internal position of the apparatus. Liem (1970) deduced movement patterns of the pharyngeal jaws of nandids from the tooth impressions on prey swallowed whole. Wainwright (1989a) applied direct stimulation of branchial muscles to link jaw displacements to muscular activity patterns. However, detailed movement analysis of the branchial apparatus is impossible without application of specialized techniques such as X-ray cinematography, either in combination with electromyography (EMG) or not (Sibbing 1982; Aerts et al. 1986; Claes and De Vree 1989, 1991a, b; Claes et al. 1991; Vandewalle et al. 1992b). Lauder (1983d) used impedance electrodes to measure relative displacements of gill arches. Recently, Sanderson et al. (1991) applied endoscopy to observe the intra-oral flows during filter feeding in the blackfish. In principle, such techniques can also be used to study aspects of pharyngeal jaw kinematics.

If one studies the kinematics of the pharyngeal jaws, one has to realize that each specific movement pattern results from *all* external forces acting on the skeletal elements. Thus, not only muscular forces, but also reaction forces from the food particle and adjacent head parts will determine the path followed by the moving jaws (presumably, inertial forces can be neglected). This implies that it is very difficult to infer jaw kinematics solely from EMG patterns, or, reversely, to predict muscular activities from observed displacements (see for instance Lauder 1983b). Such postulations, although worthwhile, must always be interpreted with caution, as stressed e.g. by Wainwright (1989a) and Claes and De Vree (1992).

In situ measurements of physical parameters like force, strain, pressure, etc. are extremely difficult. An alternative is offered by mathematical modelling, whereby morphological data and information on activity patterns (kinematics, EMG) can be used as input. Although mathematical modelling has been limited until now to statical analyses, its output can yield useful information in an ecomorphological context (e.g. Wainwright 1987; Galis 1990, 1991).

In spite of practical problems, there is an obvious need to gain insight into the mechanisms involved, because food processing by the pharyngeal apparatus (transport, mastication, reduction, etc.) forms an essential part of the feeding biology of fish. To date, most studies available concern acanthopterygians. Information on the function of the pharyngeal apparatus of the primitive teleosts is scarce.

3.1 Primitive Teleostei

Lauder (1983b) considers four phases in the feeding activity of *Esox niger* (a protacanthopterygian): initial strike, buccal manipulation, pharyngeal manipulation and pharyngeal transport. During the initial strike, several pharyngeal muscles are active (third and fourth levator, fifth adductor), but in an unpredictable fashion. Most likely, this activity brings the pharyngeal jaws into their initial position. During both manipulation phases, the same levator muscles and

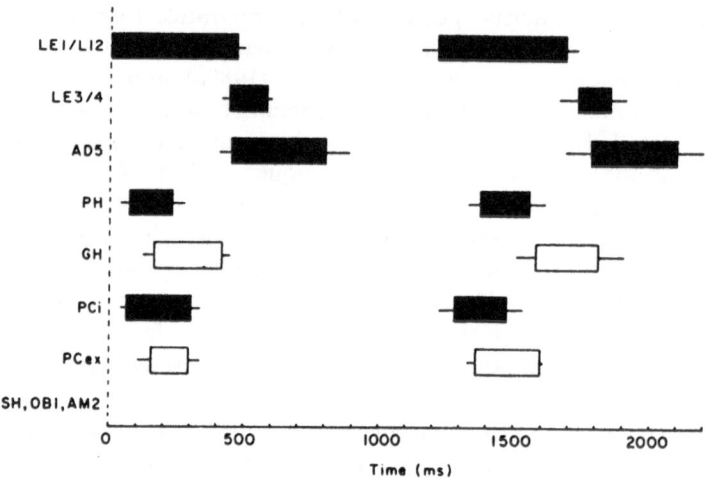

Fig. 8. Bar diagram illustrating the pattern of muscle activity during pharyngeal transport by
Esox niger. *Black bars* indicate the duration of muscle activity. *Thin lines* indicate one standard
error of the mean onset and offset times for each muscle. *White bars* indicate that activity was
observed in less than 50% of the recordings. *AD5* Adductor branchialis 5; *GH* geniohyoideus;
LE 1/LI 2 levator externus 1/levator internus 2; *LE 3/4* levatores externi 3,4 *PCex* pharyngocleithralis externus; *PCi* pharyngocleithralis internus; *PH* pharyngohyoideus; *SH, OB 1,
AM 2* sternohyoideus, obliquus *1*, adductor mandibulae *2*. (After Lauder 1983b, Fig. 11)

the pharyngocleithrales internus and externus are active. The fifth adductor is
now silent. As a result the UPJs and LPJs expand and gape rostrally, ready to
grasp and reposition the food particle. Then pharyngeal transport proceeds.

The first three phases are short and do not show any cyclical events.
Pharyngeal transport, however, can take several minutes and is the result of
cyclical, stable activity patterns of the muscles (Fig. 8). The only variability
found is an asymmetry in activity of the left and right third and fourth levators
(Lauder 1983b). During transport, the sternohyoideus is silent, whereas the
geniohyoideus and the pharyngocleithralis externus muscles are only occasionally active. The first external and the second internal levators (directed
posteriorly) and the internal pharyngocleithralis muscle contract simultaneously. In this way the UPJs and LPJs are retracted together. The third and
fourth levators protract the UPJs.

The muscles involved in the movements of the UPJs and LPJs of *Esox* suggest
that the anteroposterior movement of the LPJs dominates the pharyngeal
transport phase. Based on anatomical observations, the same assumption most
likely holds for other primitive teleosts as well. The pharyngeal jaws only ensure
the transport, without substantial processing of the prey.

Depending upon the orientation of the levator muscles, the pro- and retraction movements of the UPJs can be more or less pronounced in different species.
In *Elops saurus*, only the first levator is directed posteriorly (Winterbottom
1974), whereas in *Salvelinus fontinalis* all levators run rostrodorsally towards the

neurocranial base (Lauder and Liem 1980). In the latter species, caudal displacements of the UPJs are most likely passively induced by forces generated by the LPJs and transmitted via the prey.

In the osteoglossomorph *Notopterus chilata* the prey is also seized and transported caudally by the intermediate set of jaws formed by the parasphenoid and the basihyal bones (Sanford and Lauder 1989, 1990). Logically, only the ventral element (the basihyal) performs major anteroposterior displacements. As the LPJs are situated along the same central axis as the basihyal, both might act in synchrony to transport the prey to the oesophagus.

3.2 Acanthopterygii

In primitive acanthopterygians, Lauder (1983b) identifies four phases comparable to those described for *Esox*. However, according to Wainwright (1989a), the pharyngeal manipulation phase does not exist in the Haemulidae. In 90% of all cases the food is directly carried to the pharyngeal jaws after capture, and important trituration occurs during the subsequent transport phase. During prey capture, activity of the pharyngeal muscles can be observed obviously resulting in an initial position of the pharyngeal jaws allowing seizure of the prey.

The transport phase is characterized by regular, repetitive activity cycles of the branchial muscles (Fig. 9). The retractor dorsalis muscle is particularly active, whereas the sternohyoideus is often silent (Lauder 1983b, c; Wainwright 1989a, b). Another characteristic feature is the considerable (in some cases complete) overlap of most muscular activity (Lauder 1983b, c; Wainwright 1989a, b; Figs. 9 and 10). This is true for the main upper and lower jaw retractor muscles (respectively the retractor dorsalis and the pharyngocleithralis internus). Therefore, it is assumed that both jaws move in the same direction for the larger part of the cycle. Contrary to *Esox*, it is the UPJ which dominates the pharyngeal transport phase. Striking, however, is the nearly complete overlap of the activities of the retractor dorsalis and the levator externus III/IV (Figs. 9 and 10). Indeed, based on their fibre orientation with regard to the UPJ, one would expect them to have an antagonistic function both in the haemulids (Fig. 9) and the centrarchids (Fig. 10). The retractor dorsalis retracts and lifts the jaw, whereas the contraction of the levatores externi III and IV should result in protraction and lifting. Based on the morphological study of the generalized non-specialized perciform haemulids and on electrical stimulation of their branchial muscles, Wainwright (1989a) presents an important new model in the functioning of the levator muscles in primitive acanthopterygians. Instead of lifting and protracting the UPJs, the levator externus III and IV and the levator posterior muscle would, on the contrary, depress the lateral sides of the jaws via a simple lever system formed by the third and fourth epibranchial elements. In this system, co-contraction of the obliquus posterior muscle is very important. In this way, simultaneous activity of the retractor dorsalis and the

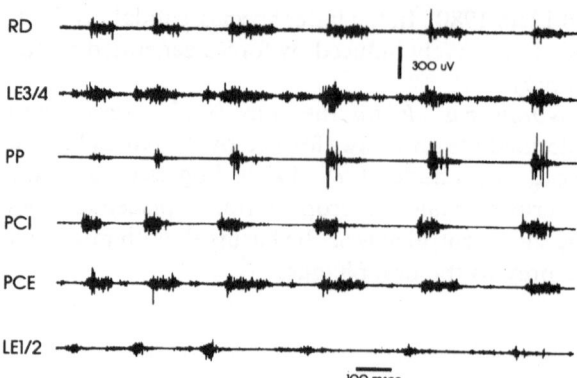

Fig. 9. Simultaneous recordings from six muscles during the pharyngeal transport phase in *Anisotremus virginicus. LE 1/2* levatores externi 1,2; *LE 3/4* levatores externi 3,4; *PCE* pharyngocleithralis externus; *PCI* pharyngocleithralis internus; *PP* protractor pectoralis; *RD* retractor dorsalis. (After Wainwright 1989a, Fig. 6)

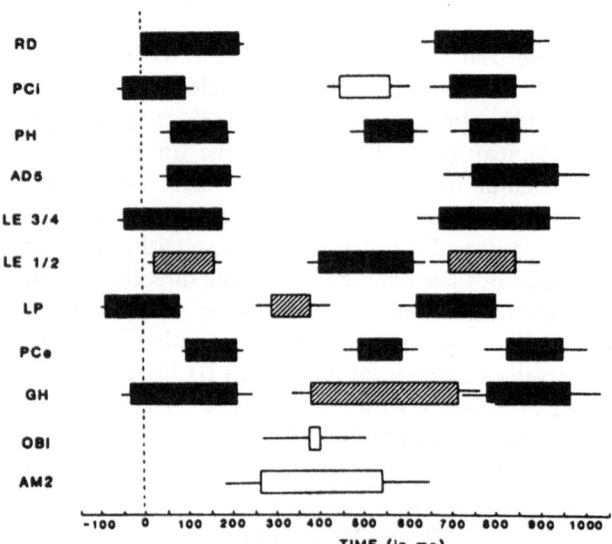

Fig. 10. Summary "block diagram" of muscle activity during pharyngeal transport in *Lepomis cyanellus.* The *left* and *right edges of the bars* mark the mean onset and offset of muscle activity and the *thin line* indicates one standard error of this mean. *Black bars* indicate activity in 67–100% of all experiments, *shaded bars* indicate activity in 34–66% of all experiments, and *white bars* indicate activity in 1–33% of all experiments. *AD5* adductor branchialis 5; *AM2* adductor mandibulae 2; *GH* geniohyoideus; *LE 1/2* levatores externi 1,2; *LE 3/4* levatores externi 3,4; *LP* levator posterior; *OB1* obliquus 1; *PCe* pharyngocleithralis externus; *PCi* pharyngocleithralis internus; *PH* pharyngohyoideus; *RD* retractor dorsalis. (After Lauder 1983c, Fig. 5)

"levator"muscles results in a retraction and depression of the UPJ, thus delivering the power stroke for transport and for prey trituration against the only slightly moving LPJ. This interpretation fits the findings of former (un-explained) experimental work on non-specialized perciforms. Liem (1970) deduces that in the Nandidae the working stroke of the UPJ is characterized by retraction and depression of its lateral side. Sponder and Lauder (1981) describe strong depression and retraction of the UPJ in the mudskipper. In the snail-crushing centrarchid, *Lepomis microlophus*, the levator posterior is extremely well developed (Lauder 1983c, e). Such a morphological specialization makes sense only when this muscle indeed does not act as a levator of the lateral sides of the UPJs, but as a forceful depressor to crush the snails. This new model might be important in the discussion of the conservation of motor patterns as well (see further).

According to Lauder (1983b, c, e), most centrarchids show only little variability in the basic activity pattern of the pharyngeal jaw muscles during the transport of different prey types (worms and fish), although asymmetrical activity of the first and second external levator could frequently be observed. Moreover, most species share the same basic pattern. In this respect the snail-crushing centrarchid *L. microlophus* might well be an exception. The crushing pattern is clearly distinct from the transport pattern, which is also peculiar in a sense that the upper and lower jaws move in opposite directions (results from X-ray recordings and confirmed by the EMG-pattern; Lauder 1983c). On the other hand, the pharyngeal repertoire of this species is also stereotyped as it "crushes" other (softer) food types as well, although other *Lepomis* species only exhibit the "transport" pattern on such prey. The conservation of the basic activity pattern is confirmed by advanced multi-variate statistical analysis applied to four haemulid species feeding on three different prey types (worms, fish and crabs) (Wainwright 1989b). Although the intensity of the electrical activity of the muscles increases with the toughness and hardness of the prey, the relative timing and the recruitment pattern are not influenced. The study of the pharyngeal jaw movements in *Serranus scriba*, on the other hand, reveals the possibility of important movement modulation during food transport (Vande-walle et al. 1992b). Movement cycles of the upper and lower pharyngeal jaws for food transport can be completely or partly out of phase (the jaws moving in opposite directions), as well as in phase with each other (same direction). In most cases the UPJ cycles are more pronounced than those of the lower elements, comparable to movements found in *Micropterus salmoides* (centrarchids; Lauder 1983b). The efficiency of the transport is variable: a movement pattern characterized by opposite cycles of upper and lower jaws results in the fastest prey transport (Fig. 11; Vandewalle et al. 1992b). However, food transport can also occur when both UPJs move in opposite directions, or even when only one of them shows displacements. This suggests that a considerable variability of the motor pattern for food processing might exist.

Vandewalle et al. (1992b) also demonstrate that in *Serranus scriba* the different elements constituting one UPJ have a certain freedom to move with

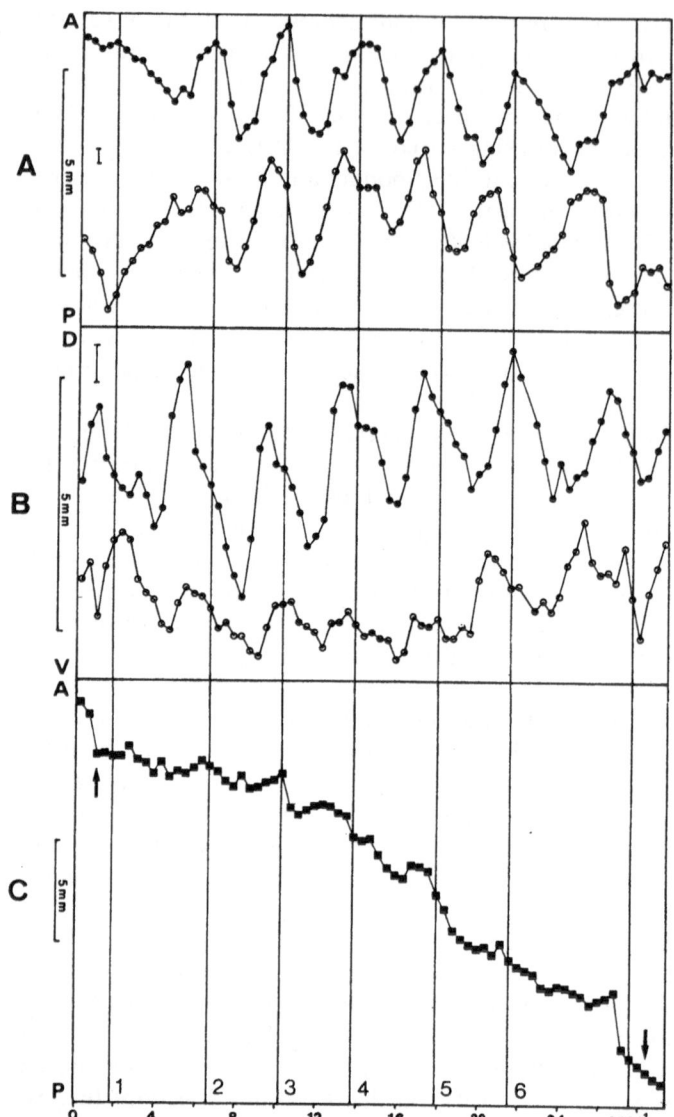

Fig. 11. Graphical representation of **A** the anteroposterior and **B** dorsoventral movements of the left pharyngeal jaws of *Serranus scriba* and **C** the displacement of the prey (a small fish, *Xiphophorus maculatus*). The error on the horizontal and vertical displacements attain maximally 0.4 and 0.8 mm respectively. The *arrows* indicate prey capture and swallowing. The *vertical lines* delimit the six successive anteroposterior movements of the left UPJ during the prey transport. *A* Anterior; *D* dorsal; *P* posterior; *V* ventral. (After Vandewalle et al. 1992b, Fig. 8)

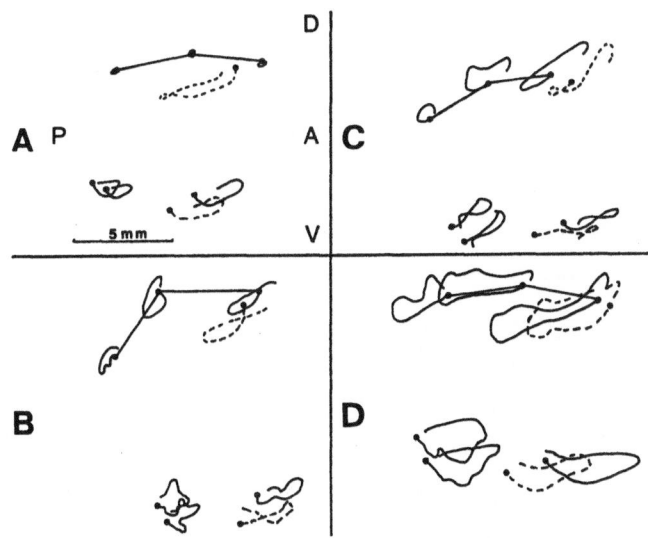

Fig. 12. Examples of food transport cycles of the pharyngeal jaws of *Serranus scriba.* The preys are **A** an earthworm; **B** a shrimp; **C** a shrimp abdomen without carapace; and **D** a fish. *Full lines* The cycles of points on the left pharyngeal jaws; *dashed lines* the cycles of points on the right pharyngeal jaws. *Straight lines* interconnect the three markers in the left UPJ. The vertical maximal error is 0.6 mm and the horizontal maximal error is 0.2 mm. *A* Anterior; *D* dorsal; *P* posterior; *V* ventral. (After Vandewalle et al. 1992b, Fig. 10)

respect to each other (Fig. 12). Such intrinsic displacements change the shape of the UPJs. The three principal elements show either a dorsally or a ventrally directed curvature. This may be related to the specific needs imposed by the prey and induced either passively by the prey shape or by muscular activity of the jaws themselves. Undoubtedly, other teleosts with UPJs built up by distinct, loosely interconnected elements exhibit the same mobility within their UPJs.

The lower pharyngeal elements of the more evolved acanthopterygians are fused, whereas the elements constituting each of the two UPJs are firmly interconnected forming, on each body side, one rigid jaw articulating with the neurocranial base. Liem (1978), Aerts et al. (1986), Claes and De Vree (1989, 1991a, b) and Claes et al. (1991) describe the cyclical pharyngeal jaw movements during feeding in cichlids. Basically, the movement pattern is consistent in a sense that the upper and lower jaws cycle in opposite directions. Nevertheless, differences related to the prey type are obvious. In *Oreochromis niloticus*, for instance, an earthworm is simply transported to the oesophagus by means of regular, opposed cycles of the pharyngeal jaws (Claes and De Vree 1991a, b; Claes et al. 1991). Actual processing of the food is not noticeable (no structural changes of the prey). On the other hand, if hard (artificial) food particles are offered to the same species, active crushing and shearing by opposite pharyngeal jaw movements can be observed (Fig. 13; Aerts et al. 1986; Claes and De Vree 1991a). However, *Haplochromis compressiceps* apparently masticates a small fish

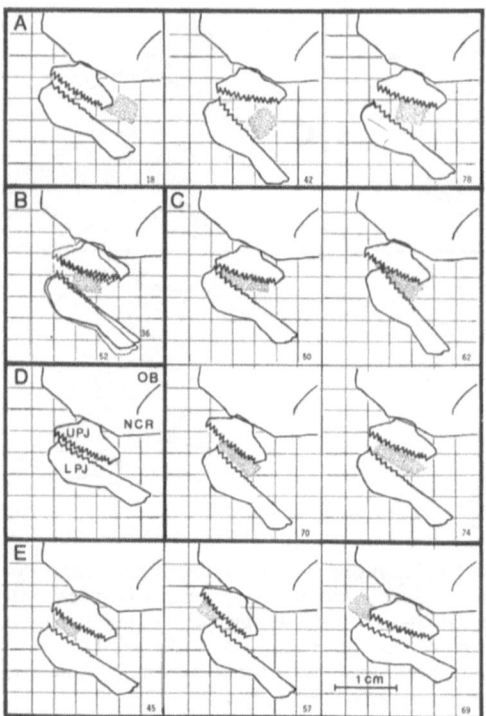

Fig. 13. Computer-generated illustrations of the pharyngeal jaw position of *Oreochromis niloticus*. Seeming overlap of the UPJ with the neurocranial base results from the schematic way of presentation. The profile contours of the jaws and the neurocranium were digitized for only one position (i.e. rest position). Thus, for this figure only the prominent anteroposterior movements can be considered (corrections for abductions and adductions of the jaws or rolling of the entire fish are impossible). Food particles are *stippled. Numbers* are those of frame numbers in the actual sequence. **A** Food uptake (soft particle); **B** squeezing of a soft particle (extreme jaw positions superimposed); **C** crushing of a soft particle; **D** rest position; **E** swallowing of a hard particle. *LPJ* Lower pharyngeal jaw; *NCR* neurocranium; *OB* orbital border; *UPJ* upper pharyngeal jaw. (After Aerts et al. 1986, Fig. 1)

by moving the upper and lower jaws in the same direction (i.e. the cycles are in phase) (Liem 1978). This contrasts with the basic pattern described above and might well be an exception. Indeed, identical prey types are processed by *Oreochromis niloticus* (a tilapiine) and *Astatotilapia burtoni* (a haplochromine) following the basic pattern (i.e. by means of opposite cycles: Liem 1986; Claes and De Vree 1989, 1991b). *Cichlasoma minckleyi* uses the same pattern to crush mollusc shells (Liem and Kaufman 1984). Mastication in *Oreochromis* can be unilateral (Claes and De Vree 1989). Liem (1978) also shows asymmetric activity patterns of pharyngeal muscles. The amplitudes of the cycles vary according to the prey type, and (as found for the primitive acanthopterygians) anteroposterior movements are more pronounced for the UPJs than for the lower pharyngeal element. According to Claes and De Vree (1992), the basic EMG pattern

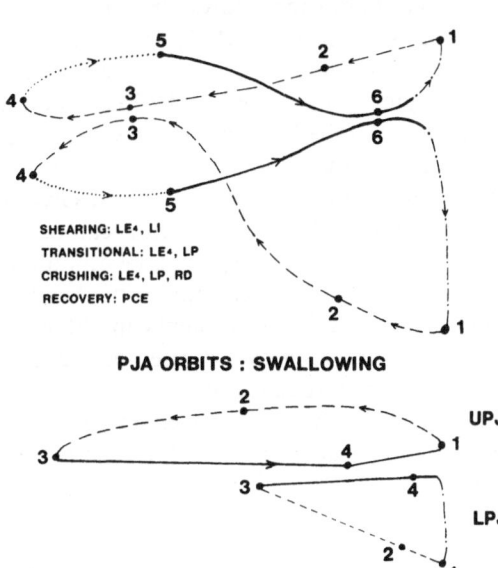

PJA ORBITS : MASTICATION

SHEARING: LE⁴, LI
TRANSITIONAL: LE⁴, LP
CRUSHING: LE⁴, LP, RD
RECOVERY: PCE

PJA ORBITS : SWALLOWING

Fig. 14. Diagrammatic representation of the pharyngeal jaw orbits derived from measurements obtained from a high-speed cineradiographic film sequence of *Embiotoca jacksoni* during mastication and swallowing. The orbits drawn represent the path of a point on the posteroventral corner of the UPJ and the posterodorsal corner of the LPJ. The masticatory cycle: *1* resting position; *2–4* shearing; *4–5* transitional phase; *5–6* crushing phase. The swallowing cycle: *1–3* protraction; *3–4* retraction. *LE4* fourth levator externus; *LI* levator internus; *LP* levator posterior; *LPJ* lower pharyngeal jaw; *PCE* pharyngocleithralis externus; *RD* retractor dorsalis; *UPJ* upper pharyngeal jaw. (After Liem 1986, Fig. 8)

during the power phase of the pharyngeal cycles is conserved in cichlids and is independent from the food type.

In the Embiotocidae, the mastication cycles are regular, describing an oblique figure eight, and are independent of the food type (Fig. 14). The involved muscular activities are particularly stable. In a rostrocaudal direction, the amplitudes of the upper and lower pharyngeal jaws are very similar throughout mastication. Crushing and shearing apparently occur by concurrent cycles of the jaws (movements in phase), whereas swallowing is characterized by a small phase shift: the (reduced) horizontal displacements of the LPJ somewhat precede those of the UPJs. Drucker and Jensen (1991) described a winnowing behaviour (separating food from debris), in which the kinematic patterns are clearly distinct from real masticatory cycles.

Apart from pharyngeal transport, cichlids and embiotocids thus clearly show important pharyngeal food processing in which the LPJ plays the dominant role. This functional shift is related to the merging of the fourth external levator and/or the levator posterior muscle with part of the obliquus posterior (Aerts

1982; Claeys and Aerts 1984; Liem 1986), forming powerful levators for the LPJ. Such a configuration allows forceful biting, as confirmed by electromyographic recordings (Liem 1978, 1986; Liem and Kaufman 1984).

In the Labridae and Scaridae, which can be considered as highly evolved acanthopterygians (Liem and Greenwood 1981; Lauder and Liem 1983), the articulation of the LPJ with the pectoral girdle implies a stabilization, but also a reduction in the kinematic possibilities of the LPJ. In fact, this LPJ now constitutes a lever system (Liem and Greenwood 1981) of which the operating muscles (like the levator posterior) can be very well developed (Yamaoka 1978). The movements of the lever can be modulated by those of the pectoral girdle. In these fish, the pharyngeal apparatus has become a forceful crushing device, in which neurocranial movements might well participate (Liem and Sanderson 1986). The crushing cycles resemble an oblique figure eight with an identical horizontal amplitude for the upper and lower jaw. Both jaws move in phase with each other. The vertical movements of the LPJ are most pronounced. Swallowing is comparable to the embiotocid pattern (Liem 1986; Liem and Sanderson 1986).

3.3 Ostariophysi

As the characiform branchial system closely resembles that of primitive teleosts, it is logically assumed that its function in feeding is also similar. Cypriniforms, however, make use of the horny chewing pad caudally on the neurocranial base for the processing of the food items. Trains of mastication cycles are followed by swallowing movements. Sibbing (1982) divides each mastication cycle of *Cyprinus carpio* into three strokes (Fig. 15): a preparatory stroke, a power stroke and a recovery stroke. During the first, the pharyngeal jaws are brought into a ventral position (contraction principally of the pharyngocleithralis externus and the rectus communis) to seize the food item between the neurocranial base and the jaws. During the power stroke, the jaws are lifted (contraction of the levator posterior, retractor posterior and transversus ventralis) to crush the food against the basioccipital chewing pad. Next to this crushing phase, grinding of the food occurs by the retraction of the jaws [contraction principally of (at least) a part of the pharyngocleithralis internus, the levator posterior and the epaxial muscles]. As a result of the simultaneous skull elevation, the basioccipital chewing pad (located ventral to the neurocranial rotation axis) moves rostrally and thus opposite to the pharyngeal jaws (grinding phase). Depending on the mastication cycle, either the crushing or the grinding phase is most prominent. Then food processing proceeds in the recovery stroke, carrying the jaws rostrally and depressing the neurocranium (relaxation of most muscles and contraction of the pharyngocleithralis externus). After one or several mastication cycles ingestion occurs, during which the constrictor pharyngis and the pharyngocleithralis externus are particularly active. Swallowing is further characterized by a constriction of the front part of the pharynx and reduced mastication movements of the jaws (Sibbing et al. 1986).

The pharyngeal movement pattern of *Cyprinus carpio* modifies with the food type. The number and frequency of the cycles, as well as the muscular activity patterns can change. Although the primary function of the pharyngeal system is reduction of the food, the pharyngeal apparatus of carp also fulfils a role in its mixing, lubrication and transport (Sibbing 1982).

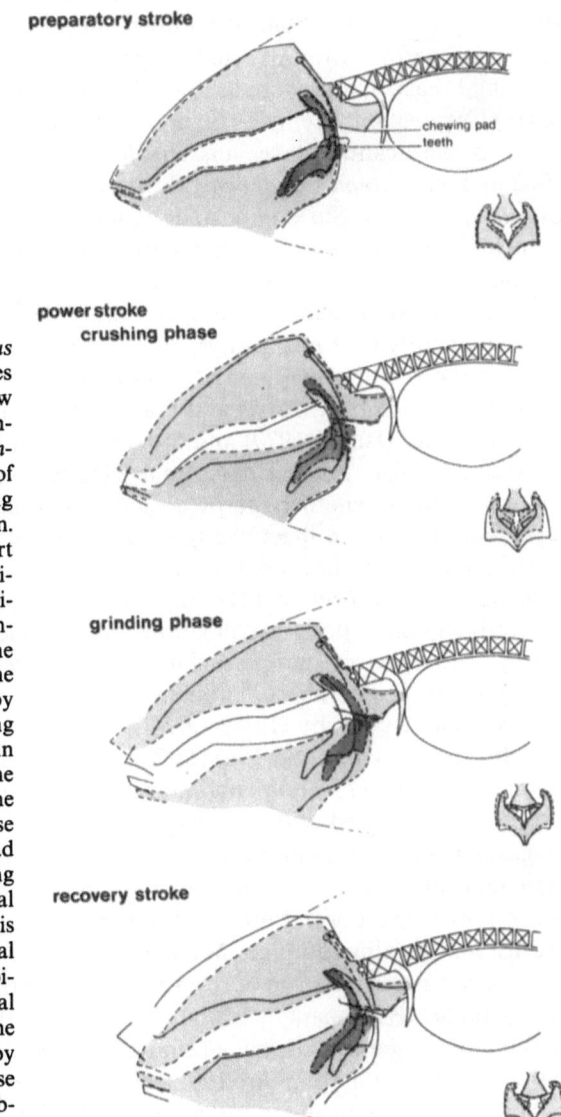

Fig. 15. A grinding cycle of *Cyprinus carpio* in lateral view. Amplitudes exaggerated about 50% to show movements more clearly (such amplitudes probably do occur). The *insets* offer a half-sized rostral view of the pharyngeal bones and chewing pad to show adduction or abduction. For each of the four strokes, start (*solid lines; white field*) and final position (*dashed lines; dark field*) are indicated. The lines marking the pharyngeal lumen indicate bony parts; the actual lumen is, at the end of the preparatory stroke, fully occupied by the pharyngeal pad. Note the lifting of the pharyngeal floor and bones in the crushing phase, whereas the chewing pad remains in about the same position. In the grinding phase the teeth as well as the chewing pad move in opposite directions, wedging the food. Rotations of pharyngeal bones are distinct; pure retraction is hardly noticed. Skull and pectoral girdle movements contribute conspicuously to grinding; the pharyngeal lumen is extensively expanded. The picture of a crushing cycle differs by domination of the crushing phase over the grinding phase. (After Sibbing 1982, Fig. 12)

4 Development of the Pharyngeal Jaws

Very few have described the early development of the branchial basket or of specialized pharyngeal jaws combining the use of closely spaced ontogenetic stages with the accuracy provided by serial sections. Indeed, most frequently cleared specimens are used, often stained either for cartilage or bone alone, and most likely missing the earliest chondrification or ossification centres. As a rule, the branchial arches differentiate (chondrify) from front to rear (e.g. de Beer 1937; Verraes 1973). Ossification does not necessarily follow this trend. Tooth plates (of dermal origin) usually appear as the first bony elements in the branchial basket, or even in the whole head (e.g. in the primitive teleost *Myrophis punctatus*, Leiby 1979; in the protacanthopterygian *Esox lucius*, Jollie 1975; in the acanthopterygians *Amphistichus argenteus*, Morris and Gaudin 1982 and *Anisotremus virginicus*, Potthoff et al. 1984; and in the ostariophysan *Clarias gariepinus*, Surlemont and Vandewalle 1991). As to ossification of the branchial elements proper (perichondral, endochondral bone), there seems to be a tendency to proceed from the rear to the front. The cartilaginous fifth ceratobranchial elements, although usually developing after the four anterior ones, frequently (but not always, Weisel 1967) ossify first. This is certainly the case for the acanthopterygians *Gasterosteus aculeatus* (Swinnerton 1902), *Amphistichus argentatus* (Morris and Gaudin 1982) and *Haplochromis* (= *Astatotilapia*) *elegans* (Ismail 1979) and for the ostariophysan *Barbus barbus* (Vandewalle et al. 1992a). Further data are needed to confirm these general tendencies and to elucidate whether the precocious differentiation of bone reflects structural requirements to meet changing demands of larval feeding.

Although the studies mentioned above deal with the appearance of skeletal elements in ontogeny, hardly any of them unravel the process by which relatively simple, purely cartilaginous visceral elements transform during ontogeny into jaws of complex shape and structure. Such an approach was attempted for a generalized haplochromine cichlid species (cf. Barel et al. 1976), *Astatotilapia elegans*, by Huysseune (1983, 1989), Huysseune and Verraes (1987) and Huysseune et al. (1988; Fig. 16). These studies indicate that the area of initial cartilage hypertrophy and perichondral bone deposition of pharyngeal elements is already established in the cartilage before overt manifestation of the process, suggesting an early commitment of the shape of the cartilaginous element. Perichondral bone restricts the growth of the elements to the distal, non-ossified portions of the cartilage. Perichondral bone also makes the cartilage susceptible to erosion. Growing tooth germs come to abut the perichondral bone collar and presumably act as inducers of resorption, first of the perichondral bone, then of the cartilage (Huysseune 1983). It is assumed that pressure exerted by the growing tooth germ is the signal that elicits resorption, as it probably is in the mandible (Huysseune and Sire 1992). Not all the tooth germs are associated with resorption cavities in the cartilage below. Whether or not a pharyngeal tooth germ in *Astatotilapia elegans* penetrates into the cartilage depends on its position, on tooth length (and stoutness) and on the depth of the dermis. Functionally, the behaviour of a given tooth germ appears to be linked to the need for all tooth tips to become aligned in the same plane to realize optimal occlusion (Huysseune 1983).

Fully differentiated teeth in *A. elegans* become attached to an annular bone of attachment and teeth do not erupt unless such a pedicel of attachment bone has

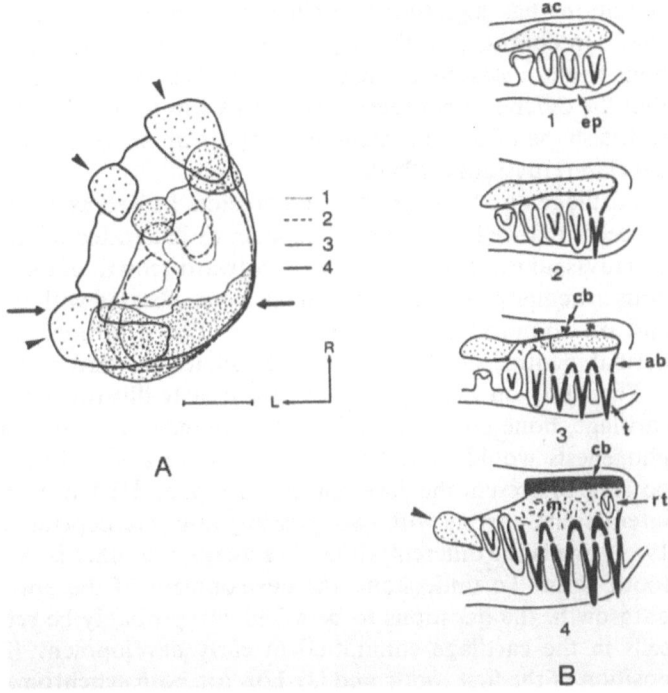

Fig. 16. A Semischematic representation of the horizontal outline of the left third/fourth pharyngobranchial (part of one UPJ) in four early postembryonic stages (*1–4*, standard length 6.5 to 10 mm) of a cichlid (*Haplochromis = Astatotilapia elegans*). Stages *1* and *2* are almost entirely cartilaginous; in *3* and *4* the cartilage to reduced to the *stippled areas*. Note the shift in position of the articular processes (*arrowheads*) (in adults the only parts persisting as cartilage). *Arrows* indicate the approximate level of sections shown in **B**. *L* Lateral; *R* rostral. Scale represents 0.1 mm (redrawn after Huysseune and Verrraes 1987). **B** Schematized transverse sections through each of these stages (*1–4*) at the level indicated in **A** to show relationships between cartilage (*stippled*), perichondral bone (*thick outline* of cartilage), chondroid bone (*cb*, *coarsely stippled*) and teeth (*t*) with their attachment bone (*ab*) (*black*). Unossified cartilage in *1* is perichondrally ossified in *2*, partially resorbed under the influence of growing teeth in *3*, and its place taken by marrow (*m*) and attachment bone of teeth in *4*. While the cartilage – except for articular processes (*arrowhead in 4*) – almost entirely disappears, chondroid bone (*cb*) develops as a skeletal support for the articular facet with the neurocranium in *3* and forms a considerable cushion in *4*. Fusion of the different pedicels of attachment bone (from *3* onwards) forms the actual dentigerous (dermal) plate. Note the presence of a first replacement tooth (*rt*), its epithelial connection to the pharyngeal epithelium (*ep*) lying posterior to this level. The arrangement as pictured in (*4*) is essentially the same as in adults. *ac* Articular cavity separating the UPJ from the neurocranial base.

been deposited below the tooth germ. On the UPJs of *A. elegans*, the dermal bone is formed solely by fusion of such pedicels of attachment bone and the dermal bone is therefore entirely tooth-mediated. Because some teeth have come to lie partially within erosion cavities in the cartilage, the dermal bone plate similarly lies partially within the former cartilage boundaries. This developmental dependence of dermal bone (tooth plates) upon teeth, which may reflect to a high degree the adaptation of a particular species to a certain trophic condition, has significant implications for phylogenetic studies (Huysseune 1983). In contrast, on the LPJ, the teeth attach by means of their attachment bone to a pre-existing dermal bone plate (Huysseune 1989). It has been proposed that the difference between UPJ and LPJ in the cichlid may be causally related to the shape of the cartilaginous support upon which the tooth germs have to develop (Huysseune 1989).

Cartilage and bone growth and erosion, as well as extension of the dentition proceed in a well-coordinated manner, as illustrated for the UPJs of *A. elegans* by Huysseune and Verraes (1987). Outward growth of the non-ossified cartilaginous articular processes in larval life is equilibrated with breakdown of cartilage and perichondral bone at their bases. The dentition, and therefore also the dermal, dentigerous bone, follow the same direction of outgrowth.

The study on *Astatotilapia elegans* clearly illustrates the interdependence of cartilage, bone and teeth during development and suggests that skeletal morphogenesis would be different if teeth were absent from these elements. It is possible to present the development of, say, an UPJ, or part of it, as a cascade of interactions starting with two (possibly also interdependent) events (Huysseune 1989); one is the differentiation of cartilage, the other is the initiation of the first tooth germ. To understand the development of the jaw, and its harmonious outgrowth, the questions to be asked can probably be reduced to (1) how are cells in the cartilage committed in early development; (2) what specifies the position of the first tooth; and (3) how are both synchronized? Processes which, at this early stage in ontogeny, speed up, truncate or in any other way perturb the cascade may well be an important mechanism for generating morphological diversification. At later stages, such diversification must be accomplished mainly through mechanisms of bone remodeling and replacement tooth recruitment. Such heterochronic processes can be partly responsible for "conversions on common phenotypes from originally different phenotypes or divergence of originally similar phenotypes into different morphologies" (Meyer 1987). Several authors have emphasized the role of heterochrony in morphological evolution (e.g. Gould 1977; Alberch et al. 1979; Alberch 1980; Hall 1984; Strauss 1990). Heterochrony has already been suggested to explain morphological variation in the oral jaw apparatus of the New World cichlid, *Cichlasoma managuense* (Meyer 1987).

All fish adjust their food choice as they grow. Changes in function concomitant with changes in prey size, velocity, etc. are reflected not only in changes in external features such as mouth size, mouth position and precocious pectoral fin development (Osse 1990) but especially in the morphology of the food transporting and processing devices such as the branchial or pharyngeal jaw apparatus (e.g. Galis 1990). Because of the important role of teeth in the morphogenesis (shaping) of the pharyngeal elements on which they occur, we want to discuss two aspects of the pharyngeal dentition which are relevant if one wants to understand how form and function of the pharyngeal jaws can change during ontogeny: (1) tooth renewal and (2) tooth morphogenesis.

1. Like in poikilotherm vertebrates in general, the dentition is polyphyodont, meaning that there is continuous tooth renewal. This allows the fish, through successive tooth generations, to adapt its dentition to a changing diet. Despite the exciting question of how changes in tooth number and arrangement and in tooth shape can be accomplished during the lifetime of an individual, pharyngeal tooth replacement has usually been studied in terms of order of tooth replacement. Often, authors have tried to fit their data with one of the two models which have been proposed to explain tooth replacement patterns in poikilotherm vertebrates, namely Edmund's (1960) Zahnreihen or Osborn's (1971) clone theory, the difference residing respectively in whether the pattern is produced by an external signal or is self-generated.

Depending on species and/or developmental stage, the pharyngeal dentition can be symmetrical or asymmetrical. Huysseune (1983) found symmetry in the first generation upper pharyngeal dentition of the cichlid *Astatotilapia elegans*, but this symmetry tends to get lost during subsequent tooth generations. A truly asymmetrical dentition, and unilateral delay of tooth germ appearance during replacement waves, occur in the cyprinid *Tribolodon hakonensis* (Nakajima 1990). Bilateral asymmetry is also found in pharyngeal tooth counts in other cyprinids (Evans and Deubler 1955; Eastman and Underhill 1973; Smith and Hocutt 1981) but the authors regard this to be due to faulty replacement. A very odd type of asymmetry is found in the carangid *Trachinotus teraia*, where the (edentulous) abrasive surfaces of the pharyngeal jaws present ridges which differ in orientation between the left and right elements (Meunier and Trébaol 1987). The literature is too scarce to tell whether a pharyngeal dentition as a rule starts symmetrically, asymmetry being the result of delay of tooth appearance during subsequent replacement waves. Neither is it known what determines asymmetry or symmetry in pharyngeal dentitions. Functionally, the presence or absence of symmetry probably highly reflects the manner in which the jaws are used for food manipulation.

2. The mechanism by which dental shape is modified during subsequent tooth generations (e.g. from slender and pointed in so-called papilliform morphs to stout and sturdy in so-called molariform morphs, cf. Meyer 1990a) to adjust to a changing diet is not yet understood. Fish teeth, like all vertebrate teeth, are assumed to require an epithelio-mesenchymal interaction for their development (Schaeffer 1977; Thesleff and Hurmerinta 1981). Replacement tooth development takes place either at the distal part of an epithelial strand emanating from the buccopharyngeal epithelium (as in cichlids, Huysseune 1989) or from the epithelial organ of its predecessor (as in trout, Bergot 1975). In some cases, the

epithelial source is unknown, as in the pharyngeal jaws of the carangid *Trachinotus teraia* (Meunier and Trébaol 1987). Teleost teeth grow in a centripetal way (Shellis and Miles 1974; Bergot 1975; Schaeffer 1977). Therefore, the shape of the tooth is determined by the shape of the epithelio-mesenchymal junction prior to matrix deposition, hence by the process of bending and folding of the epithelium. Pharyngeal replacement teeth usually develop protected from severe mechanical stresses in the medullary cavity of the jaw bone (note that in teleosts the bone marrow is not hematopoietic). If a mechanical signal triggers change in tooth shape, the message for an altered tooth shape can have been transmitted before, during or after invagination of the pharyngeal epithelium. Odell et al. (1981) provide just the sort of model which can explain the precise form of an epithelial invagination from the mechanical forces that are applied to individual cells. However, it is also possible that the mechanical forces applied to the dentigerous area are transmitted to the marrow cavity, hence to the mesenchymal cells providing the source for the dental papilla. The signal could then be transmitted to the epithelial organ. It is clear that sophisticated techniques will have to be used to unravel the processes underlying the determination of tooth shape in subsequent tooth generations.

5 Polymorphism and Plasticity of the Pharyngeal Jaw Apparatus

It has long been held by comparative anatomists, taxonomists, functional morphologists and ecomorphologists that the morphology of the pharyngeal jaw apparatus characterizes a given species and that little or no (intraspecific) variation exists. Gradually, studies are emerging which indicate that this may not be the case, and that form and function may vary to a great extent. Such a variability in pharyngeal jaw form has been reported especially for cichlids (Greenwood 1965; Sage and Selander 1975; Kornfield et al. 1982; Kornfield and Taylor 1983; Hoogerhoud 1984, 1986; Meyer 1987, 1989, 1990a, b; Witte et al. 1990). Hoogerhoud (1984) has pointed out that the magnitude of variation in morphology of the LPJ in the mollusc-crushing cichlid *Astatoreochromis alluaudi* collected from different lakes (with different snail abundance) is as high as the intergeneric differences between *Gaurochromis* and *Labrochromis*. The pharyngeal jaw apparatus of *Cichlasoma minckleyi* demonstrates an even greater variability (Kornfield and Taylor 1983). In his study of laboratory-bred *Cichlasoma managuense*, Meyer (1987) also finds cranial shapes which differ more than some intergeneric differences between African cichlids. With regard to the pharyngeal jaw apparatus, it has been pointed out that, at least for *Astatoreochromis alluaudi* (Hoogerhoud 1986) and *Cichlasoma citrinellum* (Meyer 1990b), this variability is epigenetic and related to the diet. Both the food proper and the feeding mode required to deal with a particular food item have been suggested to account for the observed shifts in morphology. E.g., in the case of mollusc-crushing cichlids, both the extent of calcium uptake (Greenwood 1965) and the muscular activity necessary to crush the prey (Greenwood 1965; Hoogerhoud 1986), have been proposed as the factor generating structural changes. It is now well established that teleosts extract hardly any calcium from their food (Meunier 1983). Moreover, Hoogerhoud (1986) has demonstrated that in a mollusc-crusher such as *A. alluaudi*, the major part of the crushed shell is ejected and ingested parts leave the body unaltered. He concluded that the observed shifts in bone shape are related to the force with which the snail has to be crushed and the effect of the resulting pressure on the bone (Hoogerhoud 1986). However, the expression of phenotypic plasticity of the pharyngeal jaw apparatus can also be constructionally constrained by surrounding anatomical structures, more particularly of the respiration apparatus (Witte et al. 1990). Clearly, a vast domain of causal morphology has yet to be explored.

6 Concluding Remarks

1. In primitive teleosts, movements of the pharyngeal jaws (predominantly of the LPJs) are capable only of transporting the food from the buccal cavity into the oesophagus (Lauder 1983b). In the acanthopterygians, the presence of a dorsal retractor muscle enhances the possibilities of retraction of the UPJs, whence their principal role in food transport (Lauder and Liem 1983). In advanced acanthopterygians, the fourth levator and the levator posterior muscles gain much importance and are related, at least partly, to the LPJ. These transformations enable the pharyngeal jaws, in addition to their function in transport, to process the food (crushing, lacerating, etc.) before it enters the oesophagus. In Ostariophysi, morphological transformations confine transport and transformation of food exclusively to the LPJs (Sibbing 1982, 1991).

The increasing specialization of the hind branchial arches, culminating in the formation of a highly specialized pharyngeal mill, has led to the introduction of the notion 'pharyngognathy'. According to Liem and Greenwood (1981), advanced euteleosts, which bite their food between the pharyngeal jaws, can be referred to as 'pharyngognath'. Among them are the Labroids, the Embiotocids, the Cichlids, the Pomacentrids, all with a single LPJ, but also the Girellidae and the Anabantoidei, which still have non-united fifth ceratobranchial elements. The Nandidae are not considered a pharyngognath because their hyal and parasphenoidal teeth participate in food uptake. In our opinion, this definition of a pharyngognathy is ambiguous for it leaves the Serranidae in an uncertain position: they possess teeth in their buccal cavity (Connes et al. 1988), yet seize their prey between their pharyngeal jaws (Vandewalle et al. 1992b). In fact, there seems to be almost no gap between the elopiform and the labrid type. Cyprinids also possess important pharyngeal jaws, which are the only buccopharyngeal elements capable of grasping and processing food particles (Sibbing 1982), yet they are not considered as being 'pharyngognath'.

The notion of pharyngognathy therefore seems to be vague and incomplete at the same time. We think it to be preferable to attribute this term to teleosts capable of processing and transforming their food with pharyngeal jaws. Such a definition would embrace Labroidei, Cichlidae, Embiotocidae and Cyprinidae sensu lato. It would represent the functional and polyphyletic outcome of a process of specializations of the pharyngeal apparatus.

2. Each specific movement pattern of the pharyngeal jaws is the combined expression of all forces deriving from muscular activity (motor pattern) and from the interactions of the jaws with the prey and other head parts. This implies that even drastically changing kinematics do not necessarily coincide with an altering motor pattern. Electromyographic recordings of pharyngeal muscles are a direct means of studying and comparing the neuromotor patterns applied by fish in pharyngeal operations. As far as the available EMG information allows far-reaching conclusions, the basic pattern is apparently conserved at least within, and even between taxa, and is in many cases relatively independent from the prey type (Lauder 1983b, c, e; Liem 1986; Liem and Sanderson 1986; Wainwright 1989a, b; Claes and De Vree 1992). Feedback modulation of the basic pattern results in changing intensity (related to force) and duration of the electrical activity, but not in the sequence of muscle recruitment and the relative times of onset of activity (Wainwright 1989b; Claes and De Vree 1992). The functions attributed to the levator muscles in primitive perciforms by the model of Wainwright (1989a; UPJ depression instead of lifting/protraction) perhaps allow one to acknowledge an extreme manifestation of the conservation of neuromotoric patterns in a biological role! In non-specialized (for instance haemulids and centrarchids) as well as specialized perciforms (e.g. cichlids, embiotocids) the muscular power for forceful food processing originates from the co-contraction of the (1) levator III/IV and the levator posterior, (2) the obliquus posterior and (3) the retractor dorsalis, in spite of the drastic morphological innovations related to the functional shift from upper to lower pharyngeal jaw in the specialized forms (i.e. merging of the levator externus IV and levator posterior with parts of the obliquus posterior to insert on the LPJ). On the other hand, it is questionable whether, for instance, the impressive diversity of pharyngeal jaw kinematics of the serranids (Vandewalle et al. 1992b), the *Lepomis* snail-crushing pattern (Lauder 1983c) or the winnowing movements observed in embiotocids (Drucker and Jensen 1991) can all be explained in terms of modulation of one and the same basic motor pattern. The present-day data set on muscular activity patterns of pharyngeal operations is still far too fragmentary to allow definite conclusions on the conservation, the evolutionary modulation to trophic specialization or the phenotypic plasticity of motor patterns.

3. It is commonplace to state that the astonishing diversity of the pharyngeal apparatus reflects the spectacular adaptive radiation among teleosts. The example of the cichlids, however, teaches one to be very cautious for such a generalization. The adaptive radiation in feeding habits and concurrent exploitation of new habitats in this speciose family has earlier been ascribed to the

fact that the pharyngeal jaws have freed the oral jaws from their dual tasks of food collection and preparation by eliminating the latter function (Liem 1973). More recently, it has been questioned whether the morphological diversity of the pharyngeal jaw apparatus is really adaptive, i.e. genetically determined, and it seems that at least some aspects of the diversity exhibited by the oral and pharyngeal jaw apparatus are epigenetic (Greenwood 1965; Witte 1984; Hooggerhoud 1986; Meyer 1987, 1990b). The concept of heterochrony has been applied to understand intraspecific epigenetic morphological variation (Meyer 1987) but the developmental mechanisms underlying heterochrony have only been partly elucidated (e.g. Hall 1984). Above all, however, these findings are crucial to taxonomic, ecomorphological and evolutionary studies.

Acknowledgments. The authors are very grateful to editors and publishing firms who have permitted the reproduction of the figures. They acknowledge grants of the FKFO 2.9005.90 and the FRFC 2.9006.90. They also wish to thank Dr. D. Kime for linguistic corrections.

References

Aerts P (1982) Development of the musculus levator externus IV and the musculus obliquus posterior in *Haplochromis elegans* Trewavas, 1933 (Teleostei: Cichlidae): a discussion on the shift hypothesis. J Morphol 173: 225–235

Aerts P, De Vree F, Vandewalle P (1986) Pharyngeal jaw movements in *Oreochromis niloticus* (Teleostei: Cichlidae): preliminary results of a cineradiographic analysis. Ann Soc R Zool Belg 116: 75–82

Alberch P (1980) Ontogenesis and morphological diversification. Am Zool 20: 653–667

Alberch P, Gould SJ, Oster GF, Wake DB (1979) Size and shape in ontogeny and phylogeny. Paleobiology 5: 296–317.

Ballintijn CM, Van Den Burg A, Egbering BP (1972) An electro-myographic study of the adductor mandibulae complex of a free-swimming carp (*Cyprinus carpio* L.) during feeding. J Exp Biol 57: 261–283

Barel CDN, van Oijen MJP, Witte F, Witte-Maas ELM (1977) An introduction to the taxonomy and morphology of the haplochromine Cichlidae from Lake Victoria. Neth J Zool 27: 333–389

Barel CDN, Witte F, van Oijen MJP (1976) The shape of the skeletal elements in the head of a generalized *Haplochromis* species:*H. elegans* Trewavas 1933 (Pisces, Cichlidae). Neth J Zool 26: 163–265

Benmouna H, Trabert I, Vandewalle P, Chardon M (1984) Comparaison morphologique du neurocrâne et du splanchnocrâne de *Serranus scriba* (Linné, 1758) et de *Serranus cabrilla* (Linné, 1758) (Pisces, Serranidae). Cybium 8(2):71–93.

Bergot C (1975) Morphogenèse et structure des dents d'un téléostéen (*Salmo fario* L.). J Biol Buccale 3: 301–324

Bishai RM (1967) Cranial muscles of *Mormyrus cashive* (L.). Anat Anz 121: 12–25

Claes G, De Vree F (1989) Asymmetrical pharyngeal mastication in *Oreochromis niloticus*. Ann Mus R Afr Centr Sci Zool 257: 69–72

Claes G, De Vree F (1991a) Kinematics of the pharyngeal jaws during feeding in *Oreochromis niloticus* (Pisces, Perciformes). J Morphol 208: 227–245

Claes G, De Vree F (1991b) Cineradiographic analysis of the pharyngeal jaw movements during feeding in *Haplochromis burtoni* (Günther, 1893) (Pisces, Cichlidae). Belg J Zool 121: 227–234

Claes G, De Vree F (1992) Pharyngeal muscle activities during food processing in *Oreochromis niloticus* (Pisces: Cichlidae). Zool Jahrb Anat 122: 173–178

Claes G, De Vree F, Vandewalle P (1991) Masticatoy operations and the functions of pharyngeal jaw movements in cichlids. Ann Mus R Afr Centr Sci Zool 262: 85–90

Claeys H, Aerts P (1984) Note on the compound lower pharyngeal jaw operators in *Astatotilapia elegans* (Trewavas), 1933 (Teleostei: Cichlidae). Neth J Zool 34: 210–214

Connes R, Grannie-Prie M, Diaz JP, Paris J (1988) Ultrastructure des bourgeons gustatifs du téléostéen marin *Dicentrarchus labrax* L. Can J Zool 66: 2133–2142

de Beer GR (1937) The development of the vertebrate skull. Clarendon Press, Oxford

Dietz PA (1912) Vergelijkende anatomie van de kaak- en kieuwboogspieren der Teleostei. PhD Thesis, University of Leiden

Drucker E, Jensen J (1991) Functional analysis of a specialized prey processing behavior: winnowing by surfperches (Teleostei: Embiotocidae). J Morphol 210: 267–287

Eastman JT, Underhill JC (1973) Intraspecific variation in the pharyngeal tooth formulae of some cyprinid fishes. Copeia 1973: 45–53

Edmund AG (1960) Tooth replacement phenomena in the lower vertebrates. R Ont Mus Life Sci Div 52: 1–190

Evans HE, Deubler EE (1955) Pharyngeal tooth replacement in *Semotilus atromaculatus* and *Clinostomus elongatus*, two species of cyprinid fishes. Copeia 1955: 31–41

Forey PL (1973) A revision of the elopomorph fishes, fossil and recent. Bull Br Mus (Nat Hist) Geol Suppl 10: 1–222

Galis F (1990) Ecological and morphological aspects of changes in food uptake through the ontogeny of *Haplochromis piceatus*. In: Hughes RN (ed) Behavioural mechanisms of food selection. NATO ASI Ser G 20. Springer, Berlin Heidelberg New York, pp 281–302

Galis F (1991) Interactions between the pharyngeal jaw apparatus, feeding behaviour and ontogeny in the cichlid fish, *Haplochromis piceatus*. PhD Thesis, University of Leiden

Gijsen L, Chardon M (1976) Muscles et ligaments céphaliques, splanchnocrâne et quelques possibilités de mouvement de la tête d'*Hoplerythrinus unitaeniatus* (Spix) (Teleostei Ostariophysi Characoidei). Ann Sc Nat Zool Biol Anim 18: 251–274

Gobalet KW (1989) Morphology of the parrotfish pharyngeal jaw apparatus. Am Zool 29: 319–331

Gould SJ (1977) Ontogeny and phylogeny. Harvard University Press, Cambridge

Greenwood PH (1965) Environmental effects on the pharyngeal mill of a cichlid fish, *Astatoreochromis alluaudi*, and their taxonomic implications. Proc Linn Soc Lond 176: 1–10

Greenwood PH (1971) Hyoid and ventral gill arch musculature in osteoglossomorph fishes. Bull Br Mus (Nat Hist) Zool 22: 1–55

Greenwood PH (1973) Interrelationships of osteoglossomorphs. In: Greenwood PH, Miles RS, Patterson C (eds) Interrelationships of fishes. London, Academic Press, pp 307–332

Greenwood PH, Rosen DE, Weitzman SH, Myers GS (1966) Phyletic studies of teleostean fishes with a provisional classification of living forms. Bull Am Mus Nat Hist 131: 339–455

Hall BK (1984) Developmental processes underlying heterochrony as an evolutionary mechanism. Can J Zool 62: 1–7

Holstvoogd C (1965) The pharyngeal bones and muscles in Teleostei, a taxonomic study. Proc K Ned Akad Wet Ser C 68: 209–218

Hoogerhoud RJC (1984) A taxonomic reconsideration of the haplochromine genera *Gaurochromis* Greenwood, 1980 and *Labrochromis* Regan, 1920 (Pisces, Cichlidae). Neth J Zool 34: 539–565

Hoogerhoud RJC (1986) Taxonomic and ecological aspects of morphological plasticity in molluscivorous haplochromines (Pisces, Cichlidae). Ann Mus R Afr Centr Sci Zool 251: 131–134

Hoogerhoud RJC, Barel CDN (1978) Integrated morphological adaptations in piscivorous and mollusc-crushing *Haplochromis* species. In: Politiek RD, Huisman EA, Oosterlee CC, Osse JWM (eds) Proc Zodiac Symposium on Adaptation. Centre for Agricultural Publishing and Documentation, Wageningen, pp. 52–56

Huysseune A (1983) Observations on tooth development and implantation in the upper pharyngeal jaws in *Astatotilapia elegans* (Teleostei, Cichlidae). J Morphol 175: 217–234

Huysseune A (1989) Morphogenetic aspects of the pharyngeal jaws and neurocranial apophysis in postembryonic *Astatotilapia elegans* (Trewavas, 1933) (Teleostei: Cichlidae). Acad Anal 51: 11–35

Huysseune A, Sire J-Y (1992) Bone and cartilage resorption in relation to tooth development in the anterior part of the mandible in cichlid fish: a light and TEM study. Anat Rec 234: 1–14

Huysseune A, Verraes W (1987) Relationship between cartilage and bone growth in pharyngeal jaw development in a cichlid fish. Biol Jaarb Dodonaea 55: 121–135

Huysseune A, Verraes W, Desender K (1988) Mechanisms of branchial cartilage growth in *Astatotilapia elegans* (Teleostei: Cichlidae). J Anat 158: 13–30

Ismail MH (1979) The ontogeny of the head parts in *Haplochromis elegans* Trewavas, 1933 (Teleostei, Cichlidae). PhD Thesis, University of Ghent, 2 vols, 228 pp

Ismail MH, Verraes W, Huysseune A (1982) Developmental aspects of the pharyngeal jaws in *Astatotilapia elegans* (Trewavas, 1933) (Teleostei: Cichlidae). Neth J Zool 32: 513–543

Johnson GD, Fritzsche RA (1989) *Graus nigra*, an omnivorous girellid, with a comparative osteology and comments on relationships of the Girellidae (Pisces: Perciformes). Proc Acad Nat Sci Phila 141: 1–27

Jollie M (1975) Development of the head skeleton and pectoral girdle in *Esox*. J Morphol 147: 61–88

Kaufman L, Liem KF (1982) Fishes of the suborder Labroidei (Pisces: Perciformes): phylogeny, ecology, and evolutionary significance. Breviora Mus Comp Zool 472: 1–19

Kornfield IL, Smith DC, Gagnon PS, Taylor JN (1982) The cichlid fish of Cuatro Cienegas Mexico: direct evidence of conspecificity among distinct trophic morphs. Evolution 36: 658–664

Kornfield IL, Taylor JN (1983) A new species of polymorphic fish, *Cichlasoma minckleyi* from Cuatro Cienegas, Mexico (Teleostei: Cichlidae). Proc Biol Soc Wash 96: 253–269

Lauder GV (1983a) Food capture. In: Webb PW, Weihs D (eds) Fish biomechanics. Praeger, New York, pp 280–311

Lauder GV (1983b) Functional design and evolution of the pharyngeal jaw apparatus in euteleostean fishes. Zool J Linn Soc 77: 1–38

Lauder GV (1983c) Functional and morphological bases of trophic specialization in sunfishes (Teleostei, Centrarchidae). J Morphol 178: 1–21

Lauder GV (1983d) Prey capture hydrodynamics in fishes: experimental test of two models . J Exp Biol 104: 1–13

Lauder GV (1983e) Neuromuscular patterns and the origin of trophic specialization in fishes. Science 219: 1235–1237

Lauder GV (1985) Aquatic feeding in lower vertebrates. In: Hildebrand M, Bramble DM, Liem KF, Wake DB (eds) Functional vertebrate morphology. Harvard University Press, Cambridge, pp 210–229

Lauder GV, Liem KF (1980) The feeding mechanism and cephalic myology of *Salvelinus fontinalis*: form, function and evolutionary significance. In: Balon EK (ed) Perspectives in vertebrate science, vol 1. Junk, The Hague, pp 365–390

Lauder GV, Liem KF (1983) The evolution and interrelationships of the actinopterygian fishes. Bull Mus Comp Zool 150: 95–197

Leiby MM (1979) Morphological development of the eel *Myrophis punctatus* (Ophichthidae) from hatching to metamorphosis, with emphasis on the developing head skeleton. Bull Mar Sci 29: 509–521

Liem KF (1970) Comparative functional anatomy of the Nandidae (Pisces: Teleostei). Fieldiana Zool 56: 166 pp

Liem KF (1973) Evolutionary strategies and morphological innovations: cichlid pharyngeal jaws. Syst Zool 22: 425–441

Liem KF (1978) Modulatory multiplicity in the functional repertoire of the feeding mechanism in cichlid fishes. 1. Piscivores. J Morphol 158: 323–360

Liem KF (1986) The pharyngeal jaw apparatus of the Embiotocidae (Teleostei): a functional and evolutionary perspective. Copeia 1986(2): 311–323

Liem KF (1990) Aquatic versus terrestrial feeding modes: possible impacts on trophic ecology of vertebrates. Am Zool 30: 209–221

Liem KF, Greenwood PH (1981) A functional approach to the phylogeny of the pharyngognath teleosts. Am Zool 21: 83–101

Liem KF, Kaufman LS (1984) Intraspecific macroevolution: functional biology of the polymorphic cichlid species *Cichlasoma minckleyi*. In: Echelle AA, Kornfield I (eds) Evolution of fish species flocks. University of Maine at Orono Press, Orono, pp 203–215

Liem KF, Osse JWM (1975) Biological versatility, evolution and food resource exploitation in African cichlid fishes. Am Zool 15: 427–454

Liem KF, Sanderson SL (1986) The pharyngeal jaw apparatus of labrid fishes: a functional morphological perspective. J Morphol 187: 143–158

Meunier FJ (1983) Les tissus osseux des Ostéichthyens. Structure, genèse, croissance et évolution. Archives et Documents, Micro-Edition, Institut d'Ethnologie, Muséum d' Histoire naturelle, Paris, SN 82-600-328, 200 pp

Meunier FJ, Trébaol L (1987) Données histologiques sur les mâchoires pharyngiennes de *Trachinotus teraia* (Cuvier 1832), Carangidae (Ostéichthyen, perciforme) d'Afrique tropicale. J Biol Buccale 15: 239–248

Meyer A (1987) Phenotypic plasticity and heterochrony in *Cichlasoma managuense* (Pisces, Cichlidae) and their implications for speciation in cichlid fishes. Evolution 41: 1357–1369

Meyer A (1989) Cost of morphological specialization: feeding performance of the two morphs in the trophically polymorphic cichlid fish, *Cichlasoma citrinellum*. Oecologia 80: 431–436

Meyer A (1990a) Morphometrics and allometry in the trophically polymorphic cichlid fish, *Cichlasoma citrinellum*: alternative adaptations and ontogenetic changes in shape. J Zool Lond 221: 237–260

Meyer A (1990b) Ecological and evolutionary consequences of the trophic polymorphism in *Cichlasoma citrinellum* (Pisces: Cichlidae). Biol J Linn Soc 39: 279–299

Miquelarena AM, Aramburu RH (1983) Osteologia y lepidologia de *Gymnocharacinus bergi* (Pisces Characidae). Limnobios 2: 491–512

Monod T (1951) Notes sur le squelette viscéral des Scaridae. Bull Soc Hist Nat Toulouse 86: 191–194

Morris SL, Gaudin AJ (1982) Osteocranial development in the viviparous surfperch *Amphistichus argenteus* (Pisces: Embiotocidae). J Morphol 174: 95–120

Muller M, Osse JWM (1984) Hydrodynamics of suction feeding in fish. Trans Zool Soc Lond 37: 51–135

Nakajima T (1990) Morphogenesis of the pharyngeal teeth in the Japanese Dace *Tribolodon hakonensis* (Pisces: Cyprinidae). J Morphol 205: 155–163

Nelson GJ (1967a) Gill arches of some teleostean fishes of the families Girellidae, Pomacentridae, Embiotocidae, Labridae and Scaridae. J Nat Hist 1: 289–293

Nelson GJ (1967b) Branchial muscles in some generalized Teleostean fishes. Acta Zool Stockholm 48: 277–288

Nelson GJ (1968) Gill arches of teleostean fishes of the division Osteoglossomorpha. J Linn Soc Zool 47: 261–277

Nelson GJ (1969) Gill arches and the phylogeny of fishes, with notes on the classification of vertebrates. Bull Am Mus Nat Hist 141: 475–552

Odell GM, Oster G, Alberch P, Burnside B (1981) The mechanical basis of morphogenesis. I. Epithelial folding and invagination. Dev Biol 85: 446–482

Osborn JW (1971) The ontogeny of tooth succession in *Lacerta vivipara* Jacquin (1787). Proc R Soc Lond Ser B 179: 261–289

Osse JWM (1990) Form changes in fish larvae in relation to changing demands of function. Neth J Zool 40: 362–385

Potthoff T, Kelley S, Moe M, Young F (1984) Description of forkfish larvae (*Anisotremus virginicus*, Haemulidae) and their osteological development. Bull Mar Sci 34: 21–59

Quignard JP (1962) Squelette et musculature branchiale des labridés. Nat Montpel Zool 4: 125–147

Ramaswami LS (1955a) Skeleton of cyprinoid fishes in relation to phylogenetic studies. 6. The skull and Weberian apparatus in the subfamily Gobioninae (Cyprinidae). Acta Zool Stockholm 36: 127–158

Ramaswami LS (1955b) Skeleton of cyprinoid fishes in relation to phylogenetic studies. 7. The skull and Weberian apparatus in the subfamily Cyprininae (Cyprinidae). Acta Zool Stockholm 36: 199–242

Roberts T (1969) Osteology and relationships of characoid fishes, particularly the genera *Hepsetus, Salminus, Hoplias, Ctenolucius*, and *Acestrorhynchus*. Proc Calif Acad Sci Ser 4, 36(15): 391–500

Rosen DE (1973) Interrelationships of higher euteleostean fishes. In: Greenwood PH, Miles RS, Patterson C (eds) Interrelationships of fishes. Academic Press, London, pp 397–513

Rosen DE (1974) Phylogeny and zoogeography of salmoniform fishes and relationships of *Lepidogalaxias salamandroïdes*. Bull Am Mus Nat Hist 153: 263–325

Sage RD, Selander RK (1975) Trophic radiation through polymorphism in cichlid fishes. Proc Nat Acad Sci USA 72: 4669–4673

Sanderson SL, Cech JJ, Patterson MR (1991) Fluid dynamics in suspension-feeding blackfish. Science 251: 1346–1348

Sanford CP, Lauder GV (1989) Functional morphology of the "tongue-bite" in the osteoglossomorph fish *Notopterus*. J Morphol 202: 379–408

Sanford CP, Lauder GV (1990) Kinematics of the tongue-bite apparatus in osteoglossomorph fishes. J Exp Biol 154: 137–162

Schaeffer B (1977) The dermal skeleton in fishes. In: Andrews SM, Miles RS, Walker AD (eds) Problems in vertebrate evolution. Linn Soc Symp Ser 4: 25–52

Shellis RP, Miles AEW (1974) Autoradiographic study of the formation of enameloid and dentine matrices in teleost fishes using tritiated amino acids. Proc R Soc Lond Ser B 185: 51–72

Sibbing FA (1982) Pharyngeal mastication and food transport in the carp (*Cyprinus carpio*): a cineradiographic and electromyographic study. J Morphol 172: 223–258

Sibbing FA (1991) Food capture and oral processing. In: Winfield IJ, Nelson JS (eds) Cyprinid fishes, systematics, biology and exploitation. Fish and fisheries series 3. Chapman & Hall, London, pp 377–412

Sibbing FA, Osse JWM, Terlouw A (1986) Food handling in the carp (*Cyprinus carpio*): its movement patterns, mechanisms and limitations. J Zool Lond A 210: 161–203

Smith RE, Hocutt CH (1981) Formulae variations of pharyngeal tooth counts in the cyprinid genus *Notropis*. Copeia 1981: 222–224

Sponder D, Lauder GV (1981) Terrestrial feeding in the mudskipper *Periophthalmus* (Pisces: Teleostei): cineradiographic analysis. J Zool Lond 193: 517–530

Stiassny M (1981) The phyletic status of the family Cichlidae (Pisces, Perciformes): a comparative anatomical investigation. Neth J Zool 31: 275–315

Strauss RE (1990) Heterochronic variation in the developmental timing of cranial ossifications in poeciliid fishes (Cyprinodontiformes). Evolution 44: 1558–1567

Surlemont C, Vandewalle P (1991) Développement postembryonnaire du squelette et de la musculature de la tête de *Clarias gariepinus* (Pisces, Siluriformes) depuis l'éclosion jusqu'à 6,8 mm. Can J Zool 69: 1094–1103

Swinnerton HH (1902) A contribution to the morphology of the teleostean head skeleton, based upon a study of the developing skull of the three-spined stickleback (*Gasterosteus aculeatus*). Quart J Microsc Sci 45: 503–597

Takahasi N (1925) On the homology of the cranial muscles of the cypriniform fishes. J Morphol 40: 1–103

Taverne L (1974) L'ostéologie d'*Elops* Linné, C., 1766 (Pisces, Elopiformes) et son intérêt phylogénétique. Acad R Belg Mém Cl Sci 41(2): 96pp

Taverne L (1977) Ostéologie, phylogenése et systématique des Téléostéens fossiles et actuels du super-ordre des Ostéoglossomorphes. Première partie. Ostéologie des genres *Hiodon, Eohiodon, Lycoptera, Osteoglossum, Scleropages, Heterotis* et *Arapaima*. Acad R Belg Mém Cl Sci 42(3): 246pp

Thesleff I, Hurmerinta K (1981) Tissue interactions in tooth development. Differentiation 18: 75–88

Thys van den Audenaerde DFE (1970) Bijdrage tot een systematische en bibliografische monografie van het genus *Tilapia* (Pisces, Cichlidae). PhD Thesis, University of Ghent

Vandewalle P (1972) Ostéologie et myologie de *Tilapia guineensis* (Bleeker, 1862). Mus R Afr Centr Ann Sc Zool 196: 50pp

Vandewalle P (1975) On the anatomy and function of the head region in *Gobio gobio* (L.) (Pisces, Cyprinidae). 3. Bones, muscles and ligaments. Forma Functio 8: 331–360

Vandewalle P (1978) Analyse des mouvements potentiels de la région céphalique du goujon, *Gobio gobio* (L.) (Poissons, Cyprinidae). Cybium 1978(3): 15–33

Vandewalle P, Focant B, Huriaux F, Chardon M (1992a) Early development of the cephalic skeleton of *Barbus barbus* (Teleostei, Cyprinidae). J Fish Biol 41: 43–62

Vandewalle P, Havard M, Claes G, De Vree F (1992b) Mouvements des mâchoires pharyngiennes pendant la prise de nourriture chez *Serranus scriba* (Linné, 1758) (Pisces, Serranidae). Can J Zool 70: 145–160

Verraes W (1973) Bijdrage tot de functioneel-morfologische studie der koponderdelen van *Salmo gairdneri* Richardson, 1836 (Pisces, Teleostei) gedurende de postembryonale ontogenie, met bijzondere aandacht voor het cranium en de kopspieren. PhD Thesis, University of Ghent, 2 vols, 242 pp

Vetter B (1878) Untersuchungen zur vergleichenden Anatomie der Kiemen- und Kiefermusculatur der Fische. II. Jena Z Naturwiss 12: 431–550

Wainwright PC (1987) Biomechanical limits to ecological performance: mollusc-crushing by the Caribbean hogfish, *Lachnolaimus maximus* (Labridae). J Zool Lond 213: 283–297

Wainwright PC (1989a) Functional morphology of the pharyngeal jaw apparatus in perciform fishes: an experimental analysis of the Haemulidae. J Morphol 200: 231–245

Wainwright PC (1989b) Prey processing in haemulid fishes: patterns of variation in pharyngeal jaw muscle activity. J Exp Biol 141: 359–375

Weisel GF (1967) Early ossification in the skeleton of the sucker (*Catostomus macrocheilus*) and the guppy (*Poecilia reticulata*). J Morphol 121: 1–18

Weitzman SH (1962) The osteology of *Brycon meeki*, a generalized characid fish, with an osteological definition of the family. Stanford Ichthyol Bull 8: 1–77

Weitzman SH (1974) Osteology and evolutionary relationships of the Sternoptychidae, with a new classification of stomiatoid families. Bull Am Mus Nat Hist 153: 329–478

Winterbottom R (1974) A descriptive synonymy of the head striated muscles of the Teleostei. Proc Acad Nat Hist Phila 125: 225–317

Witte F (1984) Consistency and functional significance of morphological differences between wild-caught and domestic *Haplochromis squamipinnis* (Pisces, Cichlidae). Neth J Zool 34: 596–612

Witte F, Barel CDN, Hoogerhoud RJC (1990) Phenotypic plasticity of anatomical structures and its ecomorphological significance. Neth J Zool 40: 278–298

Witte F, van Oijen MJP (1990) Taxonomy, ecology and fishery of Lake Victoria haplochromine trophic groups. Zool Verh 262: 1–47

Yamaoka K (1978) Pharyngeal jaw structure in labrid fish. Publ Seto Mar Biol Lab 24: 409–426

Yamaoka K (1980) Some pharyngeal jaw muscles of *Calotomus japonicus* (Scaridae, Pisces). Publ Seto Mar Biol Lab 25: 315–322

Chapter 4

Feeding in Tetrapods

F. De Vree[1] *and C. Gans*[2]

Contents

1 Introduction

The shift to tetrapody just preceded the emergence of vertebrates from the waters. This was the shift from fins to limbs, to a five-toed pattern, from support by buoying fluids to support by skeletal elements; posture becomes a biomechanical as well as a natural history phenomenon. The shift involves locomotor energetics and implies profound modification of the gas exchange and excretory systems. Similarly, terrestriality requires a fundamental change in the systems for the discovery, acquisition and reduction of food. The latter two issues are the venue of this chapter.

This transition from water to land is of particular interest as it represents one of the few generalities in the evolutionary sequence of vertebrate food acquisition systems. Predators must be opportunistic, responding both to the availability of their prey and to its continually changing defenses. This makes feeding

[1] Department of Biology, University of Antwerp (UIA), 2610 Antwerp, Belgium
[2] Department of Biology, The University of Michigan, Ann Arbor, Michigan 48109-1048, USA

Advances in Comparative and Environmental Physiology, Vol. 18
© Springer-Verlag Berlin Heidelberg 1994

(and defensive) patterns highly vagile; in contrast, locomotor patterns are constrained mainly by the relatively stable environments occupied. However, the shift from water to land was sufficiently far-reaching that it affected most aspects of the feeding patterns as well (Gans et al. 1978; Hiiemae 1978; Bramble and Wake 1985; De Vree and Gans 1989; Lauder et al. 1989). We here review the patterns following upon this transition, but omit those seen in the avian radiations.

In each case, predators face the rule that specialization for prey category is only advantageous if the category is represented by a substantial and relatively permanent biomass. Accepting this caveat one can categorize the feeding specializations in terms of their cost/benefit relationships. The benefits reflect the energetic content of the prey, as well as its content of proteins, and trace elements. The costs may be assumed to reflect not only the energy required per unit of prey captured, but also the costs associated with restructuring the feeding system to match a particular prey and exposure to predation while feeding. After all, such restructuring may also constrain the ability to feed on other prey. As the name of the game is survival and reproduction, adaptive modification for capture and reduction of new prey at the cost of an old one is only practical as long as the new resource is large and permanent. Decisions about future prey availability represent a probabilistic game (van Tienderen 1991). One can see that it is often more advantageous to select solutions that are sufficient for many prey types, rather than those that are optimum for one prey and insufficient for all others.

Water is more than 100 times as dense as air. Hence, the food objects of aquatic predators float in a medium of equivalent density, a medium that imposes substantial drag forces on their movements and on those of approaching predators. C-start prey attack modes (Webb 1985) and the development of projecting scissor jaws (Westneat 1990) represent mechanisms by which a predator may come into contact with a potentially fast-moving prey, although both are suspended in a dense medium. Other aquatic predators use suction (Muller et al. 1982; Lauder 1983), allowing the inflow of water to carry prey into the mouth. In suction patterns, the jaws may serve as secondary capture devices. They do not necessarily facilitate cutting action. The teeth may be involved in prey killing and food reduction; however, tooth placement may be on palatine or pharyngeal elements as well. Such varied possible strategies account for the spectacular diversity of actinopterygian skulls, which shows structural patterns that far transcend the range of those of the tetrapod lineages discussed below.

Finally, one must note that evolutionary experimentation (as commonly seen in experimental radiations; Gans 1990) promotes diversity and with this represents an enemy of generalizations. Although most members of an adaptive radiation may show a general condition, there remains the potential that one species, or one larger group, may specialize for a totally different aspect, utilizing a resource that happened to be available. Examples are groups such as the several kinds of anteaters; these are indeed mammals, but their specialized tongues, for instance, are hardly typical of the mammalian state.

2 Problems and Solutions for Terrestrial Feeding

2.1 Capture and Ingestion of Whole Prey

The shift to land involved primarily predatory carnivorous species, none of the transitional animals appear to have been herbivorous (Romer 1966). The need for a different kind of locomotion and support associated with the shift to land also involved the feeding system. Unlike fish, tetrapods cannot approach their prey gradually, but must propel themselves with discrete movements of individual limbs. The solution of the resulting problem has involved multiple, distinct feeding specializations.

There appear to be particular strategies, specifically for predators harvesting multiple discrete prey that are small relative to their size, predators utilizing large prey of a size equivalent to a substantial fraction of their own and predators capable of reducing large prey to smaller morsels that can be manipulated and swallowed. The benefit of small prey may be its availability in terms of total biomass, its cost may be the need repeatedly to identify and ingest individual particles. The approach of generalized filter feeding may limit the per unit cost. This cost will reflect the pumping work, which is in most fishes and whales driven by the locomotor musculature. Naturally, the method requires prey distribution of substantial density or the ability of the animals to search for and identify patches of high density. This approach is rare for terrestrial tetrapods.

Small particulate prey may be harvested effectively by reducing the cost per unit of prey. Two separate approaches are seen. The first reduces the mass of that portion of the predator that serves to catch and ingest the prey. The second is the reverse strategy of attracting the prey animal to the vicinity of the capture zone. For the first approach the benefit lies in the reduced inertia. It may be achieved by reducing the size of the head, neck, and mouth (gymnophionans, turtles, some lizards and snakes, and many birds), or by developing projectile tongues as specialized remote prey-catching elements (frogs, some salamanders, certain lizards and birds) (Gans 1991). Attractants or baits occur in many forms. They may be chemical (as in some chameleons), and visual by mimicking food objects, i.e. the lingual process of the alligator snapping turtle *Macrochelys*, or the pigmented and moving caudal tip of various small vipers. All reduce the cost per prey item. Again, prey availability determines the magnitude of potential investment; this explains the multiple specializations on termite and ant colonies.

Increases in the linear dimension of the prey that may be ingested will yield a disproportionate benefit, as the mass of nutrients ingested will rise as the third power of the linear dimension. However, each increase in the size of the prey attacked will also increase ancillary problems. Beyond the need to ingest the entire prey object, there is the task of transporting it during the process of digestion. The prey is likely to struggle, and the forces and moments it imposes on the predator will increase with its size. As the strength of cranial elements does not scale geometrically, an increase in absolute size may cause a functional conflict. Thus, the demands for prey capture may establish the need for slender low-inertia elements capable of rapid acceleration (see Chap. 6, this Vol.), whereas the demands for prey reduction may establish a need for the mandibular elements to become stouter.

This pattern of conflicting constraints is perhaps seen best in the transition from lizards to snakes and in the radiation of the latter group (Gans 1961). Capture of moving prey demands detection and sensory triangulation to establish its position in three-dimensional space; this must be followed by rapid acceleration of the head and contact with the prey. The lighter the head, the less the cost of its acceleration, but also the greater the risk of fracturing slender components upon impact. The larger the prey object, the greater its nutritional benefit, but also the greater the risk of structural failure during ingestion. Also, utilization of larger prey involves additional costs; the size of the prey constrains the locomotor capacity of the predator and conflicts with behaviors that facilitate escape from predators (Webb 1985). Many of the specializations of snakes may be viewed in this context (Gans 1974; Cundall and Gans 1979). Thus, constriction and envenomization limit the defensive movements of prey, stretching of prey objects during ingestion reduces their diameter, shift of the head over the prey (rather than pulling the prey into the esophagus) limits the forces on their slender skull and injection of venom deep into the visceral cavity of mammals accelerates digestion.

2.2 Food Reduction

In contrast to the strategies discussed thus far, large prey is commonly reduced into smaller pieces. These can be more easily ingested and swallowed than would an entire prey animal. The simplest mechanisms of prey reduction involve tearing. The prey or one of its appendages is then grasped and the predator's head withdraws, inducing shear and tensile loadings that may lead to rupture. Commonly, the prey is wedged into place; the pull then induces sudden differential acceleration rather than shift of the prey as a whole. (This has often been observed in crocodilians and also in snakes that tear snails out of their shells; Gans 1983.) Torsional tearing against wedged prey serves to tear or cut out divots (Helfman and Clark 1986). Finally, there is inertial reduction. Here, the predator bites the prey and then shakes it rapidly, causing the tissues to tear and those portions projecting from the mouth to risk being ripped off (variants of this method are seen in some aquatic feeding patterns). Such oft rather bloody performances are seen in monitor lizards and many carnivores; they may combine killing action and reduction of the prey. The major disadvantage is their intrinsic ineffectiveness as parts of the prey are likely to become lost.

The teeth, which allow the prey to be grasped, also generate stress concentration sites in its integument and facilitate subsequent tearing. In torsional tearing, the predator spins after grasping the prey and the penetrating marks of the teeth generate a zone of rupture. Teeth of different shape impose differential requirements onto the muscles that must cause their penetration into the prey. Again, tooth elongation facilitates deeper penetration at the cost of greater probability of tooth failure. Similarly, flattening of the tooth facilitates its penetration and generates a defined line of weakness that reduces the cost of removing a divot, yet also risks breakage.

Tetrapod dentitions include multiple variants of cutting systems, all facilitating removal of portions from an intact animal and later reduction of the separated portions. The biomechanics of cutting action by serrated cusps has recently been analyzed (Abler 1992). Increasing specialization involves increasing modification of the teeth of both the upper and lower tooth rows, with alternation of teeth and cusps. Initially, one sees interlocking patterns of simple conical teeth that facilitate penetration of the integument and success for the resulting tearing action. Seemingly more advanced conditions involve penetration of the integument by alternating cusps. Further modification is reflected in the specialization of matching cusps in the maxillary and mandibular dentitions. Perhaps the highest level of modification involves the parallel change in the jaw joints which permits the animals to move the teeth (propalineally) parallel to each other. Such articulations bring the teeth into specific and often unilateral occlusion, but also facilitate various horizontal power or grinding strokes. Variants are seen in the propalineal shifts that bring the rodent incisors into a position in which they may serve for cutting or self-sharpening and the lateral movements that allow the carnassials of carnivores to approach each other in order to shear prey (Turnbull 1970; Vaughan 1978).

Once large food items can be broken up to the point at which the morsels can be ingested, there may be an advantage to mastication, further trituration of food internal to the predator. The benefits likely reflect an increased digestive rate with increased surface area of the food object, thus reducing the time required for digestion. This will decrease the interval during which a predator needs to transport a gut distended with bulky food. Finally, large and unreduced food objects will tend to decay or otherwise deteriorate internally long before the process of digestion has reached their center. Whereas mastication may use the same teeth that serve the initial production of morsels, one commonly sees the development of heterodonty, structural and coordinated functional differences in the dental arcades.

2.3 Carnivory Versus Herbivory

Whereas carnivores achieve benefits from partitioning their food, reduction becomes essential for herbivores. Plant tissues contain cell walls incorporating cellulose; woody plants also contain lignin (Pough 1973). Both of these are hard materials that are notoriously difficult to digest; consequently, the walls must be ruptured in order for the digestive juices to reach the cytoplasm. Very few tetrapods have enzymes that break down cellulose directly. However, a variety of turtles and lizards, as well as some birds and mammals, avoid this difficulty by specializing on fruits, nuts, flowers and leaf buds which are more nutritious and have less cellulose. These foods also have not yet accumulated the secondary compounds which make some adult plant tissues toxic (Rosenthal and Janzen 1979).

Lizards and turtles ingest leafy materials after minimal reduction. They cannot digest these and instead rely on commensals. These may be micro-organisms or nematodes which generally pack the stomach of some turtles. It has been shown that small lizards will subsist on animal foods, whereas, with the exception of monitors (*Varanus* sp.), larger lizards specialize on plant materials (Pough 1973). A parallel change occurs during the ontogeny of a number of larger species. Whereas it may be that many larger lizards might have difficulty in encountering enough animal prey to subsist thereon, there is also the question whether there may be a minimal size (stomach volume) for neutralization of secondary compounds and for breakdown of cellulose.

In birds, herbivory commonly involves seed eating. Both for this and for feeding on other plant materials, the reduction has been shifted to the gizzard. This commonly contains a mass of hard sand grains and pebbles that are moved about by contraction of the muscular wall, grinding the interspersed plant material. However, reduction of plant materials reaches its greatest development in mammals. Several of the mammalian dentitions (see below) are specifically adjusted for grinding and rupturing cell walls as the food is compressed and then exposed to explosive decompression. Beyond this, there is rumination in which plant material is first chewed, then subjected to primary chemical digestion in the rumen, then sorted, with the floating, incompletely reduced mass regurgi-tated for secondary mastication (rumination). The secondary swallowing is followed by further digestion. The lagomorph pattern, in which all food passes the gut twice, apparently involves no secondary reduction, rather the double passage compensates for digestive processes that are located downstream of the absorption sites (Voelker 1986).

3 Amphibians: Conflict and Prey Ingestion

The food capture devices of terrestrial tetrapods are derived from the aquatic systems of their fish ancestors. These involve the first and second branchial arches, whereas the food reduction systems of fish reflect the more posterior ones. In all cases, the branchial arch system retains some association with the braincase and sensory capsules (jointly referred to as a skull), and through these with the vertebral column. However, the association of the skull and shoulder girdle, so characteristic of many fishes, is reduced to muscular rather than skeletal linkages.

The earliest tetrapods retained metamorphosis. This process presumably marked the transition between food types and modes of gas exchange as it does in the Recent amphibians. The premetamorphic stages extracted their oxygen from the water by means of gills associated with the pharyngeal skeleton; they trapped their prey in large buccal cavities (Lauder and Shaffer 1985, 1986, 1988). In adult forms, gas exchange involved manipulation of air and it has been argued that gas transport was by aspiration as in all advanced tetrapods (and possibly in rhipidistian fish as well) (Gans 1971). The phyletic reconstruction of the changing functional patterns is complicated by the situation in Recent amphibians. As far as known, these pulse-pump their air by means of the musculature of their relatively motile buccal floor (De Jongh and Gans 1969). It has been argued that their state is not transitional to that between fishes and reptiles, but rather represents a new set of functional conditions. Redevelopment of pulse pumping apparently constrains the shape and architecture of the mandibles and their bridging muscles. With this there appears to have been a functional constraint on the ability of the mandibles to support the capture of prey and its reduction and handling.

As far as known no Recent amphibian has redeveloped food reduction patterns. The closest to this one sees is the ability to tear out divots; generally, these animals swallow their prey whole. Adult gymnophionans (Bemis et al. 1983; Nussbaum 1983) and many caudates capture their prey with a forceful bite. This is also possible for some frogs; however, the method is impractical for small and agile prey. This problem of prey capture has been resolved in at least two major ways, characterizing the frogs and some terrestrial caudates. In each pattern, capture involves a projectile tongue and it appears that at least the frog condition has developed independently many times (Regal and Gans 1976; Nishikawa and Cannatella 1991; Nishikawa and Roth 1991; Nishikawa and Gans 1992).

In caudates, a novel system of prey capture could develop in those lines in which the buccal floor had been independently liberated from the demands of gas exchange. The "lungless" condition involved the shift of gas exchange to the buccal cavity and the overall integument. Relaxation of the pulse-pumping constraint on buccal architecture allowed the modification of the hyoid skeleton to provide a mechanical linkage that projected the tongue out of the mouth (Wake 1982). The projection not only can occur rapidly but can also reach far beyond the mandibular symphysis. The soft tissues on the lingual tip contain a mass of mucous glands and their inertia upon impact onto the prey causes them to deform around the prey and to bring it back into the buccal cavity as the tongue is retracted (see Chap. 6, this Vol.).

In the tongued Salientia, the ventilatory constraint is never released. Consequently, the hyoid plate ordinarily does not reach the mandibular arch, and various transverse intermandibular muscles and geniohyoid elements provide a soft tissue barrier. Its contraction can lift the buccal floor in pulse-pumping. Extensible tissues that can reach across the mandibular symphysis to pick up prey are attached to the symphysis and are only loosely connected to more posterior regions. The lingual system lacks an intrinsic skeleton; rather it involves lymph-filled chambers that can be hydrostatically stiffened by contraction of internal muscles. The demands of pulse-pumping and the slenderization of the skulls of many frogs are also reflected in the swallowing system. This involves downward bending of the skull on the head joint and depression of the eyes into the pharyngeal space then constrained by contraction of the buccal floor.

The simplest systems (not the phylogenetically earliest) already incorporate the basic mechanism (Regal and Gans 1976; Nishikawa and Cannatella 1991; Nishikawa and Roth 1991). Contraction of muscles radiating from symphyses into the tongue (the so-called genioglossal muscles) pulls the posteriorly free lingual mass anteriorly and stiffens it. As the tissues are attached to the medial aspect of the mandibles, the protrusion causes the mass to rotate and the initially dorsal surface, which is covered with mucous glands that produce a viscous fluid, then bulge over the symphysis. Thus, the pad's dorsal surface reverses and the animal impacts it downward upon the prey, complying to its surface. Retraction is then effected by contraction of hyoglossal muscles that attach and start to pull first on the lingual tips. With this, the prey enters the buccal cavity.

More advanced lingual systems have the center of the tongue occupied by various bundles of muscle fibers, the proximal portion of which is attached to the symphysial region. Contraction will transform these bundles into a stiffened median rod, the distal end of which supports the flaccid lingual mass (Gans and Gorniak 1982a, b). This genioglossal rod is generally connected to the tongue by thin muscular fascicles that insert across the field of mucous glandular crypts covering the dorsal lingual surface. A transverse submental muscle connects the mandibular tips immediately beneath the genioglossus and stiffens into a bulging rod upon stimulation; this acts to close the nostrils whenever the mouth is closed (Gans and Pyles 1983). It also facilitates the depression of the mandibular tips relative to their long axis; in many species of frogs, the mandibular tips are supported by separate mentomeckelian bones that facilitate these complex movements. Opening of the mouth coincides with downward rotation of the mandibular tips; thus, the anterior attachment of the stiffened genioglossal rod is depressed and its more posterior portion elevated. Consequently, the genioglossal rod rotates over the symphysis and allows the inertia of the soft tissues to impact onto the prey. Small prey is retracted to the midlevel of the buccal cavity; once portions of larger prey are shifted into the mouth, scraping movements of the respective forearms selectively move protruding portions inward. Retraction of the lingual mass is effected by muscles originating on the hyoid.

Whereas the basic pattern, here described, applies to numerous derived species of frogs, there is substantial diversity in the details of the muscular and apparently the neural arrangements (Trueb and Gans 1983; Gans et al. 1991; Deban and Nishikawa 1992). This suggests that the several structural arrangements still deserve much experimental attention.

4 Reptilian and Mammalian Prey Ingestion

The shift of many tetrapods to aspiration breathing allowed these to specialize in the ability to pick up prey by using the lingual system and to manipulate it within the buccal cavity. However, the tongue of many lizards and snakes shifted to a sensory role, specifically to sampling of the environment. The tongue became narrowed and its tips were transformed into rod-like tips that could be stiffened, then bent and protruded. This allows the animal to touch and sample external objects and with lingual retraction to carry the substances thus picked up to the vomeronasal organ. Various lizards show such specialization in the anterior portion of the tongue only; the lingual body and more posterior zone remain fleshy and their dorsal surface may be papillate or scale-covered (Schwenk 1988). In other species, such as the monitors (*Varanus* sp.; Smith 1986) and almost all snakes (McDowell 1972; Schwenk 1988; Smith and Mackay 1990), the posterior zone also has been modified into a rod-like pattern, thus facilitating extensive lingual protrusion. How then do these species manipulate their food?

Two basic patterns occur. The first pattern is inertial feeding (Gans 1961, 1969a; Smith 1982; Cleuren and De Vree 1992). Here, the prey is first captured between the jaws and is then shifted, laterally or posteriorly, by tossing it within the mouth. The head is jerked in the direction in which the prey object is to move and the jaws then open quickly allowing the prey momentum to determine its path to the site at which the closing jaws catch it again. This method also subserves prey manipulation in long-snouted crocodilians and some birds. Whenever it is used by aquatic species, these animals may capture their prey in the water, but always practice the inertial pattern in the air (Davenport et al. 1990; Cleuren and De Vree 1992). The second pattern is that seen in most snakes. Here, the prey is initially grasped between the dentate elements of the jaws. It is retracted into the esophagus by a combination of alternating mandibular retractions. A forward push of the neck tends to drive these bones under and around the prey (Gans 1961; Cundall and Gans 1979; Cundall 1983). Hence, the tongue is no longer required, neither for grasping the prey nor for its manipulation.

Those reptiles with soft tongues use them in a pattern superficially similar to that seen in the primitive tongues of amphibians. The body of the tongue consists of a mass of muscle fibers enclosed in a connective tissue sheath. Contraction stiffens this hydrostat (Kier and Smith 1985). Whereas the fibers are sometimes interspersed loosely amid lymph-filled spaces, the system generally differs from the amphibian one by its compartmentalization which allows controlled changes in shape. Furthermore, the fleshy reptilian tongue tends to be mounted on the anterior hyoid horn and shifting the posterior branchial elements will lift it. To the extent that the tongue is attached to the symphysial region, lifting of the lingual pad off the buccal floor tends to curve it. Protraction of the hyoid forces the tongue out of the mouth, with part of its dorsal surface facing anteriorly (or even anteroventrally); thus, movement of the head will cause the mucous membrane to contract and adhere to the prey (Gorniak et al. 1982; Schwenk and Throckmorton 1989; Bels 1990; Bels and Goosse 1990; Kraklau 1991; Chap. 7, this Vol.).

This is the pattern seen in the food acquisition of many lizards. Even if the tongue is not protruded beyond the symphysis, it may attach to prey and shift and retract it within the buccal cavity; commonly, it fixes the prey for a subsequent bite (Smith 1984). Further, the tongue becomes instrumental in shifting food objects about the buccal cavity and among the several triturating surfaces. With the shift to food breakdown, a controllably flexible tongue allows a sweeping out of the partially reduced food from the mouth so that it may be reduced further or swallowed. The manipulative and sensory capacity of the reptilian tongue will thus be important both within and beyond the cavity, an importance that is retained in mammals.

The tongue of the Old World chameleons also provides another example of a functional shift from the habitus seen in their agamid ancestors (So et al. 1992). The modification has involved both the locomotor and the feeding pattern. These slowly moving animals are found mostly in bushy and arboreal habitats. Their prey is captured by a projectile tongue that may reach as far as the length of their body. The tongue appears to represent a variation of the agamid pattern. The hyobranchial skeleton and its muscles remain remarkably similar. However, whereas the smooth and conical tip of the hyoid horn is fixed to the tongue in agamids, that of chameleons serves as a launching rod (Schwenk and Bell 1988; Bell 1989; Wainwright et al. 1991). The soft tissues of the tongue slide upon the hyoid rod. Propulsion is generated by an anterior dense sleeve of circular muscle. Its contraction deforms the mass and exerts inward forces initially on the cylindrical portion of the horn and then on its conical portion. Pressure on the conical portion only starts after the end of the exitation–contraction interval, leading to sudden outward acceleration of the mass. Coincidentally it extends the more posterior retractor muscle and the folded mucous lingual membranes. The distance of projection is limited by the length of the retractor muscle which normally rests folded on the horn and consists of fibers with hypercontracting sarcomeres. The skeletal linkage not only positions the tongue for ballistic projection, but is also able to shift prey within the buccal cavity and relative to the tooth rows (So et al. 1992).

5 Teeth, Crushing and Mastication

5.1 Dental Architecture

The shape of teeth varies enormously (Peyer 1968). This would seem confusing if one considers hard objects to be evolutionarily rigid. However, the shape of teeth is developmentally laid down by the pattern of the soft tooth bud and the position of both dentine and enamel follows this arrangement. Hence, slight modifications of the epithelial surface would simply translate into parallel modifications of dental shape.

The earliest fossil tetrapods had seemingly simple conical teeth; however, these had enameloid infoldings, hence, the name labyrinthodonts. Recent amphibians show a distinct pattern with each tooth consisting of a columnar pedicel that is capped with a dentinal cutting cusp; the two parts are joined by a collagenous connection so that the hinged tips of many species are motile (Parsons and Williams 1962; Duellman and Trueb 1986). There are relatively few species of caudates and salientians in which the adult teeth do more than provide a holdfast portion. In some frogs, the jaws include a bony projection that facilitates the generation of a deep stabbing cut.

In turtles, the edentulous jaws are covered with a heavy keratinized sheath. In crocodilians, the teeth are simply conical and set in deep sockets, but those of dentary, premaxilla and maxilla alternate. Also, there are generally a number of enlarged teeth near the front. The teeth of snakes are generally subcylindrical and elongate; they may have posterior cutting edges, although earthworm feeders have stoutly conical and rounded teeth. A few species have hinged teeth (Patchell and Shine 1986b); these are generally species that feed on cylindrical and armored prey (such as skinks). Naturally, there are the various kinds of elongate, grooved or hollow teeth associated with the venom-injecting process (Bogert 1943; Gans and Elliott 1968).

Lizards show a variety of patterns in their tooth shape (Edmund 1969). Although many textbooks claim that their dentition is homodont, it is clearly heterodont in many forms. Commonly, the anteriormost teeth are sharply conical, whereas the more posterior teeth on each side are flattened along the mandibular plane and bear one or more pairs of cusps on their edge. In a few species, such as *Gekko gecko*, the maxillary and mandibular dentitions move in the same plane and can cut effectively. In other lizards, the more posterior teeth are domed and markedly widened; this is a pattern commonly associated with crushing of hard prey (Dalrymple 1979; Estes and Williams 1984). In the rhynchocephalian *Sphenodon punctatus*, parallel rows of conical teeth lie on the maxilla and the lateral edge of the palatine. The dentary teeth fit between these (Robinson 1976; Gorniak et al. 1982) in a condition analogous to the mammalian multicusped molars.

Mammalian teeth are extremely complex and show substantial interspecific differences. Both the major mammalian groups and the species of mammals can be diagnosed by the shape, diversity and spacing of their teeth. This indicates that the present account can only deal with generalities. These patterns are discussed in terms of cranial movements.

5.2 Cranial Kinesis

The cranial bones of most ectothermic tetrapods are only loosely articulated; consequently, mandibular depression tends to be associated with movements additional to simple rotation. This cranial kinesis complicates the analysis of closing movements.

Kinesis has often been discussed and numerous papers speculate on its biological roles. Several plausible explanations exist and selection among these is confounded by taxonomic diversity. Analysis is also complicated by possible synergisms among the cited explanations. The flexibility of the skull obviously reflects its complex origins from the diverse elements of the braincase, from the independent sensory capsules, from the pharyngeal skeleton (that generated the jaws, their supports and the hyoid apparatus) and from the integumentary ossification. These elements have independent evolutionary origins and this independence reflects the phylogenetic history of the system. For instance, in most fishes, many of the elements are loosely articulated throughout life. This suggests that fusion rather than kinesis needs explanation. The second explanatory component is that the elements grow allometrically in different directions (Gans 1988). Hence, they can only fuse after they reach a size close to that of adults, which suggests that the kinesis represents a constraint on ontogeny. Finally, there are the constraints due to function.

Multiple functional (role-associated) explanations for cranial kinesis are commonly adduced (Frazzetta 1962; Bock 1964; Borsuk-Bialynicka 1984: Smith and Hylander 1985; Iordansky 1990), but tested less often, particularly in a phylogenetic framework (Condon 1987; De Vree and Gans 1987a). Mechanisms, such as those involving coordination of the strike or widening of the gape in swallowing, clearly represent possible specializations of particular groups of species. In the present context, there are at least three functional attributes of kinesis that can be documented in one or another group. These are the ability for differential movement of the cranial tooth-bearing elements, the ability for horizontal shifts of the mandibles relative to the upper jaws, and the ability to limit the buildup of stress concentrations on the nervous system, the so-called shock-absorption system.

The ability to move several cranial tooth-bearing elements relative to each other is best seen in snakes in which the maxillary and palatine arches can shift independently (Cundall and Gans 1979; Cundall 1983). As this controlled motility is fundamental to the ophidian ingestive capacity, its role is obvious, but taxonomically restricted. Hypercontraction, such as that seen in some geckos and pygopodids (Patchell and Shine 1986a; De Vree and Gans 1987b), in which the maxillary tooth row rotates relative to the base of the braincase, is another specialized example. Presumably, adjustment of the jaws permits a firm grasp on relatively large prey.

Shift of the mandibular dentition horizontally relative to that of the upper jaws is commonly seen in snakes, lizards and some amphisbaenians; it is associated with the flexibility referred to as streptostyly, in which the jugal arch is lost so that the bottom of the quadrate can rotate more freely about its dorsal articulations (directly or indirectly with the braincase). In snakes, this movement has an ingestive role in pulling prey into the mouth; it is a protoadaptation (i.e. a functional novelty based on a preexisting morphological capacity; Gans 1979). In lizards, the horizontal shifts appear to involve initial prey reduction (Hotton

1955). The prey is held forcefully between the maxillary and dentary arcades; various horizontal, generally propalineal, movements of the jaws shear it. This may be particularly beneficial in feeding on arthropods; rupture of their exoskeleton kills them and facilitates digestive access to their internal tissues. The pattern is also seen at the end of the bite of *Sphenodon*; the upper jaw holds the prey and the mandible shifts anteriorly then cutting through its soft tissues (Gorniak et al. 1982).

The occurrence of shock absorption has been debated on theoretical grounds (Frazzetta 1962). However, some lizards crush prey with heavy and brittle skeletal elements, the failure of which may generate sudden acceleration of the mandible which involves the buildup of force. Even if they have relatively solid skulls, such species tend to have connective tissue basipterygoid pads that deform substantially upon loading, thus absorbing energy (De Vree and Gans 1987a); this kind of stress reduction system has also been documented in interlocked bony joints (Jaslow and Biewener 1988). Also, animals feeding on brittle objects tend to have a stronger braincase with fusion or substantial interlocking of the cranial elements. Such construction limits excessive stress concentration during transmission of high forces through the cranial skeleton. A final argument for the significance of shock absorption patterns is seen in rabbits, the only mammals that appear to have developed an intracranial joint (Bramble 1989), and one for which shock-absorption evidence seems unequivocal.

Consequently, one may suggest that the several kinds of kinesis represent combinations of phylogenetic and ontogenetic constraints that have been utilized for additional roles. In many ectothermal tetrapods there appears to have been a set of constraints toward cranial lightening and a limitation of the mass (i.e. calcium content) of the skull. The constructions observed support the head by incorporating relatively light braces, which are joined by connective tissues. Such skulls can be relatively lighter than akinetic ones. One observes such framing patterns in lizards, and in frogs in which the skulls are relatively large. They are also seen in birds, as part of the general lightening of the animal for flight.

This requires some supplementary explanation for the patterns seen in turtles, animals which lack any obvious kinesis. However, all turtles lack teeth and cut food objects by using their variously modified keratinous sheaths. One line of turtles differs from the remainder in their capacity to protract the mandible by sliding it into the jaw joint (Gaffney 1979). It should be interesting to study the food reduction capacity of the several sheathing styles and also to consider the mechanical effect of the sliding capacity. One can then proceed to the mammalian architecture which generally lacks cranial kinesis.

5.3 Wider Roles

Before discussing the development of the complex masticatory system seen in mammals, it is necessary to digress briefly. Thus, tetrapod gape and the dentition then displayed subserve other biological roles and these must be taken into account in attempts at correlating structures and roles.

In birds and mammals, the jaws facilitate grooming. Grooming specializations appear to be associated with hair and feathers; to the best of our knowledge these patterns do not appear in any amphibian or reptile. Portions of the avian bill may be ridged in a way that cleans the feathers as these are drawn through them. Furthermore, repeated action tends to reattach the pinnules of flight feathers, maintaining their aerodynamic properties. Application of the secretion of the uropygial gland facilitates this process and also conditions the feather surface, avoiding their wetting. In some mammals, the incisors similarly allow grooming of the pellage and there have been descriptions of the association of incisive grooves with this role (Vaughan 1978).

A gape and a bite do serve as potential predator deterrents in various amphibians and reptiles (Greene 1988). This may explain the very occasional bony projections from the jaws of some large frogs (e.g. *Pyxicephalus*); these animals have been known to bite cattle, whereas the use of this system for prey capture has not been observed.

In reptiles, display gaping and associated biting are common components of social interaction and also of mating. Only snakes (and perhaps some other reduced limbed and burrowing species) do not display this pattern, substituting wrestling interactions in which the anterior body is lifted and contacts that of the antagonist (Carpenter 1977). The ubiquity of the biting patterns may be noted by some examples. Mating male turtles bite at the forequarters of their partners; perhaps this causes the females to retract the forequarters, thus bringing her cloaca into a position facilitating intromission. Male crocodilians use their jaws when fighting, and there are numerous reports of competitors being wounded or dismembered; however, crocodilians do not bite their partners during mating. In *Sphenodon* and lizards, biting is a common component of mating; the male bites the female and thus holds her, while curving his trunk in effecting intromission (Gans et al. 1984). The bite apparently serves more than holdfast action; thus, some unisexual (all female) lizards act as pseudo-males as ovulation is delayed in the absence of biting (Whittier and Tokarz 1992).

In some sexually dimorphic species of lizards, the males have substantially larger heads than the females. However, none of these mating and combat patterns appear to have been associated with the development of specialized canines or incisors. The development of protruding tusks, seen most spectacularly in suids and proboscidians, seems to be exclusively mammalian specializations, although some male plethodontid salamanders do display projecting tusks (Duellman and Trueb 1986). It might be interesting here to consider that "sabertooths" developed several times in cat-like mammals. It has been noted that such structures were unlikely to be stabbing devices (Radinsky and Emerson 1982; Miller 1984). First, modern felids do not kill large prey in this way, but break the prey's neck during a fall or cut through its throat. Also, teeth used in fighting and cutting encounter both wear and fracture (van Valkenburgh 1988). However, the tusks would be excellent disembowelling tools. The force applied to the ventral integument could be maximized, as the jaws would be applied to and could be closed on the body wall of stationary prey so that the muscle fibers would be activated while at a relatively extended position and nearly isometric – until the skin had been ruptured (Gans and Gaunt 1991).

5.4 Mastication Patterns

Mammalian heads achieve much of their final size pre-term and show a short period of growth post-term. The skulls may double in size during this period, but do not demonstrate the eightfold change in size seen in some reptiles. This may enable the limited cranial flexibility of mammals after birth. Also, because of lactation, mammalian skulls show initial hypertrophy and most heavy ossification in the dentary (Pond 1977).

The total reorganization of the mammalian jaw joint may also be important. The dentary-squamosal articulation was simpler than the dentary-quadrato-pterygoid one and all relative movements occur here. Also, the development of complex menisci allows the mammalian mandible to slide anteroposteriorly and to shift laterally on a fixed condylar surface. Thus, there is the potential for more controlled shifts of the mandible in the tooth-bearing plane, without losing the ability to apply substantial forces to the prey. Loading of the dentary-squamosal articulation and its possible limitation are discussed at the end of Sect. 6.

Except for species in which portions of the dentate arcades have become reduced, all mammals show heterodont dentitions. Whereas reptiles are also heterodont, the dentition of mammals differs profoundly in that stereotyped motor patterns bring different sets of teeth into action by the sequential shift of the mandibles (and the menisci) in the condylar fossae. Thus, many species initially break down the food by using the incisors to cut (gnaw). The food is then commonly shifted to the premolars and molars of the left or right side for unilateral food reduction, followed by distinct motor activities for the working and the balancing sides (Gorniak 1985). Other mammals show bilaterally symmetrical action, but have shifted their horizontal movements to the forward-reverse axis (Weijs and Dantuma 1975; Offermans and De Vree 1990, 1993).

Mammalian dental patterns have been classified into three major arrangements, carnivore, herbivore and rodent (Maynard Smith and Savage 1959), however, these have been further subdivided (Turnbull 1970). In the various carnivores, the jaws swing in a lateral excursion and the teeth may reach an occluded state in which they approach the interlocking state. In most ruminators and other herbivores, the jaws separate widely and the mandible then rises, shifting toward and beyond the maxillary tooth row of the working side. The food reduction movement is horizontal, and ends as the jaws again separate (De Vree and Gans 1976; Weijs and Dantuma 1981). In those rodents that use propalineal movement, occlusion is approached as the mandibles are highly retracted and the reduction stroke proceeds in the forward direction (Weijs and Dantuma 1975; Offermans and De Vree 1990).

The key to all of these movements lies in the activity of the fleshy tongue which sweeps up the food materials and deposits them at a site where they can be held between the closing teeth. Hence, the width of the opening phase is determined by the size of the food objects. Closing proceeds to the level at which the tooth surfaces can move past each other while exerting shearing forces on the food objects. The more irregular the prey, the wider the gape during the grinding movements. What should also be clear is that tooth contact, or occlusion, is an idealized state; it will only be attained when the prey has already been reduced and is presumably ready for swallowing. Selection must reflect the capacity of the dentition to facilitate reduction while the dental surfaces pass each other at some distance.

Depending upon the normal food types, one can now establish some design criteria for the shape of masticatory teeth (Lucas 1979; Lucas and Luke 1984). In carnivores, the maxillary and dentary tooth surfaces must be sufficiently coarse so that they will each maintain contact with one of the surfaces of the food object. This will place the prey into shear during the bite. Beyond shear, the interlocking cusps may be arranged so as to allow the prey to be cut during the crushing that accompanies the approach to occlusion. In herbivores, the sides of the food must be similarly held and subjected to shear loadings. Furthermore, the contour of the maxillary and dentary cusps may differ so that one set holds the prey, while the opposed set generates local failure by cutting or shredding the object along its surface.

Whereas horizontal movements may be effected by differential action of the adductor musculature, many tetrapods, particularly mammals, achieve such movements by permanent modification of the position of the jaw joint out of the plane of occlusion. Thus, elevation of the joint above this plane causes a forward shift of the mandible as it approaches the maxilla; depression of the joint below this plane causes the mandible to shift posteriorly. Both patterns induce shear during closure. Elevation of the joint is seen in herbivores such as elephants and the giant panda (Stöcker 1957; Davis 1964). Depression of the joint is seen in carnivores such as some shrews, for which it has been suggested that the closing incorporates a retractive component that pulls prey into the mouth (Gans 1969b).

The shape of the dentitional (potentially occlusal) surface establishes the movements that are possible. Hence, species with enlarged canines may not be able to engage in horizontal movements after the jaws have closed to the level at which these teeth can engage. Thus, domestic cats (Gorniak and Gans 1980) and fruitbats (De Gueldre and De Vree 1984, 1988, 1990) establish the working side of mastication with the direction in which the open jaw approaches the closing stroke, but the actual closure is almost completely vertical. Similarly, the premolar and molar cusps, for instance of microchiropteran bats (Kallen and Gans 1972) and opossums (Crompton and Hiiemae 1970), establish the direction in which the mandible may swing and the tooth cusps pass each other during food reduction.

Various mammals, most obviously species of marsupials, ungulates and rodents, environmentally modify the shape of their teeth (Rensberger 1973; Fortelius 1985; Janis and Fortelius 1988). Thus, wear matches the reduction planes. Such substantial wearing is commonly associated with open roots that allow the tooth to continue matching growth. (In proboscideans, wear matches the mean life of the teeth, so that very old animals may become edentulous.) Naturally, the ground down surface will not be protected by an enamel cover. Instead, the entire top is ground smooth, exposing the dentine and cementum. In open-rooted incisors, the enamel is concentrated entirely on the anterior surface with the dentine undercut to expose an enamel blade. On molars one commonly sees a complex folding of the tooth, producing leaves of enamel-coated dentine arranged at right angles to the direction in which the teeth approach each other. Cementum, the calcified tissue that binds the teeth to their sockets, fills the space between leaves. Wear thus produces layers of alternating hard and soft tissues, with the latter scalloping out. As two such arrays pass each other, they produce pressure and relaxation waves within the medium being masticated. This facilitates the preparation of plant tissues for subsequent chemical digestion.

6 The Adductor Muscles and Their Action Patterns

The masticatory system is powered by a combination of branchiomeric and axial muscles. The former assemblage provides the main elements for opening and closing the buccal cavity and for reducing the volume of the pharyngeal space. The axial muscles participate mainly in distending the pharynx and in activating the tongue. Initially, this appears to have been a contrast between circumferentially and longitudinally arranged muscles.

Muscles involved in the closure of the mouth during biting into prey have vertical placement. Their length will be constrained by the distance between their insertion site and their origin on the skull. As the length of the fibers determines their potential excursion (distance between longest and shortest position), short fibers will in turn constrain the degree of mandibular opening (depression) (Herring and Herring 1974). This may be the reason for the insertion of sheaves that can change the direction of tendons and hence of muscle action and allow much of the mass to be positioned horizontally rather than vertically (Schumacher 1973).

Adductor muscles essentially act within a rotating system. Furthermore, their overall angles of insertion tend to be much less acute than the angles of muscles acting within appendages. For fibers with such an insertion one can see that the excursion required for any degree of mandibular rotation will be a function of the distance from the rotational fulcrum to the insertion site. As the moment arm of the muscular insertion is also a function of this distance, one can see that the muscular sarcomeres represent units of moment, rather than units of force. In short, the mass of the active portion of the muscle, rather than its physiological cross section, will be reflected in the moment it produces (Gans and Gaunt 1991).

Implicit in this model for analyzing muscle packing is the concept of sarcomere equivalence, which is an idealization that facilitates analysis (Gans and De Vree 1987). It implies that fibers inserting closer to the fulcrum will be shorter, and will have proportionally fewer sarcomeres in series. Also, fibers inserting above the jaw line will be closer to the horizontal, which insures that in order to remain equivalent the fibers of such a muscular mass cannot lie parallel to one another. This pinnate angulation is functionally and architecturally distinct from that classically referred to as pinnation, which deals with overcoming the spatial constraints encountered in muscle packing.

The masticatory muscles of animals obviously determine the forces and displacements that the animal may exert onto its food. If the muscular volume establishes the moment, the position of the tendons of insertion may be entirely governed by packing constraints. Jaws may be kept slender, thus reducing their rotational inertia, and long, thus decreasing the length of time to closure onto prey. Slender jaws may also grasp prey hiding in narrow crevices. However, the jaws must exert forces onto the prey, moments are of internal concern. Hence, animals incur a conundrum, in that the forces that the skeletal element may exert onto the environment (i.e., food object) decrease with the length of the mandible; the moment arm (normal distance to the fulcrum) is critical for the external forces, but not for those exerted by its muscles. Hence, muscles can be packed close to the fulcrum, where their rotational inertia is minimal without affecting the moment generated. However, molar dentition, which requires the application of maximal (external) grinding forces, is placed close to the mandibular joint, so that the available moment will be divided by a minimal arm distance. This means that for any given amount of muscular contraction, animals can exert more force between their molars than between their premolars or between their canines.

The bilateral adductor muscles of akinetic lizards appear to fire almost simultaneously. However, the magnitude of activity is greater on the side on which the prey falls between the dental arcades. Not only do the bilateral muscles fire simultaneously, but so do the several adductors. This reflects again the observation that the diversity of muscles in the lizard head may mainly reflect historical and packing constraints. For example, *Sphenodon* shows differential activation only for the pterygoid muscle, in this case associated with propalineal protrusion of the mandible at the end of a cutting bite (Gorniak et al. 1982).

The masticatory motor patterns of various mammals show sequential activation and thus differ profoundly from the above. As might be expected from the observation on *Sphenodon*, the diversity seen reflects the need for horizontal movements. Commonly, muscles fire not only asymmetrically, but in alternating patterns that establish couples which will shift the condyles in the fossae (Kallen and Gans 1972; Gorniak 1985; Chap. 9, this Vol.). Consequently, examination of the sequence of muscle activation permits facile identification of the working side and the balancing one. One of the more interesting phenomena being clarified by recent experimental studies of mastication is the observation, for instance for rodents (Offermans and De Vree 1993), that morphologically similar animals often engage in distinct food reduction patterns, whereas dissimilar animals can use distinct motor regimes to achieve equivalent masticatory movements.

Modeling of muscle function from architecture has recently become more common; regrettably, it is often based on several questionable assumptions. First, it is assumed that the entire muscle acts simultaneously. If this were correct, the mass might be a good indicator of the magnitude of force applied; however, muscles are often subdivided into task groups, each reflecting one functional role (Loeb 1984, 1990). Also, one may see sequential activation of portions of large and complex muscles (Herring et al. 1979). Second, there is the assumption that force output is related to twitch, without considering the possibility of complete or partial tetanus (Gans and De Vree 1986). Finally, the force will differ depending on the portion of the length-tension surface over which shortening proceeds (Gans and De Vree 1987).

The swinging action of the mammalian mandibles incorporates substantial momentum. It has been suggested (Kallen and Gans 1972) that this system conserves energy; this idea deserves test as does the correlation of mandibular mass and masticatory frequency. Also, the increase in chewing frequency seen in smaller species may well be associated with such phenomena and it might be interesting to determine how the rates correlate with the natural frequencies of such mandibular masses.

The cutting and crushing actions on the prey by the dentition will reflect the loading of teeth. Application of forces on variably positioned food objects will generate some compensating forces within the capsule of the jaw joint. Such forces will not only appear in the vertical direction, but placement of muscles at angles to the mandible may induce propalineal sliding and risk rupture of the capsule. In mammals, the forces generated by the major adductors cross each other, balancing the load on the joint. Widely cited models have assumed that pelycosaurs showed no crossed muscle insertions and that these represent a recent invention of the mammalian line (Crompton 1963); however, these are suspect as the adductors, even of lizards, include some muscles inserting with forward open angles and others that show rearward open angles (Gans et al. 1985; De Vree and Gans 1987a).

7 Cycles

It is customary to subdivide the cycles of prey capture, possible reduction, and ingestion sequences into more or less repeatable phases. Diverse terminologies for several such phases are offered by several authors in this volume which might allow a basis of comparison. Regrettably, standardization is still lacking. What is needed is a standard, defining the start and end of phases, rather than the application of a terminology pertinent to one group of species to others studied or encountered in the literature. Even more important is that rules regarding patterns should reflect diversity; they should follow rather than precede its analysis. Phylogenetic mapping remains critical as it permits separation of components that represent plesiomorphic states from those that are clearly apomorphic.

One excellent example of this difficulty is the slow open, fast open, fast close, slow close gape cycle (Bramble and Wake 1985; Reilly and Lauder 1990). This pattern was first established for mammals and later noted to have pertinence for other tetrapods (Crompton et al. 1977). Yet, it is not clear why there should be a generality and far more species deserve attention before it can be understood (Gans et al. 1978).

The opening and closing phases of the gape cycle may represent plesiomorphic states, inherent with the possession of jaws. Even in suction feeding (and breathing), the jaws must be separated and then brought together, although the rates of opening and closing and the intervals between these appear to differ. Food transport within the buccal space will inevitably require movement of the hyoid system; at least for those species using tongues rather than inertial movements. However, some hyoid movements may well represent plesiomorphic states as they may derive from pharyngeal food reduction.

Whenever the jaws are closed onto prey items, the resistance encountered will likely slow the vertical movement. This is, of course, the basic condition of mastication, defined as the use of the dentigerous jaws to reduce food. Similarly, the opening of the jaws is likely retarded by the need to separate from the bolus. However, these two possible changes in velocity are affected by food type (De Vree and Gans 1976). Also, these pattern changes differ among animals with mainly vertical closure and those in which the closure is transverse, diagonal, or propalineal. Is the slow-fast and fast-slow transition merely an ephemeral by-product of mastication, or does it reflect a more profound division of the feeding cycle?

Just as important is that the power or reduction phase of the masticatory movement must proceed after slow closing and before slow opening. Three-dimensional plots of its path suggest that a purely vertical description is inadequate except in species, such as cats, that masticate mainly in a vertical direction (Gorniak and Gans 1980). Also, mastication always involves movements of the tongue and its supporting hyoid system which serve to reposition the food object or bolus between reducing bites. [The use of the hyoid system for transporting food in animals that do not specifically reduce their prey (Reilly and Lauder 1990) suggests that the mammalian (and even the reptilian) system represents a plesiomorphic state.] The lingual movement clearly starts during fast opening and ends prior to the end of fast closing.

Lingual repositioning must occur after the bolus has been separated from the teeth and this may mark a transition of mandibular velocity or follow it . There are no studies on the anticipatory movements of the hyoid, prior to the start of lingual motion. Also, the slow-fast transition appears less clear in many species than the discontinuity of hyoid movement. Perhaps the definition of cycle components deserves a hybrid state, following mandibular separation with transition of hyoid and then of lingual movements. In any case it will be necessary to determine whether and how often the several motions are associated and whether this association is affected by food type. Ultimately, such studies cannot continue to rely on kinematic data; we need to know the power spectrum of the driving musculature and beyond this the control circuits that coordinate the system.

Observation of regularly chewing tetrapods, particularly of ruminating ungulates, makes it plausible that the cycles are driven by simple pattern generators. However, analysis of mastication in cats feeding on several kinds of meat (Gans et al. 1990), recently confirmed for goats feeding on alfalfa, carrot and apple, or ruminating (pers. observ.), provides no evidence that the animals are generating bites of regular magnitude or that they are perhaps extrapolating from the resistance encountered during one bite to the next. Instead, there is evidence for earlier observations that the consistency (and size) of the prey is monitored during the bite. This information is immediately fed back and seems to be used to establish the forces imposed on the food object during the particular bite.

8 Overview

We hope to have documented the mechanical and biological complexity of the feeding mechanisms of tetrapods. Also, it is clear that the ways in which tetrapods acquire, manipulate, and ingest their foods involve a great amount of opportunism. Rather than a single evolutionary sequence, the processes of tetrapod feeding systems seem to represent specific responses, using the existing genetic-developmental capacity to harvest what is ecologically available. The animals have clearly responded to environmental opportunities, both by changes in the mechanical substrate and/or in the neurological (behavioral) substrate. Many experimental tools have become available recently in functional morphology. The feeding systems seem to provide ideal substrates for their application in an analysis of the evolutionary basis of such systems.

References

Abler WL (1992) The serrated teeth of tyrannosaurid dinosaurs and biting structures in other animals. Paleobiology 18: 161–183

Bell D (1989) Functional anatomy of the chameleon tongue. Zool Jahrb Anat 119: 313–336

Bels V (1990) Quantitative analysis of prey-capture kinematics in *Anolis equestris* (Reptilia: Iguanidae). Can J Zool 68: 2192–2198

Bels V, Goosse V (1990) Comparative kinematic analysis of prey capture in *Anolis carolinensis* (Iguania) and *Lacerta viridis* (Scleroglossa). J Exp Zool 255: 120–124

Bemis W, Schwenk K, Wake MH (1983) Morphology and function of the feeding apparatus in *Dermophis mexicanus* (Amphibia, Gymnophiona). Zool J Linn Soc 77: 75–96.

Bock WJ (1964) Kinetics of the avian skull. J Morphol 114: 1–42

Bogert CM (1943) Dentitional phenomena in cobras and other elapids, with notes on the adaptive modifications of their fangs. Bull Am Mus Nat Hist 81: 285–360

Borsuk-Bialynicka M (1984) Anguimorphans and related lizards from the Late Cretaceous of the Gobi desert, Mongolia. Paleontol Pol 46 (1984): 5–105

Bramble DM (1989) Cranial specialization and locomotor habit in the Lagomorpha. Am Zool 29: 303–317.

Bramble DM, Wake DB (1985) Feeding mechanisms of lower vertebrates. In: Hildebrand M, Bramble DM, Liem KF, Wake DB (eds) Functional vertebrate morphology. Harvard Univ Press, Cambridge, pp 230–261

Carpenter C (1977) Inventory of combat rituals in snakes. Smithsonian Herpetological Information Services 69. Natl Mus Nat Hist, Washington DC, pp 1–18

Cleuren J, De Vree F (1992) Kinematics of the jaw and hyolingual apparatus during feeding in *Caiman crocodilus*. J Morphol 212: 141–154

Condon K (1987) A kinematic analysis of mesokinesis in the Nile Monitor (*Varanus niloticus*). Exp Biol 47: 73–87

Crompton AW (1963) The evolution of the mammalian jaw. Evolution 17: 431–439

Crompton AW, Hiiemae KM (1970) Molar occlusion and mandibular movements during occlusion in the American opossum, *Didelphis marsupialis*. Zool J Linn Soc 49: 21–47

Crompton AW, Thexton AJ, Parker P, Hiiemae KM (1977) The activity of the jaw and hyoid musculature in the Virginia opossum, *Didelphis virginiana*. In: Gilmore D, Stonehouse B (eds) The biology of the marsupials. MacMillan, London, pp 287–305

Cundall D (1983) Activity of head muscles during feeding by snakes: a comparative study. Am Zool 23: 383–396

Cundall D, Gans C (1979) Feeding in water snakes: an electromyographic study. J Exp Zool 209: 189–208

Dalrymple GH (1979) On the jaw mechanism of the snail-crushing lizards, *Dracaena* Daudin 1802 (Reptilia, Lacertilia, Teiidae). J Herpetol 13: 303–311

Davenport J, Grove DJ, Cannon J, Ellis TR, Stables R (1990) Food capture, appetite, digestion rate and efficiency in hatchling and juvenile *Crocodylus porosus*. J Zool (Lond) 220: 569–592

Davis DD (1964) The giant panda. A morphological study of evolutionary mechanisms. Fieldiana: Zoology Memoirs, vol 3. Chicago Nat Hist Mus, Chicago, pp 5–339

Deban SM, Nishikawa KC (1992) The kinematics of prey capture and the mechanism of tongue protraction in the green tree frog *Hyla cinerea*. J Exp Biol 170: 235–256

De Gueldre G, De Vree F (1984) Movements of the mandibles and tongue during mastication and swallowing in *Pteropus giganteus* (Megachiroptera): a cineradiographical study. J Morphol 179: 95–114

De Gueldre G, De Vree F (1988) Quantitative electromyography of the masticatory muscles of *Pteropus giganteus* (Megachiroptera). J Morphol 196: 73–106

De Gueldre G, De Vree F (1990) Biomechanics of the masticatory apparatus of *Pteropus giganteus* (Megachiroptera). J Zool (Lond) 220: 311–332

De Jongh HJ, Gans C (1969) On the mechanism of respiration in the bullfrog, *Rana catesbeiana*: a reassessment. J Morphol 127: 259–290

De Vree F, Gans C (1976) Mastication in pygmy goats, *Capra hircus*. Ann Soc R Zool Belg 105: 255–306

De Vree F, Gans C (1987a) Kinetic movements in the skull of adult *Trachydosaurus rugosus*. Anat Hist Embryol 16: 206–209

De Vree F, Gans C (1987b) Intracranial movements in *Gekko gecko* (Reptilia: Sauria). Acta Anat 130: 25 (Abstr)

De Vree F, Gans C (1989) Functional morphology of the feeding mechanisms in lower tetrapods. In: Splechtna H, Hilgers H (eds) Trends in vertebrate morphology. Fortschr Zool 35. Fischer, Stuttgart, pp 115–127

Duellman WE, Trueb L (1986) Biology of amphibians. McGraw-Hill, New York

Edmund AG (1969) Dentition. In: Gans C, Parsons TS, Bellairs A d'A (eds) Biology of the Reptilia, vol 1. Academic Press, London, pp 117–200

Estes R, Williams EE (1984) Ontogenetic variation in the molariform teeth of lizards. J Vertebr Paleontol 4: 96–107

Fortelius M (1985) Ungulate cheek teeth: developmental, functional, and evolutionary inter-relations. Acta Zool Fenn 180: 1–76

Frazzetta T (1962) A functional consideration of cranial kinesis in lizards. J Morphol 111: 287–319

Gaffney ES (1979) Comparative cranial morphology of recent and fossil turtles. Bull Am Mus Nat Hist 164: 65–376

Gans C (1961) The feeding mechanism of snakes and its possible evolution. Am Zool 1: 217–227

Gans C (1969a) Comments on inertial feeding. Copeia (4): 855–857

Gans C (1969b) Nuts and bolts anatomy. Turtox News 47(9): 290–293

Gans C (1971) Strategy and sequence in the evolution of the external gas exchangers of ectothermal vertebrates. Forma Functio 3: 66–104

Gans C (1974) Biomechanics: an approach to vertebrate biology. Lippincott, Philadelphia, pp 1–261

Gans C (1979) Momentarily excessive construction as the basis for protoadaptation. Evolution 33: 227–233

Gans C (1983) Snake feeding strategies and adaptations – conclusions and prognosis. Am Zool 23: 455–460

Gans C (1988) Craniofacial growth, evolutionary questions. In: Thorogood P, Tickle C (eds) Craniofacial development. Development 103: 3–15 (Suppl)

Gans C (1990) Adaptations and conflicts. In: Mlikowsky J, Novak VJA (eds) Evolutionary biology: theory and principles. Proc Int Symp Plzen 1988, Czech Acad Sci Praha, pp 23–31

Gans C (1991) Why develop a neck? In: Berthoz A, Graf WM, Vidal PP (eds) Head-neck sensory motor system. Oxford Univ Press, New York, pp 17–21

Gans C, De Vree F (1986) Shingle-back lizards crush snail shells using temporal summation (tetanus) to increase the force of the adductor muscles. Experientia 42: 387–389

Gans C, De Vree F (1987) Functional bases of fiber length and angulation in muscle. J Morphol 192: 63–85.

Gans C, Elliott WB (1968) Snake venoms: production, injection, action. In: Staple PH (ed) Advances in oral biology, vol 3. Academic Press, New York, pp 45–81

Gans C, Gaunt AS (1991) Muscle architecture in relation to function. J Biomech 24 (Suppl 1): 53–66

Gans C, Gorniak GC (1982a) Functional morphology of lingual protrusion in marine toads (Bufo marinus). Am J Anat 163: 195–222

Gans C, Gorniak GC (1982b) How does the toad flip its tongue? Test of two hypotheses. Science 216: 1335–1337

Gans C, Pyles R (1983) Narial closure in toads; which muscles? Respir Physiol 53: 215–223

Gans C, De Vree F, Gorniak GC (1978) Analysis of mammalian masticatory mechanisms: progress and problems. Zentralbl Veterinaermed C, Anat Histol Embryol 7: 226–244

Gans C, Gillingham JC, Clark DL (1984) Courtship, mating and male combat in tuatara, Sphenodon punctatus. J Herpetol 18: 194–197

Gans C, De Vree F, Carrier D (1985) Usage pattern of the complex masticatory muscles in the shingleback lizard, Trachydosaurus rugosus: a model for muscle placement. Am J Anat 173: 219–240

Gans C, Gorniak GC, Morgan WK (1990) Bite-to-bite variation of muscular activity in cats. J Exp Biol 151: 1–19

Gans C, Nishikawa K, Cannatella D (1991) The frog Megophrys montana: specialist on large prey. Am Zool 31: 52A (Abstr)

Gorniak GC (1985) Trends in the actions of mammalian masticatory muscles. Am Zool 25: 331–337

Gorniak GC, Gans C (1980) Quantitative assay of electromyograms during mastication in domestic cats (Felis catus). J Morphol 163: 252–281

Gorniak GC, Rosenberg H, Gans C (1982) Mastication in the tuatara, *Sphenodon punctatus* (Reptilia: Rhynchocephalia): structure and activity of the motor system. J Morphol 171: 321–353

Greene HW (1988) Antipredator mechanisms in reptiles. In: Gans C, Huey RB (eds) Biology of the Reptilia, vol 16. Liss, New York, pp 1–152

Helfman GS, Clark JB (1986) Rotational feeding: overcoming gape-limited foraging in anguillid eels. Copeia 1986(3): 679–685

Herring SW, Herring SE (1974) The superficial masseter and gape in mammals. Am Nat 108: 561–575

Herring SW, Grimm AF, Grimm BR (1979) Functional heterogeneity in a multipinnate muscle. Am J Anat 154: 563–576

Hiiemae KM (1978) Mammalian mastication: a review. In: Butler PM, Josey K (eds) Studies of the development, structure and function of teeth. Academic Press, New York, pp 360–398

Hotton N (1955) A survey of adaptive relationships of dentition to diet in the North American Iguanidae. Am Midl Nat 53: 88–114

Iordansky NN (1990) The evolution of complex adaptations: the jaw apparatus of the amphibians and reptiles. Academia Nauk CCCP, Moscow, pp 3–310

Janis CM, Fortelius M (1988) On the means whereby mammals achieve increased functional durability of their dentitions, with special reference to limiting factors. Biol Rev 63: 197–230

Jaslow CR, Biewener AA (1988) Strain patterns in the cranial bones and sutures during impact loading. Am Zool 28: 175A (Abstr)

Kallen FC, Gans C (1972) Mastication in the little brown bat, *Myotis lucifugus*. J Morphol 136: 385–420

Kier WM, Smith KK (1985) Tongues, tentacles, and trunks: the biomechanics of muscular hydrostats. Zool J Linn Soc 83: 307–324

Kraklau DM (1991) Kinematics of prey capture and chewing in the lizard *Agama agama*. J Morphol 210: 195–212

Lauder GV (1983) Food capture. In: Webb PW, Weihs D (eds) Fish biomechanics. Praeger, New York, pp 280–311

Lauder GV, Shaffer HB (1985) Functional morphology of the feeding mechanism in aquatic ambystomatid salamanders. J Morphol 185: 297–326

Lauder GV, Shaffer HB (1986) Functional design of the feeding mechanism in lower vertebrates: unidirectional and bidirectional flow systems in the tiger salamander. Zool J Linn Soc 88: 277–290

Lauder GV, Shaffer HB (1988) Ontogeny of functional design in tiger salamanders (*Ambystoma tigrinum*): are motor patterns conserved during major morphological transformations? J Morphol 197: 249–268

Lauder GV, Crompton AW, Gans C, Hanken J, Liem KF, Maier WO, Meyer A, Presley R, Rieppel OC, Roth G, Schluter D, Zweers GA (1989) How are feeding systems integrated and how have evolutionary innovations been introduced? In: Roth G, Wake DB (eds) Complex organismal functions: integration and evolution in vertebrates. Dahlem Konferenzen, Life Sciences Research Report 45. Wiley, Chichester, pp 97–115

Loeb GE (1984) The control and responses of mammalian muscle spindles during normally executed motor tasks. Exercise Sport Sci Rev 12: 157–204

Loeb GE (1990) The functional organization of muscles, motor units, and tasks. In: Binder MD, Mendell LM (eds) The segmental motor system. Oxford Univ Press, New York, pp 23–35

Lucas PW (1979) The dental-dietary adaptations of mammals. Neues Jahrb Geol Paläontol Monatsh 8: 486–512

Lucas PW, Luke DA (1984) Chewing it over: basic principles of food breakdown. In: Chivers DJ, Wood BA, Bilsborough A (eds) Food acquisition and processing in primates. Plenum Press, New York, pp 283–301

Maynard Smith J, Savage JRG (1959) The mechanics of mammalian jaws. School Sci Rev 141: 289–301

McDowell SB (1972) The evolution of the tongue of snakes and its bearing on snake origins. In: Dobzhansky T, Hecht MK, Steere WC (eds) Evolutionary biology 6. Appleton-Century-Crofts, New York, pp 191–273

Miller GJ (1984) On the jaws mechanism of *Smilodon californicus* Bovard and some other carnivores. IVC Mus Soc, Occas Pap (7): 1–107

Muller M, Osse J, Verhagen J (1982) A quantitative hydrodynamical model of suction feeding in fish. J Theor Biol 95: 49–79

Nishikawa KC, Cannatella DC (1991) Kinematics of prey capture in the tailed frog, *Ascaphus truei* (Anura: Ascaphidae). Zool J Linn Soc Lond 103: 289–307

Nishikawa K, Gans C (1992) The role of hypoglossal sensory feedback during feeding in the marine toad, *Bufo marinus*. J Exp Zool 264: 245–252

Nishikawa KC, Roth G (1991) The mechanism of tongue protraction during prey capture in the frog *Discoglossus pictus*. J Exp Biol 159: 217–234

Nussbaum RA (1983) The evolution of a unique dual jaw closing mechanism in caecilians (Amphibia: Gymnophiona) and its bearing on caecilian ancestry. J Zool (Lond) 199: 545–367

Offermans M, De Vree F (1990) Mastication in springhares, *Pedetes capensis*: a cineradiographic study. J Morphol 205: 353–554

Offermans M, De Vree F (1993) Quantitative electromyography and biomechanics of the masticatory apparatus in the springhare, *Pedetes capensis*. Belg J Zool 123 (in press)

Parsons TS, Williams EE (1962) The teeth of Amphibia and their relation to amphibian phylogeny. J Morphol 110: 375–389

Patchell FC, Shine R (1986a) Feeding mechanisms in pygopodid lizards: how can *Lialis* swallow such large prey? J Herpetol 20: 59–64

Patchell FC, Shine R (1986b) Hinged teeth for hard-bodied prey: a case of convergent evolution between snakes and legless lizards. J Zool (Lond) 208: 269–275

Peyer B (1968) Comparative odontology. (ed. and tr. by Zangerl R). University of Chicago Press, Chicago

Pond CM (1977) The significance of lactation in the evolution of mammals. Evolution 31: 177–199

Pough FH (1973) Lizard energetics and diet. Ecology 54: 837–844

Radinsky L, Emerson S (1982) The late, great sabertooths. Nat Hist 91: 50–56

Regal PJ, Gans C (1976) Functional aspects of the evolution of frog tongues. Evolution 30: 718–734

Reilly SM, Lauder GV (1990) The evolution of tetrapod feeding behavior: kinematic homologies in prey transport. Evolution 44: 1542–1557

Rensberger JM (1973) An occlusion model for mastication and dental wear in herbivorous mammals. J Paleontol 47: 515–528

Robinson PL (1976) How *Sphenodon* and *Uromastyx* grow their teeth and use them. In: Bellairs A d'A, Cox CB (eds) Morphology and biology of reptiles. Linn Soc Symp Ser (3): 43–64

Romer AS (1966) Vertebrate paleontology, 3rd edn. Univ Chicago Press, Chicago

Rosenthal GA, Janzen DH (1979) Herbivores, their interaction with secondary plant metabolites. Academic Press, New York

Schumacher GH (1973) The head muscles and hyolaryngeal skeleton of turtles and crocodilians. In: Gans C, Parsons TS (eds) Biology of the Reptilia, vol 4. Academic Press, London, pp 101–199

Schwenk K (1988) Comparative morphology of the lepidosaur tongue and its relevance to squamate phylogeny. In: Estes R, Pregill G (eds) Phylogenetic relationships of the lizard families. Stanford University Press, Stanford, pp 569–598

Schwenk K, Bell DA (1988) A cryptic intermediate in the evolution of chameleon tongue projection. Experientia 44: 697–700

Schwenk K, Throckmorton GS (1989) Functional and evolutionary morphology of lingual feeding in squamate reptiles: phylogenetics and kinematics. J Zool (Lond) 219: 153–175

Smith KK (1982) An electromyographic study of the function of the jaw adducting muscles in *Varanus exanthematicus* (Varanidae). J Morphol 173: 137–158

Smith KK (1984) The use of the tongue and hyoid apparatus during feeding in lizards (*Ctenosaura similis* and *Tupinambis nigropunctatus*). J Zool (Lond) 202: 115–143

Smith KK (1986) Morphology and function of the tongue and hyoid apparatus in *Varanus* (Varanidae, Lacertilia). J Morphol 187: 261–287

Smith KK, Hylander WL (1985) Strain gauge measurement of mesokinetic movement in the lizard *Varanus exanthematicus*. J Exp Biol 114: 53–70

Smith KK, Mackay KA (1990) The morphology of the intrinsic tongue musculature in snakes (Reptilia, Ophidia): functional and phylogenetic implications. J Morphol 215: 307–324

So K-KJ, Wainwright PC, Bennett AF (1992) Kinematics of prey processing in *Chamaeleo jacksonii*: conservation of function with morphological specialization. J Zool (Lond) 226: 47–64

Stöcker L (1957) Trigeminusmuskulatur und Kiefergelenk von *Elephas maximus* L. Morph Jahrb 98: 35–76

Trueb L, Gans C (1983) Feeding specializations of the Mexican burrowing toad, *Rhinophrynus dorsalis* (Anura: Rhinophrynidae). J Zool (Lond) 199: 189–208

Turnbull WD (1970) Mammalian masticatory apparatus. Fieldiana Geol, Field Mus Nat Hist 18: 147–356

van Tienderen PH (1991) Evolution of generalists and specialists in spacially heterogeneous environments. Evolution 45: 1317–1331

van Valkenburgh B (1988) Incidence of tooth breakage among large predatory mammals. Am Nat 131: 291–302

Vaughan TA (1978) Mammalogy, 2nd edn. Saunders, Philadelphia

Voelker W (1986) The natural history of living mammals. Plexus, Medford

Wainwright PC, Kraklau DM, Bennett AF (1991) Kinematics of tongue projection in *Chamaeleo oustaleti*. J Exp Biol 159: 109–133

Wake DB (1982) Functional and developmental constraints and opportunities in the evolution of feeding systems in urodeles. In: Mossakowski D, Roth G (eds) Environmental adaptation and evolution. Fischer, Stuttgart, pp 51–66

Webb PW (1985) Locomotion and predator-prey relationships. In: Feder M, Lauder GV (eds) Predator-prey relationships. The University of Chicago Press, Chicago, pp 24–41

Weijs W, Dantuma R (1975) Electromyography and mechanics of mastication in the albino rat. J Morphol 146: 1–34

Weijs W, Dantuma R (1981) Functional anatomy of the masticatory apparatus in the rabbit (*Oryctolagus cuniculus* L.). Neth J Zool 31: 99–147

Westneat MW (1990) Feeding mechanism of teleost fishes (Labridae; Perciformes): a test of four-bar linkage models. J Morphol 205: 269–295

Whittier JM, Tokarz RR (1992) Physiological regulation of sexual behavior in female reptiles. In: Gans C, Crews D (eds) Biology of the Reptilia, vol 18. Hormones, brain and behavior. Univ Chicago Press, Chicago, pp 24–69

Chapter 5

Sensorimotor Processes That Underlie Feeding Behavior in Tetrapods

J.-P. Ewert[1], T.W. Beneke[1], E. Schürg-Pfeiffer[1], W.W. Schwippert[1]
and A. Weerasuriya[2]

Contents

1 Behavior Patterns Related to Feeding

Eating behaviors among tetrapods involve a variety of patterns that in etho-logical terms (Tinbergen 1951) can be classified as (1) *nondirected appetitive behavior*: movements seeking for a releasing stimulus situation, as the first sign of a specific internal action readiness, based on the motivational state; (2) *directed appetitive behavior*: orientational movements of head and body released by food sign stimuli that initiate the approach toward food; (3) *consummatory behavior*: grasping the food and putting it into the mouth; (4) *ingestive behavior*:

[1] Neurobiology, FB 19, University of Kassel (GhK), 34132 Kassel, Germany
[2] Basic Medical Science, Mercer University School of Medicine, Macon, Georgia 31207, USA

Advances in Comparative and Environmental Physiology, Vol. 18
© Springer-Verlag Berlin Heidelberg 1994

jaw and tongue movements that shape and position the food in the mouth, swallowing, and mouth cleaning. This general classification is obtained more or less in carnivores, herbivores, and omnivores. Behavior patterns of type 1 involve rather unspecific body movements and do not show remarkable pecularities across vertebrate groups. Various behavior patterns (type 2) are related to the strategy in acquiring food, such as orientational movements, approaching, stalking, jumping, or diving. The greatest variety occurs in grasping patterns (type 3). Most common is jaw grasping with specific patterns like pecking in birds, prey-head oriented neck biting in feline species, and death shaking in canids (Leyhausen 1965). There are also pecularities in the mode by which prey is put into the mouth: tongue flipping in amphibians and in certain reptiles, striking at prey with claws (predatory birds and felines), grasping food with a trunk (elephant), and manipulation of food with hands and/or feet eventually supplied by tools (primates). Depending on the local position of a prey and its escape strategy, predators display the directed appetitive behaviors in corresponding sequences, or in a combination with the consummatory act (e.g., jump-strike). Regarding ingestive behaviors (type 4) most common is swallowing and, if teeth are developed, mastication involving rhythmic jaw movements (Nakamura 1985; Lund and Enomoto 1988; see also Chaps. 4, 9, and 10, this Vol.).

The search for the neural bases that underlie feeding leads us to investigate types 3 and 4. In this context much progress has been made in amphibians (e.g., frogs, toads), birds (e.g., pigeons), and among mammals in rodents (e.g., rats, rabbits, guinea pigs), ruminants (e.g., sheep), and felines (e.g., cats). Since snapping, pecking, biting, mastication, and swallowing involve motoneuronal activity of the orofacial musculatuie, the trigeminal, facial, glossopharyngeal, and hypoglossal areas and the adjacent medial reticular formations are the focus of neurobiological research (e.g., see Rossignol et al. 1988). These medullary brain stem structures are rather ubiquitous across vertebrates (Gaupp 1896; Edinger 1908; Röthig 1927). Functionally, however, these structures are not specific to feeding, rather they are applicable to other behaviors, such as avoidance, aggression, and mating. This presents some problems for neurobiological research: because a striated muscle may take part in different movements, on the one hand, and since the same movement (e.g., of head, jaw, tongue, and leg) can participate in different behavioral patterns, this makes the tracing of information processing difficult to investigate. In search of the structural and functional properties of the neural network that, so to speak, mediates between a food sign stimulus and a feeding pattern, it is reasonable to focus on musculature specific to feeding. This holds for the projective tongue musculature of terrestrial amphibians. Our primary emphasis in this chapter therefore is motor pattern generation and sensorimotor function related to tongue flipping in bufonid toads in the sense of the "visual grasp reflex" according to Akert (1949). Various laboratories across the world have contributed significant data to this topic by application of different methods from ethology, neuroanatomy, neurophysiology, and neuro-information science. In addition, we discuss the integration of this behavior pattern within the entire feeding sequence involving jaw movements and swallowing. Comparable relationships between structure and function obtained in other vertebrates are also considered.

2 Definitions Referring to "Sensori-", "Motor-", and "Command-"

Sensorimotor transformation refers to the operations that translate a specific pattern of sensory input into an appropriate spatiotemporal pattern of excitation and inhibition in motoneurons – the *motor pattern* – necessary to activate and coordinate the muscle contractions for the corresponding *action pattern*.

Our concept of *motor pattern generator* (MPG) draws on Doty's (1976) definition of an internuncial network that in response to commanding trigger signals coordinates appropriate muscle contractions: (1) the network is activated if, and only if, a specific (combination of) input occurs; (2) an intrinsic pattern of neuronal connectivity in the network assures the generation of a consistent spatiotemporal distribution of excitation and inhibition, the so-called central pattern (Selverston 1980; Grillner 1985); (3) the output of the network, mediated by premotor neurons, has privileged access to the requisite motoneuronal pools; (4) sensory (e.g., proprioceptive) feedback and internal feedback from the network (Davis and Kovac 1981) can play a role in the coordination and maintenance of a motor pattern (Rossignol et al. 1988).

The neurons specified to encode the outcome of feature/space-related sensory information processing and which transmit the resulting information to an MPG are part of the *sensorimotor interface* (Capranica 1983; Scheich 1983). Sensorimotor integration points to the modes by which the processed sensory information from different sources about the same target is integrated by neurons of the MPG, for example in a manner of an AND- or an OR-gate. Wiersma and Ikeda (1964) were among the first who suggested that behaviorally relevant sensory information is fed to specific command neurons which control (i.e., trigger) a corresponding MPG. The important idea behind this concept is still valid if the following additional criteria are considered: (1) *command systems* (Kupfermann and Weiss 1978) may consist of collectively operating neurons (command elements) of the same or similar filter type, whereby these neurons are integrated in a network and thus express in their responses certain network computations. (2) An MPG can be triggered – like a safe is unlocked – according to a *sensorimotor code* (Ewert 1987) that is provided by concurrent activities of different types of command elements which evaluate various aspects of an object, such as configural features and the object's position in space; with respect to their efferent effects, these command elements together form a *command releasing system* (CRS) which is the neurophysiological correlate of the ethological concept of the releasing mechanism (RM) (Tinbergen 1951), originally called the releasing schema (Lorenz 1935). The prefix "command" points to the initiating, primarily ballistic function of the system and "releasing" refers to its coded property (concept of sensorimotor code); "code" suggests that various command elements in a combination adaptive to external and internal conditions cooperate goal-specifically (Ewert 1987). (3) *Adaptation* of a motor pattern to changing external conditions can be achieved by (a) peripheral feedback, (b) associative changes in the feature processing network that determines the properties of feature-analyzing command elements, (c) level settings of command elements, or (d) tuning of modulatory command elements (Kupfermann and Weiss 1978; Ewert 1987, 1991). (4) *Economy* in terms of "neural parsimony" probably plays a role (Comer 1987), whereby certain command elements may be shared by different CRSs. (5) *Safety* is taken into account in that the same motor pattern can be triggered by different CRS involving different sensory modalities (Grobstein et al. 1983; for command concepts see also DiDomenico and Eaton 1987).

Expressing sensorimotor concepts in the language of Arbib's (1987) schema theory, the sensorimotor code of a CRS embodies a *perceptual schema* that exists for only one purpose, namely, to determine the conditions for the activation of a specific MPG embodying a *motor schema*. The CRS must also ensure that the resultant movement is directed toward the target. A schema and its instantiation usually appear to be coextensive, i.e., instantiation of a schema appears to be identifiable with appropriate activity in certain populations of neurons of the brain, whereby each schema may involve several cell types/brain regions, while a given cell type/brain region may be involved in several schemas. The motor schemas of directed appetitive behaviors and consummatory behavior need not occur in a fixed order, rather each may proceed to completion, followed by perceptual schemas that will determine which motor schema is to be executed next. Schemas may be linked via so-called coordinated control programs. Motor schemas, for example, may take the form of "compound motor coordinations" (such as a programmed detour route to a prey object, or a programmed jump-snap-gulp sequence) which comprise a set that will proceed to completion without intervening perceptual tests in such a manner that: schema A proceeds to completion, completion of schema A triggers the initiation of schema B, or schema A passes a parameter x to schema B. In this context we also refer to Fentress's (1983) concept of *behavioral networks*, emphasizing that sequences of actions and rules of relations among these parts involve a multilayered dynamic organization, so that abstractions into "behavioral units" must be approached appropriately.

3 The Motor Part of Prey Snapping in Toads

3.1 Action Patterns and Sequential Acts

Snapping in toads, the "consummatory" act of prey catching (Fig. 1B), is usually preceded by various kinds of "directed appetitive responses" like orienting toward prey, approaching prey, and binocular fixation of prey (Fig. 1A). The

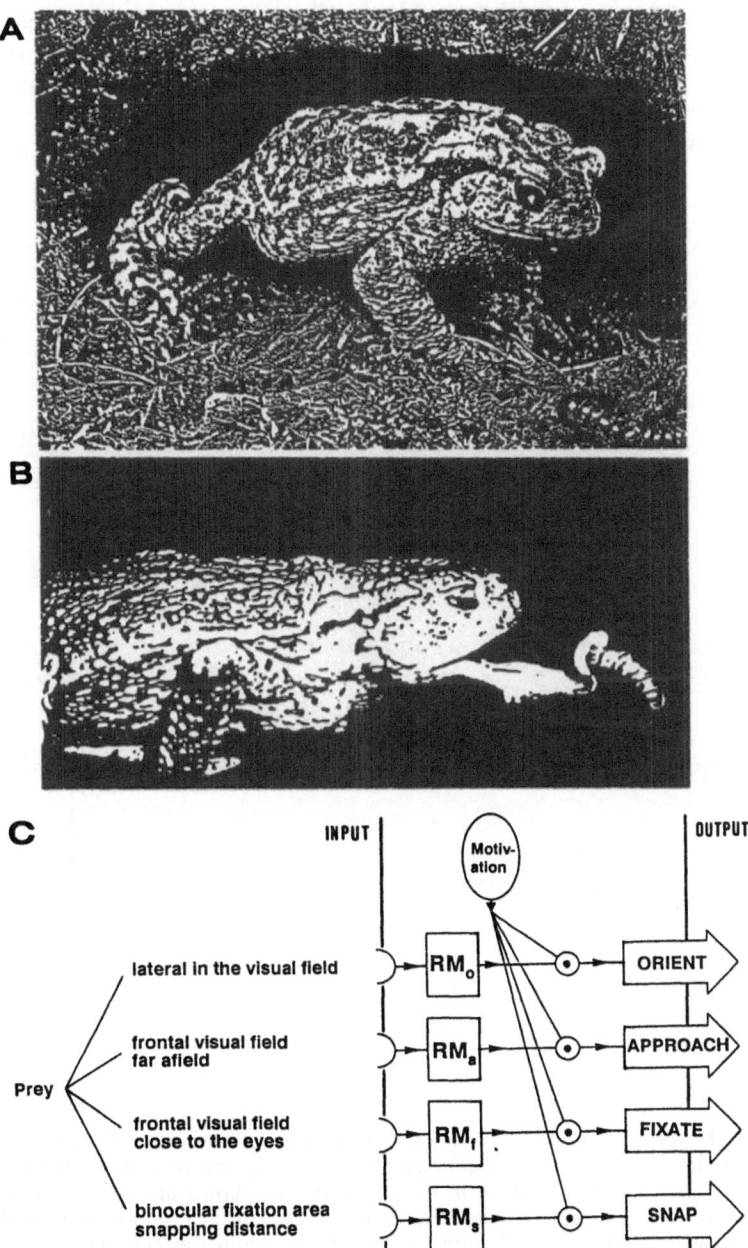

Fig. 1. A Directed appetitive and **B** consummatory behaviors of the common toad *Bufo bufo spinosus*: fixating prey and snapping it. Photos are by H. Burghagen and H.G. Meyer; with the aid of a particular electrical circuit the camera was triggered as the tongue contacted the prey, hence shutting the circuit. **C** Releasing conditions (*left*) for the action patterns related to prey-catching (*right*); depending on the prey sign stimulus in space, and in the presence of prey-catching motivation, the appropriate action pattern is selected by a corresponding releasing mechanism, *RM* (Ewert 1987)

release of each of these behavioral patterns requires prey recognition (Fig. 10A, B); the type of pattern is selected to suit the location of the prey object in space (Fig. 1C). Regarding the consummatory act, bufonid toads and ranid frogs display two main snapping patterns: biting with the jaws in response to relatively large prey objects, or grasping smaller prey with their projectile

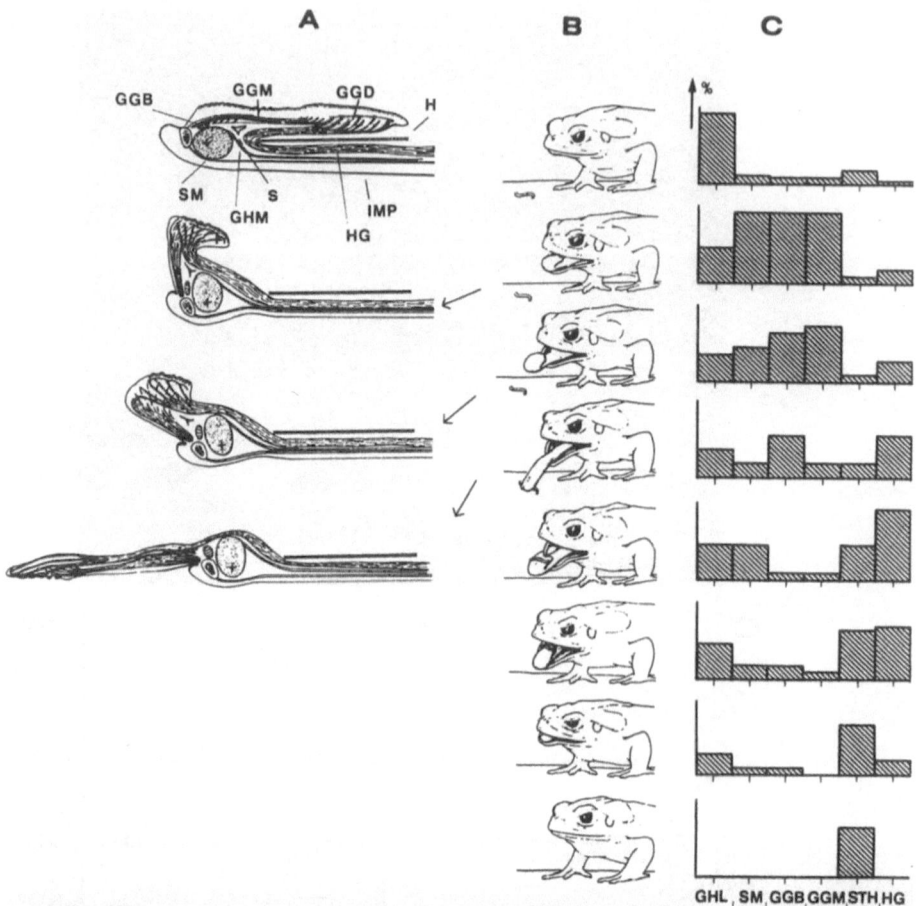

Fig. 2. Muscle contractions during tongue flipping in the marine toad *Bufo marinus*. **B** Frame-by-frame analysis of tongue-flipping phases in response to a prey object. **A** Contraction patterns of muscles involved in tongue protrusion shown in sagittal sections, from *top to bottom*: tongue in situ; tongue being lifted with the stiffened rod rotating over the symphysis; rod has proceeded beyond symphysis carrying the soft tissue; and rod projected with the soft tissues being propelled further. **C** Electromyogram values of muscles assigned to the tongue-flipping phases shown in **B**; the mean values refer to the percentage of maximum activity observed in the muscles. *GGB* Musculus genioglossus basalis; *GGD* m. genioglossus distalis; *GGM* m. genioglossus medialis; *GHL* m. geniohyoideus lateralis; *GHM* m. geniohyoideus mediâlis, *H* hyoid plate; *HG* m. hyoglossus; *IMP* m. intermandibularis posterior; *S* lymphatic sinus at the base of the tongue; *SM* m. submentalis; *STH* m. sternohyoideus. (Redrawn and combined from Gans and Gorniak 1982a, b)

tongue (Figs. 1B and 2). In toads *Bufo bufo*, the latter pattern seems to be more common, whereas the former is used mostly after unsuccessful tongue projections (Eibl-Eibesfeldt 1951; Eikmanns 1955).

Tongue snapping (Fig. 2B) consists of a sequence of behavioral acts: (1) opening the mouth, (2) projecting the tongue toward prey, (3) grasping the prey with the tongue, (4) retracting the tongue, (5) closing the mouth, and (6) gulping. Swallowing, elicited by intraoral somatosensory stimuli, is accompanied by retraction of the eye bulbs and eventually by pushing prey into the mouth by alternative movements of the forelegs. The sequence from (1) to (5) takes 200 $\pm$ 50 ms (Burghagen 1979; Gans and Gorniak 1982b). Snapping is a ballistic action pattern, which means that once triggered by an adequate visual prey stimulus, it proceeds to completion without visual control in an "open-loop" fashion. Although Gans (1961) using high-speed cinematography provided evidence of feedback-guided correctional maneuvers during a frog's leap toward an airborne prey object, he points out that at the moment of the snap, the head is elevated and this can momentarily remove the prey from the field of vision. As Fig. 1B shows, common toads often even close their eyes during snapping.

For comparison, the feeding behavior of pigeons also involves a stereotyped sequence of behavior patterns: pecking, grasping, mandibulation, and swallowing. Pecking – comparable to snapping in toads – is visually released and is a ballistic behavior executed with the eyes closed. It is introduced by two directed appetitive binocular head fixations, one at about 82 mm and the other at about 55 mm distance to the target, e.g., a grain (Goodale 1983). Mandibulation brought about by head movements and quasiperistaltic tongue movements is elicited and controlled by somatosensory stimuli arising from the grain in the pigeon's mouth (Zeigler et al. 1975; see also Chap. 8, this Vol.).

3.2 Pattern of Muscle Contractions

Applying high-speed cinematography, combined with recordings of electromyograms from the six major muscles involved in the marine toad's tongue flip (Fig. 2C), Gans and Gorniak (1982b) provided evidence for the "ballista hypothesis". It suggests that the intrinsic tongue muscles are stiffened, rotate over the symphysis, and catapult the soft tissue (Fig. 2A). More specifically, the authors concluded that tongue propulsion is due to activation of muscles connecting the tongue to the symphysis (genioglossus and submentalis muscles); protrusion depends on stiffening of these two muscle masses to form a lever system that flips the tip toward prey as a medieval ballista. Nishikawa and Gans (1990) point out that the genioglossus muscle by its contraction propels the tongue, whereas the submentalis muscle, which bends the lower jaw downward, does not seem to play a prominent role in the tongue flip. Lingual retraction – while the hyoid muscle moves – is primarily driven by the hyoglossus muscle. The hyoid muscle stabilizes the hyoid bone in the floor of the mouth during jaw and tongue movement. As Fig. 2C shows, prior to the tongue flip the sternohyoid STH muscle (and the geniohyoid muscle GHM) is active, continues during the flip, and becomes more active during tongue retraction; the genioglossus muscle GGM displays peak activities just as the tongue is flipped; the hyoglossus HG reaches its maximum as soon as the tongue starts to retract.

For the feeding mechanisms in salamanders, the reader is referred to Lauder and Reilly (1988; see also this Vol.).

3.3 Ontogenetic Aspects

In *Bufo bufo* it was observed that sensorimotor functions for feeding mature after the metamorphosis from the aquatic vegetarian tadpole to the terrestrial predatory toad (Traud 1983; Ewert 1985). With the first step of the still tailed juvenile animal to its terrestrial life, orienting toward prey is not well aimed and snapping is immature: the striking distance is not adequately estimated; the mouth is just opened in the manner of gaping and the tongue is not projected. In fact, this snapping pattern – without a projectile tongue – appears to resemble features of phylogenetically basal anuran species, e.g., from the family Discoglossidae (Burghagen 1979). Developmental studies in common toads have shown that both the visual-perceptual and the motor performances begin to improve in the first postmetamorphic week, independently of prey experience. More specifically, it was observed that during the first 2 days of terrestrial life, toads responded toward prey only with mouth opening; 6-day-old animals often showed orienting, mouth opening, and then snapping involving tongue projection; after the 10th day the toads displayed orienting and tongue flipping (Traud 1983).

The perceptual maturation (Ewert and Burghagen 1979a) – in so far as improvements in depth estimation and configural stimulus discrimination are concerned – seems to be correlated with the anatomic differentiation and parcellation of the posterior dorsal thalamus into a posterocentral (P) and a posterolateral (Lpd) nucleus (Fig. 11b) (Clairambault 1976), a process being completed 6 to 12 months after metamorphosis. In this context we are reminded of Ebbesson's (1984) "parcellation theory", which suggests that parcellation of brain nuclei results in a finer tuning of a given function (Ewert 1984b).

3.4 Localization of Tongue and Jaw Muscle Motoneurons

To localize the motoneurons that innervate the common toad's extrinsic tongue muscles – m. genioglossus (protractor) and m. hyoglossus (retractor) – horseradish peroxidase (HRP) was injected into one of these identified muscles on the one side and into the other on the opposite side, after transecting the nerves supplying the remaining muscles (Weerasuriya and Ewert 1981). The corresponding motoneurons ipsilaterally in the caudal medulla oblongata around the motor nucleus of the XIIth cranial nerve (Nieuwenhuys and Opdam 1976): genioglossal motoneurons are located caudally (Fig. 3A), whereas hyoglossal motoneurons are distributed more rostrally (Fig. 3D), and there is an area of overlap. This result was confirmed anatomically in grass frogs by soaking transected single hypoglossal nerve branches in HRP (Stuesse et al. 1983), and neurophysiologically in Japanese toads by intracellularly recording the anti-dromic activity of tongue muscle motoneurons in response to electrical stimuli applied to identified hypoglossal nerve branches (Satou et al. 1985). After intracellular injection of HRP into tongue muscle motoneurons, extensive spread of dendritic trees becomes visible (Fig. 3a, b), mainly in lateral and dorsolateral directions, even extending into the nucleus of the solitary tract, NST; some dendrites crossed to the contralateral HGN (Fig. 3a; Satou et al. 1985). Similar extensive dendritic arborizations of hypoglossal motoneurons in frogs were reported in other studies using different morphological tracers (Matesz and Székely 1977).

There is a correlation between snapping and the activity patterns of neurons of the HGN. Chronic single-cell recordings revealed H-type neurons which discharge in response to prey only if snapping is elicited (Schürg-Pfeiffer et al. 1993). H1 neurons display a burst of spikes in the phase of tongue protrusion and H2 neurons during tongue retraction. The former neurons were located in the HGN somewhat caudally, corresponding to the sites of genioglossus muscle motoneurons, and the latter rostrally, corresponding to the sites of hyoglossus muscle motoneurons. H3 neurons are active shortly before snapping, increase their discharge frequency during the tongue flip, and become more active in the phase of tongue retraction, corresponding to the activity pattern of the sterno-hyoid muscle (cf. Fig. 2B, C).

A comparable topography of tongue protractor and retractor motoneurons has been demonstrated in mammals by Kramer et al. (1979) and Uemura-Sumi et al. (1981), although here a more marked mediolateral separation of motoneu-ronal pools was observed. For a discussion of controversies on anatomic homologies between anurans and mammals regarding the XIIth cranial nerve we refer to Stuesse et al. (1983).

The jaw in toads is elevated by the temporalis, pterygoideus, and masseter muscles and lowered by the musculus depressor mandibuli. After injecting horseradish peroxidase into the respective muscles and tracing the backfilled neurons (Weerasuriya 1983), it can be shown that the depressor muscle is innervated by motoneurons of the VIIth (facial) and the elevator by motoneu-rons of the Vth (trigeminal) cranial nerve (Fig. 4B, right). Comparable data have been reported earlier for other anuran species (see Gaupp 1896; Nieuwenhuys and Opdam 1976). Intracellularly recorded and subsequently labeled visually responsive neurons of the trigeminal nucleus (Fig. 4C) revealed an extensive spread of dendritic arborizations, suggesting integrating properties (Schwippert et al. 1990).

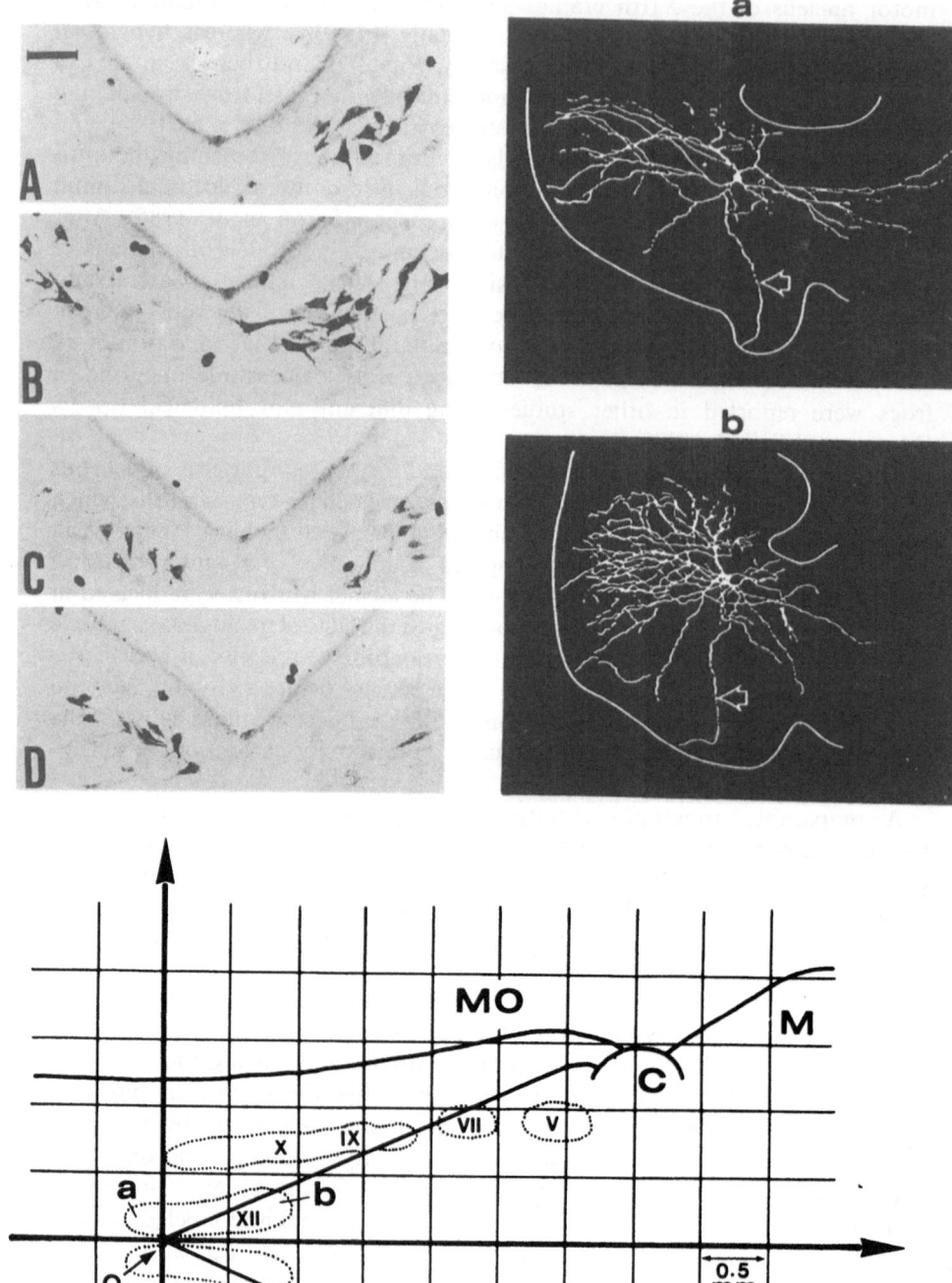

3.5 Afferents of the Hypoglossal Nucleus

After unilateral microinjections of horseradish peroxidase into the HGN, the overall pattern of retrograde cell labeling showed no backfilled cells rostral to the rhombencephalon, e.g., none in the optic tectum (Weerasuriya and Ewert 1984). However, backfilled cells were identified bilaterally in the medullary medial reticular formation (MRF) up to the level of the nucleus of the VIIIth cranial nerve, encompassing also the nucleus reticularis inferior and the nucleus reticularis medius (Fig. 5B). These cells were mostly located in the ventral gray matter (central gray) close to the border of the stratum album. A few cells in the median reticular formation (the area of the raphe nucleus) were also labeled. Bilaterally, labeling was seen also in the lateral and ventral portions of the NST with ipsilateral labeling being heavier. Backfilled cells were also found within and near the branchiomotor column (BMC), laterally to the MRF, associated with the motor nuclei of the Vth, VIIth, IXth, and Xth cranial nerves (Weerasuriya and Ewert unpubl.). A variety of visual (and tactual) properties displaying neurons have been recorded in BMC and MRF (Ewert et al. 1984, 1990b; Schwippert et al. 1990; for responses to electrical tectal stimulation see also Matsushima et al. 1989). Figure 6 shows morphologies and visual response properties of prey-selective neurons whose somata were located within or near the VIIth nucleus (Schwippert et al. 1990).

3.6 Afferents of the Medial Reticular Formation

After unilateral small injections of horseradish peroxidase into the caudal MRF (Ewert and Weerasuriya 1988) the retrogradely labeled cells were distributed bilaterally, and the dendritic trees of magnocellular neurons between gray and white matter extended into the white matter. A few cells near the ipsilateral NST and the ipsilateral lateral reticular formation were labeled, too. Backfilled cells were also seen bilaterally in and near the BMC and in the rostral spinal cord. Anterograde terminal labeling was obtained in HGN.

In the mesencephalon, labeled cells were located in the optic tectum, the torus semicircularis, and in ventral tegmental nuclei (Weerasuriya and Ewert 1981).

Fig. 3. A–D Labeling of tongue muscle motoneurons in the common toad's hypoglossal (XIIth) nucleus after injection of horseradish peroxidase (HRP) into the right hyoglossus (retractor) muscle and the left genioglossus (protractor) muscle, left/right referring to the animal; brain sections (*top to bottom*) are running from caudal to rostral: about 250 μm caudad of the obex, 200 μm rostrad, 500 μm rostrad, and 800 μm rostrad of the obex. *Calibration bar* = 100 μm (after Weerasuriya and Ewert 1981). The camera-lucida reconstructions of a protractor (**a**) and a retractor (**b**) motoneuron are from the Japanese toad after intracellular recording and subsequent intracellular injection of HRP; the *arrow* points to the axon (after Satou et al. 1985). *Bottom* Dorsal view of part of the medulla oblongata, e.g., indicating the recording sites of the retractor (*b*) and protractor (*a*) neuron; *C* cerebellum; *M* mesencephalon; *MO* medulla oblongata; *O* obex; *V-XII* cranial nerve nuclei

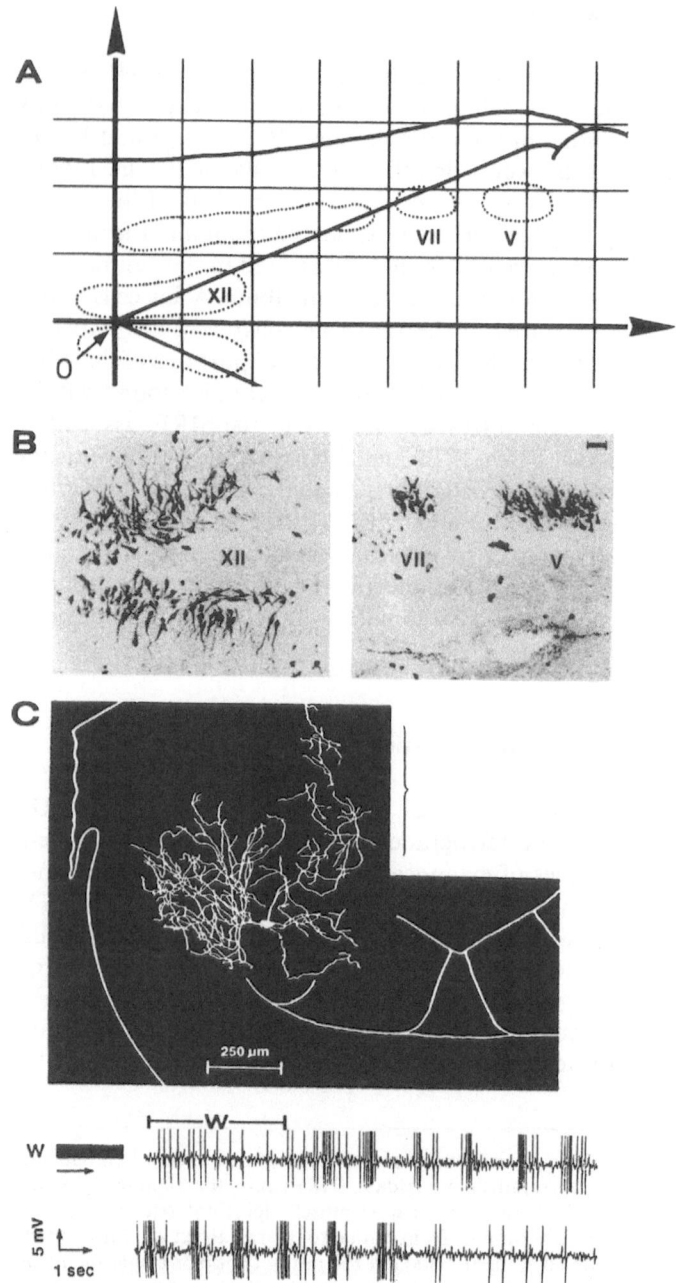

Fig. 4. A Topography and **B** labeling of tongue and jaw muscle motoneurons in the common toad. Unilateral labeling of jaw muscle motoneurons in the trigeminal (*V*) motor nerve nucleus after HRP injection unilaterally into the elevator muscles (m. temporalis, m. pterygoideus, m. masseter) and in the facial (*VII*) nucleus after HRP injection unilaterally into the depressor muscle (m. depressor mandibuli). Bilateral labeling of hypoglossal (*XII*) motoneurons follow-

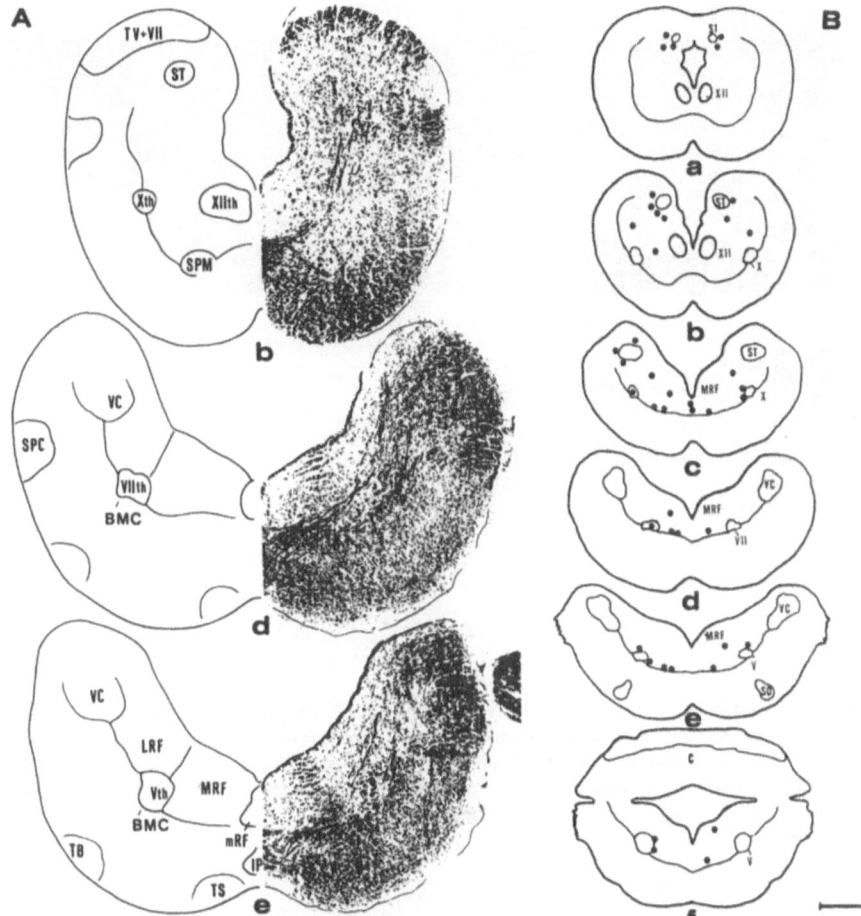

Fig. 5. Afferents to the hypoglossal nucleus in the common toad. **A** Klüver-Barrera stained transverse sections through the medulla oblongata. **B** Reconstruction of retrograde cell labeling (*black dots*) in the medulla oblongata, shown in transverse sections from caudal (*a*) to rostral (*f*), after microinjection of HRP into the left hypoglossal nucleus; *calibration bar* = 0.5 mm. *Vth-XIIth* cranial nerve nuclei; *C* cerebellum; *LRF* lateral, *MRF* medial, *mRF* median reticular formation; *SO* superior olive; *ST* solitary tract; *SPC* spinocerebellar complex; *SPM* spinal motor nucleus; *TB* tractus tecto-bulbaris directus, tr.tb.d.; *TS* tractus tecto-bulbaris et spinalis cruciatus, tr.tbs.c.; *TV + VII* descending tracts of Vth and VIIth nuclei; *VC* vestibular complex. (After results from Weerasuriya and Ewert 1984, 1988 unpubl.)

Fig. 4. *Cont.*
ing HRP injection into the tongue muscles on both sides. Sagittal sections: *Calibration bar* = 100 μm (after Weerasuriya 1983). **C** Camera-lucida reconstruction of an intracellularly recorded and iontophoretically Co^{3+}-lysine labeled trigeminal neuron; its soma was located in the Vth nucleus of the branchiomotor column (cf. Fig. 5); a branch of the dendritic tree (see *bracket*) could be traced towards the cerebellum. In response to a worm-like (*W*) moving stripe, this neuron responded with increasing spike activity that, after visual stimulation, proceeded in cyclic bursting. (Schwippert et al. 1990)

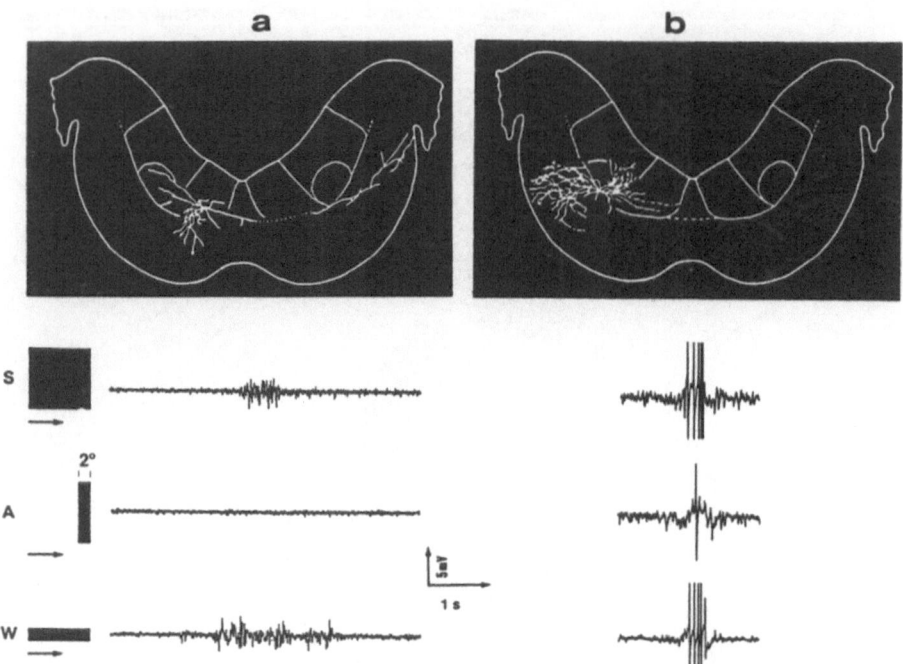

Fig. 6. Examples of two intracellularly recorded and iontophoretically Co^{3+}-lysine labeled prey-selective M5.2 cells in **a** the medial reticular formation and **b** the facial (VIIth) nucleus (cf. Fig. 5Ad) of the common toad; camera-lucida reconstructions. *Below* Records in response to visual objects moving at constant velocity (*arrows*): square (*S*) and small stripes in worm (*W*) or antiworm (*A*) configuration. (Schwippert et al. 1990)

The backfilled tectal cells were distributed bilaterally (contralateral > ipsilateral) in the ventrolateral and lateral regions, heaviest at the border between the tectal layers 6 and 7 (Fig. 7A; for histology see Fig. 11, brain section c). More specifically, about 80% of the labeled tectal cells were in the lateral half of the tectal lobe and identified as either pyramidal, ganglionic, or piriform cells (nomenclature by Székely and Lázár 1976). The axons of some of these cells could be traced in the tectal output layer 7 (see arrows in Fig. 7Ab), and fibers were observed to cross the midline of the ventral tegmentum in the ansulate commissure. The results are in agreement with previous data obtained from silver impregnation studies showing that the tectum has a bilateral projection to bulbar/spinal nuclei mediated by the tractus tecto-bulbaris directus (t.tb.d.) and the tractus tecto-bulbaris et spinalis cruciatus (t.tbs.c.) (Rubinson 1968; Lázár 1969). The horseradish peroxidase labeling study (Fig. 7A) characterizes tectal cell types and their locations (for the cells of origin of descending pathways to the frog's spinal cord, see Tóth et al. 1985).

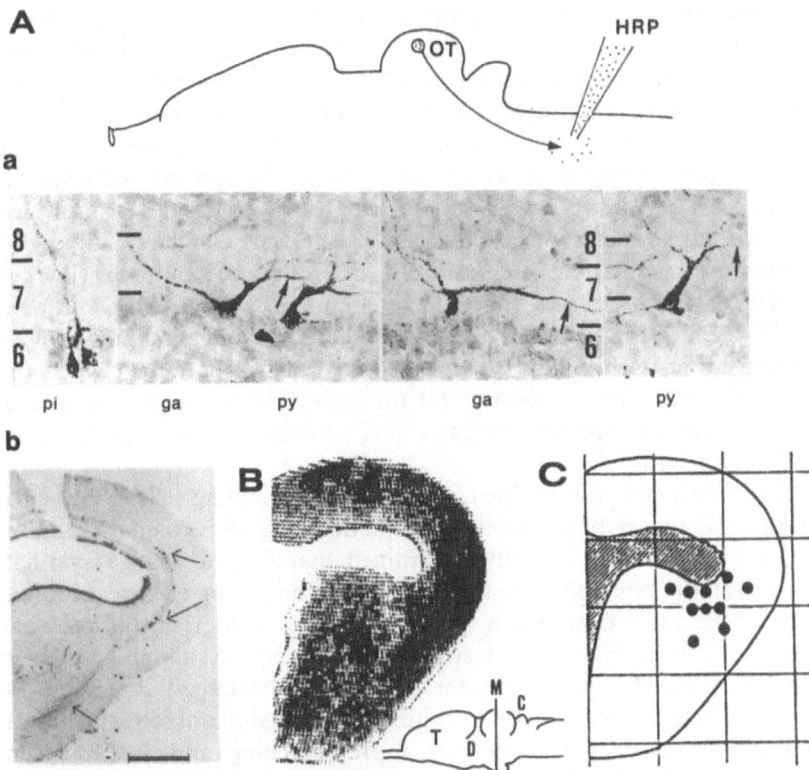

Fig. 7. A Tectomedullary efferents in common toads. **a** Different types of backfilled cells at the border of *layers 6* and *7* (cf. Fig. 11c) in the optic tectum (*OT*) following HRP injection into the contralateral caudal medial reticular formation: *ga* ganglionic, *pi* piriform, *py* pyramidal cell; *calibration bar* = 50 μm (Weerasuriya and Ewert 1981). **b** Half-section through OT; arrows on top refer to back-labeled cells (shown in a); bottom arrow indicates their axons descending in the ansulate commissure. **B** Computer-processed autoradiographic image of a ^{14}C-2-deoxyglucose (2DG)-labeled midbrain section (corresponding to Fig. 11c) from a common toad (sitting in a glass vessel) repetitively snapping toward a visual prey dummy moving (outside the vessel) to and fro; *black* indicates the strongest metabolic activity in the "snapping evoking area" of the tectum (Finkenstädt et al. 1985). **C** Sites in the midbrain which electrically stimulated elicited snapping in toads; *grid width* = 0.5 mm. *C* cerebellum; *D* diencephalon; *M* mesencephalon; *T* telencephalon (Ewert 1967)

3.7 Neurobiological Aspects of Tecto-Bulbar Information Transfer

Since Ingle (1983) has shown that cutting the crossed tecto-(bulbar)medullary tract (t.tbs.c.) at its tegmental decussation abolishes prey-orienting but not snapping, it can be concluded that visual information mediated by the ipsilaterally projecting fibers in the t.tb.d. is sufficient for triggering snapping. Electrical stimulation of the toad's optic tectum with chronically implanted electrodes has demonstrated that, depending on the stimulation site, various patterns of prey-catching behavior can be elicited, such as orientation turning, the directions of which broadly coincide with the tectal visual map (Ewert 1967, 1974; Ewert et al. 1983). Snapping was triggered by stimuli applied to the more lateral tectum that represents the frontal visual field, the "snapping evoking area". Snapping could also be released from areas extending ventrally toward subtectal and toral somatosensory regions (Fig. 7C; for anatomy, see Fig. 11, brain section c). Swallowing, eye retraction, and snout wiping were elicited mainly by toral and tegmental stimulation. Interestingly, coordinated snapping (jaw grasping, tongue flipping) was not triggered by stimulation of structures in or near the HGN (Schürg-Pfeiffer et al. 1993). These data provide two kinds of evidence: (1) motor pattern-generating systems for fixed action patterns related to feeding exist in brain-stem structures; (2) through external triggering (excitation of tecto-, toro-, or tegmento-bulbar/spinal projective cells) thier program can be called into play.

The dominance of backfilled tecto-bulbar descending cells in the ventrolateral tectal lobe after hypoglossal injection of horseradish peroxidase broadly coincides with (1) the behaviorally determined snapping-releasing visual field (Ingle 1976), (2) the "snapping evoking area" determined by electrical brain stimulation of the optic tectum (Fig. 7C), and (3) the topography of regional ^{14}C-2-deoxyglucose uptake in the tectum during repetitive snapping at a visual prey dummy (Fig. 7B). Taken together, these results suggest that there are tectal neurons in layers 6/7 which, projecting to the medulla oblongata, play a pivotal role in the sensorimotor interface of prey catching, and thus represent a link between afferent messages (prey sign stimuli) and efferent commands (snapping). Since snapping can also be elicited by touch in frogs after tectal ablation (Grobstein et al. 1983), in addition to visual command releasing systems (CRS), tactual CRS must exist whose command elements are located in somatosensory subtectal and toral regions and which have access to the same snapping-generating circuitry. Note that in the electrical brain stimulation experiments, the stimulation sites effective for snapping were found both in visual tectal and somatosensory toral areas (Fig. 7C).

Clues about the location of interneurons operating between tectal efferents and tongue muscle motoneurons have been provided by intracellular recordings of hypoglossal motoneurons (Satou et al. 1985) in response to electrical stimulation of tectal output fibers in the "snapping evoking area". Drawing on the response latencies and temporal summation properties, it could be shown that this tectomotor information transfer involves polysynaptic pathways. The fact that the amplitude of excitatory postsynaptic potentials (EPSPs) in tongue protractor motoneurons increased if both tectal lobes were stimulated simultaneously accounts for spatial facilitation, a phenomenon known from visually elicited snapping: the time interval between the onset of binocular fixation and snapping of prey could be relatively long; this period was further prolonged in the monocular animal (see also Schneider 1954), however, it was remarkably reduced, if the visually "deprived" tectal lobe was simultaneously electrically stimulated at subliminal intensities through an electrode implanted in the snapping evoking area (Ewert 1967). There are various possible substrates for convergence of information leaving both eyes at brain-stem levels, such as the ansulate commissure (Ingle 1983), the MRF that integrates inputs both from t.tbs.c. and t.tb.d., and the HGN that links tongue muscle motoneurons through dendritic commissural fibers (Fig. 3a; Satou et al. 1985).

3.8 Toward a Snapping Pattern Generating Circuit

3.8.1 Components and Functional Properties

Visual information relayed by tectal efferent neurons reaches hypoglossal motoneurons through interneurons. Drawing on the mean latency of 9 ms for the initial motoneuronal inhibitory postsynaptic potential (IPSP) to tectal stimulation (Satou et al. 1985), estimating 5 ms for conduction time (at 1 m/s for 5 mm), and assuming a synaptic delay of 1 ms (Brookhart and Fadiga 1960), then at least three interneurons seem to be sequentially interposed between tectal output and hypoglossal motoneurons; the motoneuronal EPSP had a longer latency, suggesting that there are more interneurons. If the conduction velocities are faster (the longest antidromic latencies were about 3 ms, measured by Satou and Ewert 1985), even more interneurons may be involved. In contrast, tectofugal output to neck and forelimb motoneurons in the brachial enlargement of the spinal cord crosses only one or two interneurons (Maeda et al. 1977), and glossopharyngeal afferents have monosynaptic access to hypoglossal motoneurons (Matsushima et al. 1986). This evaluation indicates how fast the tectomotor information flow for snapping could be. Recordings from freely moving toads, however, show that the actual delays between the onset of visually evoked activation in prey-selective tecto-bulbar projecting neurons and the beginning of snapping may take up to 400 ms (Schürg-Pfeiffer et al. 1993), which suggests intervening, integrative (threshold operational) and interactive (excitatory/inhibitory) loops.

Looking for the site of interneurons, we have to consider both the data from horseradish peroxidase tracings (Fig. 5B) and the preponderance of MRF cells with tectal (Ewert et al. 1984; Matsushima et al. 1989) and visual input (Ewert et al. 1984, 1990b; Schwippert et al. 1990). Hence, the tectoreceptive part of the snapping MPG appears to be located between the level of the nucleus isthmi and the rostral pole of the vagal BMC. The premotor neurons, as the components of the MPG that have monosynaptic access to the motoneurons, are probably located near the respective motor nuclei. For example, intracellular recording and cobalt-lysine staining studies in common toads have shown that stimulus-event sustaining cyclic bursting cells (Fig. 4C) are mostly situated close to or within branchiomotor nuclei (Schwippert et al. 1990). In mammals, too, it has been suggested that premotor neurons of the MRF are situated adjacent to bulbar motoneurons (Landgren et al. 1986). For example, the rhythmically active neurons of the nucleus reticularis gigantocellularis are connected to the respective cranial motor nuclei involved in the rhythmic jaw movements of mastication (Nakamura 1985; see also De Vree and Gans, this Vol.).

Regarding the functional properties of interneurons in such perimotor nuclear locations, we suggest an integrative role in recruiting their neighboring motoneurons into various motor synergies (see also Olsson et al. 1986). Interneuronal connections (Weerasuriya and Ewert 1984; Landgren et al. 1986) probably coordinate the various subsynergies of a motor pattern, whereby proprioceptive feedback may play a role. For example, Weerasuriya (1989) discovered that after transecting the hypoglossal nerves bilaterally, common toads lunge toward prey without any tongue movements and, interestingly, also without opening their jaws. The latter result is of particular importance since the jaw motoneurons were intact. The lack of mouth opening may be due to an inhibition arising from the absence of any tone in the lingual musculature detected by glossopharyngeal (Matsushima et al. 1986) or hypoglossal (Stuesse et al. 1983) afferents, hence suggesting that internal feedback mediated by hypoglossal nerves plays a role in the coordination of opening the mouth and protracting the tongue.

It appears that certain components of the snap can be uncoupled (lunge vs tongue and jaw movements), whereas others are more tightly coupled (tongue, jaw, and swallowing movements). This assumption is consistent with the following observations: if during prey fixation the prey is quickly removed from the toad's field of vision, snapping – already triggered – proceeds to completion and is often followed by repetitive swallowing, eye retraction, and mouth wiping (Hinsche 1935; Ewert 1967). A similar behavioral sequence can be observed, if snapping is elicited by electrical stimulation of the tectum. Presumably, snapping and swallowing are centrally coordinated in such a way that they are separated by a predetermined constant latency (Ewert 1967). But swallowing is also triggered by mechanical stimulation of prey in the mouth. It is difficult to test the former possibility, since contact between the tip of the returning tongue and the pharyngeal surface, following an unsuccessful strike, might be a stimulus to initiate swallowing. Suppose a toad is eating an earthworm: after the snap, which brings part of the earthworm into the toad's mouth, attempts to swallow the rest of the earthworm – aided by eye retractions and alternative use of the forelimbs – are guided probably by sensory feedback from the trigeminal-facial-glossopharyngeal-vagal afferent systems.

With respect to the kinds of cells involved in eating behavior, we have to consider both the hypoglossal afferents from the MRF and those from the NST and also from neurons in and around trigeminal and facial cranial nerve structures (Weerasuriya and Ewert 1984). In this context we refer to intracellularly recorded cobalt-lysine stained MRF neurons near the Vth and VIIth nucleus (Fig. 6a) responding to visual prey stimuli (Schwippert et al. 1990); some of these display cyclic burst activities (e.g., Fig. 4C). The connections between HGN and NST are probably important for the elementary bulbar reflexes, such as manipulation of prey within the mouth and swallowing. Again, we suggest that connections between various carnial nerve systems play a role in the coordination of the jaw and tongue musculature necessary for the timing of snapping and ingestive movements.

In pigeons that show preferences for single kernels of grain, trigeminal neurons are involved in the sensorimotor control of ingestive behavior (Zeigler and Witkovsky 1968). Furthermore, there are neurons whose somatosensory responses to intraoral stimuli participate in mandibular movements that transport the grain into the buccal cavity. Lesions to the quintofrontal structure (involving the Vth cranial nerve) left pecking intact, impaired grasping to some extent, but totally abolished the ability to retain the grain in the mouth; swallowing, on the other hand, was intact provided that the grain was put experimentally into the buccal cavity (Zeigler and Karten 1973).

In mammals – like amphibians – swallowing involves oral, pharyngeal, and esophageal actions, whereby here a pressure gradient in the mouth is achieved by a piston-like activity of the tongue that passes the food from the mouth to the pharynx (Dubner et al. 1978). Swallowing results from a centrally programmed sequence of inter- and motoneuronal activities in the medullary "swallowing center" (Doty 1968; Roman 1986; Amri and Car 1988). Investigations in sheep have shown that the interneurons responsible for the program of the spatio-temporal activity pattern in this center (Jean 1984) are located in the dorsal NST and the adjacent MRF. This structure has access to (1) premoto- and motoneurons of the ventral nucleus ambiguus and the adjacent MRF controlling the muscles of pharynx, larynx, and esophagus (Roman 1986); (2) motoneurons in the trigeminal nucleus innervating the mylohyoid, anterior digastric, and medial pterygoid muscles (Car and Amri 1982); and (3) motoneurons of the hypoglossal nucleus innervating the genioglossus, styloglossus, and hyoglossus muscles (Sumi 1969; Amri et al. 1989). As a correlate for the synchronous contraction of jaw and tongue muscles during swallowing (Doty and Bosma 1956), certain reticular interneurons with bifurcating axons seem to play a major role: they project bilaterally to the trigeminal and the hypoglossal motor nuclei, as evidenced by neurophysiological antidromic stimulation and neuroanatomical, fluorescent, double-labeling techniques (Amri et al. 1990).

3.8.2 Functional Organization

From the available evidence we envision some aspects of the functional organization of the snapping MPG and suggest certain properties of the internuncial network. The MPG receives direct input from efferent tectal neurons or other sensory analyzers (Fig. 8A). Although it appears that upon release the performance of the MPG is uninfluenced by visual feedback in toads, proprioceptive and internal feedback loops within the internuncial network probably serve to coordinate, for example, the precise timing between jaw and tongue movements. Since neuroanatomical as well as physiological studies have failed to reveal axon collaterals in hypoglossal motoneurons (Cajal 1909; Porter 1965; Satou et al. 1985), this would indicate that central programs addressed to these motoneurons do not depend on motoneuronal feedback. The interneurons must

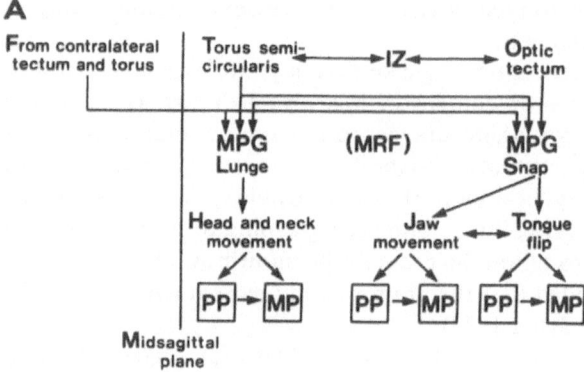

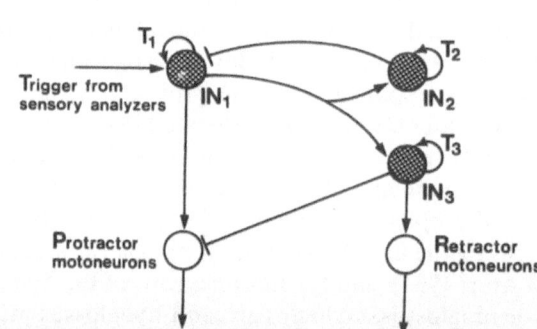

Fig. 8. A Information flow related to motor pattern generation of prey-catching action patterns; *IZ* intermediate zone between the ventrolateral tectum and far-lateral torus semicircularis; *MP* motoneuronal pools; *MPG* motor pattern-generating circuit; *MRF* medial reticular formation; *PP* premotoneuronal pools. **B** Global idea of a circuit that controls the protractor and retractor musculature for the tongue flip; *IN* interneuronal pools; *arrows*, excitatory connections; *lines with bar*, inhibitory connections; *circular arrows*, integrator circuits (involving reverberatory positive feedback loops); *T* time constants. (Weerasuriya 1991)

not function exclusively for the generation of snapping; some may take part also in other motor synergies in response to different sensory input, such as the rejection of an unpalatable object from the mouth (e.g., Sternthal 1974; Mikulka et al. 1980).

Figure 8A illustrates a global concept for the functional organization of the snapping MPG in toads. Since bilateral lesions in the MRF abolish snapping, whereas extensive ipsilateral lesions do not disrupt this action pattern (Weerasuriya 1989), we conclude that the MPG is bilaterally represented. Each of the two parts should have independent access to the motoneurons on both sides and also receive independent polysynaptic input from both tectal hemispheres. The similar latencies of postsynaptic potentials of hypoglossal motoneurons elicited by ipsi- or contralateral tectal stimulation (Satou et al. 1985) suggest that the bilateral tectal polysnaptic efferents converge. For comparison, we refer to the swallowing MPG in mammals (Doty et al. 1967) which is organized as two half-centers with each half-center controlling the bulbar motoneurons of that side. Mastication in mammals, too, is controlled by bilateral (oscillatory) circuits (Chandler and Tal 1986). The dual representation of the toad's snapping MPG, apart from the advantage of a backup system, is probably related to the fact that its optimal activation requires simultaneous impulse traffic from both tectal hemispheres. The ability of atectal (sightless) frogs to respond to tactile stimuli, and of those with toral damage to respond to visual stimuli, indicates that somatosensory and visual systems at mesencephalic levels have separate access to the snapping MPG (Grobstein et al. 1983). Furthermore, Shinn and Dole (1978) demonstrated that blind *Rana pipiens* can migrate toward a discrete source of mealworm odor and sometimes even respond with snapping, and this has been confirmed for *Bufo cognatus* (Ewert unpubl.). Whereas there is an innate basis of visual prey recognition (Ewert 1985, 1987), the evaluation of olfactory stimuli – so far investigated – has to be learned (Ewert 1968; Dole et al. 1981; Merkel-Harff and Ewert 1991).

Focusing on the visually released tongue flip (Fig. 8B), we consider different MRF interneurons receiving an adequate visual trigger input, exhibiting tonic discharges, and displaying input-sustaining reverberatory activities, i.e., proper-ties observed among medullary M5.2, M8, and M9 neurons, respectively (Ewert et al. 1990b; Schwippert et al. 1990). Once activity reaches the threshold in the protractor neurons, the retractor motoneurons are activated and the protractor motoneurons simultaneously inhibited. Comparable excitatory/inhibitory pro-cesses are known from tongue motoneurons involved in swallowing in cats; Tomomune and Takata (1988) have shown that excitation (EPSP activity) in styloglossus muscle motoneurons in combination with an excitation-suppres-sion (EPSP-IPSP) sequence in genioglossus muscle motoneurons provides a correlate of the tongue movements during the buccopharyngeal processes of swallowing. Furthermore, we must consider the fact (not shown in Fig. 8B) that tongue protractor neurons in response to tectal stimulation initially display IPSPs which in the course of stimulation – due to spatial and temporal facilitation – are then superimposed by EPSP activity (Satou et al. 1985). During swallowing in mammals, too, hypoglossal motoneurons exhibit a brief hyperpolarization before the onset of the firing that activates tongue muscles (Sumi 1969). Presumably, such IPSPs ensure that subliminal activity in motoneurons is terminated, thus "resetting" the system before the initiation of a ballistic action. We will present a concept of a neuronal circuit for the toad's tongue flip at the end of this Chapter (cf. Fig. 17).

The toad's tongue-flip controlling circuit interacts with the one responsible for jaw movements. Neurons in the Vth and VIIth nucleus capable of displaying bursts of spikes (Fig. 4C), differently phase-locked (Fig. 9C, a – c), are function-ally suitable components for such coordination. The notion of coordinating inhibitory inputs is evidenced by IPSPs that terminate each burst (Fig. 9A, B; see

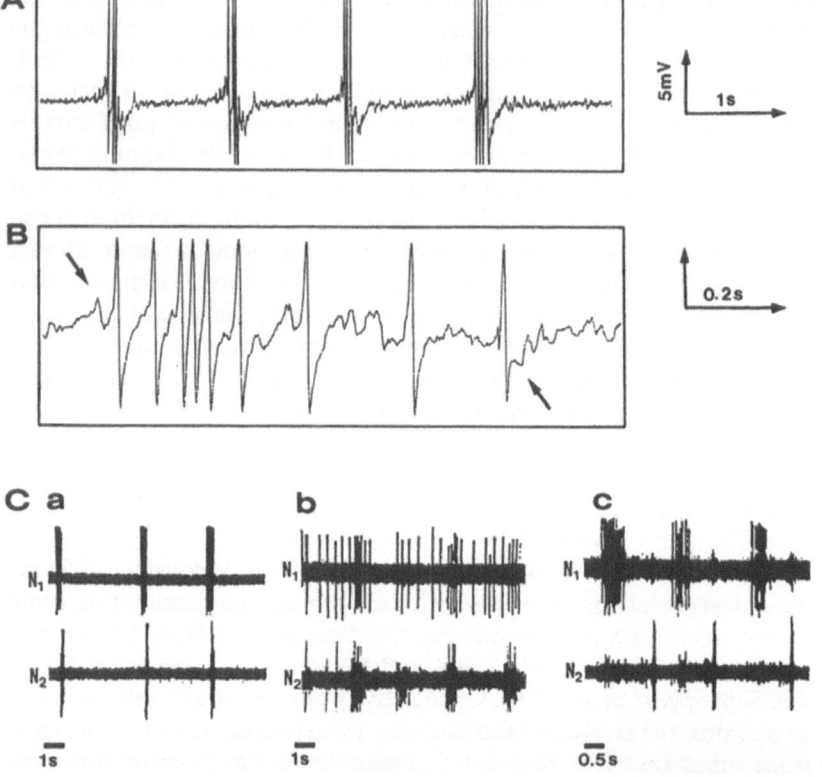

Fig. 9. Typical properties of cyclic bursts displaying medullary M10-type neurons in the common toad. **A** Intracellularly recorded sequence of bursts; each burst is introduced by depolarizing and is terminated by hyperpolarizing deflections. **B** One burst of spikes; de- and hyperpolarizing waves are indicated by *arrows* (Schwippert et al. 1990). **C** Parallel extracellular recordings of each two M10 neurons, N_1 and N_2. Neurons were recorded **a** ipsilaterally adjacent to the VIIth nucleus, **b** bilaterally from about opposite sites of the MRF adjacent to the Vth nucleus, and **c** close to the VIIth nucleus (N_1) and on the opposite side within the VIIth (N_2). The burst activities of N_1 and N_2 were differently phase-locked in **a**–**c**. (Ewert et al. 1990b)

bottom arrow). The lasting repetitive activity recorded in these examples may be due to the experimental situation, in which the toad's striated muscles were slightly relaxed after intralymphatic succinylcholine application.

3.8.3 State Control

The "state of the organism" (Hobson and Scheibel 1980), determined by the level of satiation, the season, the time of the day, etc., influences feature-analyzing sensory systems directly (Ewert 1984a; Schürg-Pfeiffer et al. 1993). But given the presence of both catechol- and indoleaminergic cells in the anuran rhombencephalon (Parent 1973) and the role of monoaminergic cells in behavioral state control (Bloom 1979), it is also possible that neurons of the MPG as well as the motoneurons are subject to neuromodulators (Harris-Warrick 1988) or are recipients of the outputs of "level setting interneurons". Serotonergic influences on anuran spinal motoneurons are reported by Soller (1977) and Cardona and Rudomin (1983). In this context the raphe nucleus of the median reticular formation, harboring serotonin-containing cells projecting to the intermediate tectal layers (Braak 1970; Parent 1973) and to the HGN, must be considered. Interactions of the hypothalamus and the reticular formation with the motor systems – taking visceral sensory information into account – are mediated by the NST. Székely et al. (1983), from electron microscope studies in frogs, imply the presence of neurotransmitters and neuromodulators (catecholamines, somatostatin, and enkephalin) in axon terminals of the NST. Visuomotor functions related to snapping in common toads appear to involve dopaminergic influences: application of the dopamine agonist apomorphine (0.4 ml/100 g) causes a strong facilitation of snapping. The snap in response to prey shows rather constant latencies, is highly stereotype, well aimed, and only elicited if the prey is presented at a short distance to the toad; orienting toward a prey object located in the lateral visual field fails to occur (Spreckelsen et al. 1993). Comparably, the snap of untreated controls displays a more variable latency and is less stereotype; for example, the variations in the snapping distance (17 – 30 mm) are surmounted by changing the amplitude of the lunge.

In pigeons, apomorphine strongly facilitates pecking (Dhawan et al. 1961) and in guinea pigs it elicits spontaneous, cyclic jaw movements (Lambert et al. 1986). Since in all these experiments apomorphine was systemically injected, the sites and the modes of its influence in the sensorimotor pathway are being presently investigated in toads (Glagow and Ewert 1992).

4 The Feature-Analyzing Part of Prey Catching

4.1 Configural Releasing Visual Features

Quantitative investigations on visual releasers of prey catching and avoidance behaviors in common toads (*Bufo bufo*) have shown that moving retinal images are essential. Prey, nonprey, and predator are not simply distinguished in terms of "small" or "big" (Eibl-Eibesfeldt 1951; Schneider 1954), but rather by an evaluation of moving configural features (Ewert 1968, 1969, 1984a). Both the release of *orienting* (the directed appetitive behavior, Fig. 10A) and *snapping* (the consummatory act, Fig. 10B) take advantage of the same features relating algorithm. Critical configural features are object expansion parallel to the direction of movement (xl_1), object expansion perpendicular to the direction of movement (xl_2), the relation between both ($xl_1 : xl_2$), and the stimulus area ($xl_1 * xl_2$). Within behaviorally relevant limits of size, $xl_1 > xl_2$ (for $xl_2 = 2.5$ mm)

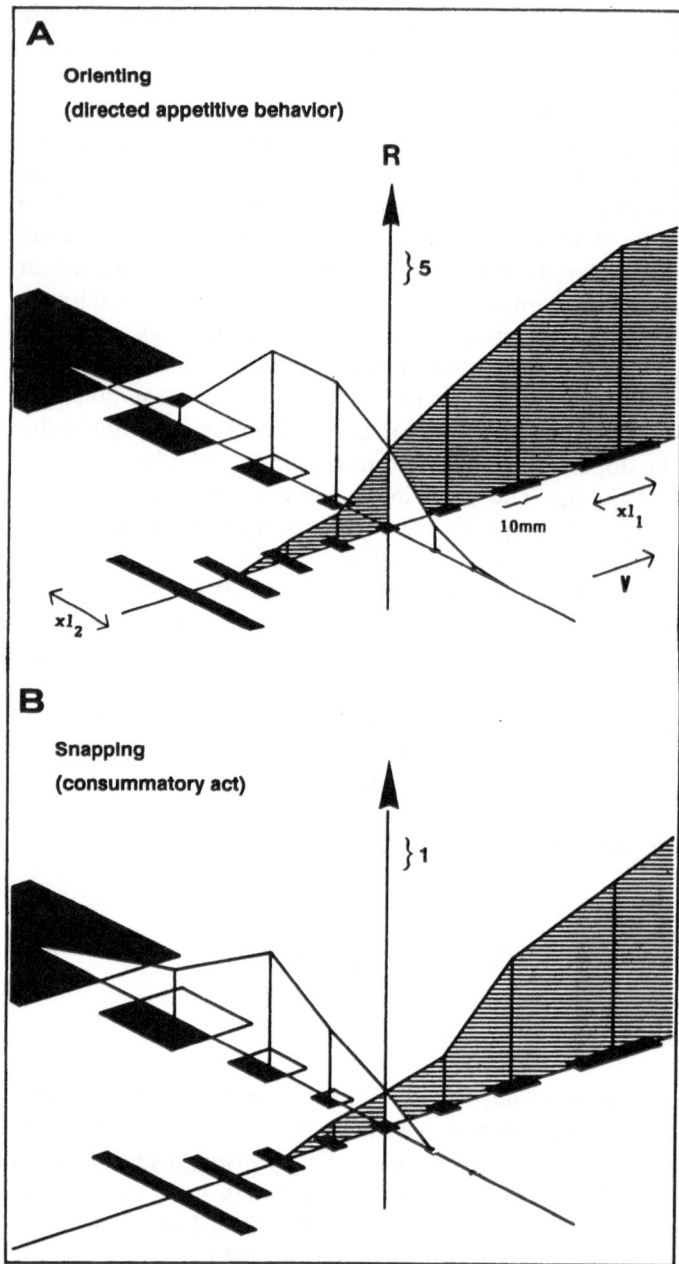

Fig. 10. Quantitative analysis of configural features of two-dimensional objects releasing **A** orientational turning and **B** snapping in common toads. During the experiment, the animal sat in a cylindrical glass vessel. The stimulus, a rectangular piece of black cardboard, was moved at an angular velocity of $v = 23°/s$ around the vessel at a distance $d = 7$ cm (**A**) or at $v = 10°/s$ and $d = 2$ cm (**B**). R refers to the average prey-catching activities: orienting responses per 60 s (**A**) and snapping responses per 30 s (**B**) from 20 toads; xl_1, edge length of the object

signals prey, $xl_1 < xl_2$ (for $xl_1 = 2.5$ mm) means nonprey (Fig. 10), whereas xl_1 ($= xl_2$) > 50 mm refers to predator. More specifically, if a small stripe of 2.5 mm width is moved in the direction of its longer axis ($xl_1 > xl_2$), we refer in laboratory jargon to "worm configuration" W, and if the longer axis of the same stripe is oriented perpendicular to the direction of movement ($xl_2 > xl_1$) to "antiworm configuration" A. The preference of W vs A in prey catching is maintained for variable velocity and directions of movement (Ewert et al. 1979a; Burghagen and Ewert 1983). The estimation of object size in terms of "size constancy" (Ewert and Gebauer 1973; Ingle 1976) requires depth measurements for which binocular vision is not essential (see also Collett 1977).

Since adult common toads, tested immediately after collection from the field, all show the same type of pattern discrimination as depicted in Fig. 10, it can be concluded that this ability is species-universal, by which we denote "universal" to the members of the same species *bufo*. The ability is based either on genetic dispositions that allow all individuals to obtain the same experiences, or on an innate property, as developmental studies suggest (Traud 1983). Generally, we can say that particularly an increased edge xl_2 of a moving object has a strong threatening effect and thus an inhibitory influence on prey-catching behavior across anuran amphibians (Ewert and Burghagen 1979b). It probably is a warning signal to the animal to "be cautious" (Ewert and Traud 1979).

4.2 Neurophysiological Correlates

Barlow (1953) and Lettvin et al. (1959) suggest that frog's retinal ganglion cells (Fig. 11) of the class R2 and R3 serve as "bug perceivers" and "fly detectors", respectively. But it could be shown that retina-filtered visual information – mediated by retinal ganglion cell classes – does not explain the feature discrimination expressed in prey catching of frogs (Grüsser and Grüsser-Cornehls 1968, 1970) or toads (Ewert and Hock 1972). The ultimate evidence against a retinal concept of prey recognition came from recordings in freely moving toads: Schürg-Pfeiffer (1989) demonstrated that retinal class R2 neurons were optimally activated if a small black stripe was moved nonprey-like in A-configuration (provided that the longer axis fits the diameter of the neuron's excitatory receptive field, ERF), while the probability of prey catching toward such a stimulus object was extremely low; if the same stripe traversed the ERF center prey-like in W-configuration, the R2 neuron's discharge frequency averaged somewhat less than to the A stimulus, but the toad readily responded with prey catching. If longer stripe stimuli were used, the inhibition provided by R2 neuron's inhibitory receptive field IRF, which surrounds the ERF (Fig. 11), was

---

Fig. 10. *Cont.*
parallel to the direction of movement; xl_2, edge length perpendicular to the direction of movement; $xl_1 > xl_2$ (for $xl_2 = 2.5$ mm): W(orm) configuration; $xl_2 > xl_1$ (for $xl_1 = 2.5$ mm): A(ntiworm) configuration; $xl_1 = xl_2$: S(quare) configuration. (**A** after Ewert 1969 and **B** from G. Herbst and J.-P. Ewert, unpubl.)

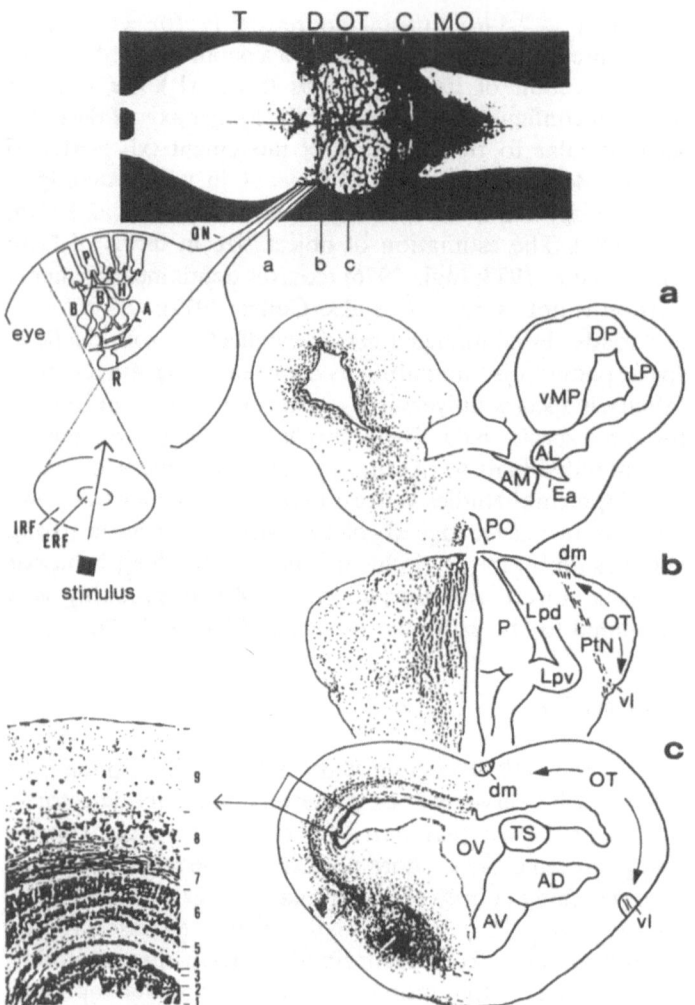

Fig. 11. Neuroanatomy of the toad's brain. *Top* Eye and brain. In the retina, one ganglion cell *R* is shown with preconnected amacrine cells *A*; bipolar cells *B*; horizontal cells *H*, and photoreceptors *P*; *ERF* excitatory receptive field of *R*; *IRF* inhibitory receptive field of *R*; the axon of the ganglion cell travels in the optic nerve *ON*. Brain regions: *T* telencephalon; *D* diencephalon; *OT* optic tectum; *C* cerebellum; *MO* medulla oblongata. *Below* Brain transverse sections at levels **a–c**; left hemispheres showing brain structures in histological Klüver-Barrera stains; right hemispheres indicating the nomenclature. **a** Telencephalon: *AL* lateral, *AM* medial nucleus amygdalae; *Ea* anterior entopeduncular nucleus; *LP* lateral, *DP* dorsal, *vMP* posterior ventromedial pallium; *PO* preoptic area (nomenclature by Kicliter and Ebbesson 1976). **b** Diencephalon: *PtN* pretectal neuropil, containing optic nerve terminals; *Lpd* lateral posterodorsal, *Lpv* lateral posteroventral, *P* posterocentral pretectal thalamic nucleus (nomenclature by Neary and Northcutt 1983). **c** Mesencephalon: *OT* optic tectum; *1–9* layers of tectal cell and fiber formations, *layer 9* containing optic nerve terminals (see Székely and Lázár 1976); *AD* anterodorsal, *AV* anteroventral tegmentum; *TS* torus semicircularis. Optic tracts: *dm* dorsomedial, *vl* ventrolateral optic tract; *OV* optic ventricle. (Ewert and IWF 1993).

not sufficient to explain the algorithm that underlies the toad's configural object discrimination for prey selection (see Ewert 1968, 1987). Nevertheless, the various retinal ganglion cell classes R1, R2, R3, and R4 express important preprocessing of visual input, e.g., drawing on the size of the ERF, the strength of the IRF, and the sensitivities to the contrast and the movement of a retinal image (for a review, see Grüsser and Grüsser-Cornehls 1976).

Our current investigations suggest that information related to the object-defining features $\{xl_1; xl_2\}$ is processed in a parallel/distributed interactive manner in mesencephalic tectal and diencephalic pretectal projection fields of the retina of the contralateral eye (Figs. 11, 12; Ewert and von Wietersheim 1974a; Tsai and Ewert 1988). Configurally, the feature xl_1 is evaluated by the excitation of tectal class T5.1 neurons, taking into account the area (xl_1*xl_2); the feature xl_2 is evaluated by excitation of pretectal thalamic class TH3 neurons, taking into account the area (xl_1*xl_2) and also responding to textured moving surfaces. These computations can be explained by tectal intrinsic mutual excitation (stimulating effect of xl_1) and by pretectal integration of R3- and R4-mediated retinal spatial summation (stimulating effect of xl_2). We suggest that the features relating algorithm takes advantage of pretectotectal inhibitory processes (Fig. 12). Such a subtractive network interaction is obviously expressed by another class of tectal neurons, called T5.2, whose activity in response to moving objects reflects the probability that the object configurally fits the prey schema (cf. Figs. 10 and 13Aa; Ewert 1974; Ewert and von Wietersheim 1974b; Ewert et al. 1974, 1979b; von Wietersheim and Ewert 1978; for network models, see Ewert and von Seelen 1974; Lara et al. 1982; Arbib 1987; Fingerling et al. 1993). The existence of T5.2 neurons has been confirmed in immobilized common toads (an der Heiden and Roth 1989), grass frogs (Schürg-Pfeiffer and Ewert 1981; Matsumoto et al. 1986), and in behaving toads (Schürg-Pfeiffer 1989; Schürg-Pfeiffer et al. 1993; Fig. 14). In freely moving toads, different from pharmacologically immobilized animals, prey-selective T5.2 neurons – in contrast to retinal ganglion cells – display a sensitivity to the real object size when objects are moved in the investigated range of 3.5 to 25 cm from the animal (Schürg-Pfeiffer et al. 1990).

Various stimulation, recording, and lesion methods provided evidence of pretecto-tectal inhibitory inputs: (1) visually elicited tectal field potentials are partly suppressed by electrical stimulation of the ipsilateral pretectum (Beneke et al. 1992). Pretectal projective TH3 cells were identified by means of the antidromic stimulation/recording technique (H. Buxbaum-Conradi, in Ewert et al. 1992). (2) Intracellular recordings from a tectal T5.2 neuron in response to electrical stimulation of the contralateral optic nerve show sequential EPSP/IPSP activity (Ewert et al. 1990a), whereas electrical stimulation in the ipsilateral pretectum elicits in the same tectal neuron predominantly an IPSP response at an appropriate shorter latency. (3) Intracellularly recorded T5.2 neurons (Fig. 13Ca, b) respond to a stripe moving in W-configuration mainly with EPSPs and spikes, but to the same stripe in A-configuration showing weak activity or mainly IPSPs. (4) After pretectal lesions, extracellularly recorded

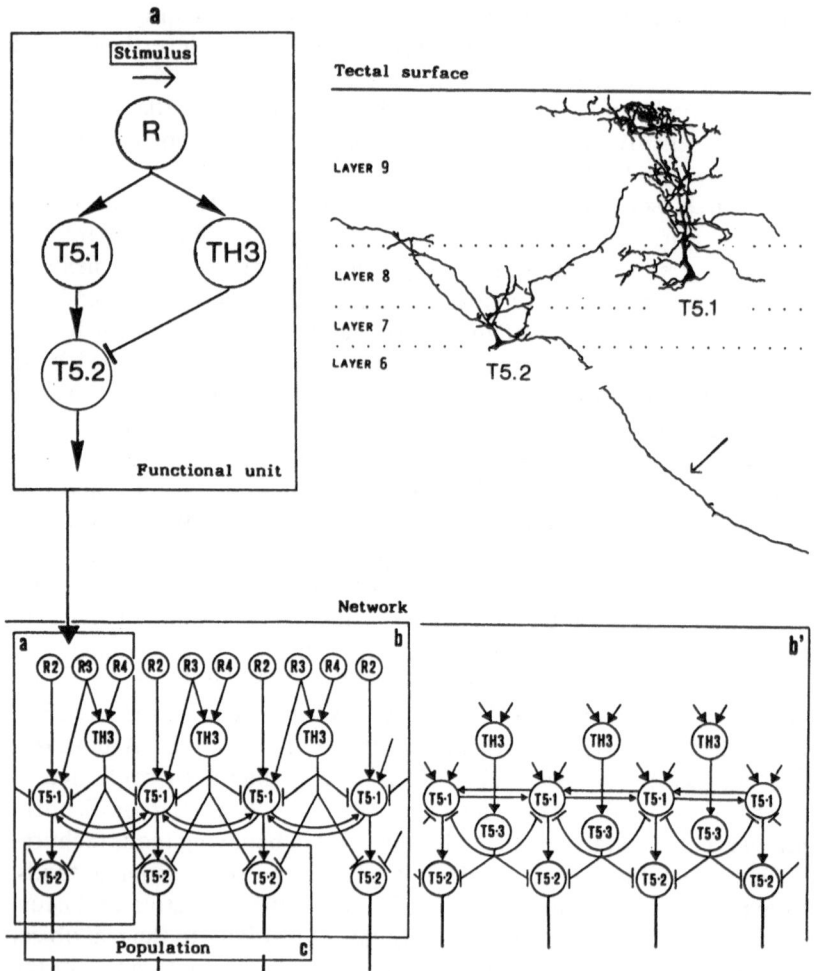

Fig. 12. Concept of a feature-analyzing retino-pretectal/tectal network. *Top left* In a functional unit (*a*) tectal T5.1 neurons and pretectal thalamic *TH3* neurons receive inputs of different combinations of retinal *R2* and *R3* ganglion cells; the outputs converge on a tectal T5.2 neuron which expresses in its response the feature relations (xl_1;xl_2) that determine the prey schema; *arrows* indicate excitatory and *lines with bars* inhibitory influences (after Ewert 1974). *Top right* Camera-lucida reconstructions of an intracellularly recorded and iontophoretically Co^{3+}-lysine labeled T5.1 pear cell and a T5.2 pyramidal cell; *arrow* points to the axon of the pyramidal cell projecting to the medulla (Matsumoto et al. 1986); *TH3* cells are located in the *Lpd* and the lateral *P* nucleus (see Fig. 11b). *Below* The *functional unit* (*a*) is integrated in a pretectal/tectal *network* (*b*) whose output is mediated by a *population* (*c*) of T5.2 neurons. ERF diameters of the monocularly driven neurons: *R* [4–6°], *R3* [8–10°], T5.1, T5.2 [27°], *T5.3*, *TH3* [35–45°]. The lateral excitatory and inhibitory connections are not restricted to immediately adjacent neurons. Pretectal influences inhibit T5-type neurons at two levels, hence T5.1 and T5.2, respectively. The network *b'* considers tectal interneurons *T5.3* which in response to excitatory *TH3* input may inhibit T5.1 and T5.2 correspondingly; networks *b* and *b'* are suggested to be integrated. (Ewert 1987)

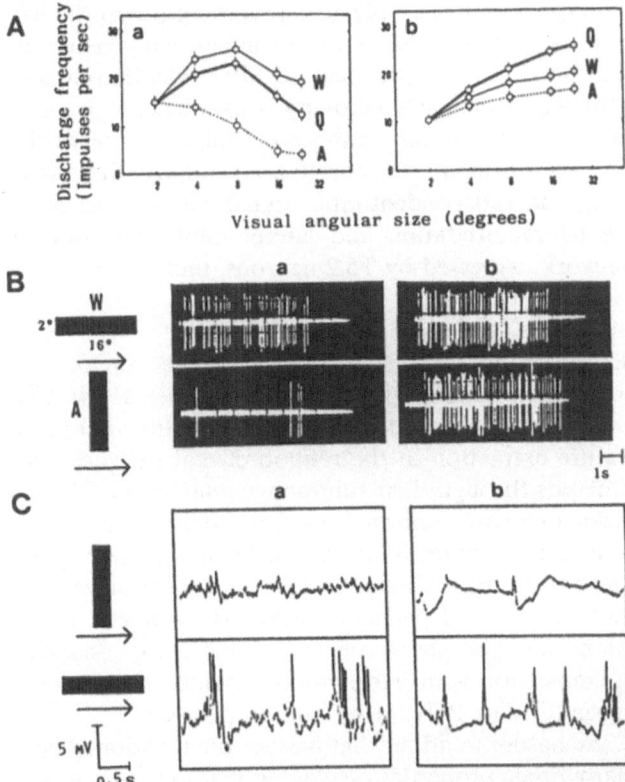

Visual angular size (degrees)

Fig. 13. Aa Discharge activities of T5.2 neurons (n = 20) extracellularly recorded between tectal layers 6 and 8 from immobilized common toads in response to black worm (*W*)-, antiworm (*A*)-, or square (*Q*)-configural moving objects corresponding to Fig. 10; v = 7.6°/s (Ewert and von Wietersheim 1974a). **b** Responses of T5 neurons (n = 20) of the same tectal layer in toads after an ipsilateral pretectal lesion (Ewert and von Wietersheim 1974b). **B** Extracellular responses of a toad's T5.2 neuron toward the *W*- or *A*-configuration of a black moving stripe **a** before and **b** after a pretectal lesion applied ipsilaterally to the recording site (T. Finkenstädt, in Ewert 1987). **C** Intracellularly recorded T5.2 cells (**a, b**) of the grass frog in response to the *W*- or *A*-configuration of a black moving stripe; *arrows* indicate the direction of movement; v = const. (Matsumoto et al. 1986)

tectal neurons lose their configural object discrimination (Fig. 13Ab), and the animal's prey recognition becomes so to speak agnostic, i.e., prey, predator, or moving background structures are all responded to with prey catching (Ewert 1968; Ewert and von Wietersheim 1974b; Schürg-Pfeiffer 1989). (5) Extracellular recordings from a T5.2 neuron before (Fig. 13Ba) and after an ipsilateral pretectal lesion (Fig. 13Bb) demonstrate that postlesion the neuron's firing in response to moving visual objects is increased ("disinhibited") and the configural *W* vs *A* discrimination almost completely abolished.

Regarding the role of neuron types for preception (Barlow 1985), our hypothesis suggests that T5.2 neurons – integrated in a neural network (Fig. 12) – fulfill the function of a prey-feature filter which compares information regarding configural features of a moving visual object with stored information inherent in the property of the retino-pretectal/tectal network. Such a comparison is possible through cross-correlation: suppose that $y(r, s, t)$ is the space (r, s)- and temporal (t)-dependent input signal, $H(r, s, t)$ the coupling function of neurons via lateral excitation and lateral inhibition, and $z(r, s, t)$ the output of the network expressed by T5.2 neurons, then

$$F[z(r, s, t)] = F[H(r, s, t)]\{F[y(r, s, t)]\}$$

describes the Fourier transform of the convolution integral of the cross-correlation function (Ewert and von Seelen 1974). Given that an operation H_1 characterizes the feature extraction in the retinotectal network and H_2 the feature extraction in the retinopretectal network, then feature discrimination proceeds through their subtractive interaction. Thus, the above-described filter picks out prey information from the visual environment, weights it, and suppresses nonprey information. An appropriate tuning of $\{H_1, H_2\}$ and weighting of the subtractive interaction in connection with threshold operations could explain species-dependent variations in object discrimination. Such computations may also play a role in the object size-constancy phenomenon, probably in connection with information provided by the lense accommodation mechanism (Collett 1977) or by motion parallax.

We hasten to admit that besides the mentioned classes TH3, T5.1, and T5.2, many other properties displaying neurons exist (e.g., see Figs. 16, 17; Grüsser and Grüsser-Cornehls 1970; Ewert 1971, 1984a, 1987). For example, there are cells sensitive to large moving (classes TH4, T5.4), looming (TH6) objects, or those (classes T1, T3) capable of mediating information about depth. There are also T4 neurons whose ERFs encompass the entire field of vision, so to speak for a "general view".

The retino-pretectal/tectal network described so far (Fig. 12) is integrated in a macro-network. Intracellular recording studies have shown that TH3 cells of the pretectal Lpd/P region receive inhibitory (and excitatory) inputs from the ipsilateral ventrocaudal striatum (Matsumoto et al. 1991). Such a pathway is suitable (1) for gating a visual response to prey in a disinhibitory manner using two inhibitory projections, a striatopretectal and a pretectotectal one, and (2) for keeping tectal reverberatory excitatory processes within behaviorally relevant limits, a prerequisite for object recognition. Furthermore, recent learning studies in toads suggest loop-operated processes by which the properties of feature-sensitive/selective cells may be modulated (modified) through pretectal and/or hypothalamic influences involving various telencephalic structures (Finkenstädt and Ewert 1988a, b; Ewert 1991, 1992; Merkel-Harff and Ewert 1991; Ewert et al. 1992; Dinges and Ewert 1993). In this context, the ventromedial pallium (Fig. 11, brain section a) – the homologous structure of the mammalian hippocampus (Herrick 1933) – seems to play a prominent role.

4.3 Premotor Properties of Tectal Neurons

In freely moving toads, tectal T5.2 neurons have been chronically recorded whose ERFs were located laterally in the ventrohorizontal visual field (Schürg-Pfeiffer 1989; Schürg-Pfeiffer et al. 1993). When a small stripe traversed the ERF in W-configuration, a strong increase in the discharge rate preceded and predicted an orienting response toward the stimulus. This warming up increase in the firing rate was only displayed when the toad was responsive to prey behaviorally (cf. Fig. 14 A, B). Presented in A-configuration, the stripe elicited a weak neuronal response and no prey catching. If the frontal ERF of a T5.2 neuron was traversed by the prey stimulus at snapping distance (Fig. 14C), here too, a premotor warming up introduced snapping; however, the firing immediately ceased during snapping. After snapping, the neuron sometimes showed a short, weak rebound activity. Strong warming up firing was only displayed when the stimulus elicited snapping; spontaneous prey-catching movements or

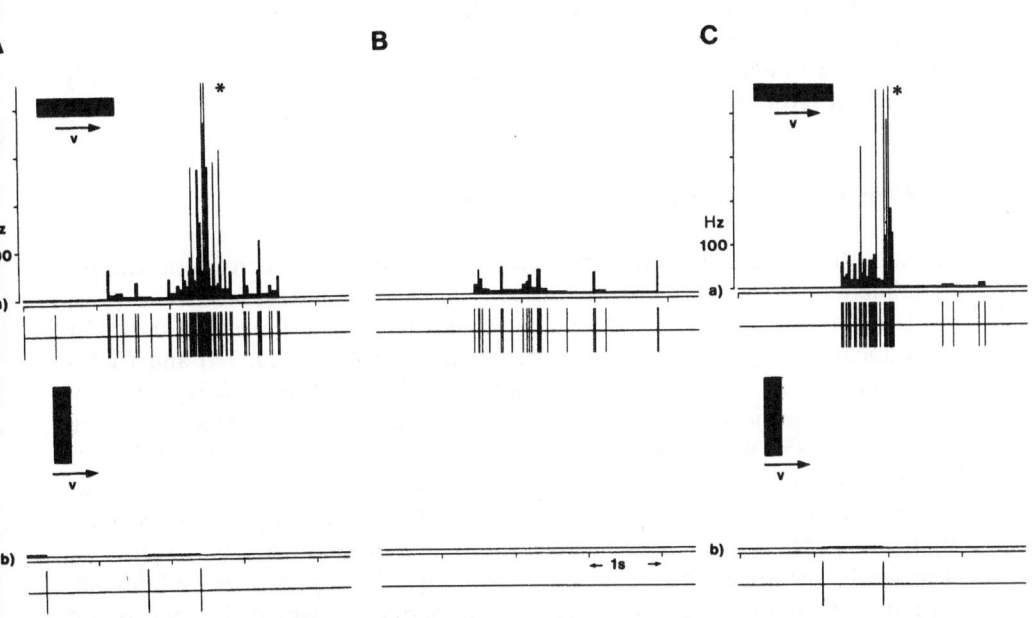

Fig. 14. Premotor and state-dependent properties of prey-selective tectal T5.2 neurons chronically recorded in freely moving common toads. **A** Interspike frequency time histogram (*upper trace*) calculated from the discharges (*lower trace*) of a representative tectal T5.2 neuron in response to a 2×20 mm² stripe traversing the laterally located ERF at 10°/s either *a* worm-like or *b* in antiworm configuration; a strong increase in the neuronal activity preceded the onset of orienting (indicated by the *asterisk*) toward the prey stimulus; the neuron was active during the turning movement. **B** Activity of the same T5.2 neuron to the stimuli while the toad was unresponsive to prey. **C** Activity of another T5.2 neuron whose ERF – located in the toad's frontal visual field – was traversed by the stimulus at snapping distance; strong increase in the neuronal activity preceded the onset of snapping (indicated by an *asterisk*); during snapping, the neuron was silent. (Schürg-Pfeiffer et al. 1993)

responses to prey outside the ERF were not introduced by any activity of the respective T5.2 neuron under investigation.

In the deep tectal and subtectal layers the activity of two types of tonically discharging T8 neurons was correlated with any kind of behavioral movement (Borchers 1982; Megela et al. 1983; Schürg-Pfeiffer et al. 1993). *Premotor activity* of T8.1 neurons: they increased their firing before a visual stimulus elicited orienting, snapping, avoiding, or another visual behavioral response; but also any kind of spontaneous movement (e.g., of head, trunk, or legs) was introduced by an increasing spike activity, even when the toad's environment was absolutely dark. *Premotor inhibition* of T8.2 neurons: they stopped their firing before a visual stimulus elicited a behavioral response; but also any spontaneous movement was introduced by a firing pause.

There are various explanations for T5.2 neuron's premotor warming up. For example, it could be the expression of a striato-pretecto-tectal disinhibitory function in order to gate a response to prey (Ewert et al. 1991). Alternatively, or in addition, T5.2 prey-selective cells might be optimally activated only if activity of T8.1 movement cells overlaps in time with their visual responses (cf. Fig. 17). The latter is supported by the observation that at times of spontaneously fluctuating T8.1 activity, visual behavioral responses can be elicited more readily at a certain level of firing.

4.4 Combinatorial Control of MPGs

In toads, the concept of a command releasing system (CRS) is consistent with the experiments in which electrostimulation of small tectal areas elicits prey-catching patterns (Fig. 7C). Tecto-bulbar/spinal projections have been shown by anatomical tracing studies using horseradish peroxidase (Fig. 7A) and Co^{3+}-lysine (Tóth et al. 1985). Furthermore, we know that prey signal-carrying tecto-bulbar/spinal-projecting T5.2 cells exist (Figs. 12 top right and 15; see also Saton and Ewert 1985). We hence suggest that T5.2 neurons in various combinations with cells sensitive to local properties of an object (e.g., T1, T3) yield CRSs for prey-catching MPGs (cf. Figs. 1C and 16). The notion of state-dependent modulatory inputs to CRS shows a neurophysiological correlate in the fact that chronically recorded T5.2 neurons display significantly weaker (subliminal) activity in response to a prey object during periods in which the toad is not ready (motivated) to feed (Fig. 14, cf. B and A).

In and adjacent to the motor nuclei for head, jaw, and tongue movements, we have recorded various types of medullary (M-) neurons which receive different inputs, among these are T5.2 property-displaying M5.2 cells. The integrative character of such cells suggested from intracellular recording/labeling studies is consistent with their extensive dendritic trees and stimulus-dependent EPSP/IPSP activities suitable for "perceptual sharpening". The branches of one cell may spread toward medullary regions (Figs. 4C and 6), so that adjacent neurons literally form a "reticular" network. Some of these fulfill criteria of

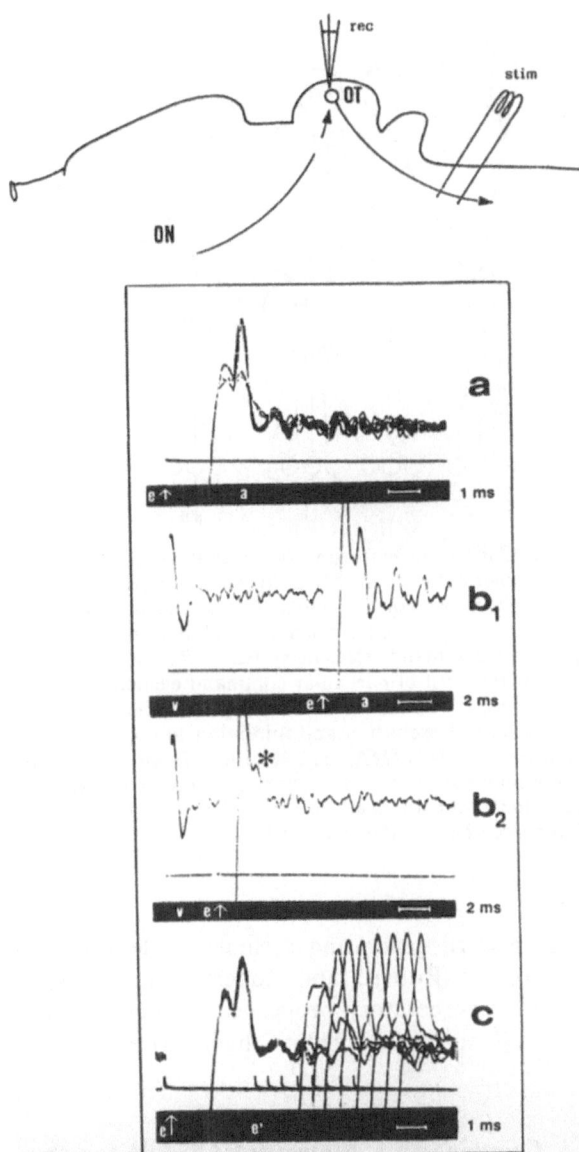

Fig. 15. Evidence of the tectomedullary projective character of a T5.2 neuron in the common toad. **a** An antidromic spike *a* was recorded (*rec*) extracellularly from the top of tectal layer 6 in response to a 0.1-ms electrical impulse *e* (see *vertical arrow*) applied bipolarly to the tecto-bulbar tract tr.tbs.c. (*stim*) in the contralateral medulla oblongata at the level of the VIIth nucleus. Constant-latency response of the antidromic spike *a* at L = 2.3 ms (several super-imposed traces); two traces are just below response threshold to determine the all-or-nothing reaction. $\mathbf{b}_{1,2}$ Collision test in which a visually elicited spike *v* triggered the electrical impulse *e* that evoked an antidromic spike *a* at a spike (*v*)-stimulus (*e*) delay of 9.5 ms ($\mathbf{b}_1$); the antidromic spike was extinguished ($\mathbf{b}_2$) at a critical delay, D = 3.3 ms, that meets the collision criterion (E-D)/E < 0.5; expected delay E = L + A. **c** The following ability and the absolute refractory period (A = 3.2 ms) were measured with electrical double impulses *e-e'* of variable intervals (eight superimposed traces); *ON* optic nerve; *OT* optic tectum. (Ewert et al. 1990b)

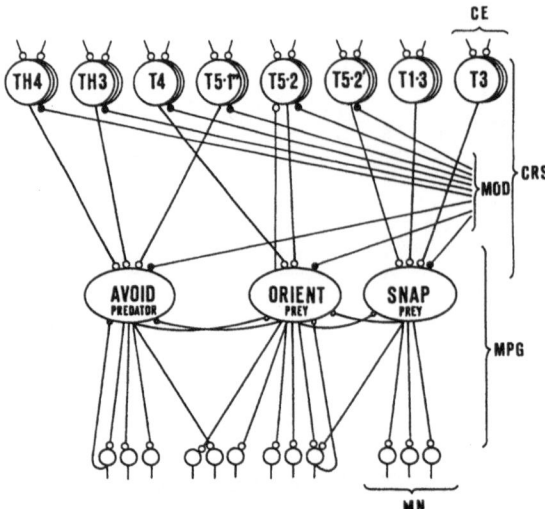

Fig. 16. Concept of sensorimotor codes in command-releasing systems. Triggering a motor pattern-generating circuit (*MPG*) for an action pattern (e.g., orient toward prey, snap at prey, or avoid predator) requires inputs of a certain combination of specified neurons functioning as command elements (*CE*). Such a combination provides the sensorimotor code for the appropriate command-releasing system (*CRS*). Modulatory influences (*MOD*) probably enter a CRS at the level of particular command elements and the MPG. Upon activation by the CRS, a specific "routine" is called into play in the appropriate MPG to produce a spatiotemporal pattern of excitation and inhibition to coordinate the activities in the corresponding motoneuronal pools (*MN*) e.g., see Fig. 17. Among the various CRSs, and MPGs, certain neurons may be shared; a *circle* stands for a set of neurons belonging to the same class, an *oval symbol* for a circuit; *small circles* indicate excitatory or inhibitory inputs which are not differentiated here. (After Ewert 1987)

motor pattern generating systems as described in other vertebrates and in invertebrates (Roberts and Roberts 1983), such as tonically discharging M8 neurons, reverberatory properties displaying M9 neurons, and bursting M10 neurons (Ewert et al. 1990b; Schwippert et al. 1990).

5 A Concept of a Neuronal Circuit for the Tongue Flip

Drawing on the neurophysiological data regarding *premotor warming up* of prey-selective T5.2 cells, *premotor activity* of T8.1 movement cells, *snap-correlated inhibition* of T5.2, *premotor inhibition* of T8.2, *perceptual sharpening* of medullary prey-selective M5.2 cells, and *resetting inhibition* of tongue muscle motoneurons, Fig. 17A proposes a neuronal circuit for the tongue flip. This circuit is suggested to coordinate the genioglossus, hyoglossus, and hyoid muscle contractions involved during tongue projection/retraction elicited by a

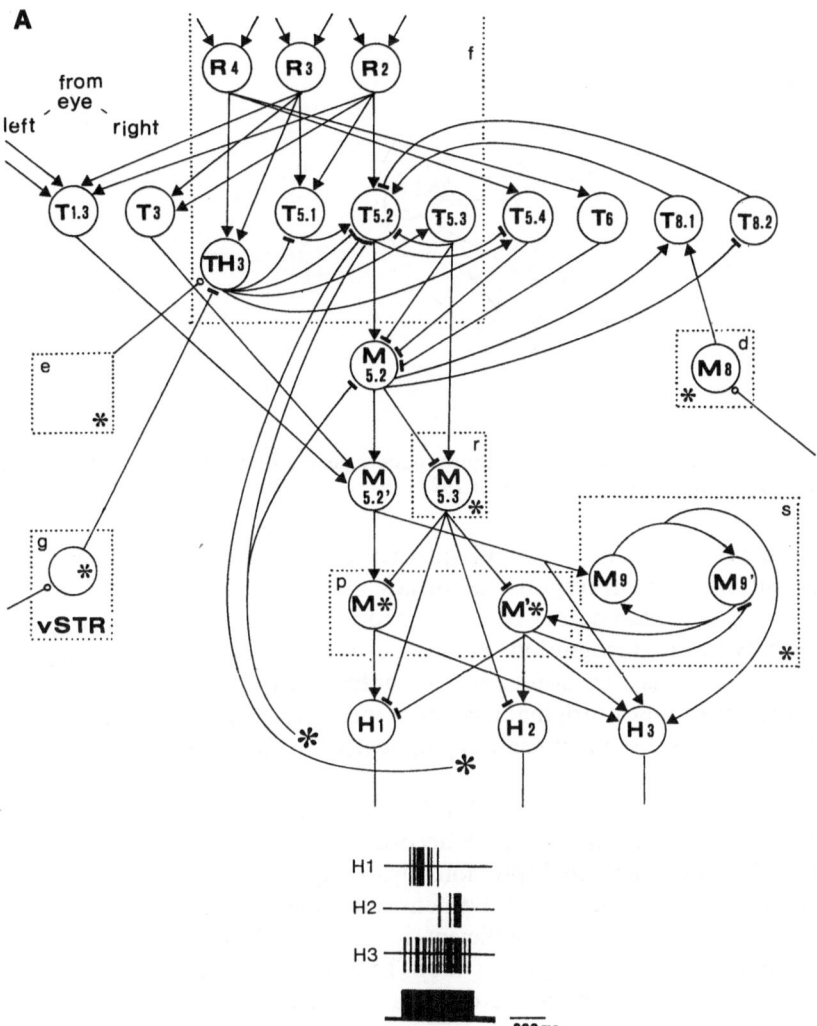

Fig. 17. A Concept of a neuronal circuit (cf. Fig. 8B) that controls the toad's tongue flip in response to a visual prey stimulus (see also Fig. 16). *Dotted boxes* refer to integrated functional systems related to prey feature detection (*f*), depth estimation (*e*), priming/driving (*d*), gating (*g*), sustaining (*s*), premotoneuronal controlling, (*p*), and resetting (*r*); *circles* stand for sets of neurons; *arrows* indicate excitatory, *lines with bars* inhibitory, and *lines with small circle* not yet differentiated influences; *asterisks inside a circle* mark neurons not yet recorded, *outside the circles* hypothetical systems. *Bottom* Representative records of H-type neurons during snapping (*black bar*) from lunging through backmoving to the starting position.

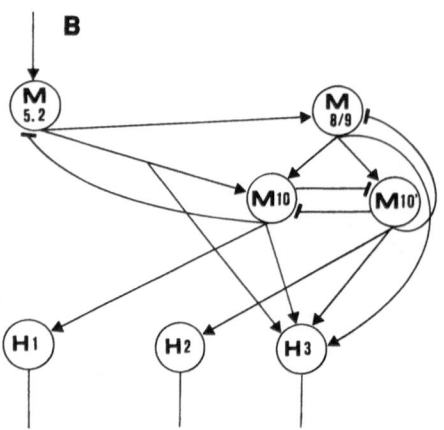

Fig. 17. B Alternative circuit for motor pattern generation. For further explanations see text. (Schürg-Pfeiffer, Spreckelsen, Ewert 1993.)

visual prey stimulus. The circuit interacts through internal feedback loops with a circuit (not shown) responsible for jaw muscle contraction.

In Fig. 17A, the dotted boxes refer to various integrated systems responsible for prey feature recognition (f), depth estimation (e), priming/driving (d), gating (g), sustaining (s), premotoneuronal controlling (p), and resetting (r). The traces at the bottom show original records of three hypoglossal H-type neurons during lunging and subsequent snapping indicated by the black bar (Schürg-Pfeiffer et al. 1993).

More specifically, the R2, R3, R4, TH3, T5.3, and T5.2 neurons belong to the prey feature-analyzing network according to Fig. 12. Prey-selective T5.2 cells are adequately activated (premotor warming up of T5.2), if appropriate visual input and input from T8.1 movement cells (premotor activity of T8.1) converge; then the T5.2 output is sufficient to drive medullary M5.2 cells which are components of an integrating loop T5.2-M5.2-T8.1-T5.2. The output of this loop, cooperatively with the outputs of space monitoring T1.3 and T3 cells (command releasing system), triggers medullary M5.2′ cells which, via M* premotoneurons, activate genioglossus (tongue protractor) muscle motoneurons, suggested to be H1. Shortly before H1 cells discharge, M5.2′ collateral output activates hyoid (hyoid stabilizing) muscle motoneurons, suggested to be H3; collateral output of M* keeps H3 cells further firing. Feedback from the motor system (premotoneurons or muscle) inhibits M5.2′ and T5.2 (snap-correlated inhibition of T5.2). Further collateral output of M5.2′ feeds into an integrating medullary loop involving M9-type neurons which, via M′* premotoneurons, activate hyoglossus (tongue retractor) muscle motoneurons, suggested to be H2. The collateral outputs of M* and M′* increase the activity of H3, inhibit H1, and terminate at a delay first the firing of H2 and then of H3 by inhibition of M9′ (Fig. 17B shows an alternative coordination principle). It is further suggested that T8.1 neurons obtain inputs from M8 neurons whose tonic activity fluctuates depending on (attentional, arousal) states. In parallel to the integrating T5.2-M5.2-T8.1-T5.2 loop there may be a disinhibitory loop M5.2-T8.2-T5.2 (premotor inhibition of T8.2). Also, a striato-pretecto-tectal vSTR-TH3-T5.2 disinhibitory pathway should be considered (response gating of T5.2). Furthermore, it is suggested that T5.3 interneurons and predator-sensitive T5.4 and T6 cells deliver inhibitory inputs to M5.2 (perceptual sharpening). There may also be inhibitory influences from T5.3, mediated via M5.3 to premoto- and motoneurons (resetting property).

Another, computer-simulated model is presented by Liaw et al. (1993). Of course, the precise organization and coordination of different motor pattern-coordinating systems, the sharing of components, and the triggering of these systems by neurons collectively transmitting "sensorimotor codes" remain current topics for further experimental research in the dialogue with models of brain function (Arbib and Ewert 1991).

Acknowledgments. This paper is dedicated to Dr. Hasso Kuczka on the occasion of his 65th birthday. The work is supported by the Bundesministerium für Forschung und Technologie (BMFT), Deutsche Forschungsanstalt für Luft- und Raumfahrt e.V., Projektträger des BMFT für Informationstechnik (Neuroinformatik), Verbundprojekt "SEKON" (Sensomotoric Co-ordination of Robotic Movements with Neuronal Nets) No. 413-5839-01 IN 104 C/4 (Ewert).

References

Akert K (1949) Der visuelle Greifreflex. Helv Physiol Pharmacol Acta 7: 112–134

Amri M, Car A (1988) Projections from the medullary swallowing center to the hypoglossal motor nucleus: a neuroanatomical and electrophysiological study in sheep. Brain Res 441: 119–126

Amri M, Lamkadem M, Car A (1989) Activity of extrinsic tongue muscles during swallowing in sheep. Brain Res 503: 141–143

Amri M, Car A, Roman C (1990) Axonal branching of medullary swallowing neurons projecting on the trigeminal and hypoglossal motor nuclei: demonstration by electrophysio-logical and fluorescent double labeling techniques. Exp Brain Res 81: 384–390

Arbib MA (1987) Levels modeling of mechanisms of visually guided behavior. Behav Brain Sci 10: 407–465

Arbib MA, Ewert J-P (eds) (1991) Visual structures and integrated functions. Research notes in neural computing 3. Springer, Berlin Heidelberg New York

Barlow HB (1953) Summation and inhibition in the frog's retina. J Physiol (Lond) 173: 377–407

Barlow HB (1985) The twelfth Barlett memorial lecture: the role of single neurons in the psychology of perception. Q J Exp Biol 37(A): 121–145

Beneke TW, Schwippert WW, Ewert J-P (1992) Pretectal inputs to the optic tectum of toads – changes of field potentials. Verh Dtsch Zool Ges (short comm) 85: 57

Bloom FE (1979) Chemical integrative processes in the central nervous system. In: Schmidt FO, Worden FG (eds) The neurosciences: fourth study program. MIT Press, Cambridge

Borchers H-W (1982) Correlation between behavior patterns and single unit responses from the optic tectum in the freely moving toad (*Bufo bufo* L.). In: Trappl R, Pask G, Ricciardi L (eds) Progress in cybernetics and systems research, vol 9. Hemisphere Publ Corp, Washington, pp 109–117

Braak H (1970) Biogene Amine im Gehirn vom Frosch (Rana esculenta). Z Zellforsch 106: 269–308

Brookhardt JM, Fadiga E (1960) Potential field initiated during monosynaptic activation of frog motoneurons. J Physiol 150: 633–655

Burghagen H (1979) Der Einfluß von figuralen, visuellen Mustern auf das Beutefangverhalten verschiedener Anuren. PhD Thesis, Univ of Kassel

Burghagen H, Ewert J-P (1983) Influence of the background for discriminating object motion from self-induced motion in toads Bufo bufo (L.). J Comp Physiol 152: 241–249

Cajal SR y (1909) Histologie du systeme nerveux de l'homme et des vertebres. Maloine, Paris

Capranica RR (1983) Sensory processing of key stimuli. In: Ewert J-P, Capranica RR, Ingle DJ (eds) Advances of vertebrate neuroethology. Plenum Press, New York, pp 3–6

Car A, Amri M (1982) Etude des neurones déglutileurs pontiques chez la brebis. I. Activité et localisation. Exp Brain Res 48: 345–354

Cardona A, Rudomin P (1983) Activation of brain stem serotonergic pathways decreases homosynaptic depression of monosynaptic responses of frog spinal motoneurons. Brain Res 280: 373–378

Chandler SH, Tal M (1986) The effects of brain stem transections on the neuronal networks responsible for rhythmical jaw muscle activity in the guinea pig. J Neurosci 6: 1831–1842

Clairambault P (1976) Development of the prosencephalon. In: Llinás R, Precht W (eds) Frog neurobiology. Springer, Berlin Heidelberg New York, pp 924–945

Cohen AH, Rossignol S, Grillner S (eds) (1988) Neural control of rhythmic movements in vertebrates. Wiley, New York

Collett TS (1977) Stereopsis in toads. Nature 267: 349–351

Comer CM (1987) Sensorimotor functions: what is a command, that a code may yield it? A commentary. Behav Brain Sci 10: 372

Davis WJ, Kovac MP (1981) The command neuron and the organization of movement. TINS 4: 73–76

Dhawan B, Saxena PN, Gupta GP (1961) Apomorphine-induced pecking in pigeons. Br J Pharmacol 15: 285–295

DiDomenico R, Eaton RC (1987) Toward a reformulation of the command concept. A commentary. Behav Brain Sci 10: 374–375

Dinges AW, Ewert J-P (1993) Interocular transfer of visual associative memory in toads Bufo bufo spinosus. Naturwissenschaften 80: 285–286

Dole JW, Rose RB, Tachiki KH (1981) Western toads (Bufo boreas) learn odor of prey insects. Herpetologia 37: 63–68

Doty RW (1968) Neural organization of deglutition. In: Field J, Magoun HW, Hall VE (eds) Handbook of physiology, sect VI, vol IV. Alimentary canal. Am Physiol Soc, Washington, pp 1861–1902

Doty RW (1976) The concept of neural centers. In: Fentress JC (ed) Simpler networks and behavior. Sinauer Assoc, Sunderland, MA, pp 251–265

Doty RW, Bosma JF (1956) An electromyographic analysis of reflex deglutition. J Neurophysiol 19: 44–60

Doty RW, Richmond WH, Storey AT (1967) Effect of medullary lesions on coordination of deglutition. Exp Neurol 17: 91–106

Dubner R, Sessle BJ, Storey A (1978) The neural basis of oral and facial function. Plenum Press, New York

Ebbesson SOE (1984) Evolution and ontogeny of neural circuits. Behav Brain Sci 7: 321–366

Edinger L (1908) Vorlesungen über den Bau der nervösen Zentralorgane des Menschen und der Tiere, vol 2. Vogel, Leipzig

Eibl-Eibesfeldt I (1951) Nahrungserwerb und Beuteschema der Erdkröte (Bufo bufo L.). Behaviour 4: 1–35

Eikmanns K-H (1955) Verhaltensphysiologische Untersuchungen über den Beutefang und das Bewegungssehen der Erdkröte (Bufo bufo L.). Z Tierpsychol 12: 229–253

Ewert J-P (1967) Aktivierung der Verhaltensfolge beim Beutefang der Erdkröte (Bufo bufo L.) durch elektrische Mittelhirnreizung. Z Vergl Physiol 54: 455–481

Ewert J-P (1968) Der Einfluß von Zwischenhirndefekten auf die Visuomotorik im Beute- und Fluchtverhalten der Erdkröte (Bufo bufo L.). Z Vergl Physiol 61: 41–70

Ewert J-P (1969) Quantitative Analyse von Reiz-Reaktions-Beziehungen bei visuellem Auslösen der Beutefang-Wendereaktion der Erdkröte (Bufo bufo L.). Pflügers Arch Ges Physiol Menschen Tiere 308: 225–243

Ewert J-P (1971) Single unit response of the toad (Bufo americanus) caudal thalamus to visual objects. Z Vergl Physiol 74: 81–102

Ewert J-P (1974) The neural basis of visually guided behavior. Sci Am 230: 34–42

Ewert J-P (1984a) Tectal mechanisms that underlie prey-catching and avoidance behaviors in toads. In: Vanegas H (ed) Comparative neurology of the optic tectum. Plenum Press, New York, pp 247–416

Ewert J-P (1984b) Behavioral selectivity based on thalamo-tectal interactions: ontogenetic and phylogenetic aspects in amphibians. A commentary. Behav Brain Sci 7: 337–338

Ewert J-P (1985) The Niko Tinbergen lecture 1983: concepts in vertebrate neuroethology. Anim Behav 33: 1–29

Ewert J-P (1987) Neuroethology of releasing mechanisms: prey-catching in toads. Behav Brain Sci 10: 337–405

Ewert J-P (1991) A prospectus for the fruitful interaction between neuroethology and neural engineering. In: Arbib MA, Ewert J-P (eds) Visual structures and integrated functions. Research notes in neural computing 3. Springer, Berlin Heidelberg New York, pp 31–56

Ewert J-P (1992) Neuroethology of an object features relating algorithm and its modification by learning. Rev Neurosci 3: 45–63

Ewert J-P, Burghagen H (1979a) Configurational prey selection by Bufo, Alytes, Bombina, and Hyla. Brain Behav Evol 16: 157–175

Ewert J-P, Burghagen H (1979b) Ontogenetic aspects on visual 'size-constancy' phenomena in the midwife toad Alytes obstetricans (Laur.). Brain Behav Evol 16: 99–112

Ewert J-P, Gebauer L (1973) Größenkonstanzphänomene im Beutefangverhalten der Erdkröte (Bufo bufo L.). J Comp Physiol 85: 303–315

Ewert J-P, Hock FJ (1972) Movement sensitive neurones in the toad's retina. Exp Brain Res 16: 41–59

Ewert J-P, IWF (1993) Image processing in the visual system of the toad: behavior – brain function – artificial neuronal net. Institut für den Wissenschaftlichen Film (IWF), Göttingen

Ewert J-P, Traud R (1979) Releasing stimuli for antipredator behaviour in the common toad Bufo bufo (L.). Behaviour 68: 170–180

Ewert J-P, von Seelen W (1974) Neurobiologie und System-Theorie eines visuellen Muster-Erkennungsmechanismus bei Kröten. Kybernetik 14: 167–183

Ewert J-P, von Wietersheim A (1974a) Musterauswertung durch tectale und thalamus/praetectale Nervennetze im visuellen System der Kröte (Bufo bufo L.). J Comp Physiol 92: 131–148

Ewert J-P, von Wietersheim A (1974b) Der Einfluß von Thalamus/Praetectum-Defekten auf die Antwort von Tectum-Neuronen gegenüber bewegten visuellen Mustern bei der Kröte (Bufo bufo L.). J Comp Physiol 92: 149–160

Ewert J-P, Weerasuriya A (1988) The motor pattern generator for snapping in toads. Soc Neurosci Abstr 14: 312

Ewert J-P, Hock FJ, von Wietersheim A (1974) Thalamus/Praetectum/Tectum: retinale Topographie und physiologische Interaktionen bei der Kröte (Bufo bufo L.). J Comp Physiol 92: 343–356

Ewert J-P, Arend B, Becker V, Borchers H-W (1979a) Invariants in configurational prey selection by Bufo bufo (L.). Brain Behav Evol 16: 38–51

Ewert J-P, Borchers H-W, Wietersheim A von (1979b) Directional sensitivity, invariance, and variability of tectal T5 neurons in response to moving configural stimuli in the toad Bufo bufo (L). J Comp Physiol 132: 191–201

Ewert J-P, Burghagen H, Schürg-Pfeiffer E (1983) Neuroethological analysis of the innate releasing mechanism for prey-catching behavior in toads. In: Ewert J-P, Capranica RR, Ingle DJ (eds) Advances in vertebrate neuroethology. Plenum Press, New York, pp 413–475

Ewert J-P, Schürg-Pfeiffer E, Weerasuriya A (1984) Neurophysiological data regarding motor pattern generation in the medulla oblongata of toads. Naturwissenschaften 71: 590–591

Ewert J-P, Schwippert WW, Beneke TW (1990a) Parallel distributed processing of configural moving objects in the toad's visual system. In: Eckmiller R, Hartmann G, Hauske G (eds)

Parallel processing in neural systems and computers. North-Holland, Amsterdam, pp 109–112

Ewert J-P, Framing EM, Schürg-Pfeiffer E, Weerasuriya A (1990b) Responses of medullary neurons to moving visual stimuli in the common toad. I. Characterization of medial reticular neurons by extracellular recording. J Comp Physiol A 167: 495–508

Ewert J-P, Matsumoto N, Schwippert WW, Beneke TW (1991) Striato-pretecto-tectal connections: a substrate for arousing the toad's response to prey. In: Arbib MA, Ewert J-P (eds) Visual structures and integrated functions. Research notes in neural computing 2. Springer, Berlin Heidelberg New York

Ewert J-P, Beneke TW, Buxbaum-Conradi H, Dinges AW, Fingerling S, Glagow M, Schürg-Pfeiffer E, Schwippert WW (1992) Adapted and adaptive properties in neural networks for visual pattern discrimination: a neurobiological analysis toward neural engineering. Adaptive Behav 1: 123–154

Fentress JC (1983) The analysis of behavioral networks. In: Ewert J-P, Capranica RR, Ingle DJ (eds) Advances in vertebrate neuroethology. Plenum Press, New York, pp 939–968

Fingerling S, Ewert J-P, Menzel R, Pfeiffer F (1993) From the toad to a robot: implementation of neurobiological principles of object discrimination in neural engineering. Naturwissenschaften 80: 321–324

Finkenstädt T, Ewert J-P (1988a) Stimulus-specific long-term habituation of visually guided orientation behavior toward prey in toads: a ^{14}C-2DG study. J Comp Physiol A 163: 1–11

Finkenstädt T, Ewert J-P (1988b) Effects of visual associative conditioning on behavior and cerebral metabolic activity in toads. Naturwissenschaften 75: 95–97

Finkenstädt T, Adler NT, Allen TO, Ebbesson SOE, Ewert J-P (1985) Mapping of brain activity in mesencephalic and diencephalic structures of toads during presentation of visual key stimuli: a computer assisted analysis of ^{14}C-2DG autoradiographs. J Comp Physiol 156: 433–445

Gans C (1961) A bullfrog and its prey. Nat Hist 70: 26–37

Gans C, Gorniak GC (1982a) Functional morphology of lingual protrusion in marine toads (*Bufo marinus*). Am J Anat 163: 195–222

Gans C, Gorniak GC (1982b) How does the toad flip its tongue? Test of two hypotheses. Science 216: 1335–1337

Gaupp E (1896) A. Ecker's und R. Wiedersheim's Anatomie des Frosches. Viehweg, Braunschweig

Glagow M, Ewert J-P (1992) The influence of apomorphine on prey-catching and visually evoked excitability of retinal class-R2-neurons in toads *Bufo bufo*. Verh Dtsch Zool Ges (short comm) 85: 66

Goodale MA (1983) Visuomotor organization of pecking in the pigeon. In: Ewert J-P, Capranica RR, Ingle DJ (eds) Advances in vertebrate neuroethology. Plenum Press, New York, pp 349–357

Grillner S (1985) Neurobiological bases of rhythmic motor acts in vertebrates. Science 228: 143–149

Grobstein P, Comer C, Kostyk SK (1983) Frog prey capture behavior: between sensory maps and directed motor output. In: Ewert J-P, Capranica RR, Ingle DJ (eds) Advances in vertebrate neuroethology. Plenum Press, New York, pp 331–347

Grüsser O-J, Grüsser-Cornehls U (1968) Neurophysiologische Grundlagen visueller angeborener Auslösemechanismen beim Frosch. Z Vergl Physiol 59: 1–24

Grüsser O-J, Grüsser-Cornehls U (1970) Die Neurophysiologie visuell gesteuerter Verhaltensweisen bei Anuren. Verh Dtsch Zool Ges Köln 64: 201–218

Grüsser O-J, Grüsser-Cornehls U (1976) Neurophysiology of the anuran visual system. In: Llinás R, Precht W (eds) Frog neurobiology. Springer, Berlin Heidelberg New York, pp 297–385

Harris-Warrick R (1988) Chemical modulation of central pattern generators. In: Cohen AH, Rossignol S, Grillner S (eds) Neural control of rhythmic movements in vertebrates. Wiley, New York

Heiden U an der, Roth G (1989) Retina and optic tectum in amphibians: a mathematical model and simulation studies. In: Ewert J-P, Arbib MA (eds) Visuomotor coordination: amphibians, comparisons, models, and robots. Plenum Press, New York, pp 243–267

Herrick CJ (1933) The amphibian forebrain. VIII. Cerebral hemispheres and pallial primordia. J Comp Neurol 58: 737–759

Hinsche G (1935) Ein Schnappreflex nach "Nichts" bei Anuren. Zool Anz 111: 113–122

Hobson JA, Scheibel AB (eds) (1980) The brain stem core: sensorimotor integration and behavioral state control. Neuro Sci Res Prog Bull Vol 18. MIT Press, Cambridge

Ingle D (1976) Spatial vision in anurans. In: Fite KV (ed) The amphibian visual system: a multidisciplinary approach. Academic Press, New York, pp 119–140

Ingle DJ (1983) Brain mechanisms of visual localization by frogs and toads. In: Ewert J-P, Capranica RR, Ingle DJ (eds) Advances in vertebrate neuroethology. Plenum Press, New York, pp 177–226

Jean A (1984) Brain stem organization of the swallowing network. Brain Behav Evol 25: 109–116

Kicliter E, Ebbesson SOE (1976) Organization of the "nonolfactory" telencephalon. In: Llinás R, Precht W (eds) Frog neurobiology. Springer, Berlin Heidelberg New York, pp 946–972

Kramer EB, Rath I, Lischka MF (1979) Somatotopic organization of the hypoglossal nucleus: HRP study in the rat. Brain Res 170: 533–537

Kupfermann I, Weiss KR (1978) The command neuron concept. Behav Brain Sci 1: 3–39

Lambert RW, Goldberg LJ, Chandler SH (1986) Comparison of mandibular movement trajectories and associated patterns of oral muscle electromyographic activity during spontaneous and apomorphine-induced rhythmic jaw movements in the guinea pig. J Neurophysiol 55: 301–319

Landgren S, Olsson KA, Westberg KG (1986) Bulbar neurones with axonal projections to the trigeminal motor nucleus in the cat. Exp Brain Res 65: 98–111

Lara R, Cervantes F, Arbib MA (1982) Two-dimensional model of retinal-tectal-pretectal interactions for the control of prey-predator recognition and size preference in amphibia. In: Amari S, Arbib MA (eds) Competition and cooperation in neural nets. Springer, Berlin Heidelberg New York, pp 371–393

Lauder GV, Reilly SM (1988) Functional design of the feeding mechanism in salamanders: causal bases of ontogenetic changes in function. J Exp Biol 134: 219–233

Lázár G (1969) Efferent pathways of the optic tectum in frog. Acta Biol Acad Sci Hung 20: 171–183

Lettvin JY, Maturana HR, McCulloch WS, Pitts WH (1959) What the frog's eye tells the frog's brain. Proc Inst Radio Eng 47: 1940–1951

Leyhausen P (1965) Über die Funktion der relativen Stimmungshierarchie, dargestellt am Beispiel der phylogenetischen und ontogenetischen Entwicklung des Beutefangs von Raubtieren. Z Tierpsychol 22: 412–498

Liaw J-S, Weerasuriya A, Arbib MA (1993) A neural network model for snapping in frogs and toads (submitted)

Lorenz K (1935) Der Kumpan in der Umwelt des Vogels. Der Artgenosse als auslösendes Moment sozialer Verhaltensweisen. J Ornithol 83: 137–213, 289–413

Lund JP, Enomoto S (1988) The generation of mastication by the mammalian central nervous system. In: Cohen A, Rossignol S, Grillner S (eds) Neural control of rhythmic movements in vertebrates. Wiley, New York, pp 41–72

Maeda M, Magherini PC, Precht W (1977) Functional organization of vestibular and visual inputs to neck and forelimb motoneurons in the frog. J Neurophysiol 40: 225–243

Matesz C, Székely G (1977) The dorsomedial nuclear group of cranial nerves in the frog. Acta Biol Acad Sci Hung 28: 461–474

Matsumoto N, Schwippert WW, Ewert J-P (1986) Intracellular activity of morphologically identified neurons of the grass frog's optic tectum in response to moving configurational visual stimuli. J Comp Physiol A 159: 721–739

Matsumoto N, Schwippert WW, Beneke TW, Ewert J-P (1991) Forebrain-mediated control of visually guided prey-catching in toads: investigation of striato-pretectal connection with intracellular recording/labeling methods. Behav Processes 25: 27–40

Matsushima T, Satou M, Ueda K (1986) Glossopharyngeal and tectal influences on tongue muscle motoneurons in the Japanese toad. Brain Res 265: 198–203

Matsushima T, Satou M, Ueda K (1989) Medullary reticular neurons in the Japanese toad: morphology and excitatory inputs from the optic tectum. J Comp Physiol A 166: 7–22

Megela A, Borchers H-W, Ewert J-P (1983) Relation between activity of tectal neurons and prey-catching behavior in toads *Bufo bufo*. Naturwissenschaften 70: 100–101

Merkel-Harff C, Ewert J-P (1991) Learning-related modulation of toad's responses to prey by neural loops involving the forebrain. In: Arbib MA, Ewert J-P (eds) Visual structures and integrated functions. Research notes in neural computing 3. Springer, Berlin Heidelberg New York, pp 417–426

Mikulka P, Hughes J, Aggerup G (1980) The effect of pretraining procedures and discriminative stimuli on the development of food selection behaviors in the toad (*Bufo terrestris*). Behav Neurol Biol 29: 52–62

Nakamura Y (1985) Localization and functional organization of masticatory rhythm generator in the lower brain stem reticular formation. Neurosci Lett 20: 53–54

Neary T, Northcutt RG (1983) Nuclear organization of the bullfrog diencephalon. J Comp Neurol 213: 262–278

Nieuwenhuys R, Opdam P (1976) Structure of the brain stem. In: Llinás L, Precht W (eds) Frog neurobiology. Springer, Berlin Heidelberg New York, pp 811–855

Nishikawa K, Gans C (1990) Neuromuscular control of prey capture in the marine toad, *Bufo marinus*. Am Zool 30: 141A

Olsson KA, Landgren S, Westberg KG (1986) Location and peripheral convergence on the interneurone in the disynaptic path from the coronal gyrus of the cerebral cortex to the trigeminal motoneurones in the cat. Exp Brain Res 65: 83–97

Parent A (1973) Distribution of monoamine-containing neurons in the brainstem of the frog, *Rana temporaria*. J Morphol 139: 67–78

Porter R (1965) Synaptic potentials in hypoglossal motoneurones. J Physiol (Lond) 180: 209–244

Roberts A, Roberts BL (eds) (1983) Neural origin of rhythmic movements. Cambridge Univ Press, Cambridge

Roman C (1986) Nervous control of swallowing and oesophageal mobility in mammals. J Physiol (Paris) 81: 118–131

Rossignol S, Lund JP, Drew T (1988) The role of sensory inputs in regulating patterns of rhythmical movements in higher vertebrates. In: Cohen AV, Rossignol S, Grillner S (eds) Neural control of rhythmic movements in vertebrates. Wiley, New York, pp 201–283

Röthig P (1927) Beiträge zum Studium des Zentralnervensystems der Wirbeltiere. 11. Über die Faserzüge im Mittelhirn, Kleinhirn und der Medulla oblongata der Urodelen und Anuren. Z Mikrosk Anat Forsch 10: 381–472

Rubinson K (1968) Projections of the tectum opticum of the frog. Brain Behav Evol 1: 529–561

Satou M, Ewert J-P (1985) The antidromic activation of tectal neurons by electrical stimuli applied to the caudal medulla oblongata in the toad *Bufo bufo* L. J Comp Physiol A 157: 739–748

Satou M, Matsushima T, Takeuchi H, Ueda K (1985) Tongue-muscle-controlling motoneurons in the Japanese toad: topography, morphology and neuronal pathways from the 'snapping-evoking area' in the optic tectum. J Comp Physiol A 157: 717–737

Scheich H (1983) Sensorimotor interfacing. In: Ewert J-P, Capranica RR, Ingle DJ (eds) Advances in vertebrate neuroethology. Plenum Press, New York, pp 7–14

Schneider D (1954) Beitrag zu einer Analyse des Beute- und Fluchtverhaltens einheimischer Anuren. Biol Zentralbl 73: 225–282

Schürg-Pfeiffer E (1989) Behavior-correlated properties of tectal neurons in freely moving toads. In: Ewert J-P, Arbib MA (eds) Visuomotor coordination: amphibians, comparisons, models, and robots. Plenum Press, New York, pp 451–480

Schürg-Pfeiffer E, Ewert J-P (1981) Investigation of neurons involved in the analysis of gestalt prey features in the frog *Rana temporaria*. J Comp Physiol 141: 139–152

Schürg-Pfeiffer E, Spreckelsen C, Ewert J-P (1990) Tectal small-field neurons recorded in prey-catching toads are sensitive to the real object size. Eur J Neurosci Suppl 3: 186

Schürg-Pfeiffer E, Spreckelsen C, Ewert J-P (1993) Temporal discharge patterns of tectal and medullary neurons chronically recorded during snapping toward prey in toads *Bufo bufo spinosus*. J Comp Physiol A 173: 363–376

Schwippert WW, Beneke TW, Ewert J-P (1990) Responses of medullary neurons to moving visual stimuli in the common toad: II. An intracellular recording and cobalt-lysine labeling study. J Comp Physiol A 167: 509–520

Selverston AI (1980) Are pattern generators understandable? Behav Brain Sci 3: 535–571

Shinn EA, Dole JW (1978) Evidence for a role for olfactory cues in the feeding response of leopard frogs *Rana pipiens*. Herpetologia 34: 167–172

Soller RW (1977) Monoaminergic inputs to frog motoneurons: an anatomical study using fluorescence histochemical and silver degeneration technique. Brain Res 122: 445–458

Spreckelsen C, Schürg-Pfeiffer E, Ewert J-P (1993) Apomorphine-influenced visuomotor functions in toads *Bufo bufo spinosus* (L.). (submitted)

Sternthal DE (1974) Olfactory and visual cues in the feeding behavior of the leopard frog (*Rana pipiens*). Z Tierpsychol 34: 240–246

Striker EM (1983) Brain neurochemistry and the control of food intake. In: Satinoff E, Teitelbaum P (eds) Handbook of behavioral neurobiology, vol 6. Motivation. Plenum Press, New York, pp 329–366

Stuesse SL, Cruce WLR, Powell KS (1983) Afferent and efferent components of the hypoglossal nerve in the grass frog. J Comp Neurol 217: 432–439

Sumi T (1969) Synaptic potentials of hypoglossal motoneurons and their relation to reflex deglutition. Jpn J Physiol 19: 68–79

Székely G, Lázár G (1976) Cellular and synaptic architecture of the optic tectum. In: Llinás R, Precht W (eds) Frog neurobiology. Springer, Berlin Heidelberg New York, pp 407–434

Székely G, Levai G, Matesz K (1983) Primary afferent terminals in the nucleus of the solitary tract of the frog: an electron microscopic study. Exp Brain Res 53: 109–117

Tinbergen N (1951) The study of instinct. Clarendon Press, Oxford

Tomomune N, Takata M (1988) Excitatory and inhibitory postsynaptic potentials in cat hypoglossal motoneurons during swallowing. Exp Brain Res 71: 262–272

Tóth P, Csank G, Lázár G (1985) Morphology of the cells of origin of descending pathways to the spinal cord in *Rana esculenta*. A tracing study using cobalt-lysine complex. J Hirnforsch 26: 365–383

Traud R (1983) Einfluß von visuellen Reizmustern auf die juvenile Erdkröte (*Bufo bufo* L.). PhD Thesis, Univ of Kassel

Tsai H-J, Ewert J-P (1988) Influence of stationary and moving textured backgrounds on the response of visual neurons in toads (*Bufo bufo* L.). Brain Behav Evol 32: 27–38

Uemura-Sumi M, Mizuno N, Iwahori N, Tackeuchi Y, Matsushima R (1981) Topographical representation of the hypoglossal nerve branches and tongue muscles in the hypoglossal nucleus of macaque monkeys. Neurosci Lett 22: 31–35

von Wietersheim A, Ewert J-P (1978) Neurons of the toad's (*Bufo bufo* L.) visual system sensitive to moving configurational stimuli: a statistical analysis. J Comp Physiol 126: 35–42

Weerasuriya A (1983) Snapping in toads: some aspects of sensorimotor interfacing and motor pattern generation. In: Ewert J-P, Capranica RR, Ingle DJ (eds) Advances in vertebrate neuroethology. Plenum Press, New York, pp 613–627

Weerasuriya A (1989) In search of the motor pattern generator for snapping toads. In: Ewert J-P, Arbib MA (eds) Visuomotor coordination: amphibians, comparisons, models, and robots. Plenum Press, New York, pp 589–614

Weerasuriya A (1991) Motor pattern generators in anuran prey capture. In: Arbib MA, Ewert J-P (eds) Visual structures and integrated functions. Research notes in neural computing Vol. 3. Springer, Berlin Heidelberg, New York, pp 255–270

Weerasuriya A, Ewert J-P (1981) Prey-selective neurons in the toad's optic tectum and sensorimotor interfacing: HRP studies and recording experiments. J Comp Physiol 144: 429–434

Weerasuriya A, Ewert J-P (1984) Afferents of the hypoglossal nucleus in the common European toad, *Bufo bufo*. Am Assoc Anat Abstr 84: 192A

Wiersma CAG, Ikeda K (1964) Interneurons commanding swimmeret movements in the crayfish, *Procambarus clarkii* (Girard). Comp Biochem Physiol 12: 509–525

Zeigler HP, Karten HJ (1973) Brain mechanisms and feeding behavior in the pigeon (*Columba livia*). I. Quintofrontal structures. J Comp Neurol 152: 59–82

Zeigler HP, Witkovsky P (1968) The main sensory trigeminal nucleus in the pigeon: a single unit analysis. J Comp Neural 134: 255–264

Zeigler HP, Miller MG, Levine RR (1975) Trigeminal nerve and eating in the pigeon (*Columba livia*): neurosensory control of the consummatory response. J Comp Physiol Psychol 89: 845–858

Chapter 6

Amphibian Feeding Behavior: Comparative Biomechanics and Evolution

G.V. Lauder[1] and *S.M. Reilly*[2]

Contents

1 Introduction

The clade Amphibia is critical for our understanding of vertebrate evolution. Because of their position as a basal lineage of tetrapods, nearly all aspects of amphibian biology are of special interest to those interested in the origin of terrestrial life and in the morphological, physiological, ecological, and behavioral changes involved in aquatic to terrestrial transitions. In addition, amphibian taxa illustrate with particular clarity the phenomenon of metamorphosis, allowing the experimental study of aquatic-to-terrestrial transitions on a single individual during ontogeny. Although phylogenetic relationships among the three extant amphibian clades (and among fossil amphibian taxa) are still a matter of debate (Bolt 1977; Carroll and Holmes 1980; Duellman and

[1] School of Biological Sciences, University of California, Irvine, California 92717, USA
[2] Dept. of Zoology, Ohio University, Athens, Ohio 45701, USA

Advances in Comparative and Environmental Physiology, Vol. 18
© Springer-Verlag Berlin Heidelberg 1994

Trueb 1986; Trueb and Cloutier 1991), the relevance of amphibian clades to problems in vertebrate biology is not at issue. Extant amphibian lineages, because of their phylogenetic position near the base of the tetrapod radiation and because of the mosaic nature of character distribution in these taxa (many amphibian taxa retain large numbers of primitive features in the musculoskeletal system, while at the same time displaying numerous derived characteristics), are prime candidates for comparisons to both fish and amniote clades.

One key aspect of a transition (evolutionary or ontogenetic) between an aquatic and a terrestrial environment is the problem of obtaining food. How do animals manage to modify behavioral, morphological, or physiological patterns associated with food acquisition to permit function in both environments? Extant amphibian clades allow this question to be addressed from a number of different perspectives. First, individual amphibians may undergo an ontogenetic transformation from an aquatic feeding mode as a larva to a terrestrial feeding mode after metamorphosis. This provides an experimental opportunity to study directly form and function in the feeding mechanism across environments. Second, some terrestrial amphibians may be induced to feed in the water, providing an opportunity to examine how well a terrestrial feeding mechanism functions biomechanically in an aquatic environment. Third, the three major extant amphibian clades are widely divergent in their cranial morphology, providing an opportunity to examine how different lineages of amphibians have (perhaps independently) solved biomechanical problems (such as terrestrial prey capture using tongue projection). Fourth, many features of larval amphibian feeding mechanisms are very similar to those of outgroup clades such as lungfishes and ray-finned fishes (Actinopterygii), facilitating evolutionary analyses via comparisons of homologous morphologies, functions, and behaviors.

In this chapter, our aim is to summarize the current state of knowledge about amphibian feeding biomechanics, with the general objective of placing the biomechanics of amphibian feeding within the framework of vertebrate evolution. Our specific goals are first, to review the biomechanics of prey capture in outgroup clades to provide historical and phylogenetic background; second, to analyze the biomechanics of feeding in the two extant amphibian lineages for which there are the most complete data (salamanders and frogs); third, to compare feeding mechanisms in fish, amphibian and amniote taxa to search for general historical patterns and relationships; finally, to assess needed future directions for research in the functional morphology of amphibian feeding.

2 Overview of Feeding Mechanics in Fishes

2.1 Initial Prey Capture

We begin our consideration of amphibian feeding biomechanics with an overview of functional patterns that have been established for outgroup clades such as lungfishes and ray-finned fishes. Analyses of tetrapod feeding mechanics have often proceeded without considering the functional patterns present in outgroup taxa. Biomechanical analyses benefit in many ways from an historical perspective on the evolution of function (Gans 1980; Lauder 1986b, 1990, 1991), and aquatic feeding systems in fishes have much to teach us about both amphibian and amniote feeding patterns. If models of tetrapod feeding behavior are to be properly generated and interpreted, then a firm phylogenetic foundation for hypothesized functional patterns is essential.

A key conclusion from past research on the functional morphology of the feeding mechanism in fishes is that, despite considerable specialization among lineages of fishes [consider the differences between a lungfish (Dipnoi) and a largemouth bass (Actinopterygii: Centrarchidae)], there are many common functional patterns that are of general occurrence (Lauder 1985a, b). This is a fortunate result because it allows us to identify several primitive biomechanical features of fish feeding systems and then to use these as a basis for evaluating amphibian biomechanics. These features are hypothesized to be primitive for tetrapod feeding systems (Lauder 1985a; Reilly and Lauder 1990a; Lauder and Shaffer 1993).

Figure 1 summarizes the pattern of muscle activity and gape and hyoid kinematics common to feeding mechanisms in many species of ray-finned fishes. Similar functional patterns have been noted in lungfishes (Bemis and Lauder

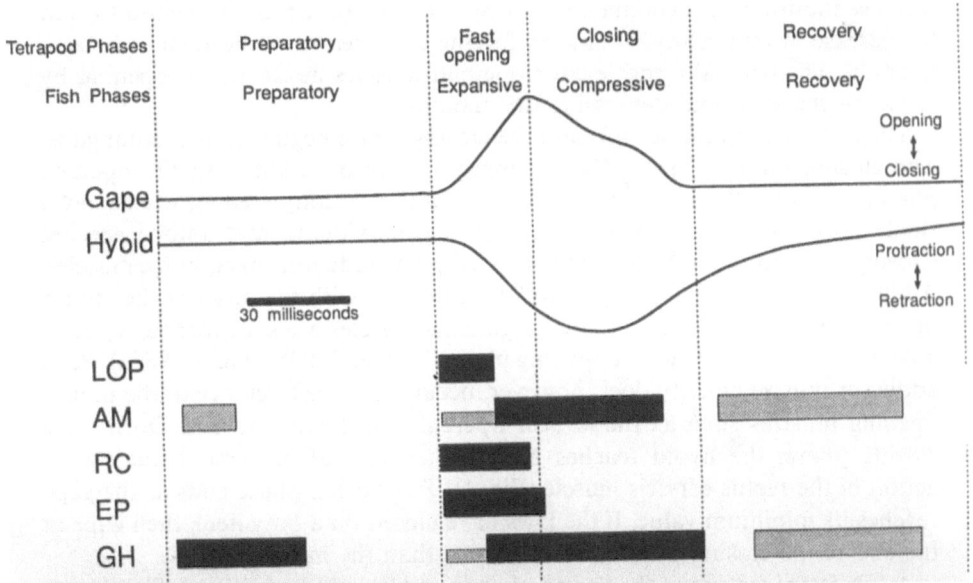

Fig. 1. Generalized feeding pattern for actinopterygian fish and lungfish, hypothesized to represent the primitive condition for tetrapods. The distinct kinematic phases of the strike are labeled at the *top* of the figure. *Black bars* indicate muscle activity that is consistently present, *gray bars* indicate frequent activity, and changes in height of the bars show large-scale changes in the amplitude of muscle activity. *AM* Adductor mandibulae; *EP* epaxial; *GH* geniohyoideus; *LOP* levator operculi; *RC* rectus cervicis

1986; Bemis 1987). Although experimental data are currently not available on coelacanths, mechanical linkages hypothesized from anatomical analysis (Lauder 1980a) also suggest that similar patterns existed in the Actinistia.

The process of initial prey capture in fishes was formerly divided into four phases: preparatory, expansive, compressive, and recovery (Lauder 1985a). These terms are similar to those proposed for amphibian (and some amniote) clades: preparatory, fast opening, closing, and recovery, respectively. To be consistent, we will use the tetrapod phase names shown at the top of Fig. 1 throughout this work (see also Reilly and Lauder 1990a). Ray-finned fishes and lungfishes lack a slow opening phase (which is present in many amniote taxa) and thus this phase is not discussed here. Most fishes also lack a preparatory phase. When the preparatory phase is present, the buccal cavity is compressed to force out water and reduce intraoral volume. Activity of the jaw adductor muscles and ventral throat muscles (such as the geniohyoideus) is often seen during this phase.

The preparatory phase ends and the fast opening phase begins with the onset of mouth opening. As the gape increases rapidly at the onset of the strike, the hyoid arch moves posteroventrally from its initial protracted position (Fig. 1). Three main muscles effect these actions: the levator operculi, epaxialis, and rectus cervicis (= sternohyoideus in the ichthyological literature). The rectus cervicis is a major muscle of the fast opening phase and acts to (1) increase the gape via ligamentous connections between the hyoid and mandibular arches, (2) increase mouth cavity volume by moving the floor of the mouth ventrally, and (3) increase mouth cavity volume by forcing the sides of the head (suspensoria) laterally. The epaxialis and levator operculi muscles assist mouth opening by elevating the skull and depressing the mandible.

The fast opening phase ends and the closing phase begins at maximum gape. The closing phase is generally of slightly longer duration than the opening phase, although there is considerable variability among feeding events on a single prey type, among prey types, among individuals, and among species. Closing of the mouth is initiated by activity in the adductor mandibulae muscles. Adductor muscle activity may begin at a low level with the onset of the mouth opening muscles, indicating that antagonistic muscles are synchronously activated with the onset of the fast opening phase (Lauder 1980b, 1983b, 1985a). Peak adductor muscle activity does, however, occur after peak activity in the mouth opening muscles such as the levator operculi and rectus cervicis. During the closing phase, the hyoid reaches peak posteroventral movement due to the action of the rectus cervicis muscle (Fig. 1). The closing phase ends as the gape reaches its minimum value. If the jaws have closed on a prey item, then gape at the end of the closing phase will be greater than the initial gape.

A consistent feature of the timing of gape and hyoid kinematic profiles during initial prey capture in fishes is the delay in peak hyoid excursion relative to peak gape: the hyoid reaches its maximum posteroventral excursion after maximum gape has been reached. In most fishes, lateral opercular expansion reaches its peak after the hyoid, providing an anteroposterior timing in peak excursions:

first maximum gape, then maximum hyoid retraction, and finally maximal opercular abduction.

The fast opening and closing phases together constitute the gape cycle, which in fishes is roughly bell-shaped: no gape cycles have been published which show a plateau during the fast opening phase.

The recovery phase is that phase during which the gape remains relatively constant and the hyoid moves anterodorsally due to synchronous activity in the jaw adductor muscles and ventral mouth musculature such as the geniohyoideus (Fig. 1). The jaw adductor muscles act to stabilize the mandible so that the geniohyoideus muscle, with its origin at the mandibular symphysis, can act to protract the hyoid. Completion of the recovery phase may take up to several seconds, and is frequently interrupted by the onset of transport sequences (or buccal manipulation events) to move prey posteriorly in the oral cavity.

The function of mouth cavity expansion and the role of cranial muscle activity is to cause a pressure reduction within the oral cavity which draws water and prey into the mouth. This mode of feeding in fishes is called suction feeding because of the negative (or "suction") pressure generated within the mouth cavity. Experimental measurement of these negative pressures (Lauder 1980c, d, 1983a, 1986a; Liem 1980) has established several features of the suction feeding mechanism. First, pressure measurements made simultaneously both anterior and posterior to the gill bars (which are composed of hypobranchial, ceratobranchial, and epibranchial elements supporting the gill filaments) in sunfishes (Centrarchidae) have demonstrated that there is a clear pressure differential across the gill bars, with the pressure anteriorly in the mouth two to six times more negative than the pressure in the opercular cavity. Second, the timing of gill bar, opercular, and hyoid movement effects the predominantly unidirectional flow of water through the mouth cavity (a small reverse flow from the opercular into the buccal cavity may also occur at the onset of the fast opening phase). Third, gill bar adduction is timed so that the gill bars are maximally adducted near the time of peak negative pressure, preventing significant water flow from posterior to anterior into the buccal cavity. Overall, the process of initial prey capture by suction feeding involves the initiation of a rapid (and unsteady) flow of water into the mouth that results from the rapid opening of the mouth and depression of the hyoid.

2.2 Prey Manipulation and Transport

Once the prey has been captured, a series of mouth movements is usually observed that manipulates the prey into a position for swallowing (Liem 1970; Lauder 1979, 1983b). Although many strikes involving suction result in the prey being brought directly into the buccal cavity, unless the prey is very small, buccal manipulation is still employed to position prey for swallowing. Such prey manipulation within the mouth cavity occurs by creating currents of water flow via jaw, hyoid, and opercular movements, and Bemis and Lauder (1986) referred to this as a hydraulic mechanism of prey transport. Prey are moved anteriorly and posteriorly, as well as mediolaterally, by currents of water. Hydraulic prey transport is accomplished via similar mechanisms to those described above for initial prey transport: prey are moved posteriorly in the mouth toward the pharyngeal jaws by a current of water resulting from mouth opening and delayed hyoid and opercular expansion (Lauder 1980b, 1983b). Anterior flows of water are created by changing the timing of opercular, hyoid, and gape movements, and this results in an altered pressure profile during prey manipulation (Lauder 1980b).

The kinematic patterns involved in prey transport toward the esophagus are very similar to those seen during initial prey capture by suction feeding. The timing of gape and hyoid profiles is similar, with maximal hyoid retraction occurring after peak gape (Bemis and Lauder 1986).

3 Salamander Feeding Mechanics

3.1 Comparative Framework

Due to the variety of possible comparisons that may be made among taxa, behaviors, and between the aquatic and terrestrial environments, it is useful to begin a consideration of the biomechanics of feeding in salamanders with the general comparative scheme outlined in Fig. 2. The two key behaviors of interest are (1) the initial strike in which the prey is first captured and brought into the mouth cavity, and (2) the transport of prey to the esophagus following the initial strike. Given these two behaviors, there are four levels of comparison (Fig. 2).

First, strike and transport behaviors may be compared within each species or clade being studied (Fig. 2: comparison A). Thus, within the species *Ambystoma tigrinum*, measured features of strike and transport skeletal movement and muscle activity patterns may be compared to determine (quantitatively) what differences exist between the two behaviors. Second, the two behaviors may be compared across metamorphosis (Fig. 2: comparison B) to determine how skeletal movement and muscle activity change during ontogeny. Third, post-metamorphic salamanders may be studied feeding in both aquatic and terrestrial environments and the strike and transport behaviors compared (Fig. 2: comparison C). Fourth, strike and transport behaviors in salamander taxa may be compared to outgroup clades to determine which kinematic or motor pattern attributes have a general distribution within lower vertebrates (comparison D). In addition, different clades of salamanders may be compared with each other (not shown in Fig. 2) to determine the extent of phylogenetic variation within salamanders (both aquatic and terrestrial) in kinematic and motor patterns used during both initial prey capture and subsequent prey transport.

The goal of the comparative tests outlined in Fig. 2 is to provide a comprehensive picture of the ontogeny and phylogeny of the feeding mechanism in

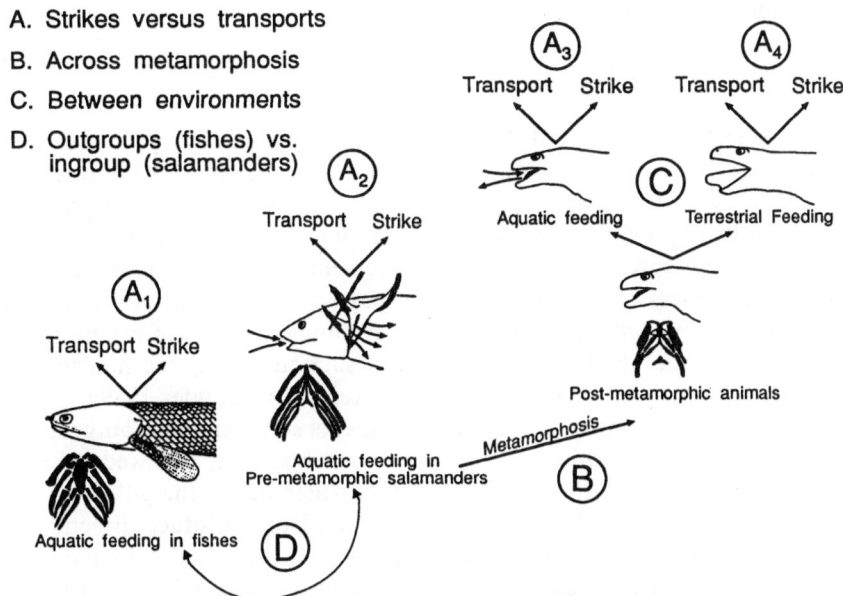

A. Strikes versus transports

B. Across metamorphosis

C. Between environments

D. Outgroups (fishes) vs. ingroup (salamanders)

Fig. 2. Chart of the feeding behaviors of fish and salamanders and possible comparisons to be made among taxa, metamorphic conditions, and behaviors. Comparisons A_{1-4} contrast strike and transport behaviors within taxa: ray-finned fishes, larval salamanders, and postmetamorphic salamanders feeding in the water and on land. Aquatic feeding in premetamorphic salamanders occurs by suction feeding with unidirectional flow through the mouth cavity (indicated by *arrows* showing water flow in the mouth and out the gill openings posteriorly in comparison A_2). Aquatic feeding in postmetamorphic salamanders also occurs by suction feeding, but water flow is bidirectional (as indicated by *arrows* showing water flow in and out of the mouth in comparison A_3. Comparison B compares feeding function across metamorphosis during salamander ontogeny. Comparison C contrasts prey capture (both strikes and transports) in postmetamorphic salamanders feeding in the water and on land. Comparison D compares prey capture (both strikes and transports) in salamanders and outgroup taxa. Another possible comparison contrasts feeding behavior in different salamander taxa (e.g., Reilly and Lauder 1992). Schematic diagrams of the hyobranchial apparatus are shown for outgroup taxa (fishes, *Polypterus*), and for pre- and postmetamorphic salamanders

salamanders. To date, only a few of these comparisons have been conducted, and there are many gaps in our knowledge. Many studies of salamander feeding kinematics have been conducted without considering potentially primitive aspects of feeding kinematics present in fishes, and electromyographic data on muscle function are available only for two species. Furthermore, just one species has been studied across metamorphosis, and limited data are available only for one species comparing strike and transport behaviors quantitatively. A great deal of work remains to be done.

3.2 Initial Prey Capture

3.2.1 Aquatic Feeding

The biomechanics of initial prey capture in the water has been studied most extensively in the genus *Ambystoma*: Lauder and Shaffer (1985) described the morphology of the feeding apparatus in conjunction with high-speed cinematography of prey capture, electromyography, buccal pressure recordings, and impedance measurements of gill bar movement. Despite a number of morphological differences from outgroup taxa, patterns of skeletal movement during prey capture by *Ambystoma* are extremely similar to that of ray-finned fishes: most aquatic salamanders and fishes use a suction feeding mechanism (Lauder 1985a, b; Lauder and Shaffer 1985, 1993, Reilly and Lauder 1989a, 1992). The general pattern of timing of skeletal movements so common in outgroup clades is preserved in *Ambystoma*: peak gape is reached first, followed by maximal hyoid depression, and then by the efflux of water out of the gills.

An impedance recording technique was used to transduce directly the distance between adjacent gill bars and this showed that maximal gape coincides with maximal gill bar adduction which prevents water influx from the region posterior to the head as the mouth opens (Lauder and Shaffer 1985). The gill bars in larval *Ambystoma* possess interlocking gill rakers which form an effective barrier to water flow in a similar manner to the gill bars of fishes. Maximal gill bar adduction coincides also with maximal negative pressure in the mouth cavity. As the mouth closes on the prey, the gill bars are abducted and water flows out posteriorly between the abducted gill bars: water flow during prey capture is unidirectional (from anterior to posterior).

Recordings of muscle activity during prey capture made in conjunction with film records of head movement (at 200 frame/s) show that the mouth opens as a result of combined activity in the epaxialis, depressor mandibulae, rectus cervicis, and geniohyoideus muscles. Activity in anatomical antagonists, the depressor mandibulae, and adductor mandibulae begins simultaneously just prior to the onset of mouth opening, but the depressor mandibulae reaches peak activity (as measured by spike number and amplitude in electromyograms) 5 to 20 ms before the adductor mandibulae.

Comparative data from other salamander taxa are only available for patterns of head movement (Reilly and Lauder 1992); no electromyographic data are available for aquatic feeding to allow comparisons across salamander families. Reilly and Lauder (1992) studied initial prey capture in species in the families Ambystomatidae, Dicamptodontidae, Amphiumidae, Sirenidae, Proteidae, and Cryptobranchidae. By measuring seven variables from high-speed videos of prey capture, they showed that there was highly significant differentiation among these taxa in feeding kinematics. In particular, *Cryptobranchus* and *Siren* showed the most divergent patterns of feeding kinematics from the other taxa. Reilly and Lauder (1992) did find clear kinematic correlates of the reduction in posterior gill openings. In both *Cryptobranchus* and *Amphiuma* (which possess restricted posterior gill openings for the exit of water during feeding) the angle of the head is depressed well below its starting value. Also, hyoid depression is delayed for up to 15 ms after the start of mouth opening in both taxa. However, despite showing significant quantitative differences in initial prey capture, all taxa showed the general patterns outlined above of the delay in maximum hyoid depression relative to maximum gape. All taxa also showed the characteristic bell-shaped gape profile seen in fish.

3.2.2 Terrestrial Feeding

In contrast to aquatic feeding in salamanders, which has just begun to attract attention from experimental zoologists, the process of prey capture by terrestrial salamanders has interested morphologists for nearly a century (Druner 1902, 1904; Francis 1934; Edgeworth 1935; Lombard and Wake 1976, 1977; Reilly and Lauder 1989a). Most functional research on terrestrial feeding has been conducted on the genus *Ambystoma*, although comparative functional data are now available on the families Plethodontidae (Severtsov 1971; Thexton et al. 1977; Larsen et al. 1989) and Salamandridae (Findeis and Bemis 1990; Miller and Larsen 1990).

Terrestrial prey capture occurs by projection of the tongue out of the mouth toward the prey, and involves the coordinated movement of the skull and hyobranchial apparatus. The strike of a terrestrial salamander differs considerably from an aquatic prey capture event. Figure 3 shows five fields from a high-speed video sequence of prey capture in *Ambystoma tigrinum* which illustrate the general sequence of head movements used to capture prey by tongue projection on land.

As the mouth opens (Fig. 3: 0 ms), the hyobranchial apparatus is moved anterodorsally (protracted) and the tongue base lifted. The protracted tongue base serves as a platform from which the tongue flips forward to contact the prey. Twenty-five milliseconds after the start of mouth opening, maximum gape has been almost reached and the plateau phase of the gape cycle begins. Maximal tongue projection is reached at about 35 ms near the middle of the gape plateau, and the tongue then contacts the prey. Peak tongue projection separates the projection and retraction phases of hyobranchial movement (Fig. 3). The retraction phase of tongue movement involves posteroventral hyobranchial movement to pull the base of the tongue back into the mouth with the prey attached. During the closing phase, the gape decreases and the jaws close on the prey (Fig. 3: 90 ms).

The plateau in the gape cycle appears to be a consistent feature of salamanders feeding on land, and may be related to the necessity of projecting the tongue out of the mouth: a near-constant gape occurs while the tongue is extended beyond the plane of the gape toward the prey (Reilly and Lauder 1989a). As a result of the distinct plateau, discrete fast opening and closing phases may be difficult to define. Although a tongue-based feeding system in salamanders

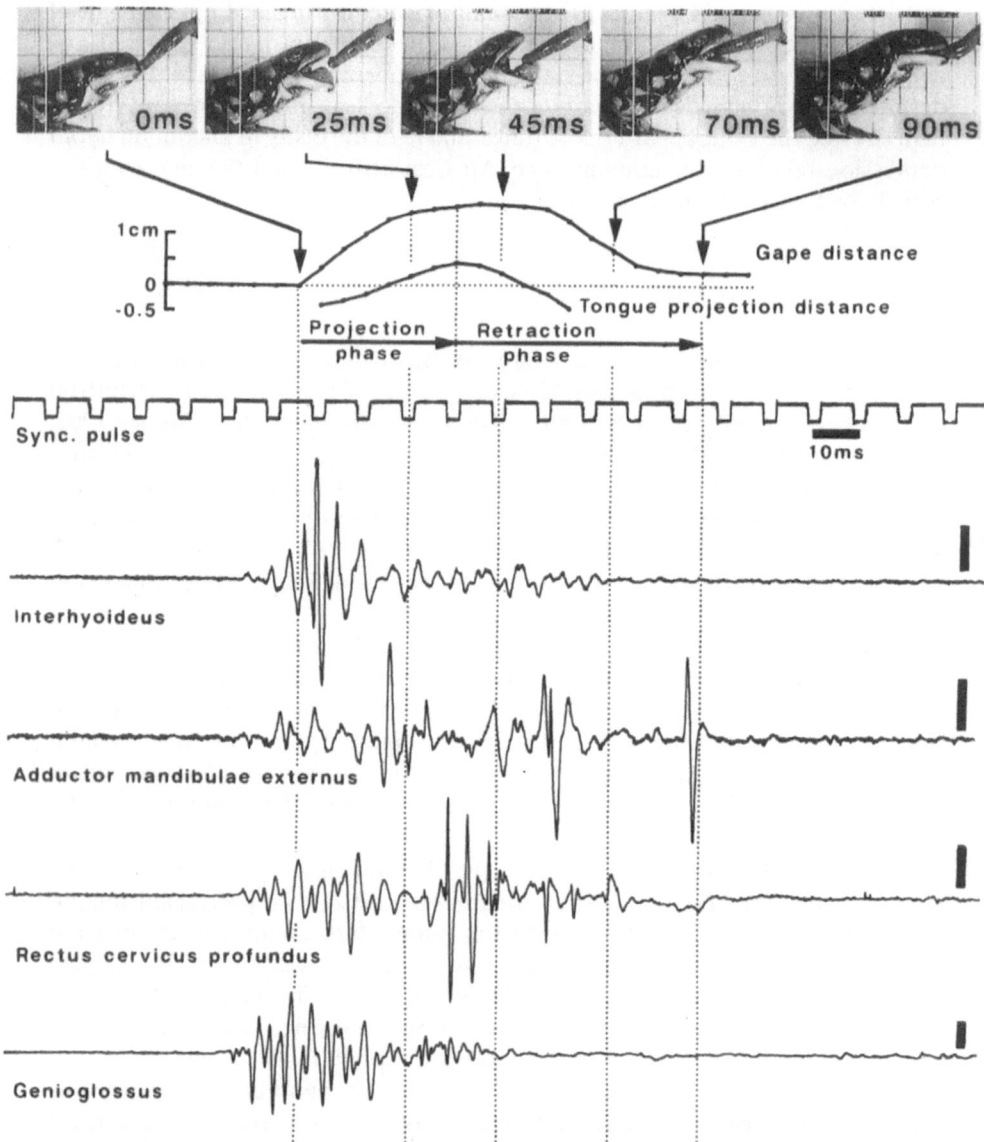

Fig. 3. Synchronized kinematic and electromyographic patterns during the strike with five representative video fields from this feeding sequence illustrated at the *top*. The time between peaks in the synchronization pulse illustrated below the kinematic plots is 10 ms. Note that activity in the interhyoideus and genioglossus muscles rises rapidly to a peak, while the adductor mandibulae and rectus cervicis profundus peak activity times are delayed until the plateau phase of the gape profile. Electrode positions in the illustrated muscles were confirmed by dissection following the experiment. *Vertical bars* to the *right* indicate 0.1 mV. (Reilly and Lauder 1990b)

would appear to function best in a terrestrial setting, Schwenk and Wake (1988) did find a case of a plethodontid salamander that feeds underwater using tongue projection. If nothing else, this illustrates that functional mechanisms thought to work only in one medium can be used elsewhere, and provides rather convincing evidence for the independent evolution of aquatic feeding in this species.

The major morphological components of the feeding system in a transformed tiger salamander (*Ambystoma tigrinum*) are illustrated in Fig. 4. Figure 4A shows the tongue in an elevated position as it is after hyobranchial protraction (roughly the position depicted in Fig. 3: 25 ms). The position of the tongue at rest is shown (equivalent to Fig. 3: 0 ms video frame) in Fig. 4B. The major muscles acting to protract and elevate the hyobranchial apparatus are the geniohyoideus (GH), genioglossus (GG), intermandibularis (IM), and subarcualis rectus one (SAR). These muscles act synergistically to move the branchial apparatus (Fig. 4: horizontal hatching) anterodorsally relative to the hyoid and lower jaw thus protracting the tongue. In particular, the SAR acts to slide the first ceratobranchial element anterior relative to the hyoid which is restrained by its ligamentous attachment to the quadrate bone (Reilly and Lauder 1988b, 1991a;

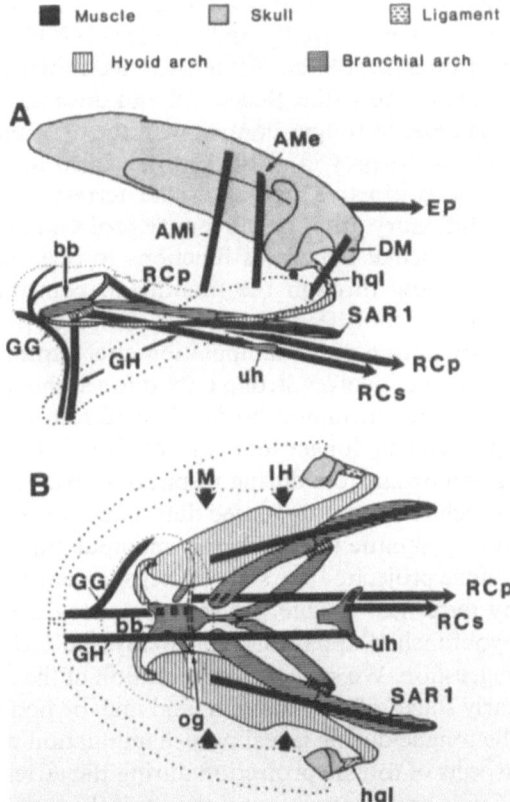

Fig. 4. Schematic diagram of representative muscles used during feeding in transformed salamanders (based on *Ambystoma tigrinum*). **A** The tongue is shown in a partially projected state, corresponding to the second video field (25 ms) in Fig. 3. The hyobranchial apparatus has been protracted, but the tongue pad is not flipped. A *black dot* marks the position of the jaw joint. *Small dots* outline the tongue, while *large dots* outline the lower jaw. **B** The tongue, lower jaw, and hyobranchial apparatus are shown in rest position as in the first video field of Fig. 3. *AMe* Adductor mandibulae externus; *AMi* adductor mandibulae internus; *bb* basibranchial; *DM* depressor mandibulae; *EP* epaxial; *GG genio glossus*; *GH* geniohyoideus; *hql* hyoquadrate ligament; *IH* interhyoideus; *IM* intermandibularis; *og* otoglossal cartilage; *RCs* rectus cervicis superficialis; *RCp* rectus cervicis profundus; *SAR* subarcualis rectus one; *uh* urohyal

Fig. 4). As the branchial apparatus is protracted, the tongue is also flipped forward by contraction of the genioglossus muscle fibers which reach up into the tongue pad. The flipping action of the tongue extends the tip of the tongue forward to contact the prey (Reilly and Lauder 1989a, 1990b).

Electromyographic recordings during prey capture reveal a complex pattern of activity in head muscles (Lauder and Shaffer 1988; Reilly and Lauder 1990b). Nearly all muscles are synchronously active (Fig. 5A), but display different patterns of amplitude and spike frequency following initial activity. Onset times alone reveal very little about muscle function, and it is important to consider the pattern of activity *within* the muscle burst. The depressor mandibulae, genioglossus, interhyoideus, rectus cervicis superficialis, and geniohyoideus all show a rapid rise to a single dominant peak in activity within 25 ms from the start of mouth opening (Reilly and Lauder 1990b). The adductor mandibulae muscles tend to peak later, but show some variability both among muscle divisions and among feedings and individuals in the precise timing of peak activity. The SAR and epaxial muscles show a double-peak pattern with the first period of activity reaching a maximum at 5 to 10 ms after the mouth starts to open, and a second period of intense activity at 60 to 80 ms.

The process of prey capture in terrestrial tiger salamanders appears to be a relatively stereotyped one based on comparisons of successful and unsuccessful strikes at prey. Reilly and Lauder (1990b) found that 66 out of 77 variables measured from 11 cranial muscle electromyograms did not differ with success or failure of the strike. Successful and unsuccessful strikes also had indistinguishable times to tongue contact with the prey and gape cycle times (the tongue did contact the prey in all strikes, but failed to adhere in unsuccessful strikes).

The distinctive nature of the terrestrial strike invites hypotheses on the evolutionary origin of the tongue projection behavior. In *Ambystoma* larvae, the hyobranchial apparatus functions to open and close the gill slits (controlling water flow through the mouth), to contribute to intraoral volume changes during suction feeding, and to compress prey against the roof of the buccal cavity during prey manipulation after capture (Reilly and Lauder 1989a). This latter role involves strong anterodorsal movements of the hyobranchial apparatus. After metamorphosis when terrestrial feeding occurs, the hyobranchial apparatus no longer functions to move water, and buccal volume changes play an important role during respiration. However, anterodorsal movement of the branchial apparatus is used during tongue projection, and hyobranchial motion during aquatic intraoral prey manipulation and the early phases of terrestrial tongue projection are similar. Regal (1966) proposed the hypothesis that feeding by terrestrial tongue projection evolved from a manipulative function of the hyobranchial apparatus, and Reilly and Lauder (1989a) elaborated on Regal's suggestion. We suggest that elevation of the hyobranchial apparatus during the early stages of the strike on land may be homologous to the dorsal elevation of the tongue during larval prey manipulation within the mouth. Novel functional aspects of tongue projection during the terrestrial strike (such as flipping of the tongue and protraction of the ceratobranchial relative to the ceratohyal) are a

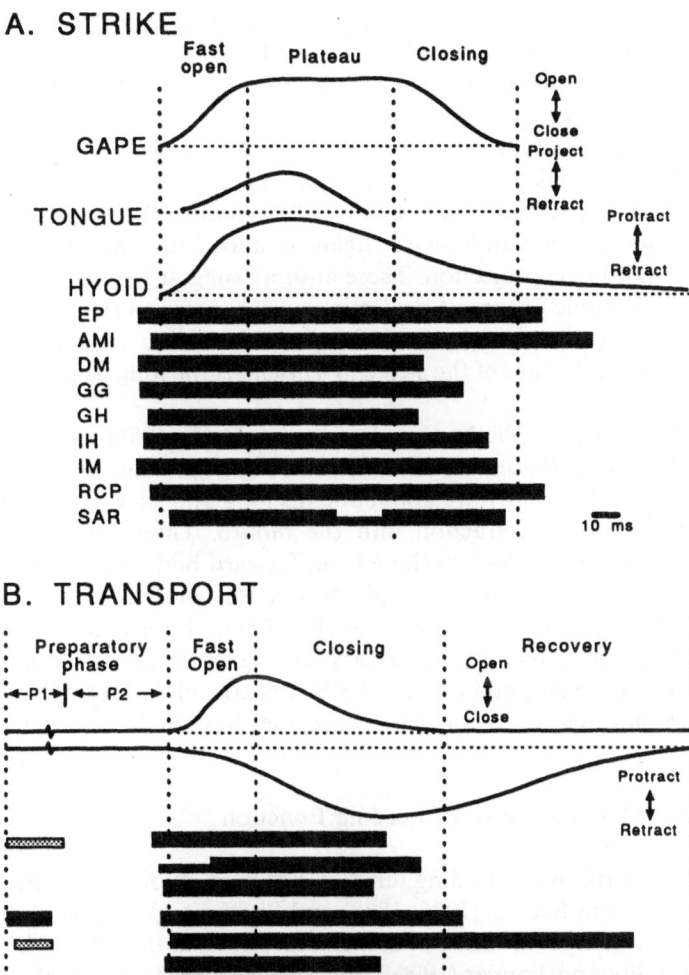

Fig. 5. Generalized terrestrial salamander feeding patterns during **A** the strike and **B** prey transport. Kinematic patterns for the gape and hyoid are shown for each behavior and a summary of the motor pattern is illustrated below: *black bars* indicate the time during which each muscle is active, *gray bars* indicate occasional activity, and changes in bar height reflect consistent patterns of amplitude variation. In **B** (transport), the duration of the *P1* and *P2* portions of the preparatory phase are shortened in this schematic figure from their real durations. *P1* may last up to 5 s, while *P2* generally is of 100 to 150 ms duration. During most of *P2*, there is no muscle activity until about 10 ms before the mouth begins to open. Note the presence of a plateau in the gape profile during the strike, but the lack of such a plateau during prey transport, and the similarities between transport kinematics and the strike kinematics for fish are shown in Fig. 1. Data for this figure were taken from Reilly and Lauder (1990b, 1991b)

consequence of novel morphological features acquired at metamorphosis. Thus, the terrestrial strike may have evolved via the addition of morphological and functional novelties at metamorphosis onto a primitive pattern of hyobranchial elevation present in larvae.

The most comprehensive comparative data on feeding kinematics in salamanders are available for the families Salamandridae and Plethodontidae. In salamandrids, Findeis and Bemis (1990) studied *Taricha* in detail and found that the lack of a mandibulohyoid ligament allows the ceratohyal to move anteriorly during tongue projection. These authors suggest that salamandrids may have a derived mode of tongue projection as compared to the primitive mechanism in which the hyoid arch is relatively stable during tongue projection. Another interesting feature of the strike in *Taricha* is the long gape cycle (on the order of 200 ms).

In the family Plethodontidae, Larsen et al. (1989) have analyzed data from *Bolitoglossa, Ensatina, Plethodon,* and *Desmognathus.* Their data show that gape cycles in this family are about 100 ms long and that peak gape occurs near the end of tongue retraction into the mouth. These authors also note that in *Bolitoglossa occidentalis* there is no forward body movement during the strike, which they explain as an adaptation to arboreal feeding, noting that "...a sudden forward lunge could cause it to fall to the ground". However, *Ambystoma tigrinum* also remains stationary during both the strike and prey transport behaviors (Reilly and Lauder 1989a, 1990a) and the lack of body movement may have nothing to do with an arboreal feeding mode.

3.2.3 Metamorphosis of Feeding Function

Metamorphosis of feeding function has been studied in *Ambystoma tigrinum* by Lauder and Shaffer (1986, 1988) and Shaffer and Lauder (1988), and morphological changes in the head at metamorphosis in this species have been analyzed by Reilly and Lauder (1990c) and Lauder and Reilly (1990).

A useful starting point for analyses of metamorphosis and the functional changes that take place is to consider three "stages" of ontogeny: larvae, metamorphosed salamanders feeding on land, and metamorphosed salamanders feeding in the water. By comparing feeding function in larvae (which feed in the water) to metamorphosed animals feeding in the water, the effect of changes in morphology across metamorphosis may be analyzed: the environment is held constant. By comparing feeding function of salamanders after metamorphosis in the water and on land, the effect of the environment alone may be seen: morphology is held constant (e.g., Fig. 2, comparison C). For analyses of the ontogeny of function in salamanders, it is vital to be able to separate the effects of environment and morphology.

The major conclusion of the studies of Lauder and Shaffer (1986, 1988) and Shaffer and Lauder (1988) is that the process of metamorphosis does not carry with it obligatory changes in muscle activity patterns. Measurements of muscle activity patterns in larvae and metamorphosed *Ambystoma* feeding in the water showed that there was no difference in motor output (Lauder and Shaffer 1988). However, there is a dramatic drop in feeding performance after metamorphosis that appears to be due to a decrease in the mass of cranial muscles that power the fast opening phase (depressor mandibulae and rectus cervicis; Lauder and Reilly 1990) and the change from a unidirectional feeding mode in larvae to a bidirectional mode after metamorphosis (Lauder and Reilly 1988). The capture of elusive prey is less effective when water must reverse course and exit from the mouth anteriorly during the closing phase (Lauder and Shaffer 1986; Reilly and Lauder 1988a).

Feedings in the water by both larvae and metamorphosed individuals showed a characteristic bell-shaped gape profile with no plateau phase. Only minimal and occasional tongue elevation was observed in metamorphosed animals when feeding in the water.

Both muscle activity and kinematic aspects of feeding differed significantly between metamorphosed tiger salamanders feeding in the water and on land (Lauder and Shaffer 1988; Shaffer and Lauder 1988). On land, feedings showed a characteristic plateau in the gape profile and muscle activity durations tended to be longer than during aquatic feedings. Terrestrial feedings also involve tongue projection toward the prey and concomitant kinematic and motor pattern novelties in those muscles that arise at metamorphosis.

3.3 Prey Transport

Although nearly all studies of head function in salamanders have focused on the initial strike, a second vital function of head muscles and skeletal elements is the manipulation and transport of prey from the mouth to the esophagus and stomach after prey are captured. Quantitative studies of prey transport behavior have been conducted only for terrestrial feedings in *Ambystoma tigrinum*, and yet some intriguing patterns have been found. The study of prey transport is an area very much in need of further investigation, especially in nonamniote tetrapod taxa.

Terrestrial prey transport in *Ambystoma tigrinum* involves repeated cycles of jaw and hyoid motion that move the prey toward the esophagus. There is little "chewing" or reduction of the prey and captured food is usually swallowed whole. Distinct preparatory, fast opening, closing, and recovery phases are present. Reilly and Lauder (1990a, 1991b) have divided the preparatory phase in *Ambystoma tigrinum* into two parts: P1 and P2. During the first part of the preparatory phase (P1) which may last up to 5 s (Fig. 5B: P1), the mouth is closed (gape distance is zero or nearly so), and the hyobranchial apparatus is elevated pressing the prey against the roof of the mouth. Muscles that may be active during this phase include the genioglossus, interhyoideus, geniohyoideus, adductor mandibulae externus, and epaxial muscles (Reilly and Lauder 1991b). The subarcualis rectus one, rectus cervicis, and depressor mandibulae muscles are all silent. Activity in the genioglossus and buccal elevating muscles acts to press the prey against the roof of the mouth during this phase. During some prey transport sequences, the gape opens slightly (about 1 mm) during P1 phase, but this is a variable occurrence.

During the P2 phase, all muscles are silent until the last 10 to 15 ms prior to the onset of the fast opening phase, and the gape remains constant. The fast opening phase begins with the onset of gape increase due to a rapid rise in activity of the depressor mandibulae, epaxial muscles, and rectus cervicis. Interestingly, the subarcualis rectus one muscle is strongly active during prey transport even though the tongue is not projected from the mouth (Reilly and Lauder 1991b). This muscle has been thought to function primarily in tongue projection, but electromyographic data indicate that it is strongly active during the fast opening phase of prey transport. During this phase, the hyobranchial apparatus moves posteroventrally, pulling the tongue and the attached prey into the oral cavity (Fig. 5B: hyoid curve). Hyoid movement continues into the closing phase, reaching a peak as the gape closes. Thus, it is the posteroventral movement of the hyoid, tongue, and attached prey that mechanically pulls the food posteriorly.

Activity of the epaxial muscles, adductor mandibulae, rectus cervicis, genioglossus, geniohyoideus, and intermandibularis continues through the fast opening phase and into the closing phase (Fig. 5B). Activity in the subarcualis rectus one and interhyoideus is usually completed by the end of the closing phase.

Kinematic analyses of prey transport show that the gape profile (Fig. 5B) is bell-shaped and does not have a plateau (Reilly and Lauder 1990a). In addition, *Ambystoma tigrinum* does not appear to use inertial transport to any significant extent, as the position of the body relative to a fixed background remains nearly constant throughout prey transport. Posteroventral hyobranchial movement during each transport cycle (defined as the time between P1 onset times) moves the prey from 4 to 8 mm toward the esophagus. A number of transport cycles are thus necessary for prey to be completely swallowed.

The recovery phase is characterized by a closed gape with continued muscle activity in the geniohyoideus, intermandibularis, genioglossus, and low level activity in the jaw adductor muscles. During the recovery phase, the tongue and hyobranchial apparatus move anteriorly to a new position under the prey, "resetting" the hyoid for another event. A new transport cycle may then begin with the onset of the P1 phase.

The transport of prey to the esophagus during aquatic feedings has not yet been analyzed in detail. High-speed video records of aquatic prey transport show that the captured prey are moved posteriorly within the buccal cavity toward the esophagus by an anterior-to-posterior flow of water. This water flow is created by rapid jaw movements similar to those used during the initial aquatic strike. Fast opening, closing, and recovery phases are all present, and the relative timing of peak bone excursions is the same as during the strike. Quantitative comparisons between aquatic strike and transport behaviors remain to be conducted.

3.4 Comparisons Among Behaviors in Salamanders

Analyses of the process of initial prey capture and transport both in the water and on land provide data for an overall hypothesis about the phylogenetic and ontogenetic relationships among these behaviors and their underlying physiological mechanisms (Fig. 2). We propose the hypothesis that the terrestrial transport and aquatic prey capture behaviors together are distinct from initial prey capture on land, and that the kinematic and motor patterns used during terrestrial prey transport are derived from and are homologous to the process of prey transport in the water. Specifically, we suggest that when homologous muscles are considered, motor output during terrestrial prey capture will be significantly different from aquatic prey capture, aquatic transport, and terrestrial transport. In addition, the kinematic patterns associated with these three behaviors will be more similar to each other than to the kinematics of terrestrial strikes.

There are some data to support the above hypothesis. Kinematically, only terrestrial strikes possess a plateau in the gape cycle, while the gape cycles of aquatic capture and transport are similar both to each other and to terrestrial transport. The timing and pattern of hyoid movement are similar in the two transport behaviors and in aquatic strikes. Only during terrestrial strikes is the hyobranchial apparatus protracted and elevated during fast opening to serve as a platform from which the tongue is projected. Tongue projection from the mouth may also explain the gape plateau: the mouth must be held open as the tongue moves out and then back into the mouth.

Although there are few quantitative data, initial prey capture in the water and the process of prey transport both in the water and on land appear to be similar in many ways. As noted by Reilly and Lauder (1991b), these behaviors share similar profiles and timings of jaw bone movements, similar sequences of muscle activity, similar durations (aquatic strikes and both transport behaviors show relatively short duration bursts), and similar electromyographic activity profiles of muscles such as the depressor mandibulae (which shows a single peak during aquatic strikes and transports in both environments).

This hypothesis would benefit from an explicit quantitative test among all four behaviors. To date, only the terrestrial strikes and transports have been compared (Reilly and Lauder 1991b), and this comparison revealed many differences between the two behaviors in muscle activity pattern.

The similarity of terrestrial prey transport in *Ambystoma tigrinum* to aquatic feeding strikes and transports in fishes led Reilly and Lauder (1990a) to propose the hypothesis that many features of the terrestrial prey transport cycle in salamanders may be primitive characters retained from nontetrapods and to suggest homologies between the phases of the gape cycle observed in *Ambystoma tigrinum* with those proposed previously for amniote gape cycles. In amniotes, salamanders (aquatic feedings, prey transport in both environments), and fishes, the hyoid undergoes posteroventral excursions during the fast opening phase. During the closing phase, hyoid retraction reaches a peak. In amniotes, hyoid protraction occurs during the slow opening phase, a time that corresponds to the recovery and preparatory phases in salamanders and fishes. Thus, one possible hypothesis is that the preparatory and recovery phases of fishes and salamanders are together homologous with the slow opening phase in the amniote transport cycle.

According to this view, the relationship between gape and hyoid cycles during the fast opening and closing phases in amniotes is a primitive and complex functional character that has been retained from outgroup taxa such as ray-finned fishes and lungfishes (Bemis and Lauder 1986). These features of the transport cycle have thus been little modified with the origin of terrestrial vertebrate life. On the other hand, the transformation of the recovery and preparatory phases in salamanders into a slow opening phase in amniotes represents a significant functional shift that may represent a key novelty in amniote feeding systems.

4 Frog Feeding Mechanics

4.1 Background

The mechanism by which frogs capture their prey has been the subject of speculation since the 1820s (Gans and Gorniak 1982a). The most obvious feature of the frog prey capture system, tongue projection from the mouth, has been the focus of several divergent hypotheses purporting to provide a mechanical explanation for observed movements. As Gans and Gorniak (1982a) note in their comprehensive review of the history of prey capture mechanisms in frogs, proposals for the cause of tongue projection have ranged from elevated pressure in the lymphatic sublingual sinus to various hypotheses about muscular mechanisms of tongue projection. One possible reason for the number of divergent hypotheses is the considerable phylogenetic diversity that exists in feeding behavior and morphology in anurans (Trueb 1973; Emerson 1976; Regal and Gans 1976; Gans and Gorniak 1982a; Horton 1982; Trueb and Gans 1983).

In 1977, Emerson proposed a hypothesis for projection of the tongue in *Bufo marinus* that involved the hyoid apparatus, intrinsic tongue muscles such as the genioglossus, and the geniohyoideus muscles. Emerson proposed that the hyoid moves anteriorly during mouth opening as a result of relaxation of the sterno-hyoideus muscle, which contracts to hold the hyoid in place during initial mouth opening. Anterior movement of the hyoid during subsequent mouth opening then assists in anterior movement of the tongue out of the mouth. This hypothesis was tested by Gans and Gorniak (1982a, b), who conducted both high-speed filming and electromyographic studies of prey capture in *Bufo marinus*. They concluded that anterior hyoid movement does not play a role in tongue projection by *Bufo*.

The Gans and Gorniak papers proposed another mechanical hypothesis for tongue projection. They suggested that the submentalis muscle contracts to depress the symphysis of the mandible and to adduct the dentary bones. Contraction of the genioglossus medialis muscle causes the tongue to stiffen into a rigid rod that is then rotated anteriorly by the rising wedge formed by the submentalis and genioglossus basalis. The work of Gans and Gorniak corroborated the notion that muscle activity (and not hydraulic pressure) is responsible for tongue projection in *Bufo*, and showed that the hyoid probably plays only a minor role in tongue projection.

The Gans and Gorniak papers have stimulated new research on frog feeding systems in the 10 years since their publication, and data on the neural control and phylogenetic diversity of anuran feeding behavior are beginning to appear (Trueb and Gans 1983; Ewert 1984; Grobstein et al. 1985; Matsushima et al. 1985; Anderson 1990; Deban and Nishikawa 1990; Nishikawa and Gans 1990; Nishikawa and Cannatella 1991; Nishikawa and Roth 1991; Nishikawa et al. 1991). Earlier seminal studies include those by Comer and Grobstein (1978, 1981) and Ingle (1968).

4.2 Data and Current Hypotheses

The morphology of the feeding system in anurans differs considerably from that of the salamanders analyzed above. Although there is also a great deal of variation among anurans in morphology, Fig. 6 presents some basic features of the feeding mechanism in an anuran such as *Bufo* as a general basis for comparison to salamander feeding systems (Fig. 4) and as an aid to understanding the kinematic patterns of prey capture discussed below in frogs. Both mandibular depressor and adductor muscles are present, originate on the skull, and insert on the lower jaw (Fig. 6A; Duellman and Trueb 1986). The intermandibularis posterior muscle forms a sheet extending broadly between the mandibular rami (this is a primitive feature of gnathostomes; Lauder 1980b), while the submentalis muscle extends between the mandibular rami just posterior to the mentomeckelian bones at the symphysis (Fig. 6B). The geniohyoideus medialis muscle extends posteriorly from its origin at the symphysis to insert near the laryngeal cartilages (Gans and Gorniak 1982a). Dorsal to this muscle, the complicated genioglossus muscle, with many separate slips, extends from the tip of the mandible into the tongue. Many other muscles involved in the anuran feeding mechanism are described by Gans and Gorniak (1982a) and Duellman and Trueb (1986).

While recent research has shown that there may be differences in the feeding mechanisms of neobatrachian frogs and more plesiomorphic taxa (Nishikawa and Gans 1990; Nishikawa and Roth 1991; Smith and Nishikawa 1991), the overall kinematics of prey capture are qualitatively similar among the taxa studied to date. Figure 7 illustrates gape and tongue profiles measured from prey

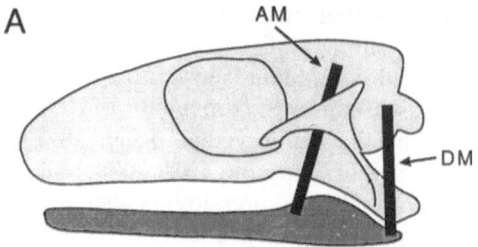

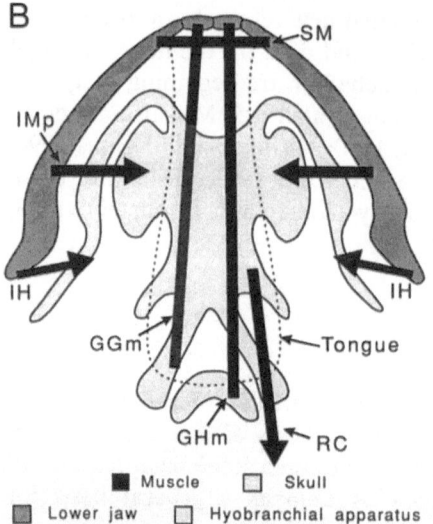

Fig. 6. Schematic diagram of representative muscles used during feeding in frogs
(based on *Bufo*) in **A** lateral view and
B ventral view. Note the expanded hyobranchial plate that contrasts with salamanders which possess separately articulating ceratobranchials. Several muscles are
not shown for clarity (see Gans and Gorniak 1982a). These include the genioglossus basalis located deep to the submentalis, the hyoglossus, and the geniohyoideus lateralis. Note that the genioglossus
medialis passes beneath the hyoid and
into the tongue (*dotted outline*) in this
view. *AM* Jaw adductor complex; *DM*
depressor mandibulae; *FFm* genioglossus
medialis; *GHm* geniohyoideus medialis;
IH interhyoideus; *IMp* intermandibularis
posterior; *RC* rectus cervicis or sternohyoideus; *SM* submentalis

capture sequences in *Ascaphus, Discoglossus,* and *Bufo* (Gans and Gorniak
1982a, b; Nishikawa and Cannatella 1991; Nishikawa and Roth 1991). The gape
profile may exhibit considerable variation from feeding to feeding (Nishikawa
and Cannatella 1991), but peak tongue projection usually occurs during the
plateau phase of the gape cycle. Gape cycles are often bimodal and the valley
between the two peaks roughly corresponds to the time of peak tongue
projection.

The gape cycle may vary in duration from 80 to over 300 ms (Nishikawa and
Cannatella 1991). In primitive anurans (e.g., *Discoglossus, Ascaphus;* Nishikawa
and Cannatella 1991; Nishikawa and Roth 1991), the tongue is projected only
about 3 mm beyond the margin of the lower jaw. Key characteristics of the
feeding mechanism in primitive anurans are (1) forward movement of the body
which begins at or before the mouth starts to open and reaches a peak near
maximum gape, which moves the head (and tongue) closer to the prey, (2)
ventral bending of the mandible at the mentomeckelian joint as the mouth
opens, and (3) considerable variation in feeding kinematics depending on the
success of the strike.

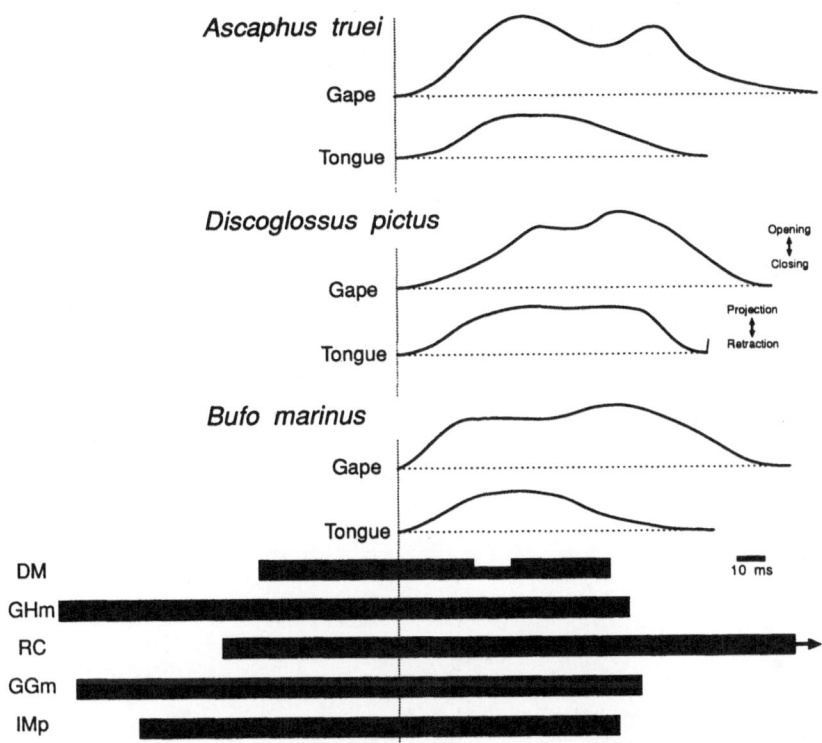

Fig. 7. Generalized anuran feeding patterns during the strike. Gape, hyoid profiles and muscle activity patterns were constructed from the data in Nishikawa and Cannatella (1991), Nishikawa and Roth (1991), and Gans and Gorniak (1982a, b). *Black bars* indicate the duration of muscle activity in *Bufo marinus* based on data in Gans and Gorniak (1982a). Note the variability in gape profiles among taxa; each taxon also shows considerable intraspecific variability (not shown)

As described for *Ascaphus* by Nishikawa and Cannatella (1991), the mouth opens and the tongue pad rotates anteriorly around the mandibular symphysis from an initial dorsal position to a ventral position at peak tongue extension. In this position, prey contact is made, and forward body and head movement continues past the prey as the tongue adheres to the prey item. Retraction of the tongue with the prey attached begins after peak gape and is completed as the mouth closes (Fig. 7).

Nishikawa and Roth (1991) conducted an experiment on the feeding system of *Discoglossus* to examine the hypothesis that the submentalis muscle is an important part of the tongue projection mechanism. As proposed by Gans and Gorniak (1982a), the submentalis muscle acts as an elevating wedge when it contracts, and forms a fulcrum around which the stiffened genioglossus muscle can rotate with the tongue. In *Discoglossus*, denervation of the submentalis

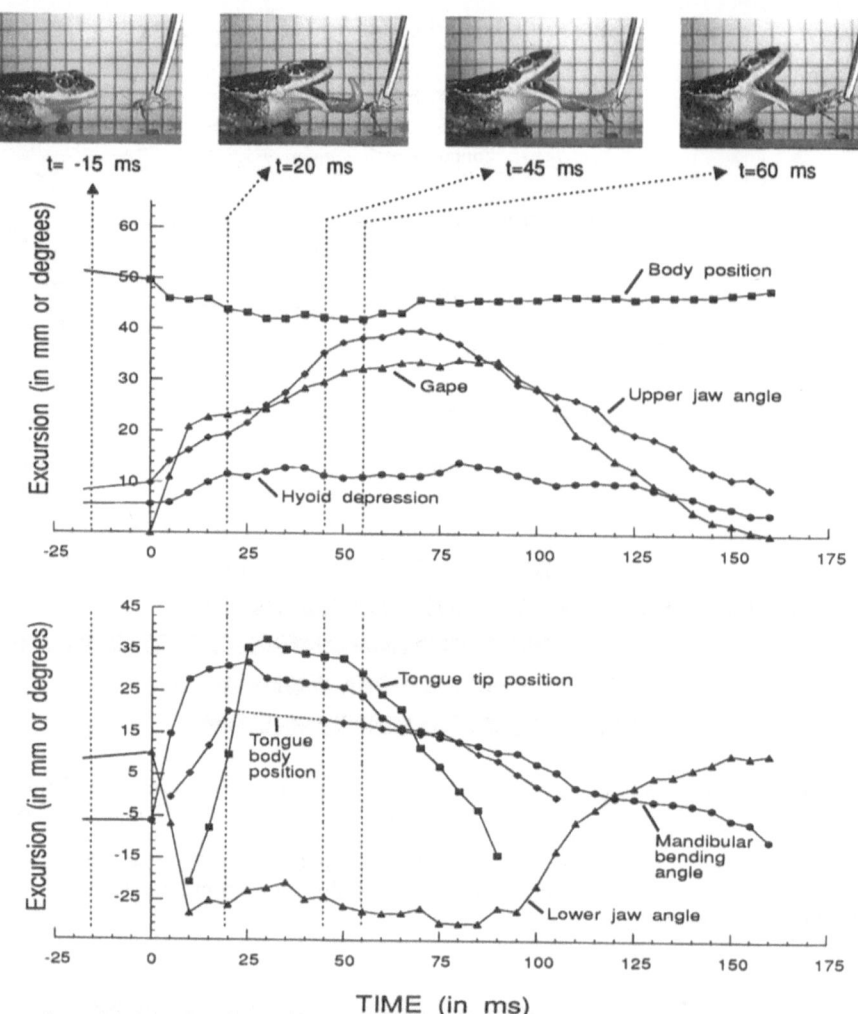

Fig. 8. Graphs of feeding kinematics in *Bufo cognatus*, with four representative video fields from this sequence shown at the *top* for reference. The video fields were obtained with a NAC HSV-400 high-speed video system in a manner similar to that described in Reilly and Lauder (1990b, 1991b, 1992). The *top graph* shows gape distance (*triangles*, in mm) measured between the tips of the upper and lower jaw; body position (*squares*, in mm) relative to a fixed reference line in the background located in front of the head; upper jaw angle (*diamonds*, in degrees) measured relative to a fixed external horizontal line; and hyoid depression (*circles*, in mm) measured as the distance from the angle of the jaw to the ventral-most buccal depression below the mandible. The *bottom graph* shows lower jaw angle (*triangles*, in degrees) measured relative to a fixed external horizontal line, mandibular bending angle (*circles*, in degrees) measured along the ventral margin of the mandible; tongue tip position (*squares*, in mm; negative values indicate that the tip of the tongue is located inside the plane of the gape, positive values show that the tongue has moved out of the mouth); and tongue body position (*diamonds*, in mm), a measurement that tracks the anterior position of the rigid body of the tongue. Note that positive angles indicate that the bone is elevated above the horizontal, while negative angles indicate that the bone is angled below the horizontal. The four video fields at the top of the

eliminated mandibular bending, but did not affect tongue projection. Denervation of the genioglossus muscle did significantly affect tongue morphology and projection during the strike, indicating that this muscle is critical to successful tongue protraction. Similar results have been obtained in *Hyla cinerea* (Deban and Nishikawa 1990). However, denervation of the submentalis in *Spea* had little effect on either tongue protrusion or mandibular bending during feeding, and both the geniohyoideus and genioglossus appear to be necessary for tongue protrusion in this genus (Smith and Nishikawa 1991).

In order to depict the overall kinematic pattern during the strike in anurans in a manner similar to that which we have previously used on ambystomatid salamanders, we present in Fig. 8 an analysis of a prey capture sequence in *Bufo*. These data were obtained from high-speed video sequences (200 fields/s) of prey capture in *Bufo cognatus*. Four representative video fields are reproduced at the top of Fig. 8 and correspond to four times during the strike. These video fields may be compared to similar high-speed video data for ambystomatid salamanders shown in Fig. 3.

The data presented in Fig. 8 reveal that the mouth opens rapidly for 10 ms and then the rate of gape increase declines as the gape increases to its peak at 80 ms after the time when the mouth first begins to open. Gape velocity thus varies considerably within the fast opening phase (time 0 to peak gape), and the fast opening phase is longer than the closing phase. The gape increase is achieved both by elevation of the head (and thus the upper jaw) and by depression of the mandible. Lower jaw angle (Fig. 8) declines precipitously during the first 10 ms of the feeding, and remains at an extended plateau throughout the rest of the fast opening phase and the beginning of the closing phase. Comparison of the gape, lower jaw, and upper jaw profiles in Fig. 8 shows that the initial rapid increase in gape is due mostly to depression of the lower jaw. For the remainder of the fast opening phase, the increase in gape is due to elevation of the head and thus the upper jaw. During the closing phase, both the lower jaw and head move to close the gape distance. During the first 10 ms of the fast opening phase, the mandible bends dramatically as the lower jaw is depressed (Fig. 8) and the peak in mandibular bending occurs only 25 ms into the strike. Mandibular bending decreases steadily during the last two-thirds of the fast opening phase and throughout the closing phase.

Forward body movement increases as the mouth begins to open and maximal forward movement occurs during the fast opening phase. Body position remains relatively stable during the closing phase. Ventral hyoid movement also begins with the onset of mouth opening, and peaks at maximum gape, before returning to the initial rest position at the end of the closing phase.

Fig. 8. *Cont.*
figure correspond to the positions on the graphs indicated by the *dashed lines*. Note that the video field at t = 20 ms shows the tongue being flipped out of the mouth. The stiff body of the tongue is held nearly horizontal in this field, while the more flexible distal portion is nearly vertical as it rotates over the base of the tongue toward the prey. Time = 0 ms indicates the field before the mouth first begins to rapidly open: prior to time 0, the gape distance is 0

The tip of the tongue, first visible within the opening mouth 10 ms after time 0 (Fig. 8), moves rapidly out of the mouth during the fast opening phase: peak tongue tip extension occurs at 30 ms in this feeding, and has returned to nearly its initial position by the onset of the closing phase. The position of the anterior extent of the stiffened body of the tongue follows a similar course, with a rapid anterior movement followed by a slow return during the last half of the fast opening phase.

There are very few comparative electromyographic data on anurans to correlate with the kinematic patterns described above, although the studies of Gans and Gorniak (1982a, b) provide excellent data on *Bufo*. Matsushima et al. (1985) do present electromyographic data on *Bufo* that resulted from snapping behavior elicited by brain stimulation. However, we believe that these data should be treated with considerable caution until quantitative kinematic analyses show that the "snapping" behavior is the same as unrestrained strikes at live prey.

Some of the electromyographic data presented by Gans and Gorniak (1982a, b) are summarized in Fig. 7 for muscles that are comparable to those of salamanders. A key feature of the strike in *Bufo* is the early onset of muscle activity relative to the onset of mouth opening. During the equivalent of the salamander preparatory phase 2 (100 ms immediately preceding the onset of mouth opening), many muscles are active in *Bufo*. The depressor mandibulae muscle shows a tendency toward a double burst pattern (or at least a biomodal spike amplitude distribution) with the second burst occurring during the plateau or dip in the gape profile.

5 Comparative Analysis of Amphibian Feeding

The data available to date on the feeding systems of anurans and salamanders, although limited in comparative scope, do point strongly to substantial differences in the physiological mechanisms underlying feeding behavior in these two clades of amphibians (Roth et al. 1990). The comparative morphological and neurobiological data, however, are of much greater comparative breadth than functional data. Kinematic patterns are just now becoming available from a variety of salamander and anuran taxa to permit quantitative comparative analyses (Erdman and Cundall 1984; Cundall et al. 1987; Larsen et al. 1989; Reilly and Lauder 1989a, 1990b, 1992; Findeis and Bemis 1990; Nishikawa and Cannatella 1991; Nishikawa and Roth 1991). Quantitative electromyographic data are only available for two taxa: *Bufo marinus* (Gans and Gorniak 1982a, b) and *Ambystoma tigrinum* (Lauder and Shaffer 1985, 1988; Shaffer and Lauder 1985b; Reilly and Lauder 1989b, 1990b, 1991b), and this greatly restricts the generality of comparisons between anurans and salamanders.

Even with the limited comparative data now available, there appear to be two key differences between anurans and salamanders in feeding function. First, if electromyographic data on *Bufo* are corroborated by future studies, then anurans would appear to use a fundamentally different motor pattern during feeding than salamanders (also see Roth et al. 1990). Second, although there may be a diversity of feeding systems within anurans, the role of the hyoid in the feeding mechanism seems to differ substantially between salamanders and anurans. The salamander feeding system is fundamentally hyoid-based. This is true for both aquatic and terrestrial feedings, as well as for strike and transport behaviors. In most species of anurans, the hyoid appears to play a relatively small role in tongue projection, and this may be related to larval specializations in feeding behavior.

An overall hypothesis of historical patterns to the major functional features of the feeding mechanism in amphibians is presented in Fig. 9 within the phylogenetic context of other vertebrate clades. Three key conclusions emerge from this analysis, and each will be considered seriatim.

First, many features of the feeding mechanism of amphibians are primitive characteristics that are retained from nontetrapod taxa. The extent to which both amphibian and amniote feeding kinematics retain primitive functions from nontetrapods has not been widely recognized, and yet some characters are conserved throughout many vertebrate clades. A prime example of functional conservatism is hyoid retraction during the fast opening phase (Fig. 9). Movement of the hyobranchial apparatus in a posteroventral direction occurs during the time in which the gape is rapidly increasing in a wide diversity of vertebrates during either the initial strike, prey transport within the mouth cavity, or both. Thus, taxa as divergent as turtles, mammals, lizards, salamanders, lungfishes, and sunfishes all exhibit hyoid retraction during the fast opening phase of prey transport.

Second, the functional patterns (both kinematic and electromyographic) involved in prey capture by amphibians do not fit the general pattern proposed for tetrapods by Bramble and Wake (1985). Bramble and Wake (1985) proposed a generalized model of jaw function during prey transport in lower tetrapods. This model predicts four distinct components of the gape cycle and associated muscle activities.

However, results from experimental studies in salamanders (Reilly and Lauder 1991b) have revealed neither the predicted muscle activity patterns nor the predicted kinematic features of the gape cycle. Rather than a single "generalized feeding cycle", tetrapods appear both to have retained primitive features from fishes and to have acquired distinct functional novelties at several phylogenetic levels (Fig. 9). In particular, amniote feeding systems possess several novelties in feeding function (Fig. 9) including extensive intraoral food processing, the presence of a slow opening phase in the gape cycle (Crompton 1989; Schwenk and Throckmorton 1989), and the common use of inertial feeding. These attributes are primitively lacking in nonamniote taxa. Specifically, salamanders and frogs lack a slow opening phase of the gape cycle and this is a primitive character inherited from nontetrapod ancestors. Indeed the slow opening phase of amniotes appears to be homologous to the preparatory and

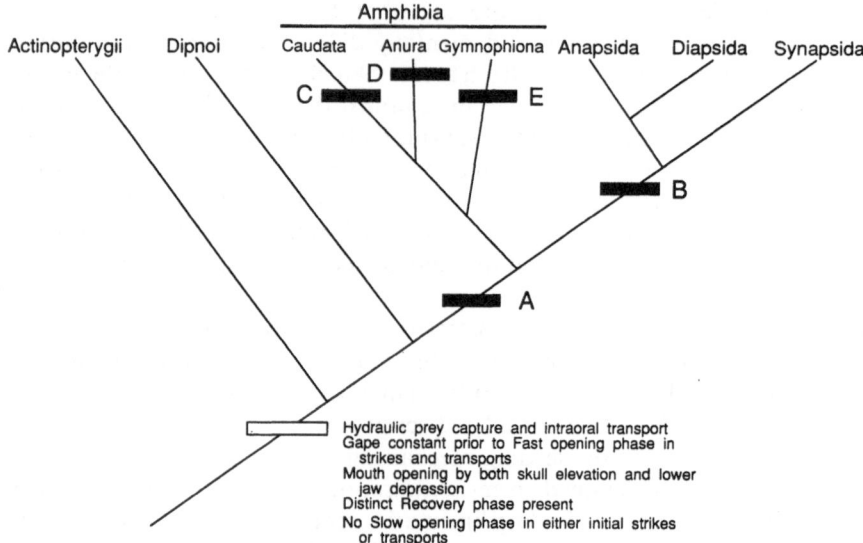

Fig. 9. Tetrapod phylogeny [with representative outgroups: ray-finned fishes (Actinopterygii) and lungfishes (Dipnoi)] to show one hypothesized historical pattern to the transformation of functional characteristics in the feeding mechanism. This figure is a modification and expansion of the hypothesis presented in Reilly and Lauder (1990a; Fig. 6). Features described next to the *open bar* at the stem of the phylogeny reflect functional characters as follows. (Note that due to the scarcity of comparative data, functional characteristics of the feeding mechanism proposed to be primitive for the entire clade. *Letters next to black bars* indicate groups of functional characters as follows. (Note that due to the scarcity of comparative data, functional characteristics listed here may turn out to be restricted to a derived set of taxa within the clade or may be more general and thus primitive for salamanders and frogs together or even for the Amphibia.) *A* Long preparatory phase prior to fast opening, tongue-based terrestrial intraoral prey capture and transport; *B* extensive intraoral food processing, inertial feeding present, recovery phase compressed into gape cycle, gape increase mostly by lower jaw depression, short slow opening phase just prior to fast opening; *C* tongue projection based on a mechanism involving the branchial skeleton and the subarcualis rectus one muscle, relatively little influence of sensory feedback on the feeding motor pattern; *D* submentalis muscle present, mandibular bending present, extensive modulation of the feeding kinematic pattern, novel strike motor pattern with jaw muscle activity in several muscles beginning well before the onset of mouth opening, lack of a distinct recovery phase; *E* novel jaw closing mechanism involving the interhyoideus posterior muscle

recovery phases together of anamniotes (Reilly and Lauder 1990a, b, 1991b), a time when the hyoid is protracting (moving anterodorsally).

Third, frogs and salamanders appear to have acquired many novelties in the feeding mechanism, both morphological and functional, that reflect both the biomechanical requirements of terrestrial feeding and phylogenetic diversification into distinct feeding systems. As emphasized by Roth et al. (1990), frogs appear to exhibit considerable diversification in feeding function, with several different methods of tongue projection used within the Anura (Trueb and Gans 1983; Roth et al. 1990; Nishikawa and Roth 1991; Smith and Nishikawa 1991).

The anuran feeding mechanism displays much more extensive diversification in function of the hyobranchial apparatus, variability in the feeding mechanism (Anderson 1990; Nishikawa and Cannatella 1991; Nishikawa and Roth 1991), and involvement in sensory feedback pathways (Nishikawa and Gans 1990) than do salamander taxa that have been studied so far. Indeed, the salamander feeding system seems to retain many more primitive (nontetrapod) morphological and functional features than the anuran feeding system which has acquired numerous functional and morphological apomorphies.

Comparison of the data presented here for salamanders (Figs. 4 and 5) and frogs (Figs. 6, 7, and 8) illustrates the many differences between these two taxa. Most larval salamanders retain the basic morphological configuration of the nontetrapod feeding mechanism with a hyoid arch, branchial elements, and associated muscles, as well as a large number of functional similarities in motor pattern and kinematics to outgroup taxa (Lauder and Shaffer 1985; Lauder and Reilly 1988). Even transformed salamanders, while acquiring both morphological and functional novelties at metamorphosis for terrestrial tongue projection (Lauder and Shaffer 1988, 1993), retain outgroup functional characteristics during prey transport behaviors (Reilly and Lauder 1990a, 1991b). Thus, the gape cycle during prey transport lacks the plateau phase, and is similar to the gape cycle of larval salamanders, transformed salamanders feeding in the water, and outgroup taxa like lungfishes and ray-finned fishes. In addition, analyses of variation in the feeding mechanism of ambystomatid salamanders have shown that, while there is often significant variation among individuals, there is very little variation in either kinematic or electromyographic patterns within an individual when different prey are caught or if prey are captured or missed (Shaffer and Lauder 1985a, b; Reilly and Lauder 1989b, 1990a, 1991b).

When compared to both salamanders and outgroup taxa (Fig. 9), frogs possess a highly modified hyoid apparatus, jaws and skull, novel features of the buccal musculature (such as the submentalis muscle), relatively variable strike kinematics that are altered based on strike success and prey size, and many novel features of the kinematic pattern of the strike. In addition, frogs also possess a highly modified larval stage with a feeding mechanism that also exhibits a large number of novel features and considerable diversity across taxa (De Jongh 1968; Wassersug and Hoff 1979; Ruibal and Thomas 1988).

Bone movement and muscle function in frogs differ from salamanders and nontetrapod outgroups in at least four major attributes evident from currently available data. First, there is no recovery phase (this may be due in part to the many morphological, behavioral, and ecological novelties in anuran larvae). In *Bufo* for example (Fig. 8), hyoid depression and tongue projection return to their initial rest positions by the end of the closing phase. Data for *Ascaphus* and *Discoglossus* (Nishikawa and Cannatella 1991; Nishikawa and Roth 1991) also show that these taxa lack a recovery phase comparable to the plesiomorphic condition for tetrapods (Fig. 9). Second, patterns of hyoid movement during the strike are quite different in both fish (in which the hyoid moves posteroventrally) and terrestrial salamanders (in which the hyobranchial apparatus is protracted).

Third, the patterns of jaw movement in frogs may be considerably different from outgroup taxa. For example, in *Bufo* (Fig. 8), the lower jaw drops rapidly as the mouth opens, and maintains a plateau until the start of the closing phase. Not all frogs show this specific pattern (Nishikawa and Roth 1991) and many frog species show considerable intraspecific (as well as interspecific) variation in upper and lower jaw kinematics not found in outgroup clades. Fourth, the electromyographic patterns at the strike in frogs appear to differ (for homologous muscles) from those in salamanders. Jaw muscles in frogs may be active for a significant period of time prior to the onset of mouth opening, something not yet seen in salamanders.

In salamanders, quantitative analyses of electromyographic patterns at the strike are only available for *Ambystoma tigrinum* (larvae: Lauder and Shaffer 1985, 1988; transformed individuals: Lauder and Shaffer 1988; Reilly and Lauder 1990b, 1991b, in prep.), and these data show that the onset of muscle activity is indeed nearly synchronous (within 5 ms, even for antagonistic muscles). Reilly and Lauder (1990b) also showed that each muscle in the feeding mechanism possesses a distinctive pattern of activation and amplitude variation during the strike.

Although this discussion has focused on anurans and salamanders, caecilians form a third clade that is important to future discussions of the evolution of form and function in the amphibian feeding mechanism. As yet, the only data available on caecilians are those of Bemis et al. (1983) who studied prey transport, Nussbaum (1983), O'Reilly (1990), and O'Reilly and Deban (1991). Bemis et al. (1983) and Nussbaum (1983) showed that caecilians possess many novelties in the feeding mechanism, including a new mechanism of jaw closing, and involving the interhyoideus posterior muscle. Further data on all aspects of caecilian feeding mechanisms (initial strike and transports in both aquatic and terrestrial environments) are badly needed.

6 Recommendations and Future Directions

In describing and comparing feeding mechanisms of anurans and salamanders, several directions for future research have become apparent. In addition, we would like to make a number of recommendations to facilitate comparative analyses among amphibian taxa.

Our first recommendation is that a standard set of terminology be developed for comparing feeding function that is phylogenetically based. We suggest, therefore, that the terminology used to describe the plesiomorphic feeding pattern for tetrapods be extended to all amphibian taxa as the basis for initial descriptions of feeding behavior. Thus, the preparatory, fast opening, closing, and recovery phases with their associated definitions based on gape and hyoid cycles (Reilly and Lauder 1990a, b; Lauder and Prendergast 1992) could be used as a foundation of primitive kinematic features present in outgroups. Not all taxa will possess all these phases (e.g., frogs appear to lack a recovery phase), and new kinematic features will certainly be described that require new terms (the frog fast opening phase may need to be subdivided into distinct parts). These apomorphic behaviors should be given apomorphic terms, and not be confusingly described using terminology with well-established meanings in plesiomorphic taxa. Additional descriptive terms for tongue movements should be agreed on so that kinematic phases such as projection, protraction, and retraction have accepted meanings to all workers. The precise definition of all kinematic terms will be of considerable value when studies of the feeding motor pattern are expanded to include more taxa.

Our second recommendation stems from the unfortunate tendency of workers investigating one amphibian taxon to proceed independently from functional research being done on other taxa. We suggest that specific collaborative research utilizing frogs, salamanders, and caecilians be conducted to permit direct, quantitative, statistical comparisons in feeding system function across taxa. Functional research on amphibian feeding mechanisms has proceeded to date via analyses of behavior within each major amphibian clade. This is appropriate to establish a baseline of data, but does not encourage quantitative comparisons of feeding systems across taxa.

Third, we suggest that when comparisons among major amphibian clades are conducted, that homologous morphological and functional components be compared using quantitative statistical methods. It is all too easy to make qualitative generalizations based only on a few species and on the analysis of a limited number of variables. Analyses of motor patterns, for example, should involve appropriate signal filtering, quantification, and statistical analysis of measured variables before conclusions regarding differences among taxa are announced. While uncertainties in the homology of morphological features across taxa may limit the breadth of functional comparisons, there are a number of muscles (such as the intermandibularis, geniohyoideus, and depressor mandibulae) which may be suitable subjects for comparative studies.

Based on the research described in this chapter, there are three areas which we view as being in particular need of investigation. First, there are currently no data on prey transport in anurans. Transport behaviors are of special importance because they may provide evidence that primitive motor and kinematic patterns, thought to be absent when only the initial strike is studied, are in fact present. Second, functional data on caecilians are lacking and, in particular, comparative analyses of aquatic and terrestrial prey capture will be of interest to contrast with currently available data on salamanders. Third, a broadly quantitative comparison among strike and transport behaviors in all three major clades of amphibians and nontetrapod outgroups will test the extent of phylogenetic conservatism in both strike and transport behaviors. It is somewhat shocking to discover, given the rather large number of published generalizations about amphibian feeding behavior, that quantitative data on muscle function are available only for two species of Amphibia. We suggest that further generalizations should await the production of experimental data on which to base functional and historical hypotheses. Such comparative functional data will also contribute to our understanding of the patterns of diversification in amphibian feeding mechanisms, a key theme especially in the Anura.

The feeding system of amphibians offers an excellent study system within which to investigate problems of structural and functional evolution. The three major clades of extant amphibians have diversified structurally to a remarkable extent, ecologically, environmentally, and morphologically. The examination of how functional patterns have been transformed in this clade will be of great assistance in our attempts to understand major features of vertebrate evolution such as the transition from water to land, and the nature of historical patterns to functional characters.

Acknowledgments. We thank Bruce Jayne for assistance with the *Bufo* recordings, and Miriam Ashley-Ross, Peter Wainwright, and Gary Gillis for comments on the manuscript, Kiisa Nishikawa and Jim O'Reilly kindly provided extensive reviews of the chapter. This research was supported by NSF grants DIR 8820664, BSR 8520305, and DCB 8710210 to George Lauder.

References

Anderson CW (1990) The effect of prey size on feeding kinematics in two species of ranid frogs. Am Zool 30: 140A

Bemis WE (1987) Feeding systems of living Dipnoi: anatomy and function. J Morphol Suppl 1: 249–275

Bemis WE, Lauder GV (1986) Morphology and function of the feeding apparatus of the lungfish, *Lepidosiren paradoxa* (Dipnoi). J Morphol 187: 81–108

Bemis WE, Schwenk K, Wake MH (1983) Morphology and function of the feeding apparatus in *Dermophis mexicanus* (Amphibia: Gymnophiona). Zool J Linn Soc Lond 77: 75–96

Bold JR (1977) Dissorophoid relationships and ontogeny, and the origin of the Lissamphibia. J Paleontol 51: 235–249

Bramble DM, Wake DB (1985) The feeding mechanisms of lower tetrapods. In: Hildebrand M, Bramble DM, Liem KF, Wake DB (eds) Functional vertebrate morphology. Havard Univ Press, Cambridge, pp 230–261

Carroll RL, Holmes R (1980) The skull and jaw musculature as guides to the ancestry of salamanders. Zool J Linn Soc Lond 68: 1–40

Comer C, Grobstein P (1978) Prey acquisition in atectal frogs. Brain Res 153: 217–221

Comer C, Grobstein P (1981) Tactually elicited prey acquisition behavior in the frog, *Rana pipiens*, and a comparison with visually elicited behavior. J Comp Physiol A 142: 141–150

Crompton AW (1989) The evolution of mammalian mastication. In: Wake DB, Roth G (eds) Complex organismal functions: integration and evolution in vertebrates. Wiley, London, pp 23–40

Cundall D, Lorenz-Elwood J, Groves JD (1987) Asymmetric suction feeding in primitive salamanders. Experientia 43: 1229–1231

Deban SM, Nishikawa K (1990) The mechanism of tongue protrusion in *Hyla cineria* and its evolutionary implications. Am Zool 30: 141A

De Jongh HJ (1968) Functional morphology of the jaw apparatus of larval and metamorphosing *Rana temporaria* L. Neth J Zool 18: 1–103

Druner L (1902) Studien zur Anatomie der Zungenbein-, Kiemenbogen-, und Kehlkopfmuskelen der Urodelen, I Theil. Zool Jahrb Anat 15: 435–622

Druner L (1904) Studien zur Anatomie der Zungenbein-, Kiemenbogen-, und Kehlkopfmuskelen der Urodelen, II Theil. Zool Jahrb Anat 19: 361–690

Duellman WE, Trueb L (1986) Biology of amphibians. McGraw Hill, New York

Edgeworth FH (1935) The cranial muscles of vertebrates. Cambridge Univ Press, Cambridge

Emerson S (1976) A preliminary report on the superficial throat musculature of the Micro-
hylidae and its possible role in tongue action. Copeia 1976: 546–551

Emerson S (1977) Movement of the hyoid in frogs during feeding. Am J Anat 149: 115–120

Erdman S, Cundall D (1984) The feeding apparatus of the salamander *Amphiuma tridactylum*:
morphology and behavior. J Morphol 181: 175–204

Ewert JP (1984) Tectal mechanisms that underlie prey-catching and avoidance behaviors in
toads. In: Vanegas H (ed) Comparative neurology of the optic system. Plenum Press, New
York, pp 247–416

Findeis EK, Bemis WE (1990) Functional morphology of tongue projection in *Taricha torosa*
(Urodela: Salamandridae). Zool J Linn Soc Lond 99: 129–157

Francis ETB (1934) The anatomy of the salamander. Oxford Univ Press, London

Gans C (1980) Biomechanics: an approach to vertebrate biology. University of Michigan
Press, Ann Arbor

Gans C, Gorniak GC (1982a) Functional morphology of lingual protrusion in marine toads
(*Bufo marinus*). Am J Anat 163: 195–222

Gans C, Gorniak GC (1982b) How does the toad flip its tongue? Test of two hypotheses.
Science 216: 1335–1337

Grobstein P, Reyes A, Zwanziger L, Kostyk SK (1985) Prey orienting in frogs: accounting for
variations in output with stimulus distance. J Comp Physiol A 156: 775–785

Horton P (1982) Diversity and systematic significance of anuran tongue musculature. Copeia
1982: 595–602

Ingle D (1968) Visual releasers of prey-catching behavior in frogs and toads. Brain Behav Evol
1: 500–518

Larsen JH, Beneski JT, Wake DB (1989) Hyolingual feeding systems of the Plethodontidae:
comparative kinematics of prey capture by salamanders with free and attached tongues. J
Exp Zool 252: 25–33

Lauder GV (1979) Feeding mechanisms in primitive teleosts and in the halecomorph fish *Amia
calva*. J Zool (Lond) 187: 543–578

Lauder GV (1980a) The role of the hyoid apparatus in the feeding mechanism of the living
coelacanth, *Latimeria chalumnae*. Copeia 1980: 1–9

Lauder GV (1980b) Evolution of the feeding mechanism in primitive actinopterygian fishes: a
functional anatomical analysis of *Polypterus*, *Lepisosteus*, and *Amia*. J Morphol 163:
283–317

Lauder GV (1980c) The suction feeding mechanism in sunfishes (*Lepomis*): an experimental
analysis. J Exp Biol 88: 49–72

Lauder GV (1980d) Hydrodynamics of prey capture in teleost fishes. In: Schenck D (ed)
Biofluid mechanics, vol 2. Plenum Press, New York, pp 161–181

Lauder GV (1983a) Prey capture hydrodynamics in fishes: experimental tests of two models. J
Exp Biol 104: 1–13

Lauder GV (1983b) Food capture. In: Webb PW, Weihs D (eds) Fish biomechanics. Praeger,
New York, pp 280–311

Lauder GV (1985a) Aquatic feeding in lower vertebrates. In: Hildebrand M, Bramble DM,
Liem KF, Wake DB (eds) Functional vertebrate morphology. Harvard University Press,
Cambridge, pp 210–229

Lauder GV (1985b) Functional morphology of the feeding mechanism in lower vertebrates. In:
Duncker H-R, Fleischer G (eds) Functional morphology of vertebrates. Springer, Berlin
Heidelberg New York, pp 179–188

Lauder GV (1986a) Aquatic prey capture in fishes: experimental and theoretical approaches. J
Exp Biol 125: 411–416

Lauder GV (1986b) Homology, analogy, and the evolution of behavior. In: Nitecki M, Kitchell
J (eds) The evolution of behavior. Oxford University Press, Oxford, pp 9–40

Lauder GV (1990) Functional morphology and systematics: studying functional patterns in an
historical context. Annu Rev Ecol Syst 21: 317–340

Lauder GV (1991) Biomechanics and evolution: integrating physical and historical biology in the study of complex systems. In: Rayner JMV, Wooton RJ (eds) Biomechanics in evolution. Cambridge Univ Press, Cambridge, pp 1–19

Lauder GV, Prendergast T (1992) Kinematics of aquatic prey capture in the snapping turtle, *Chelydra serpentina*. J Exp Biol 164: 55–78

Lauder GV, Reilly SM (1988) Functional design of the feeding mechanism in salamanders: causal bases of ontogenetic changes in function. J Exp Biol 134: 219–233

Lauder GV, Reilly SM (1990) Metamorphosis of the feeding mechanism in tiger salamanders (*Ambystoma tigrinum*): the ontogeny of cranial muscle mass. J Zool (Lond) 222: 59–74

Lauder GV, Shaffer HB (1985) Functional morphology of the feeding mechanism in aquatic ambystomatid salamanders. J Morphol 185: 297–326

Lauder GV, Shaffer HB (1986) Functional design of the feeding mechanism in lower vertebrates: unidirectional and bidirectional flow systems in the tiger salamander. Zool J Linn Soc 88: 277–290

Lauder GV, Shaffer HB (1988) The ontogeny of functional design in the tiger salamander *Ambystoma tigrinum*: are motor patterns conserved during major morphological transformations? J Morphol 197: 249–268

Lauder GV, Shaffer HB (1993) Design of the aquatic vertebrate skull: major patterns and their evolutionary interpretations. In: Hanken J, Hall B (eds) The vertebrate skull. University of Chicago Press, Chicago Vol. 3, pp 113–149

Liem KF (1970) Comparative functional anatomy of the Nandidae (Pisces: Teleostei). Fieldiana Zool 56: 1–166

Liem KF (1980) Acquisition of energy by teleosts: adaptive mechanisms and evolutionary patterns. In: Ali MA (ed) Environmental physiology of fishes. Plenum Press, New York

Lombard RE, Wake DB (1976) Tongue evolution in the lungless salamanders, family Plethodontidae. I. Introduction, theory and a general model of dynamics. J Morphol 148: 265–286

Lombard RE, Wake DB (1977) Tongue evolution in the lungless salamanders, family Plethodontidae. II. Function and evolutionary diversity. J Morphol 153: 39–80

Matsushima T, Satou M, Ueda K (1985) An electromyographic analysis of electrically-evoked prey-catching behavior by means of stimuli applied to the optic tectum in the Japanese toad. Neurosci Res 3: 154–161

Miller BT, Larsen JH (1990) Comparative kinematics of terrestrial prey capture in salamanders and newts (Amphibia: Urodela: Salamandridae). J Exp Zool 256: 135–153

Nishikawa K, Cannatella DC (1991) Kinematics of prey capture in the tailed frog, *Ascaphus truei* (Anura Ascaphidae). Zool J Linn Soc Lond 103: 289–307

Nishikawa K, Gans C (1990) Neuromuscular control of prey capture in the marine toad, *Bufo marinus*. Am Zool 30: 141A

Nishikawa K, Roth G (1991) The mechanism of tongue protraction during prey capture in the frog, *Discoglossus pictus*. J Exp Biol 159: 217–234

Nishikawa K, O'Reilly JC, Cannatella DC (1991) Biomechanical and behavioral transitions in the evolution of frog feeding. Am Zool 31(5): 52A

Nussbaum RA (1983) The evolution of a unique dual jaw-closing mechanism in caecilians (Amphibia: Gymnophiona) and its bearing on caecilian ancestry. J Zool (Lond) 199: 545–554

O'Reilly JC (1990) Aquatic and terrestrial feeding in caecilians (Amphibia: Gymnophiona): a possible example of phylogenetic constraint. Am Zool 30: 140A

O'Reilly JC, Deban SM (1990) The evolution of aquatic prey capture in amphibians: phylogenetic constraints and exaptations. Am Zool 31(5): 17A

Regal PJ (1966) Feeding specializations and the classification of terrestrial salamanders. Evolution 20: 392–407

Regal PJ, Gans C (1976) Functional aspects of the evolution of frog tongues. Evolution 30: 718–734

Reilly SM, Lauder GV (1988a) Ontogeny of aquatic feeding performance in the eastern newt *Notophthalmus viridescens* (Salamandridae). Copeia 1988: 87–91

Reilly SM, Lauder GV (1988b) Atavisms and the homology of hyobranchial elements in lower vertebrates. J Morphol 195: 237–245

Reilly SM, Lauder GV (1989a) Kinetics of tongue projection in *Ambystoma tigrinum*: quantitative kinematics, muscle function and evolutionary hypotheses. J Morphol 199: 223–243

Reilly SM, Lauder GV (1989b) Physiological bases of feeding behavior in salamanders: do motor patterns vary with prey type? J Exp Biol 141: 343–358

Reilly SM, Lauder GV (1990a) The evolution of tetrapod feeding behavior: kinematic homologies in prey transport. Evolution 44: 1542–1557

Reilly SM, Lauder GV (1990b) The strike of the tiger salamander: quantitative electromyography and muscle function during prey capture. J Comp Physiol A 167: 827–839

Reilly SM, Lauder GV (1990c) Metamorphosis of cranial design in tiger salamanders (*Ambystoma tigrinum*): a morphometric analysis of ontogenetic change. J Morphol 204: 121–137

Reilly SM, Lauder GV (1991a) Experimental morphology of the feeding mechanism in salamanders. J Morphol 210: 33–44

Reilly SM, Lauder GV (1991b) Prey transport in the tiger salamander (*Ambystoma tigrinum*): quantitative electromyography and muscle function in tetrapods. J Exp Zool 260: 1–17

Reilly SM, Lauder GV (1992) Morphology, behavior and evolution: comparative kinematics of aquatic feeding in salamanders. Brain Behav Evol 40: 182–196

Roth G (1976) Experimental analysis of the prey catching behavior of *Hydromantes italicus* Dunn (Amphibia, Plethodontidae). J Comp Physiol 109: 47–58

Roth G (1978) The role of stimulus movement patterns in the prey catching behavior of *Hydromantes genei* (Amphibian, Plethodontidae). J Comp Physiol A 123: 261–264

Roth G (1982) Responses in the optic tectum of the salamander *Hydromantes italicus* to moving prey stimuli. Exp Brain Res 45: 386–392

Roth G, Nishikawa K, Wake DB, Dicke U, Matsushima T (1990) Mechanics and neuromorphology of feeding in amphibians. Neth J Zool 40: 115–135

Ruibal R, Thomas E (1988) The obligate carnivorous larvae of the frog, *Lepidobatrachus laevis* (Leptodactylidae). Copeia 1988: 591–604

Schwenk K, Throckmorton GS (1989) Functional and evolutionary morphology of lingual feeding in squamate reptiles: phylogenetics and kinematics. J Zool (Lond) 219: 153–176

Schwenk K, Wake DB (1988) Medium-independent feeding in a plethodontid salamander: tongue projection and prey capture under water. Am Zool 28: 115A

Severtsov AS (1971) The mechanism of food capture in tailed amphibians. Dokl Biol 197: 185–187

Shaffer HB, Lauder GV (1985a) Patterns of variation in aquatic ambystomatid salamanders: kinematics of the feeding mechanism. Evolution 39: 83–92

Shaffer HB, Lauder GV (1985b) Aquatic prey capture in ambystomatid salamanders: patterns of variation in muscle activity. J Morphol 183: 273–284

Shaffer HB, Lauder GV (1988) The ontogeny of functional design: metamorphosis of feeding behavior in the tiger salamander (*Ambystoma tigrinum*). J Zool (Lond) 216: 437–454

Smith SR, Nishikawa KC (1991) The mechanism of tongue protrusion in the spade foot toad *Spea multiplicatus*. Am Zool 31(5): 52A

Thexton AJ, Wake DB, Wake MH (1977) Tongue function in the salamander *Bolitoglossa occidentalis*. Arch Oral Biol 22: 361–366

Trueb L (1973) Bones, frogs, and evolution. In: Vial J (ed) The evolutionary biology of the Anura. Univ of Missouri Press Columbus, pp 65–132

Trueb L, Cloutier R (1991) A phylogenetic investigation of the inter- and intra-relationships of the Lisamphibia (Amphibia: Temnospondyli). In: Schultze H-P. Trueb L (eds) Origins of the higher groups of tetrapods: controversy and consequences. Cornell Univ Press, Ithaca, pp 223–313

Trueb L, Gans C (1983) Feeding specializations of the Mexican burrowing toad, *Rhinophrynus dorsalis* (Anura: Rhinophrynidae). J Zool (Lond) 199: 189–208

Wassersug RJ, Hoff K (1979) A comparative study of the buccal pumping mechanism of tadpoles. Biol J Linn Soc 12: 225–259

Chapter 7

Biomechanics of the Hyolingual System in Squamata

V.L. Bels[1], M. Chardon[2] and K.V. Kardong[3]

Contents

1 Introduction

The hyolingual system of Squamata is a highly versatile system used in different feeding, drinking, chemoreception, and social behaviors. In each of these activities, either the entire hyolingual system or one of its elements is used. For instance, in the majority of lizards, the tongue acts as the main element for liquid

[1] Agronomic Centre of Applied Researches, C.A.R.A.H., rue Paul Pastur 11, B-7800 Ath, Belgium

[2] Laboratory of Functional Morphology, University of Liège, Quai Van Benden 22, B-4020 Liège, Belgium

[3] Department of Zoology, Washington State University, Pullman, Washington 99164-4236, USA

Advances in Comparative and Environmental Physiology, Vol. 18
© Springer-Verlag Berlin Heidelberg 1994

uptake, intraoral food and liquid transport, and in chemoreception, whereas the hyoid apparatus plays a major role during social interactions by acting on the ventral floor of the throat. In varanids, the hyoid apparatus is involved in both deglutition of foods and liquids, and during social displays.

Early morphological studies noted the differences in morphology and function of the tongue and the hyoid apparatus (see, for example, Bell 1826; Houston 1828; Duvernoy 1836; Brücke 1872; Kent 1895; Chemin 1899; Gandolfi 1908; Zavattari 1911; Willard 1915; Von Geldern 1919; Camp 1923; Gnanamuthu 1930a, 1937; Richter 1933; Kesteven 1944; El-Toubi 1947; Oelrich 1956; Sondhi 1958; and see Table 1 in Avery and Tanner 1982).

Based upon two tongue functions (feeding and chemoreception), Underwood (1971) separated lizards into ascalabotans (Iguanidae, Agamidae, and Gekkonidae) and autarchoglossans (all other families). Although the tongue may be specialized for chemosensory behavior in varanids and snakes, the general use of the tongue for gathering chemical cues has proven to be widely distributed among lizard families (Cooper 1989, 1990a,b).

Based primarily on morphological characters of the tongue, the division of lizards into two sister groups (Estes and Pregill 1988) has been supported (Schwenk 1987, 1988). This taxonomic division emphasized that iguanians (Agamidae, Iguanidae, and Chameleontidae) caught food with their tongues (tongue protrusion), whereas the scleroglossans (all other families) used only their jaws.

Within the hyolingual apparatus, these evolutionary trends in squamates were based only on the morphology of the tongue and its role in prey capture. However, the tongue serves other roles besides feeding; the hyoid apparatus not only contributes to various lingual functions but supports specialized functions not shared with the tongue. In fact, the hyoid apparatus may act simultaneously or not at all with the tongue even during feeding activities (Smith 1984). Therefore, our purpose is to review the hyolingual apparatus in a broader context of form and function, and then to evaluate the significance of this system to evolutionary events within squamates.

2 Functional and Biomechanical Studies of the Hyolingual System

2.1 Feeding

A feeding episode includes up to four successive phases: capture (food intake into the buccal cavity), reduction, transport (to the esophagus through the buccal cavity), and swallowing or deglutition. Food acquisition involves different behavioral strategies (Greene 1982, 1983; Mushinsky 1987). Before swallowing food, killing, crushing, and reducing facilitate the swallowing and digestion. These feeding acts are clearly separated into successive phases (Bels and Baltus 1989), or overlap during food ingestion (Smith 1984; Schwenk and Throckmorton

1989). In snakes, these phases are absent (reduction) or included in the jaw transport cycles (Frazzetta 1966; Kardong 1974, 1977, 1986; Cundall and Gans 1979; Cundall 1987).

In this chapter, the term "reduction" is used for all the jaw cycles that crush or break up the captured prey, the term "transport" for the intraoral displacements of the prey to the pharynx and the entrance of the esophagus, and the term "swallowing" for pharyngeal packing, pharyngeal compression, passage into, then along the esophagus, and "cleaning".

2.1.1 Kinematics and Biomechanics

2.1.1.1 Capture

Squamates capture various types of food that pose different biomechanical problems. In general, lizards are insectivorous, but some species, particularly in a xeric environment, eat plants (Iverson 1982). Carnivory has been reported for a few species (Auffenberg 1972, 1978, 1981). Diet specializations are often accompanied by morphological and physiological changes in the jaw apparatus, such as in the teeth and the mandibular mechanics (Presch 1974; Rieppel and Labhardt 1979) and the digestive tract which may deploy symbiotic microbes and valves in the colon of herbivorous lizards (Greene 1982; Iverson 1982). Snakes are carnivores that can feed on large prey. Greene (1983) hypothesized "that very early snakes used constriction and powerful jaws to feed on elongated, heavy prey". This would have permitted a shift from feeding often on small items to feeding infrequently on heavy items, without initially requiring major changes in jaw structure relative to a lizard-like predator (Greene 1983, p. 431).

The relationships between hyolingual morphology and diet specializations are not yet understood in lizards or in snakes. For instance, herbivory is restricted to Agamidae, Iguanidae, and Scincidae that possess a completely different tongue morphology. Smith (1986) emphasized that there is little evidence that varanid morphology is related to consumption of large prey. In his extensive study of the tongue morphology in lizards, Schwenk (1988) concluded, "Hence, phylogeny is a better predictor of tongue morphology than is ecology".

The prey capture phase involves bringing the jaws over the prey so that it enters the buccal cavity. Kinematic data of displacements of the jaws, tongue, and hyoid apparatus do not show large differences during food capture between specialist and generalist iguanian lizards. For instance, a herbivorous agamid, *Uromastyx aegyptius*, uses its tongue in the same way as an insectivorous agamid, *Agama agama* (Schwenk and Throckmorton 1989; Kraklau 1991). The only quantitative differences to emerge are in some properties of the cycles of tongue displacement, which in herbivorous species are slower than in insectivores or carnivores (Schwenk and Throckmorton 1989; Bels 1990a). The use of the tongue implies functional consequences in displacements of the jaw apparatus that are illustrated in this chapter.

The primitive mode of prey capture in lizards is based on tongue protraction (Schwenk and Throckmorton 1989). In iguanians, the tongue is protruded (Agamidae, Iguanidae) or projected (Chameleontidae) to the prey. Then the jaws are opened, allowing the retracting tongue to carry the food to the buccal cavity. Tongue protrusion or projection is accompanied by jaw opening characterized by a slow opening (SO) stage and fast opening (FO) stage (Fig. 1).

In scleroglossans, which do not use their tongues, such a two-stage division is completely absent (Bels and Goosse 1990). Instead, some scleroglossans, such as *Tiliqua scincoides* feeding on small prey (i.e., mealworms), exhibit a complex prey capture sequence that consists of a fixation of the prey between the tips of the jaws without the use of the tongue. The tongue is then used for the first time following fixation during the posterior displacements of the prey into the buccal cavity (Fig. 2). The gape increase during tongue transport is also divided into slow (SO) and fast (FO) stages. In snakes, jaw displacements lack SO and FO stages (Kardong and Bels, pers. observ., but see Cundall and Gans 1979).

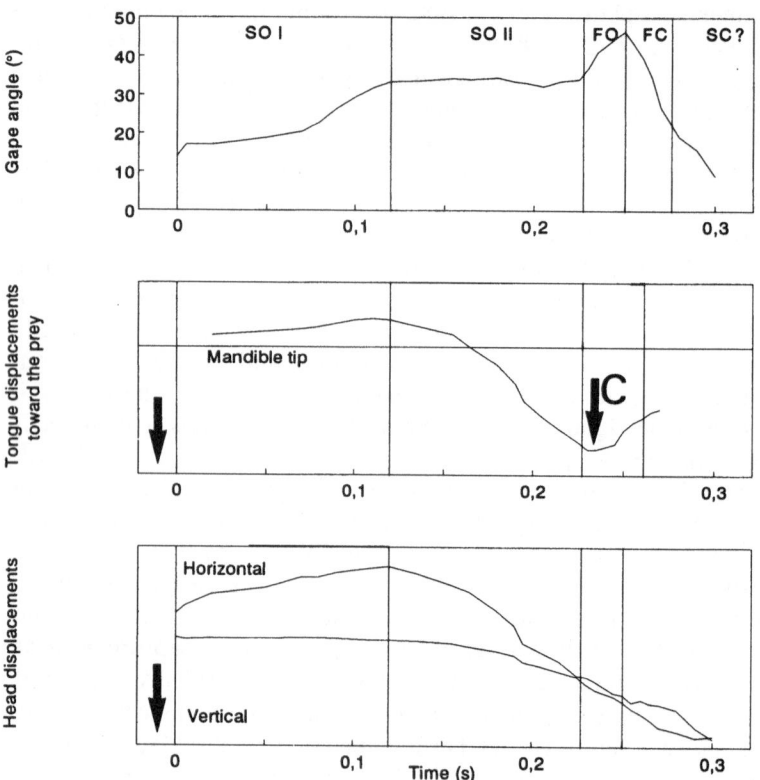

Fig. 1. Generalized kinematic diagrams of gape, tongue, and cyclic head displacements during prey capture of a typical iguanian lizard. These diagrams are based on data from agamids and iguanids (Schwenk and Throckmorton 1989; Bell 1990; Bels 1990a; Kraklau 1991; Wainwright et al. 1991; Delheusy and Bels 1992). The lizards move their head toward the prey simultaneously with tongue protrusion. *C* Contact between the prey and the tongue

In lizards, the differences in tongue use are correlated with differences in the microanatomy. For instance, the presence of serous and seromucous secretory cells in the tongues of iguanians may function adhesively to hold prey on the tongue as in Chameleontidae (Bell 1989).

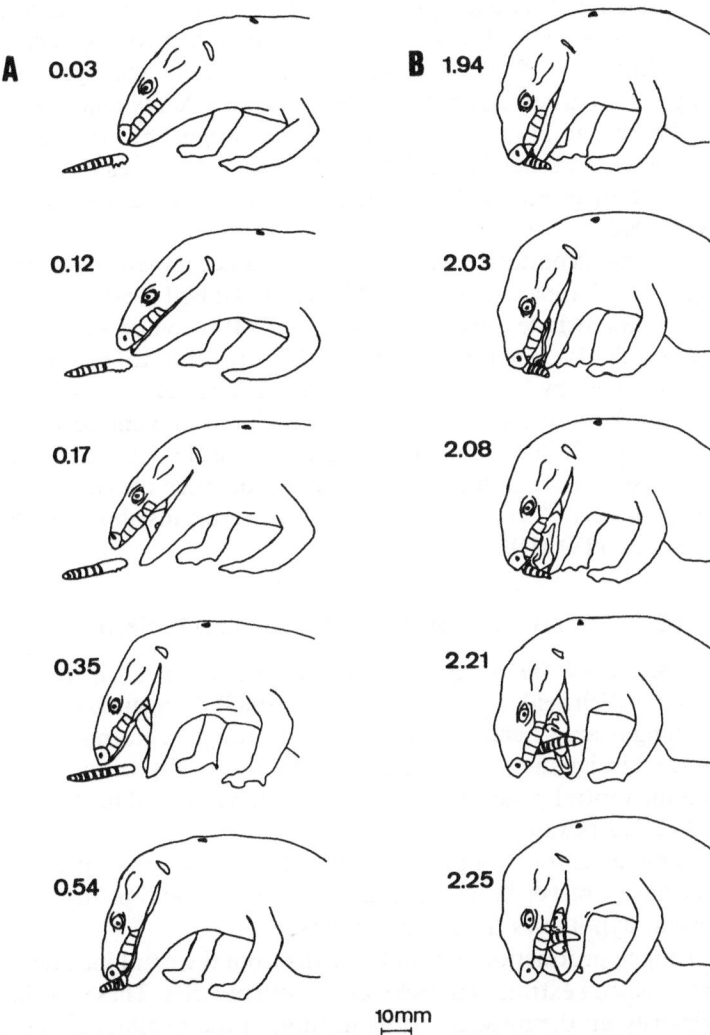

Fig. 2. A Prey capture (mealworm) by the scincid, *Tiliqua scincoides* (from time = 0.03 to time − 0.54 s). **B** The second gape-tongue cycle used for the posterior displacement of the prey item in the buccal cavity. The prey is firmly held between the tip of the jaws during the capture cycle, and then it is moved posteriorly into the buccal cavity by the tongue. The gape cycle of capture is divided in slow opening (SO), fast opening (FO) and fast closing (FC) stages. The slow opening stage consists of a regular increase in the gape angle before the beginning of fast opening. SO: time 0.03–0.15 and 1.94–2.08; FO: time 0.15–0.35 and 2.08–2.21; FC: 0.35–0.54 and 2.21–2.25

During feeding, tongue displacements can be divided into three successive stages: the tongue is moved forward in the buccal cavity (TO I), projected out of the buccal cavity toward the prey (TO II), and retracted (TO III). During TO I, the tongue tip is moved to just behind the mandibular symphysis (Schwenk and Throckmorton 1989; Bels 1990a), or advanced slightly in front of this symphysis (Kraklau 1991). Tongue protrusion of *Pogona barbata* is different because it is accompanied by the forward displacement of the hyoid apparatus (Schwenk and Throckmorton 1989). Although displacement of the hyoid apparatus during tongue protrusion has not been confirmed by X-ray analysis, a slight elevation of the throat always accompanies the tongue protrusion in the iguanians (Delheusy and Bels 1992). This elevation could result from a slight hyoid protraction produced by contractions of M. mandibulohyoidei I, II, and III (Bels 1990a).

In Chameleontidae, tongue projection is accompanied by large displacements of the hyoid apparatus (Bels and Baltus 1987; Bell 1990; Wainwright et al. 1991). The stages include tongue protrusion (TO I) accompanied by hyoid protraction (stage HY I), tongue projection (TO II), and tongue retraction (TO III) accompanied by hyoid retraction (stage HY II). During the end of TO I (about the last 0.5 s), the tongue protrudes about 1 to 3 cm beyond the mandibular symphysis (Wainwright et al. 1991). The stage HY I, which begins with stage TO 1, corresponds to an increase in the angle between the ceratobranchial I and the lingual process. The hyoid retraction (HY II) begins about 50 ms prior to tongue retraction (TO III).

2.1.1.2 Mechanism of Tongue Protrusion and Projection in Iguanians

The shape of the tongue during its protrusion is different in agamids and iguanids (Schwenk and Throckmorton 1989). In agamids, the ventral pallets of the tongue are oriented ventrally and the tongue deforms relative to the lingual process; in iguanids, the tongue is curled ventrally within the buccal cavity so that the ventral pallets become oriented dorsally, and the tongue is protruded in this curled position.

Tongue protraction at prey involves contractions of the hyolingual muscles that have not yet been experimentally studied except in the Chameleontidae. Several hypotheses have been proposed.

Tongue protraction probably results from the combined action of the tongue (intrinsic and extrinsic) muscles and hyoid muscles. These displacements depend primarily on the muscular organization of the hyolingual system (Smith 1984, 1988; Schwenk 1986; Smith and Kier 1989). The hyoid musculature and the external musculature of the tongue are similar for all the iguanians, divided functionally into sets of protractor and retractor muscles that are similar for all agamids and iguanids (see, for example, Gnanamuthu 1930a,b; Oelrich 1956; Avery and Tanner 1971, 1982; Smith 1984, 1988; Throckmorton et al. 1985; Schwenk 1988; Bels and Goosse 1989; Bels 1990a; Font and Rome 1990). Smith (1986) provides a summarizing table of the terminology used in the literature for these muscles (Table 4 in Smith 1986).

In contrast to extrinsic musculature, the intrinsic musculature of the iguanians is different in agamids (Smith 1988). The tongue morphology of the Iguanidae is most similar to that of *Sphenodon* (Schwenk 1986, 1988). In her study of tongue morphology of Agamidae, Smith (1988) observed (1) the absence of circular bundles around the M. hyoglossus, and (2) the presence of dorsal and ventral septa in the tongue.

Schwenk provided a hypothesis about the role of the hyoid and tongue muscles in tongue protrusion of *Sphenodon* that occurs when catching small prey such as insects (Gorniak et al. 1982). For agamids (Smith 1988), the muscular arrangement of the posterior ring muscle around the lingual process plays a major role in tongue protrusion. Contractions of this circular muscle on the lingual process may impart forward displacement of the tongue mass. Tongue protrusion is further aided by the inclined angle of the lingual process that encourages the tongue mass to move forward. This model of tongue protrusion may be accepted only if the coefficient of friction between the lingual process and the tongue musculature is small enough to allow the tongue mass to slide on this hyoid element (Smith 1988).

In iguanids, the mechanics of tongue protrusion has not been studied. But in *Oplurus cuvieri*, the M. verticalis is similar to the ring muscle of agamids: muscular fibers pass under the lingual process in the hindtongue (Fig. 3).

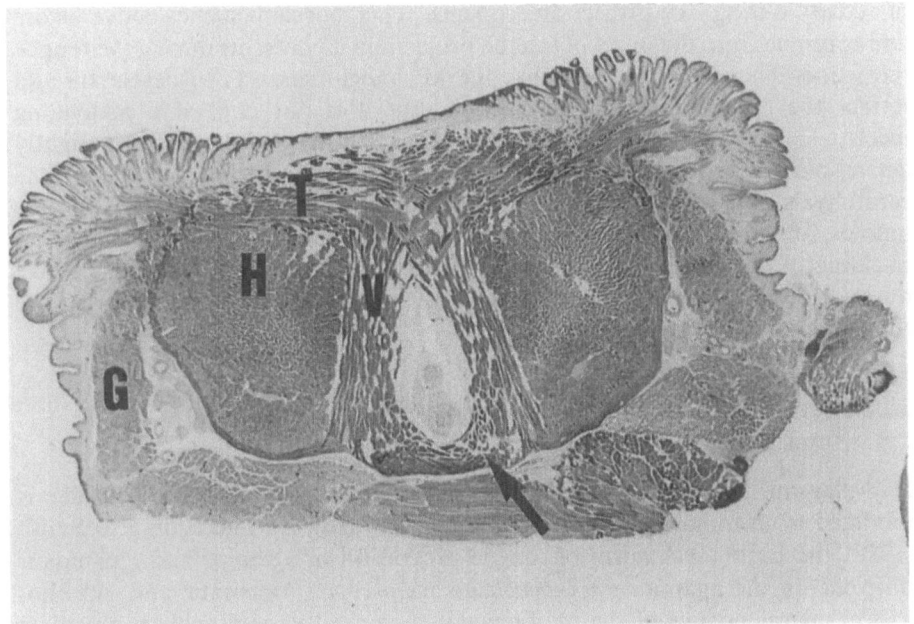

Fig. 3. Transverse section of the posterior tongue in *Oplurus cuvieri*. The fibers of the M. verticalis (*black arrow*) pass under the lingual process. Such muscular organization is similar to that described previously for agamids (Smith 1988). *G* M. genioglossus; *H* M. hyoglossus; *T* M. transversalis; *V* M. verticalis

Muscular septa are not present as in other iguanids, and the lingual process is anteriorly tapered as in agamids (Avery and Tanner 1971). The contraction of this muscle should thus produce a forward sliding movement of the tongue mass relative to the lingual process.

In Chameleontidae, tongue projection and retraction have drawn attention (Mivart 1870; Katharinarer 1894; Gans 1967; Bell 1984, 1987, 1989; Bels and Baltus 1987; Wainwright et al. 1991), and Bell (1989) proposed a hypothetical sequence of tongue muscle activity during projection. Muscle functions during prey capture have been experimentally tested by Wainwright and Bennett (1992a,b). The M. accelerator of the tongue is a sphincter muscle, but it contracts about 180 ms before the onset of tongue projection and stops about 10 ms prior to tongue projection. During this period of contraction, the M. accelerator exerts compressive forces against the parallel sides of the lingual process (Wainwright and Bennett 1992b). The contraction of this muscle does not begin to produce an anterior vector of these forces until it moves onto the tapered distal portion of the process. This delay provides a preloading mechanism that substantially enhances maximal acceleration and velocity of the projectile tongue.

Activity of some hyobranchial muscles has been recently investigated by Wainwright and Bennett (1992a) for tongue projection. In contrast to the models that predict a biphasic activity (Zoond 1933; Altevogt and Altevogt 1954; Bell 1989, 1990), the activity of the M. hyoglossus presents only one burst of activity during one projection sequence. This muscle becomes active before and continues into the onset of tongue projection; it contracts during the tongue retraction. These data suggest that the M. hyoglossus acts to decelerate and retract the tongue following projection, but it is not part of a preloading mechanism as suggested by previous models. The M. geniohyoideus is slightly active during protraction of the hyoid apparatus. During projection, the ceratohyals and ceratobranchials I move anteriorly and the hyoid apparatus unfolds. Activity of the M. sternothyroideus during this protraction may prevent buckling of the hyobranchial system. Finally, the M. sternohyoideus is active during the retraction of the hyoid apparatus which actually begins during the tongue projection phase.

2.1.1.3 Evolutionary Comparisons Between Tongue Protrusion and Projection in Iguanians

Schwenk and Bell (1988) suggest that tongue projection in Chameleontidae is modified from tongue protrusion in Agamidae. However, according to Smith (1988), the basic mechanism of tongue protrusion in agamids and iguanids is different. In the agamid, *Phrynocephalus helioscopus*, kinematic and morphological characteristics of the hyolingual design are "a functional intermediate between the plesiomorphic condition found in *Sphenodon*, Iguanidae, and Agamidae (most of species), and the highly derived lingual projection of Chameleontidae" (Schwenk and Bell 1988, p. 699). The evolutionary transformation of the hyolingual system in iguanians proceeds: (1) protrusion associated with hyoid protraction and limited lingual translation caused by extrinsic muscles (little or activities in M. verticalis); (2) additional lingual translation produced by activity of M. verticalis so that the glandular portion of the tongue now wraps around the end of the lingual process; and (3) forceful action of M. verticalis transformed into the M. accelerator producing an anterior translation of the tongue along the lingual process. There are two key differences between tongue *protrusion* and tongue *projection*. First, ballistic projection protracts the tongue at greater distances (3 vs 35 cm for agamid compared to chameleontids). Second, the hyoid apparatus is displaced during projection. Other unique features occur in gape angle and in the use of the tongue (Bels and Baltus 1987; So et al. 1992).

Primitively, tongue protrusion in iguanids is produced by the action of M. genioglossus (external tongue protractor muscle). In agamids, two biomechanical novelties are significant for tongue protrusion: a tapered lingual process and acquisition of a ring muscle around the lingual process. Smith (1988) suggests that tongue morphology in agamids and iguanids has proceeded in two divergent directions. In one direction, agamids retained the *Sphenodon* condition (a circular muscle around the lingual process, and septa in the hindtongue). Also, the adults have an M. genioglossus internus which contributes to the ring muscle. In the other phylogenetic direction, the iguanid tongue, *Oplurus cuvieri*, muscular fibers also pass under the lingual process in the hindtongue and could act as in agamids (Delheusy and Bels 1992).

2.1.2 Intraoral Manipulation and Transport of Food in Lizards

2.1.2.1 Kinematics and Morphological Specializations

Once in the buccal cavity, the food is next transported to the esophagus. Usually, living food is killed and reduced (biting, crushing) in the buccal cavity. Both this reduction and the subsequent transport of the food consist of cyclic displacements of the jaws and the hyolingual system or the hyoid apparatus alone (i.e., Varanidae). Reduction and transport may occur together or in sequence (Smith 1984; Bels and Baltus 1989). Reduction may be interrupted by rearrangement of the prey between the jaws (Bels and Baltus 1989). It is not yet clear whether kinematic differences between food reduction and transport are only a functional consequence of the position of the food relative to the jaws or whether they are produced by a different motor pattern. Consequently, reduction cycles are best seen as subsets of transport cycles (Schwenk and Throckmorton 1989; Delheusy and Bels 1992).

Two main functional properties may be documented in the actual data: (1) in all lizards (except *Varanus*), the tongue plays a major role during all parts of intraoral food manipulation and transport; (2) the kinematics and motor patterns of hyolingual displacements do not differ during their successive roles

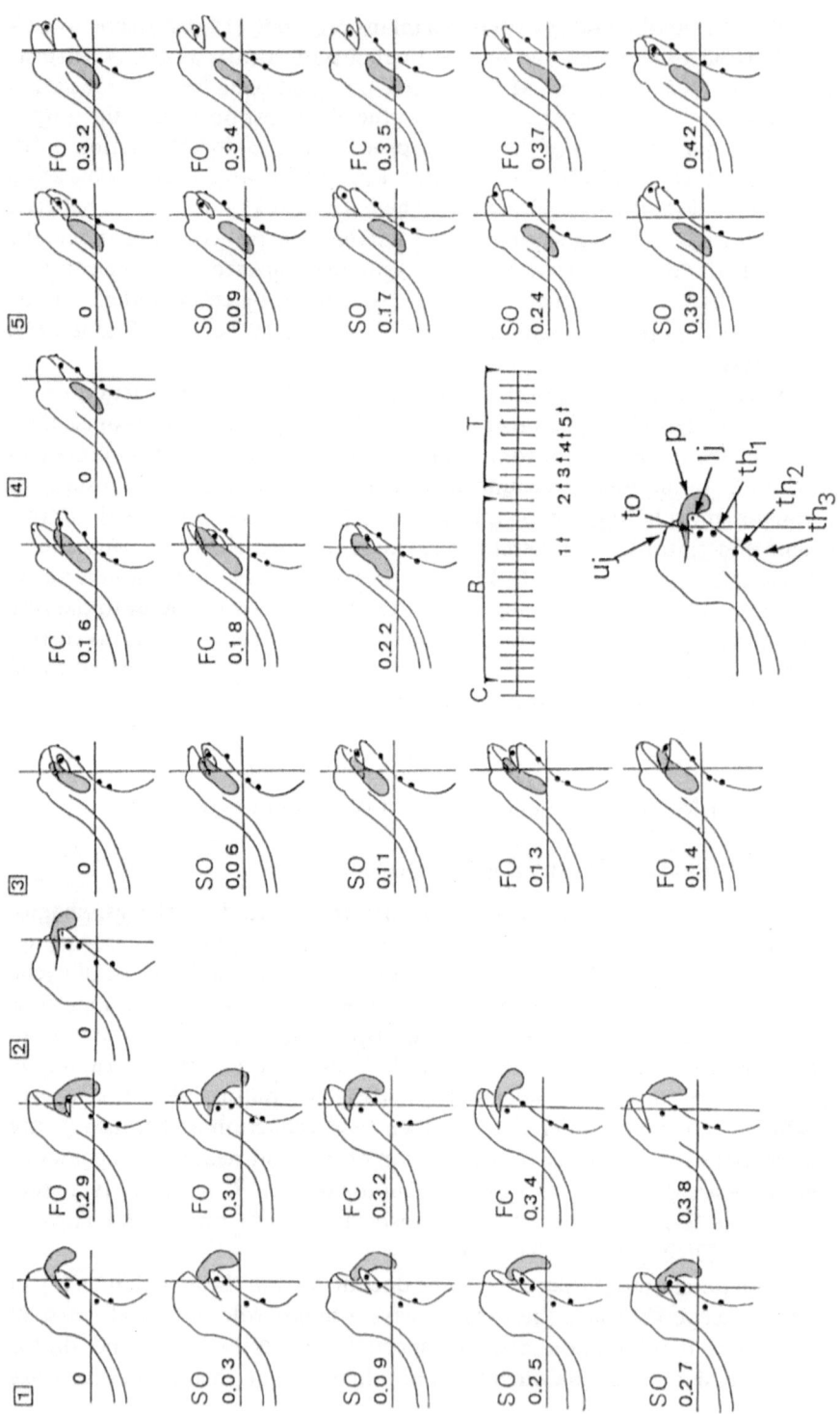

(Fig. 4). For instance, based on variables depicting the jaw and tongue displacements, the reduction and transport cycles in *Oplurus cuvieri* form a kinematically homogeneous group in a multivariate space (Delheusy and Bels 1992). However, So et al. (1992) showed that the reduction (chewing) and transport cycles are clearly separated in such a multivariate space in the chameleontid. These authors note that they chose *unambiguous* cycles in their categorization, whereas Delheusy and Bels (1992) based their distinction between reduction and transport cycles on the position of the prey relative to the jaws.

For iguanians, kinematic analyses of food reduction have been presented for three iguanids only (Smith 1984; Bels and Baltus 1989; Delheusy and Bels 1992), one agamid (Kraklau 1991), and the chameleontids (So et al. 1992). Tongue displacements are still poorly known, except in *Ctenosaura* (Smith 1984) and *Anolis* (Bels and Goosse 1989). During reduction, the tongue and the hypobranchium move in a forward-backward displacement in relationship to gape angle (Fig. 6). The tongue moves forward and bulges during SO and the beginning of FO stages; at the end of FO and during FC stages it retracts and flattens (Bels and Goosse 1989; Delheusy and Bels 1992).

Tongue and related hyoid displacements during food transport are poorly known. For lizards, only two studies provide detailed data on the hyolingual displacements in food transport for iguanians (Smith 1984; Delheusy and Bels 1992). Functionally, transport of the prey to the esophagus is a complex phase (Fig. 5): the food bolus moves posteriorly to the pharynx, and enters the esophagus (Delheusy and Bels 1992). For an iguanian, *Ctenosaura similis*, Smith (1984) describes three types of hyolingual displacements: transport, pharyngeal packing, and pharyngeal compression. In *C. similis*, the "difference in tongue cycles in pharyngeal packing and transport reflects the differences in the position of the food bolus relative to the tongue" (Smith 1984). The food is on the tongue during transport, and behind the tongue during pharyngeal packing. Two major differences are reported (Smith 1984). (1) During transport, the posterior region of the tongue moves upward and forward during SO I, and during pharyngeal packing, it moves upward and backward during SO I, and (2) the orbital-shaped displacements described by the tongue become elongated along the horizontal axis during pharyngeal packing.

Fig. 4. A typical reduction and transport sequence in *Oplurus cuvieri* feeding on cricket taken from successive cycles of X-ray analysis. The cycles illustrate three phases: reduction (*cycle 1*), transport cycle to the esophagus through the buccal cavity (*cycle 2-4*), and pharyngeal packing cycle (*cycle 5*). Frames 2 and 4 illustrate, respectively, the position of the prey at the beginning of the transport and the pharyngeal packing phases. The prey is *gray* and the *points* correspond to lead markers placed into the tongue (*to*) and on the throat (*th 1-3*). During the reduction phase (*cycle 1*), the prey is maintained between the upper (*uj*) and lower (*lj*) jaws. During the transport of the prey to the esophagus, the prey passes under the tongue (*cycle 2*) or is placed behind the tongue (*cycle 3*) into the pharynx. The cycles are positioned in the entire feeding sequence. *C* Capture phase; *R* reduction phase; *T* transport phase; *FC* fast closing; *FO* fast opening; *SO* slow opening; *lj* lower jaw; *p* prey; *th 1-3* throat; *to* tongue; *uj* upper jaw

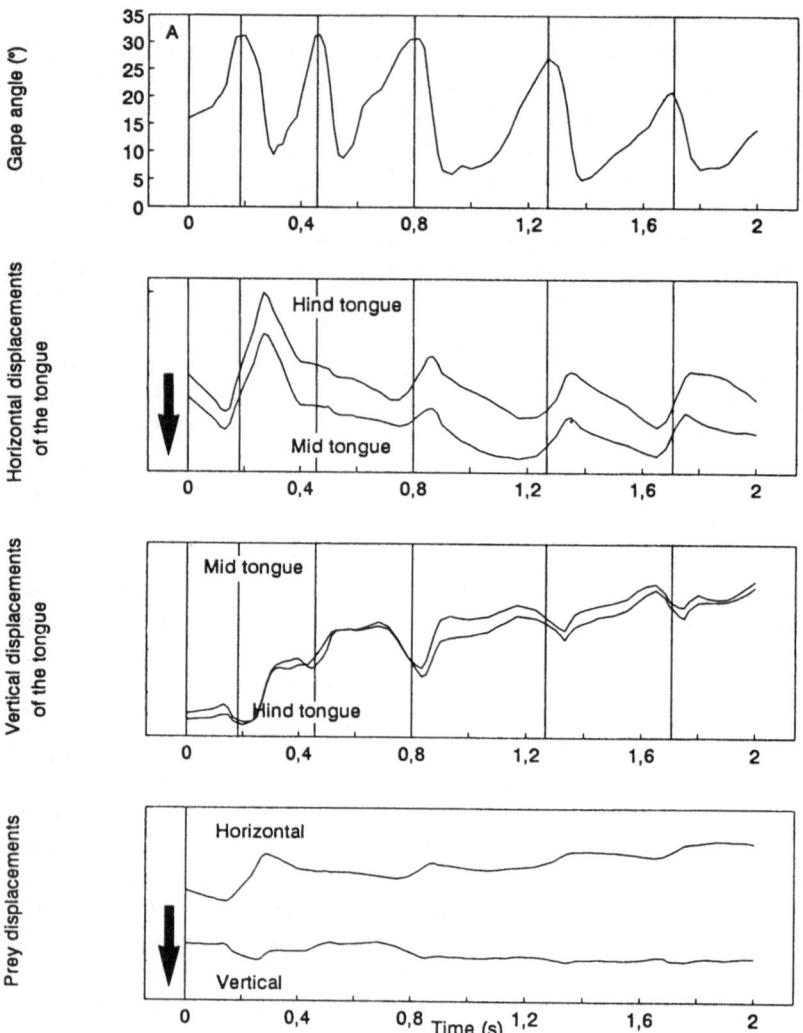

Fig. 5. Kinematic diagrams showing the jaws, tongue, and hyoid displacements during reduction and transport cycles in iguanian lizards

In the iguanid, *O. cuvieri*, displacements of the tongue and hyoid apparatus do not exhibit such differences during the transport and pharyngeal packing: (1) the tongue moves upward and forward, then backward and downward during the two phases, (2) elongation of the tongue displacements along the horizontal direction occurs only during some cycles involving a small tongue protrusion (Delheusy and Bels 1992).

In all the studied scleroglossans, the tongue also plays a major role in the transport cycles (Smith 1984; Goosse and Bels 1992a). There is only a few difference between *Tupinambis* and *Ctenosaura*: "the difference between the two animals in tongue use in feeding are of degree rather than in general pattern"

(Smith 1984, p. 134). This difference mainly concerns the use of the hyoid system. Several morphological specializations of the hyoid body fail to be explained by the behavioral constraints of the feeding and/or drinking activities. Except in varanids, which have a highly specialized feeding behavior (inertial thrusts), the scleroglossan hyoid apparatus and its elements follow the anterior-posterior cyclic movements of the tongue during all phases of feeding and are not used during lapping cycles (Smith 1984; Goosse and Bels, pers. observ.). Even if ceratobranchials I move faster to act on tongue movements during mechanical reduction, these elements follow a cyclic anterior-posterior displacement associated with that of ceratohyals and ceratobranchials II (Bels and Goosse 1989). In varanids, the relationship between jaw and hyoid cycles is very close: the hyoid apparatus moves ventrally and slightly caudally during FO and retracts during FC (Fig. 6, in Smith 1986). For food transport and pharyngeal packing, a ventral depression of the ceratobranchials II should produce a larger opening of the pharyngeal cavity to aid swallowing of the prey. Such an increase in the cavity anterior to the esophagus is produced in varanids by relative displacements of the hyoid elements (Smith 1988). In the other scleroglossan lizards, such depression should be produced by the action of the bolus itself entering into the esophagus.

During pharyngeal packing and pharyngeal compression in scleroglossans, the tongue moves out of the mouth, even in varanids, in a "lip-licking" cycle (Smith 1986). In iguanians, some pharyngeal packing cycles and cleaning cycles, which appear very late in the whole feeding bout (Delheusy and Bels 1992), are also accompanied by slight tongue protrusion. In such cycles, the tongue may help to force the food into the esophagus by increasing the anterior-posterior force acting on the food entering the esophagus. Because it passes beneath the prey, forward elongation of the tongue could store elastic energy in the laryngohyoid ligament (Owasa 1898; Oelrich 1956; Schwenk 1986; Delheusy and Bels 1992). This energy could be returned during retraction of the tongue, thus adding to its force against the food in the esophagus. During the later hyoid cycles of varanids, which are roughly equivalent to the hyolingual cycles used in pharyngeal packing of other lizards, Smith (1986) notes that the buccal floor is lifted during jaw opening and then pulled back, moving posteriorly against the food item.

In conclusion, some differences may appear in timing (Schwenk and Throckmorton 1989; Bels 1990a) or in movements of the entire hyolingual system relative to the jaws (Schwenk and Throckmorton 1989), but *not* in the relative movements of the hyoid elements through the cycles (Fig. 6). Therefore, we suggest that the primitive mode of movements to reduce and transport the food to the esophagus includes the following: (1) successive anterior-posterior cyclic activities of the hyolingual apparatus; (2) this cyclic activity is produced by successive contractions of the hyolingual protractor muscles (M. genioglossus, M. mandibulohyoidei I, II, III) and retractor muscles (M. hyoglossus, M. sternohyoideus, M. omohyoideus, and M. branchiohyoideus); and (3) this cyclic activity is produced without highly specialized displacements of the hyoid elements. During the evolution of scleroglossans, the specialization of the

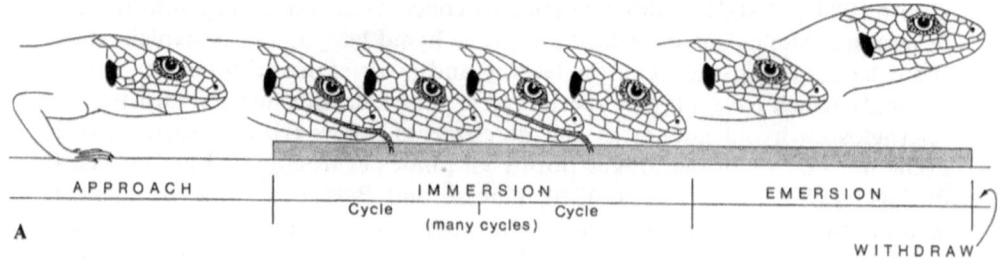

A

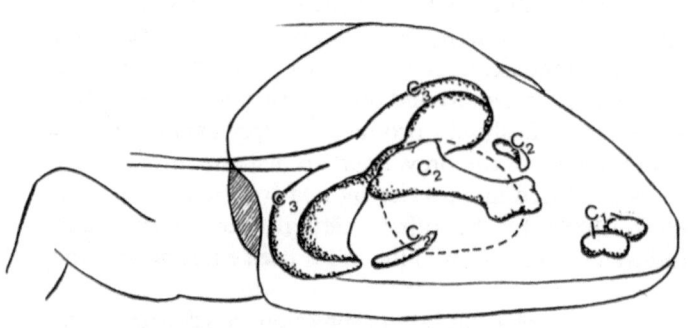

B

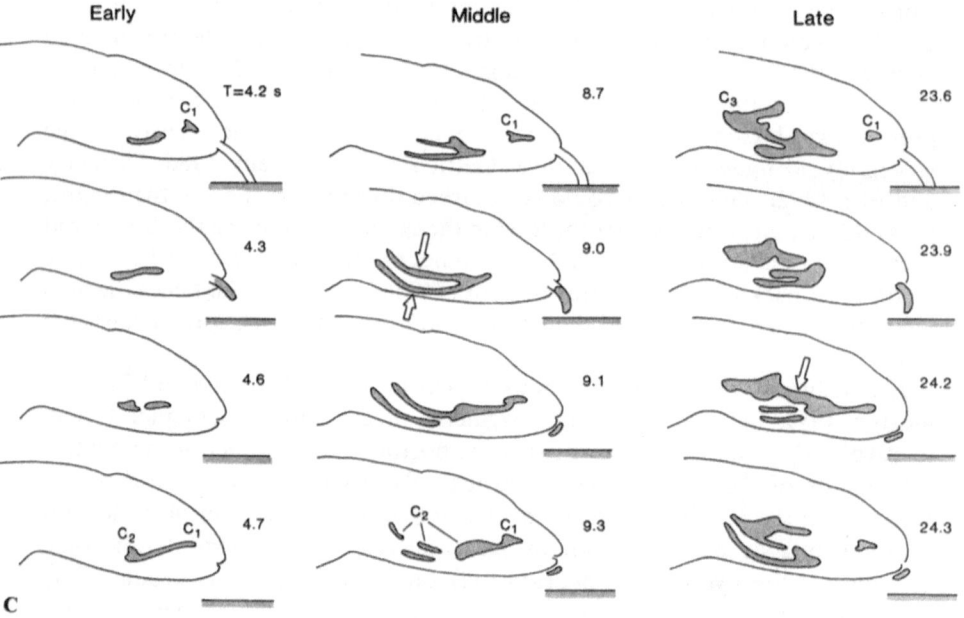

C

Fig. 6. A–C

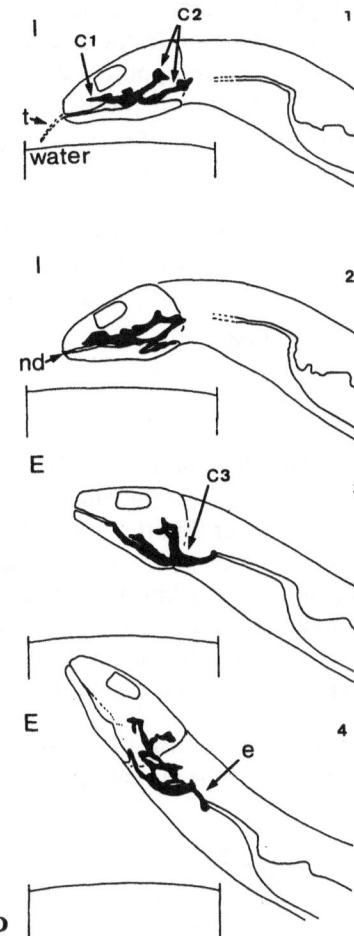

Fig. 6. Drinking behavior in *Lacerta viridis*. **A** Drinking is divided into four phases. **B** The buccal cavity of *Lacerta viridis* includes three successive compartments. Compartment 1 (C_1) is most anterior and smallest. It lies beneath the recessed vomeronasal fenestra. Compartment 2 (C_2) is larger and positioned in the middle of the buccal cavity. It is connected by dorsal and ventral channels to C_1 when filling with water. Dorsally, it includes the posterior nasopharynx, a single, medial space behind and above the posterior edge of the secondary palate. It connects with the spaces adjacent to the sides of the tongue along arching lateral channels located within the sides of the buccal cavity. Compartment 3 (C_3) is the largest and includes all of the posterior buccal cavity situated within the throat. The esophagus departs from the center of its posterior wall. **C** During the immersion phase, the tongue plays a double role. In early and middle tongue cycles, the water is brought by the foretongue from the dish or the drop to compartment C_1. At the same time, the water of C_1 is carried back to C_2 by the hindtogune. In late tongue cycles, water moves from C_2 to C_3. The *arrows* indicate the lateral channels. **D** The water moves from C_1 to C_2 and to C_3 and enters the esophagus during the emersion phase. The lizard ends the immersion phase (*I*) and tilts the head during the emersion phase (*E*). *nd* New drop; C_1–C_3 are defined in **B**. (Bels et al. 1992)

tongue for chemoreception (i.e., Teiidae and Varanidae) has been accompanied by specialization in the use of the hyoid apparatus in the feeding function.

Certainly, the hypothesized effects of hyolingual specializations on the feeding function remain to be tested. The food itself and its resistance may be two important constraints on hyolingual displacements. The food item that completely fits the buccal cavity should produce strong effects on hyolingual displacements. Differences in the lingual morphology (i.e., surface and intrinsic structure) may be related to variation in details of tongue displacements related to food entry into the buccal cavity.

2.2 Drinking

2.2.1 Lizards

General Observations. So far, two modes of drinking have been described in lizards. In the first (iguanians and scleroglossans), the tongue is used directly for lapping water (Smith 1986; Schwenk and Greene 1987; Bels et al. 1992). In the second (*Varanus*), the tongue is not used to collect water but instead the snout is inserted directly into the liquid (Auffenberg 1981; Smith 1986). A scleroglossan drinking session typically includes four defined phases: approach, immersion, emersion, withdrawal (Bels et al. 1992). The approach and withdrawal phases, respectively, began and ended the drinking session (Figs. 7 and 8).

Drinking Mechanism. Displacement of water into and through successive compartments of the buccal cavity is based primarily upon active tongue movements

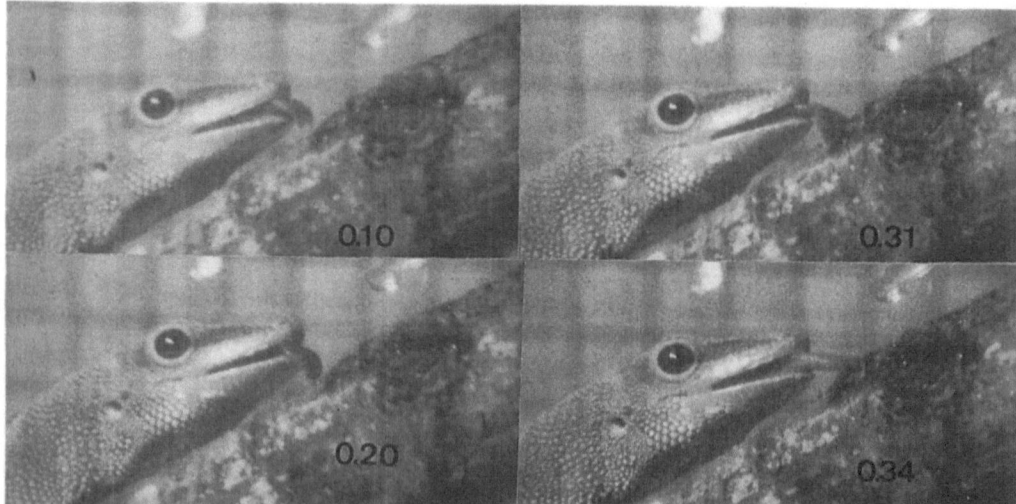

Fig. 7. Successive frames of drinking water from a branch by *Phelsuma madagascariensis* (Scleroglossa, Gekkon). The time of each frame is indicated in ms

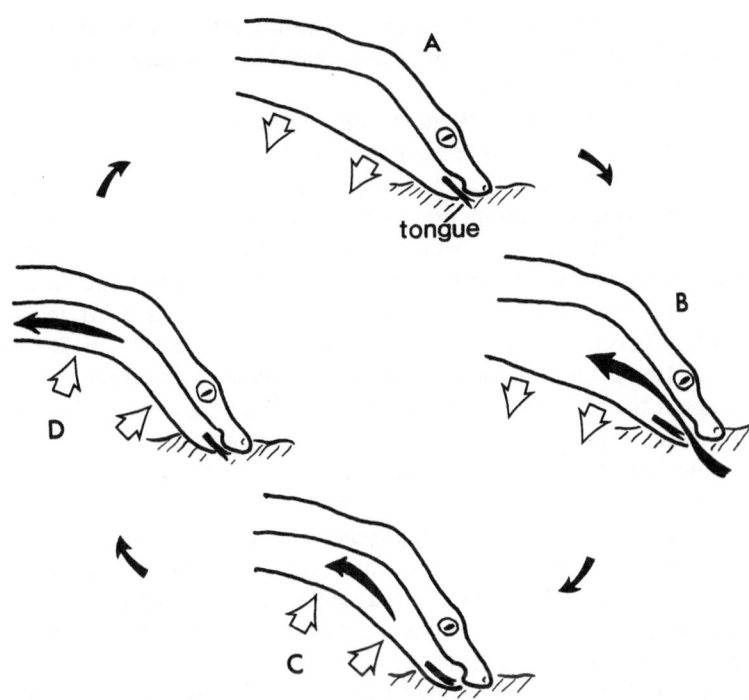

tongue

Fig. 8. Buccal-pump mechanism of drinking as illustrated with *Boa constrictor*, one cycle. **A** Depression of the floor of the oropharynx (*open arrows*) creates a negative pressure within the mouth compared to ambient. **B** Depression of the jaws and retraction of the tongue break the seal around the margins of the mouth and allow water to be aspirated into the oropharynx. **C** Elevation of the floor of the oropharynx begins to raise the pressure along with initial elevation of the jaws. **D** By this point in the cycle, the margins of the jaws have closed and the tongue has completely entered the lingual canal to close the mouth. Elevated pressure forces the aspirated water into the esophagus, past the esophageal sphincter where each cycle of water collects until the snake lifts its head and water moves to the stomach

(Fig. 7) and gravity flow. The tongue in these lizards performs a dual role (Fig. 7). First, it participates in collection (protrusion) and delivery (retraction) of water into the anterior mouth, compartment 1. Second, during retraction, the hindtongue participates in intraoral water transport from this compartment (C1) to a posterior compartment 2. Thus, the foretongue functions to collect and transport water into the buccal cavity; the hindtongue functions to transport water intraorally (Fig. 7). As successive tongue cycles bring water into the buccal cavity, receiving areas within it fill. Swallowing of this water into the stomach is based upon gravity flow, and the opening and action of the esophagus. Lifting of the head brings the pool of collected buccal water above the level of the largest, and most posterior of the buccal spaces (compartment 3). Water flows to this third compartment primarily under the action of gravity. Occasionally, movements of the throat are evident during this emersion phase, but these are not synchronized with intraoral fluid movements. They are part of the breathing mechanism and if they contribute to swallowing, they do so by flattening pleats within the oral epithelial lining that may have captured small rivulets of water held by surface tension.

During lapping in *Ctenosaura similis* and *Tupinambis nigropunctatus* (Smith 1984), the gape profile is similar to that of drinking in scleroglossans. The hyoid apparatus in lacertids moves very slightly during the cyclic immersion and emersion tongue movements. Such hyoid movement is related to hindtongue movement. The throat movements may be produced by forward and backward movements of the hindtongue itself or by slight movements of the entire hyoid apparatus produced by hyoid protractor muscles, such as *M. mandibulohyoidei* I and II (Throckmorton et al. 1985; Bels and Goosse 1989; Bels 1990b; Font and Rome 1990). Lateral tongue expansion seems to be absent in *Ctenosaura*, an iguanian lizard, but present in *Tupinambis* and *Lacerta*, which are scleroglossans. Such expansion, which broadens the tongue, occurs after protrusion of the tongue from the mouth and may be related to the mechanism of loading with water.

Drinking in varanids (scleroglossans) is completely different. Varanids drink by inserting the snout into the water. Smith (1986) reports that "the tongue is usually but not always protruded and withdrawn slightly". She also observed that throat movements were associated with tongue displacements. As a mechanism of loading water, Smith (1986) hypothesized that the tongue, transversely narrower than in *Lacerta*, may pick up some water and bring it to the buccal cavity. However, the head elevation reported in *L. viridis* does not occur in varanids which use continuous "sucking" movements to fill the digestive tract with water.

2.2.2 Snakes

In snakes, drinking is not characterized by frequent head-back tipping or tongue scooping as occurs in, for instance, some birds (Heidweiller and Zweers 1990). There is instead in snakes one basic drinking mechanism, a buccal-pump mechanism, with several variations.

Oral Morphology. The upper and lower arcades of teeth borned by the maxillae and dentaries, respectively, arch forward but do not meet at the midline. Instead, a gap remains, the lingual canal, at the anterior midline of the skull. This lingual canal, together with slight parting of the lips, offers an opening through which the tongue protracts from the mouth during chemosensory activity. The lateral seal to the mouth is opened or closed by parting or pressing together complementary soft tissue along the upper and lower margins of the mouth. These soft tissues include the oral mucosae, embracing the tooth arcades, that extend far back towards the corners of the mouth. Upper and lower lips also fit into one another when the jaws close. Consequently, elevation of the mandibles engages these soft tissues along the upper and lower margins of the buccal cavity, sealing the mouth laterally. Closure of the lingual canal seals the mouth at its anterior midline. Forming and breaking this seal is an important mechanical component of the drinking mechanism of snakes.

Drinking Phases. Four drinking phases of unequal length are recognised – approach, immersion, emersion, and withdrawal. During the approach phase, movement of the snake brings its head into the vicinity of the water. The immersion phase begins upon contact of the tongue or snout with the water. Rhythmic cycles of depression and elevation of the mandibles begin immediately. Depending upon disturbance and thirst, the snake may remain in contact with the water for several minutes during which time many such rhythmic cycles occur without interruption and without changing the position of the head. Sometimes during the immersion phase, the snout is pushed deep into the water, submerging the nostrils. But usually, the nostrils are out of the water and only the anterior tip of the snout makes contact with the water. In fact, during some drinking bouts, snakes may have only their chin in contact with the water; the capillarity of the water draws up the meniscus to the front of the mouth where water then enters through the lingual canal. When drinking drops of water, this capillarity is especially important in bringing water to the lingual canal.

Eventually, the head is lifted and contact with the water is broken, i.e., emersion phase, although rhythmic movements of the jaws and throat continue for several cycles. Sometimes, snakes will again make contact with water and continue drinking cycles. When the snake is finished, it departs from the vicinity of the water, i.e., the withdrawal phase.

Drinking Mechanics. Unlike swallowing of food (Frazzetta 1966), in which unilateral reciprocating motions (*sensu* Gans 1961) of the jaws occur, drinking by snakes instead involves the movement of the jaws together and in synchrony. During the first stage in a drinking cycle, the expansion phase, the jaws and especially the floor of the buccal cavity are depressed, expanding the volume of the mouth and throat; the tips of the jaws part to anteriorly break the seal around the margins of the mouth and permit entry of water. During the second stage, the compression stage, the jaws close, reforming the oral seal. The expansion phase produces a buccal pressure lower than ambient, thus encouraging the entry of water into the mouth when the oral seal is broken; the compression stage raises buccal pressure, but since the oral seal is reformed (and the nasal passages are plugged by the larynx), the elevated pressure forces water just drawn into the mouth to pass next into the esophagus. At about the level of the quadrado-mandibular joint, a constriction, the esophageal sphincter, is present. Water is forced past this sphincter during the compression phase, but its closure prevents the retrograde flow of water back into the mouth. Consequently, during prolonged drinking, water accumulates behind this sphincter in the anterior esophagus, so the snake does not need to raise its head frequently to allow gravity to move this water into the esophagus (cf. lizards, Bels et al. 1992).

Mechanically, the most important element in snake drinking is the floor of the buccal cavity. It rises and falls together with the lower jaws, producing super-ficially a "gulping"-like action. However, it is a very precise mechanism syn-chronized with the opening and closing of the seal to the buccal cavity. This cyclic movement of the buccal cavity produces timely negative and positive buccal pressures to bring water first into the mouth and then to transport water intraorally, past the sphincter, and into the esophagus. Such a *buccal-pump mechanism* (Kardong and Haverly 1993) seems to be part of the drinking pattern in all snakes.

So far two variations of the buccal-pump mechanism of snake drinking are known. One variation occurs in the *Boa constrictor* (Kardong and Haverly 1993), and perhaps in most henophidians (pers. observ.). When establishing a seal around the buccal cavity, the boa protrudes its tongue into the lingual canal to help plug this gap at the front of the mouth. When water is to enter, the tongue is withdrawn. This retraction of the tongue, together with parting of the adjacent soft tissues at the front of the mouth, result in opening this route for aspiration of water into the mouth (Kardong and Haverly 1993). Low, rounded papillae are present along the foretongue (McDowell 1972) that possibly capture some water and deliver it to the front of the mouth upon retraction. However, the major mechanism for water entry is based on the buccal pump.

The second variation of the buccal-pump mechanism occurs in *Boiga irregu-laris* (Berkhoudt, Kardong and Zweers, pers. observ.) and possibly in most caenophidians (pers. observ.). The tongue does not protrude from the mouth during drinking in a rhythmic pattern correlated with drinking cycles. This loss of tongue protrusion represents a derived condition. The lingual canal is closed instead by jaw adduction that tightly presses soft tissues together and by the elevation of the anterior, lower labial and mental scales.

This buccal-pump mechanism produces negative and positive pressures relative to ambient to draw water in, transport it intraorally, and force it past the esophageal sphinctor to the anterior esophagus where it temporarily collects. Consequently, frequent head lifting to employ gravity is not required. Parting and forming of the oral seal are such that water intake occurs primarily at the front of the mouth. This control of the oral seal together with the pressures generated mean that water in small quantities, such as droplets that might form as morning dew, can be gathered as well as water in large standing pools.

2.3 Displaying in Lizards

2.3.1 Definition of the Displays

Based on literature description (Carpenter and Ferguson 1977; Greenberg 1977; Carpenter 1978; Shine 1990), throat displays can be divided into four categories: (1) throat extension, (2) dewlap flap, (3) dewlap and throat expansion, and (4) frillneck extension (Fig. 9).

2.3.2 Morphological Diversification of the Hyoid Apparatus

The morphological diversification of the hyoid apparatus in lizards may be illustrated by five characteristics of the elements: (1) presence or absence of paired elements; (2) spatial orientation of the elements relative to another; (3) shape, thickness, or length of paired and unpaired elements; (4) relative orientation of the elements within a pair; and (5) mode of articulation between elements.

In lizards, some of the hyoid elements are always present (the lingual process, the hyoid body, the ceratohyals, and the ceratobranchials I), whereas other elements such as the ceratobranchials II or the small paired elements (epihyals and epibranchials) may be absent. Except in some agamids (*Pogona* and *Chlamydosaurus*, Throckmorton et al. 1985; Smith 1988) and in chamaeleontids, the hyoid apparatus of iguanian lizards includes all the basic elements (see, for example, Camp 1923; Gnanamuthu 1930b; Tilak 1964; Avery and Tanner 1971, 1982; Smith 1988; Bels 1990b; Kraklau 1991). In some groups (i.e., *Phrynosoma*), the ceratobranchials II are extremely reduced (see, for example, Camp 1923; Avery and Tanner 1982). When present (in agamids and iguanids), the ceratobranchials II are separated or fused along their longitudinal axis. When fused, the longitudinal axis of both elements are parallel (i.e., *Anolis, Iguana, Draco*) or the elements are separated proximally but become joined distally (i.e., *Oplurus cuvieri, Pogona muricata*). In *P. barbata*, the hyoid apparatus is highly modified: the ceratobranchials II are absent, but the ceratobranchials I and the ceratohyals are present and even longer than in other iguanians (Throckmorton et al. 1985; Smith 1988). Based on histological analysis in adult *Anolis carolinensis* (Bels 1990b), the hyoid body is the beginning of the ceratobranchials II (Fig. 10A).

The degree of ossification of the hyoid elements has been described in *Anolis* lizards (Bels 1990b; Font and Rome 1990). As in all lizards, the ceratobranchials I are ossified. The cortex of the ceratohyals and hypohyals is ossified (Fig. 10B); their cores remain hyaline cartilage in living cells. The ceratobranchials II are less ossified than the other elements (Fig. 10A–B). The semicircular ring of ossification is ventrally opened. This opening extends to the mid-diaphysis of each element (*Anolis carolinensis*) or extends along the entire element (*Anolis equestris*). In *Lacerta viridis*, the cortexes of ceratohyals, hypohyals, and ceratobranchials II are ossified.

In scleroglossans, the shape of the hyoid body and the hypohyals is variable (see, for example, Table 1 in Avery and Tanner 1982). The presence or absence of the ceratobranchials II is extremely variable in scleroglossans (Camp 1923; Gnanamuthu 1937; Avery and Tanner 1971). When present, the ceratobranchials II are *always* separated. Both elements may be either parallel such as in *Cabrita* (Gnanamuthu 1930) or sharply diverge such as in *Gerrhonotus* and *Coleonyx* (Camp 1923), *Eumeces* and *Xantusia* (Avery and Tanner 1982). The hyoid apparatus of Varanidae is very different from that of other scleroglossans (Sondhi 1958; Smith 1986). For instance, the hypohyals and the ceratohyals

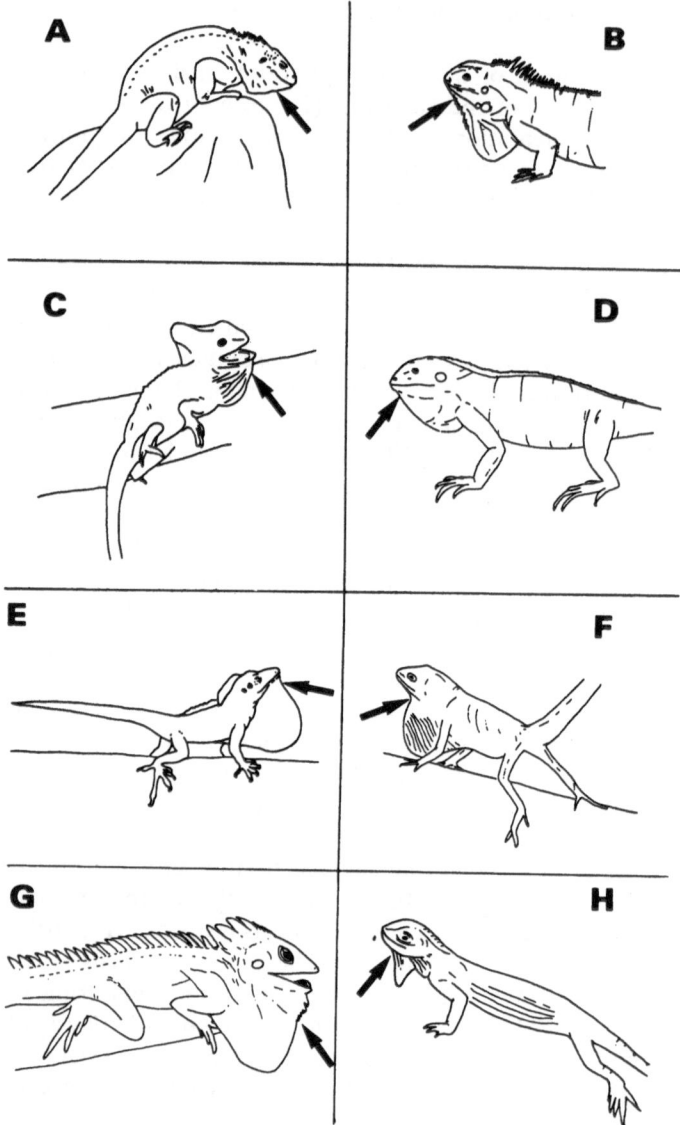

Fig. 9A–J. Display behaviors in iguanians with various types of hyoid apparatus. **A** *Colonophus subcristatus* (Carpenter 1978); **B** *Iguana iguana* (from photograph); **C** *Corythophanes hernandezii* (Carpenter 1978); **D** *Brachylophus fasciatus* (from photograph); **E** *Anolis marcanoi* (Losos 1985); **F** *Anolis roquet* (Gorman 1968); **G** *Gonocephalus* sp. (from photograph); **H** *Draco volans* (from photograph)

which are attached by flexible, ligamentous sheets (Smith 1986) form a kind of horizontal fan used during display behavior (Bels Renous and Gasc, pers. observ.). Rieppel (1981) proposed that the reduction in the hyoid elements (the hypohyals and the ceratohyals might be correlated with the morphological transformations of the body suited to burrowing activities.

Fig. 9I–J. Throat extension and tongue flicking in *Varanus griseus*

The form of the hyoid body seems to be more variable in scleroglossans than in iguanians. In agamids and iguanids with parallel joined or separated ceratobranchials II, the hyoid body does not show large lateral expansions. Such expansions occur in agamids with highly separated ceratobranchials I and II such as *Uromastyx aegyptius* (Carpenter and Ferguson 1977). The lateral expansions of the hyoid body are very common in the scleroglossans. Because

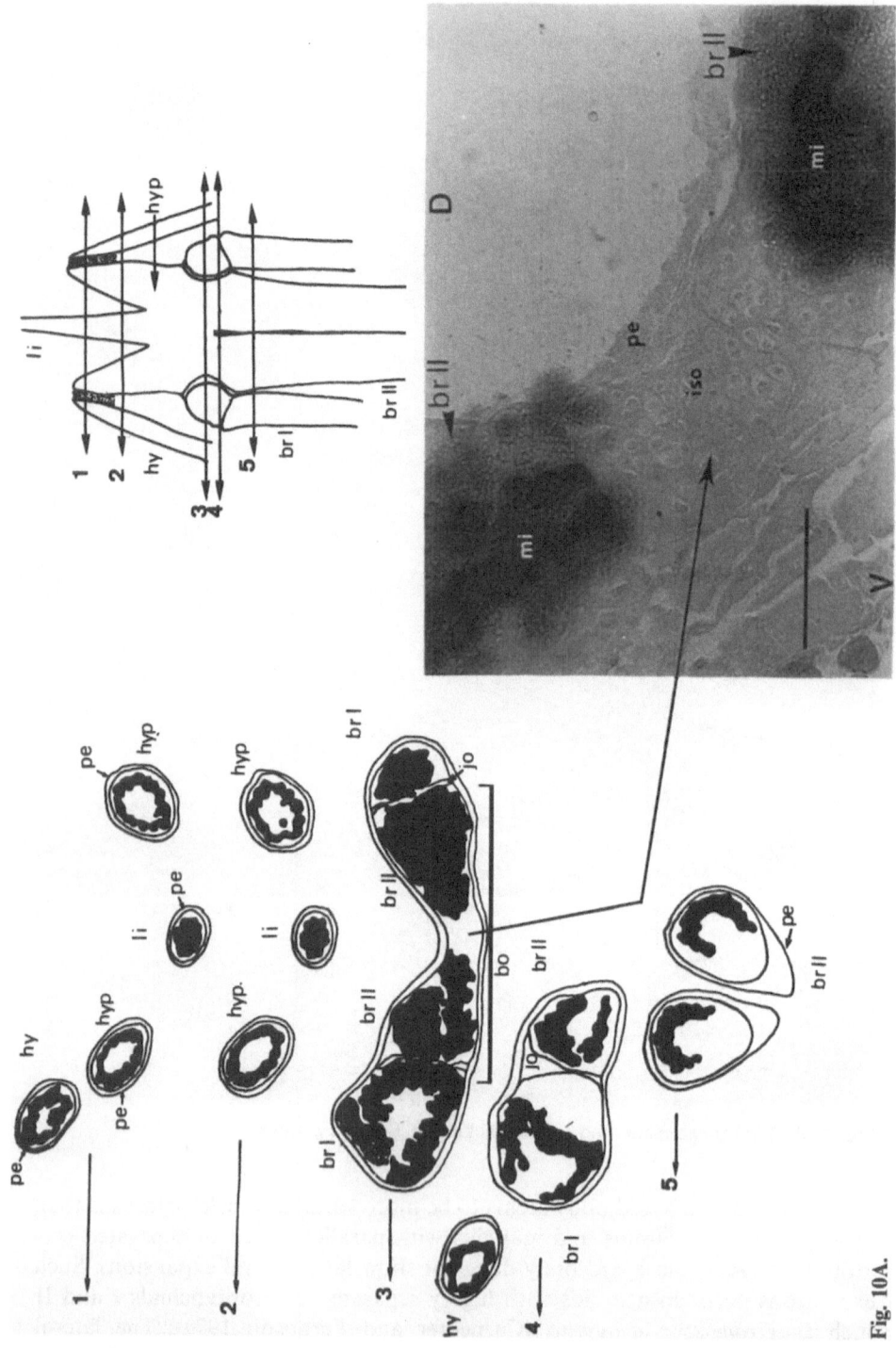

Fig. 10A.

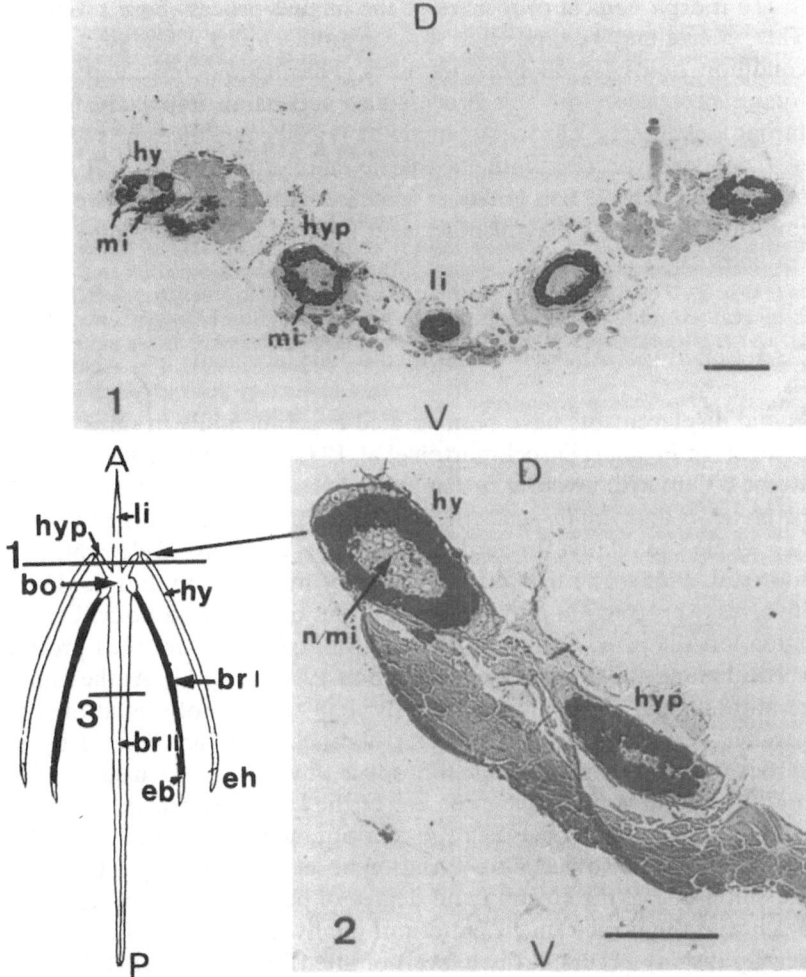

Fig. 10. A Drawing of successive sections of the hyoid body stained with alizarin red. The sections are numbered from the anterior to posterior regions of the hyoid apparatus. The mineralized areas are *black*. The mid-cartilage within the hyoid body is not mineralized. *Scale bar* = 1 mm. **B** Transverse sections showing the mineralization of the hyoid elements stained with alizarin red. 3: section in br II not illustrated in this paper. *D–V* Dorsoventral axis of the hyoid body; *Bo* hyoid body; *br I* ceratobranchial I: *br II* ceratobranchial II; *eb* epibranchial; *eh* epihyal; *hy* ceratohyal; *hyp* hypohyal; *iso* isogenic group of cells within the cartilage; *li* lingual (entogossal) process; *mi* mineralized area; *n/mi* no mineralized area; *pe* perichondrium (Bels 1990b)

the ceratobranchials I articulate with the hyoid body by a synovial joint, lateral separation of these ossified elements produces lateral expansions of the hyoid body as illustrated by *Gerrhonotus* (Camp 1923), *Mabuya* (Gnanamuthu 1937), *Tupinambis nigropunctatus* (Smith 1984), and *Varanus* (Gnanamuthu 1937; Smith 1986).

The morphological properties of the lingual process pose several questions. This process may be tapered in some agamids, but it is not yet known if such a condition holds for all lizard families. The tapered lingual process may aid tongue protrusion, but this process also acts as an important fulcrum during throat movements. Throckmorton et al. (1985) correlated the short, stubby size of the lingual process with its role as a fulcrum during frill erection in *A. barbatus*. However, Bels (1990b) pointed out that because it is ossified, a long, thin lingual process may also act as a fulcrum during dewlap extension in *A. carolinensis*.

2.3.3 Functional Analysis of the Hyoid Displacements and Motor Pattern

Hyoid displacements have been studied experimentally in threat and agonistic contexts in *Pogona* (Throckmorton et al. 1985) and *Anolis* (Bels 1990b; Font and Rome 1990). Frill erection in *Pogona barbata*, and dewlap extension in *Anolis carolinensis* and *Anolis equestris*, involve displacements of the hyoid apparatus as a whole and relative forward displacements of the hyoid elements (Throkmorton et al. 1985). The first display phase of frill erection and dewlap extension involves a protraction and ventral rotation of the hyoid body. During dewlap extension the ceratohyals and the ceratobranchials I are thus protracted in a vertical plane and ventrally oriented (Bels 1990b; Fig. 11). At the same time, the ceratobranchials I move more rapidly than the ceratohyals in their respective parallel, vertical plane and the ceratobranchials II are erected by increasing vertically the gular skin. A signal produced by dewlap extension is *always* a by-product of the vertical displacement of the ceratobranchials II resulting from relative displacements between the ceratobranchials I and the ceratohyals (Bels 1990b; Font and Rome 1990). Variation observed in dewlap extension occurs in the amplitude of the erection and degree of bending of the ceratobranchials II, but not in the displacement pattern of the hyoid elements. In *Anolis marcanoi*, dewlap display occurs at three levels of intensity (Losos 1985). These levels are related to the body elevation and the degree of bending of the ceratobranchials II (Fig. 1 in Losos 1985). Lizard dewlap display may be either a "round" signal produced in gular skin by bending of the ceratobranchials II, such as in *Anolis carolinensis* or *Anolis chlorocyanus*, or a "teardrop" signal, an extended gular skin without a "round shape" as in the majority of dewlapping iguanians, such as *Lyrioephalus scutatus* (Bels 1992). During frill displays, the relative displacement of the ceratobranchials I and the ceratohyals in their respective horizontal plane directly produces the expansion of the throat skin (Throckmorton et al. 1985). However, Throckmorton et al. (1985) suggest that "the ceratohyals undergoes less abduction than the ceratobranchial because its posterior tip is attached to the mandible". In contrast to dewlap extension, which occurs mainly in a vertical plane, frill erection occurs mainly in a horizontal plane; but protraction and ventral rotation of the hyoid body also involve vertical hyoid displacements.

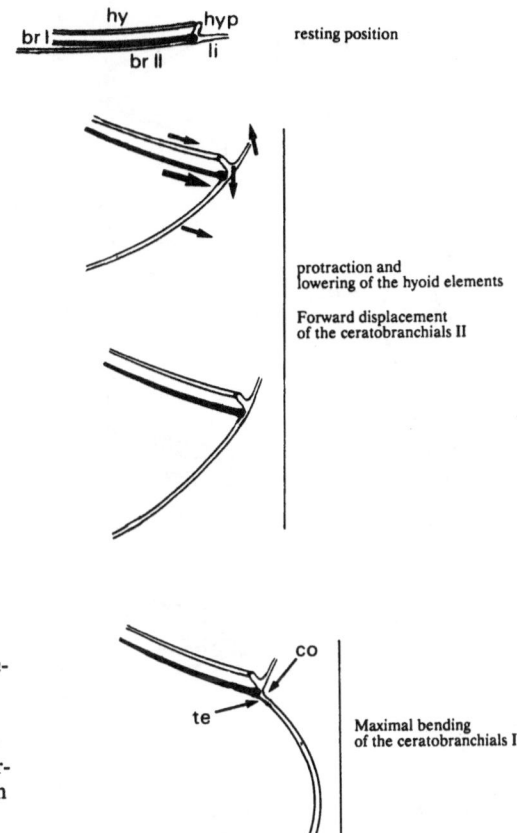

Fig. 11. Schematic diagrams of the dewlap mechanism in *Anolis carolinensis*. The compression (*co*) and tension (*te*) occurring within the cartilage are indicated in the last diagram. Ceratobranchial I, which acts powerfully during dewlap extension, is represented in *black* (see Fig. 10 for abbreviations) (Bels 1990)

Throat extension occurs in all the lizards at different body orientations (Murphy and Mitchell 1974; Carpenter and Ferguson 1977) but its kinematics is variable. For instance, in *A. carolinensis* (Bels 1990b) and *A. equestris* (Font and Kramer 1989), the hyoid apparatus is strongly protracted during throat extension (Bels 1990b), whereas in *Varanus*, the throat extends ventrally without such a great protraction during threat displays. In many scleroglossans such as *Lacerta viridis*, protraction of the throat is very slight. In *Varanus griseus*, throat extension is a complex motor pattern. This behavior mainly involves depression of the hyoid body accompanied by a small protraction (Bels 1990b; Font and Rome 1990). The distances between the hyoid body, neck of the lizard, and the ceratobranchials I increase simultaneously during extension. The angle between the ceratohyals and the hypohyals increases drastically during the extensive phase (Bels 1992). By contrast, the angle between the hypohyals, which rotate backward relative to the ceratobranchials I, decreases slightly, and the angle between the ceratobranchials I and the lingual process also decreases. The position of the hyoid body during throat extension in *A. equestris* (Font and Kramer 1989) is more anterior than in *Varanus*.

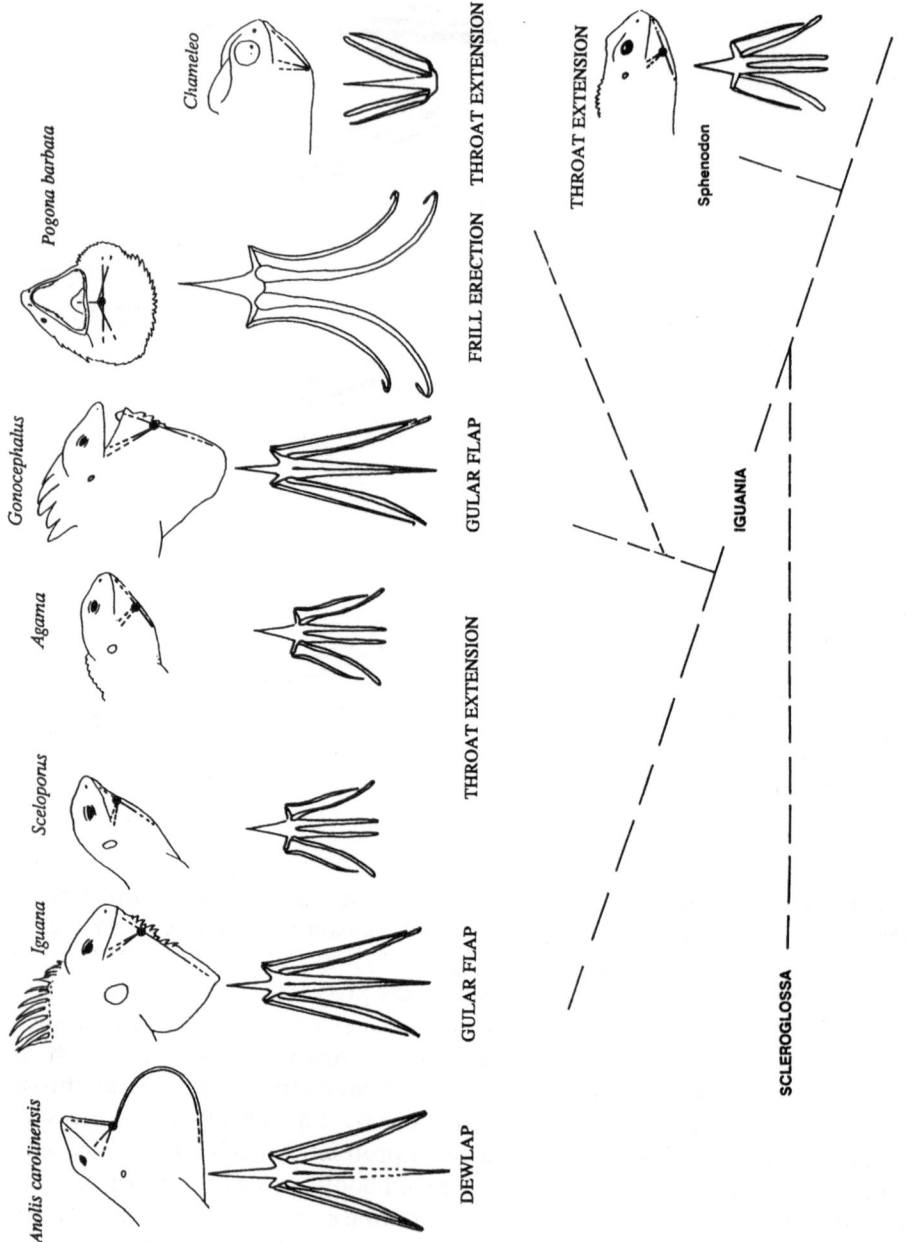

The only functional data on motor display patterns concern the activity of the M. branchiohyoideus which contributes to the extension of the ceratobranchials II during dewlap extension and lateral expansion during frill erection. In *P. barbata* (Throckmorton et al. 1985), *A. carolinensis* (Bels 1990b), and *A. equestris* (Font and Rome 1990; Bels and De Vree in prep.), electrical stimulation and preliminary EMG recordings of this muscle in *A. equestris* during dewlap pulses (Font and Rome 1990) show that contraction of M. branchiohyoideus is *always* associated with the social display. This muscle moves the ceratobranchial I relative to the ceratohyal. The role of the other throat muscles in behavioral display patterns remains to be studied by electromyography. Throckmorton et al. (1985) reported that several muscles may act during frill erection: "During movements of the two horns, the ceratobranchials (= ceratobranchials I) are abducted and rotate around their long axes by the action of the mm. ceratomandibularis externus (= M. mandibulohyoideus I), hyoglossus/genioglossus and certohyoideus (= branchiohyoideus)".

The activities of throat muscles during vertical throat displays remain to be studied by electromyography. Oelrich (1956) and Font and Rome (1990) propose that contractions of M. mandibulohyoideus I, II, and III in *Ctenosaura pectinata* and *A. equestris* produce displacements of the hyoid body during this display. However, hyoid displacements in *A. equestris* and *C. pectinata* in social displays are rather different.

2.3.4 Evolutionary Hypothesis

Based on the structural properties of the hyoid body and signaling displays, lizards may be divided into two groups: (1) lizards with *separated*, short ceratobranchials II displaying ventral throat depression, and (2) lizards with *joined*, long ceratobranchials II displaying dewlap extension, gular flapping, and ventral throat depression. Based on their display behaviors, all the scleroglossans without ceratobranchials II belong to this first group (Fig. 12). Carpenter (in Carpenter and Ferguson 1977) does not make clear the distinction between a dewlapper and a gular flapper, but he does distinguish between both movements and inflated throat (increase in the size of the throat region). Extension of the dewlap or gular region has no differential effect on social interactions. However, it is different from a structural and functional point of view. In dewlappers, a large gular region is vertically extended and the distal portion of the ceratobranchials II may be bent. Gular flapping requires a forceful enlargement of

Fig. 12. Evolutionary hypothesis of the form function of the hyoid apparatus in iguanian lizards in relationship to its role in displays. Scleroglossans exhibit throat extension, whereas iguanians exhibit throat extension, gular flapping, dewlap extension, and frill erection. Frill erection is only displayed by the agamid, *Pogona barbata*. The ceratobranchials II of dewlapping agamids and iguanids are always joined along their long axis, whereas the iguanians displaying throat extension, the ceratobranchials II are separated. The hyoid apparatus has been drawn after Avery and Tanner (1971), Throckmorton et al. (1985), and Smith (1988)

the posterior region of the throat and should be restricted to iguanians with ceratobranchials II that are anteriorly separated but caudally joined. We suggest that dewlappers *Sensu stricto* include iguanids such as *Anolis, Iguania,* and *Basiliscus*; and agamids such as *Gonocephalus* and *Draco.* Gular flappers include iguanids such as *Sceloporus, Crotaphytus,* and *Tropidurus*; and *Agama* among the agamids. *Amphibolurus* lizards are particularly interesting because *A. muricatus* has short ceratobranchials II and performs only ventral throat depression (Carpenter and Ferguson 1977; Throckmorton et al. 1985).

One parsimonious hypothesis may be (1) that display in ancestors of lizards, particularly in threat display patterns, involve contractions of several throat muscles, and (2) that the patterns in modern species retain this pattern: if the pattern has been specialized, this was accomplished by using a small number of muscles together with morphological behavior and specializations. Differences in the pattern of threat, challenge, and/or sexual displays may be related to difference in the functional characteristics of the display behavior such as amplitude or velocity. This could then influence the responses of the receivers (i.e., predators and conspecifics). From such a perspective, the display may be considered to be a response related to the recruitment of many or just one throat muscle. In this evolutionary hypothesis, the ventral throat depression is suggested to be the plesiomorphic *behavioral* condition, whereas short, separated ceratobranchials II are the plesiomorphic *morphological* condition within squamates. Romer (1956) assumed that the hyoid apparatus of Gekkones is primitive. Our suggestion confirms Romer's finding because the hyoid apparatus of Gekkones has short, separated ceratobranchials. The proposed behavioral and morphological conditions seem also to be plesiomorphic for all the reptiles which ventrally depress the throat during display behaviors (i.e., turtles and crocodiles; Carpenter and Ferguson 1977). The transformation series related to simultaneous diversification of the hyoid apparatus and throat display may include the following characteristics:

1. The primitive hyoid apparatus is characterized by short, *separated* ceratobranchials II. This hyoid conformation is present in rhyncocephalians (Romer 1956), scleroglossan, and some iguanian lizards such as *Oplurus.* The ceratohyals articulate at the distal end (most cases) or in the middle of the lateral anterior hypohyals (e.g., *A. muricatus*; Throckmorton et al. 1985).
2. The ceratobranchials II increase in length and their association along the mid-axis changes (decrease in their maximal distance along the transverse axis). The ceratohyals never articulate at the distal tips of the hypohyals. This conditions is only known in iguanian lizards. Dewlap extension is thus performed during social interactions, but the ceratobranchials II may not bend relative to the hyoid body (e.g., *Draco, Iguana, Gonocephalus*).
3. The ceratobranchials II lengthen to the level of the pectoral girdle. This lengthening is associated with novel bending properties of the cartilage. This condition occurs in some *Anolis* lizards (e.g., *A. carolinensis, Anolis chlorocyanus, Anolis sagrei,* and *Anolis roquet*). The modification of the cartilaginous properties consists of (a) a decrease in the mineralization of the cartilage

matrix, and (b) a change in shape of the mineralized area within the ceratobranchials II as shown in *A. carolinensis* lizards (Bels in prep.).

2.4 Chemosensory Activities

In snakes and lizards, chemical cues are primarily detected through three chemosensory systems: taste buds, nasal epithelium, and the vomeronasal epithelium. Taste buds occur throughout the tongue region and oral epithelium of most lizards but are reduced or absent in *Varanus* and snakes (Schwenk 1985). The nasal and vomeronasal epithelia house chemoreceptor cells whose axons project to the olfactory bulb. The receptor axons from the nasal epithelium project to the main olfactory bulb; receptor axons of the vomeronasal epithelium project to the accessory olfactory bulb (Kubie et al. 1978; Halpern and Kubie 1984). Volatile, airborne chemicals are primarily gathered by the nasal epithelium; chemical cues, often of high molecular weight, sampled from the air or substrates related to prey and social events of reproduction are primarily gathered by the vomeronasal system (Burghardt 1970; Duvall et al. 1986). To distinguish these two categories of chemical cues and the two distinct but parallel systems devoted to their detection, the chemicals are termed odors or vomodors, and the processes termed olfaction and vomerolfaction, respectively (Cooper and Burghardt 1990).

The mechanics of gathering the chemicals are different for the two systems as well. Airborne chemicals arriving at the olfactory eipthelium are carried in through the nostrils in the respiratory air. When interest is aroused, the rate of breathing and therefore of chemosensory sampling by the nasal epithelium increases. Vomerolfaction depends upon the hyolingual system to gather and deliver chemicals to the vomeronasal epithelium through the mechanical mechanism of tongue flicking. In both snakes and lizards, elevated rates of tongue flicking, termed tongue-flick attack scores (TFAS), accompany the stimulus of prey vomodors (Burghardt 1990; Cooper and Burghardt 1990); a stereotypic high rate of tongue flicking, termed strike-induced chemosensory searching (SICS), is stimulated by striking prey and is exhibited when relocating dispatched prey (Chiszar et al. 1977, 1983, 1990).

During tongue flicks, the tongue changes shape by elongation and bending (Fig. 13). Tongue elongation in scleroglossans such as *Tupinambis* and *Varanus* reaches more than 100% (Smith and Kier 1989). One tongue flick, or "flick cluster" (Ulinski 1972), consists of up to three phases: tongue protrusion phase, during which the tongue is initially protracted from the mouth; oscillation phase, during which the protruded tongue sweeps up and down in front of the mouth; and tongue withdrawal phase returning the tongue to the mouth, ending the tongue flick. Generally, during the tongue flick, the tongue moves out of the mouth via the lingual canal to sweep the air and even make contact with the substrate directly in front of the snout. However, once protruded from the lingual canal, the tongue may be bent laterally toward the direction of interest or movement (Ulinski 1972). The oscillation phase is especially variable, and four oscillation patterns are recognized: simple downward extension (SDE), single oscillation (SOC), multiple oscillation (MOC), and sub-multiple oscillation (SMOC) have been described (Gove 1979; Goosse and Bels 1992b). Actually, one, two, or all of these oscillation patterns – SDE, SOC, SMOC, MOC – are performed in different behavioral contexts by scleroglossans and snakes, whereas only one behavioral unit, SDE, has been reported for iguanians. Three morphological elements (the jaw apparatus, the hyoid apparatus, and the

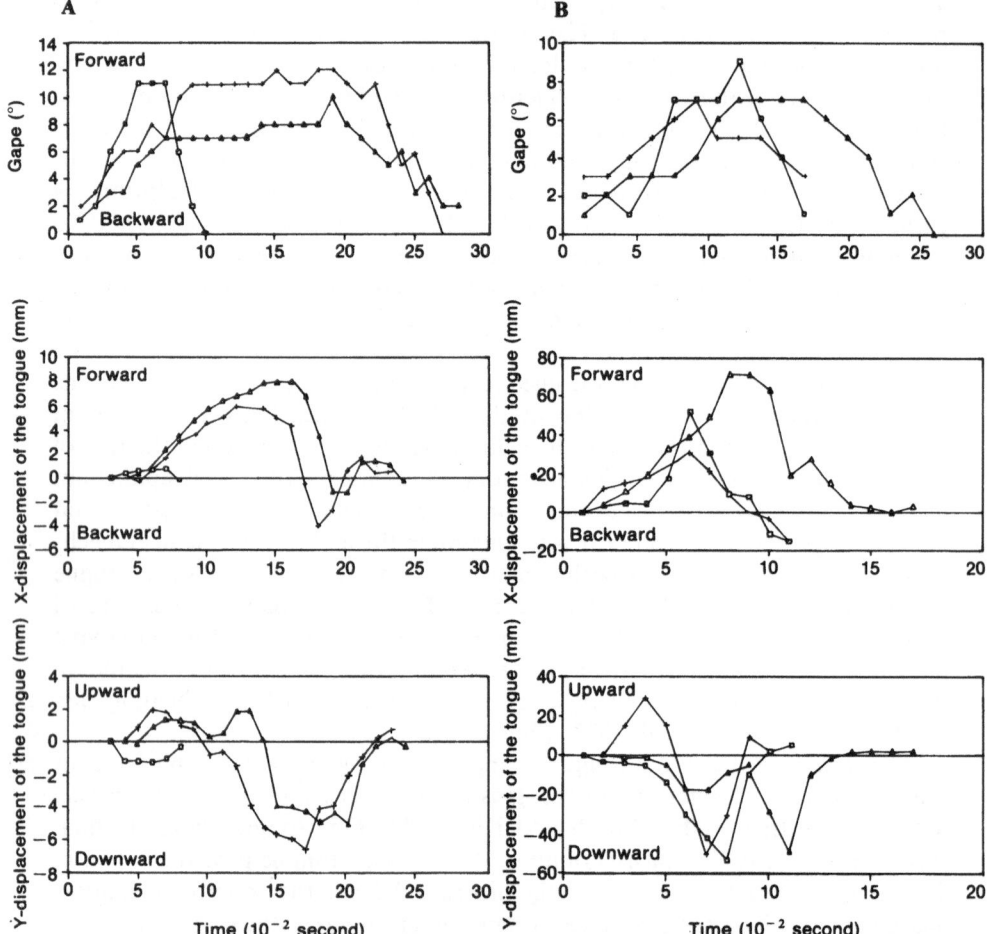

Fig. 13. Tongue flicking in lizards. **A–B** Representative kinematic diagrams of the tongue flicking units in *Lacerta viridis* (**A**) and *Varanus griseus* (**B**) illustrating the simple downward extension (+), the single oscillation (△), and the submultiple oscillation (□). **C** Representative kinematic diagrams representing the horizontal and vertical displacement of the tongue (△) and the hyoid apparatus (□) during single oscillation in *Varanus griseus*. Displacements of the tongue are independent of the displacements of the hyoid apparatus

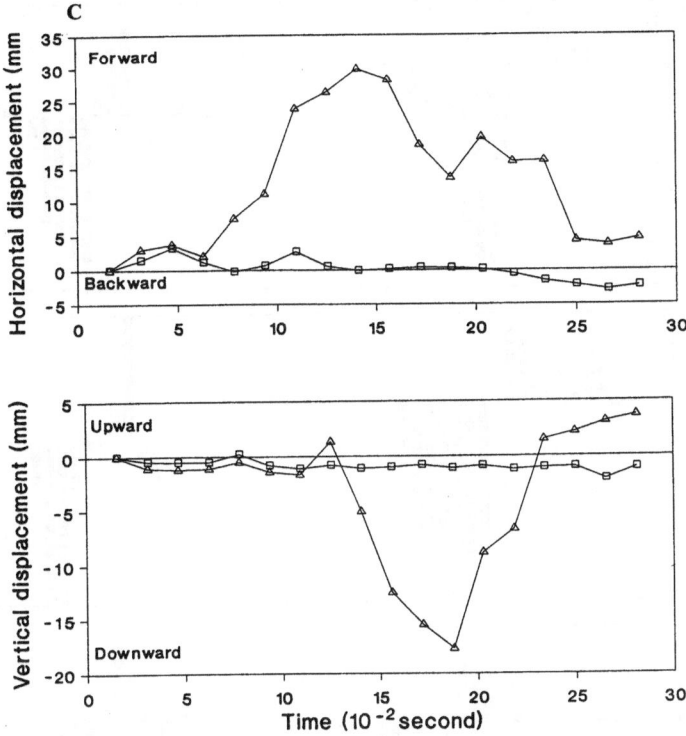

Fig. 13. *Cont.*

tongue) may be involved in tongue flicking. The hyoid apparatus, on which the posterior portion of the tongue inserts by the retractor M. hyoglossus, does not move at all during flicking in *Varanus* in which the tongue elongates (Fig. 13).

During the oscillation phase, the protruded tongue is generally elevated and lowered, exhibiting one or more of the oscillation patterns before being retracted. The variability in the patterns of these tongue oscillations may suggest various sequential organizations and different muscle recruitments in tongue-flicking patterns (Goosse and Bels 1992b). In the SOC oscillation pattern, the tongue crosses the horizontal plane twice, once during lowering of the tongue, and then a second time during an immediate, full elevations of the tongue. In the SMOC oscillation patterns, the tongue does not cross the horizontal plane between two successive upward movements (Fig. 14). This SMOC pattern is considered to result from activation of a second elevation of the tongue before its initial descent carries it across the horizontal plane. The flicking category (SMOC) may be an evolutionary intermediate between a single oscillation (SOC) and multiple oscillations (MOC) as described for some lizards (e.g., Scincidae and Anguidae) and serpents (Ulinski 1972; Gove 1979).

Gove (1979) proposed that multiple oscillations arose independently in snakes and lizards because the relative area swept becomes greater from one

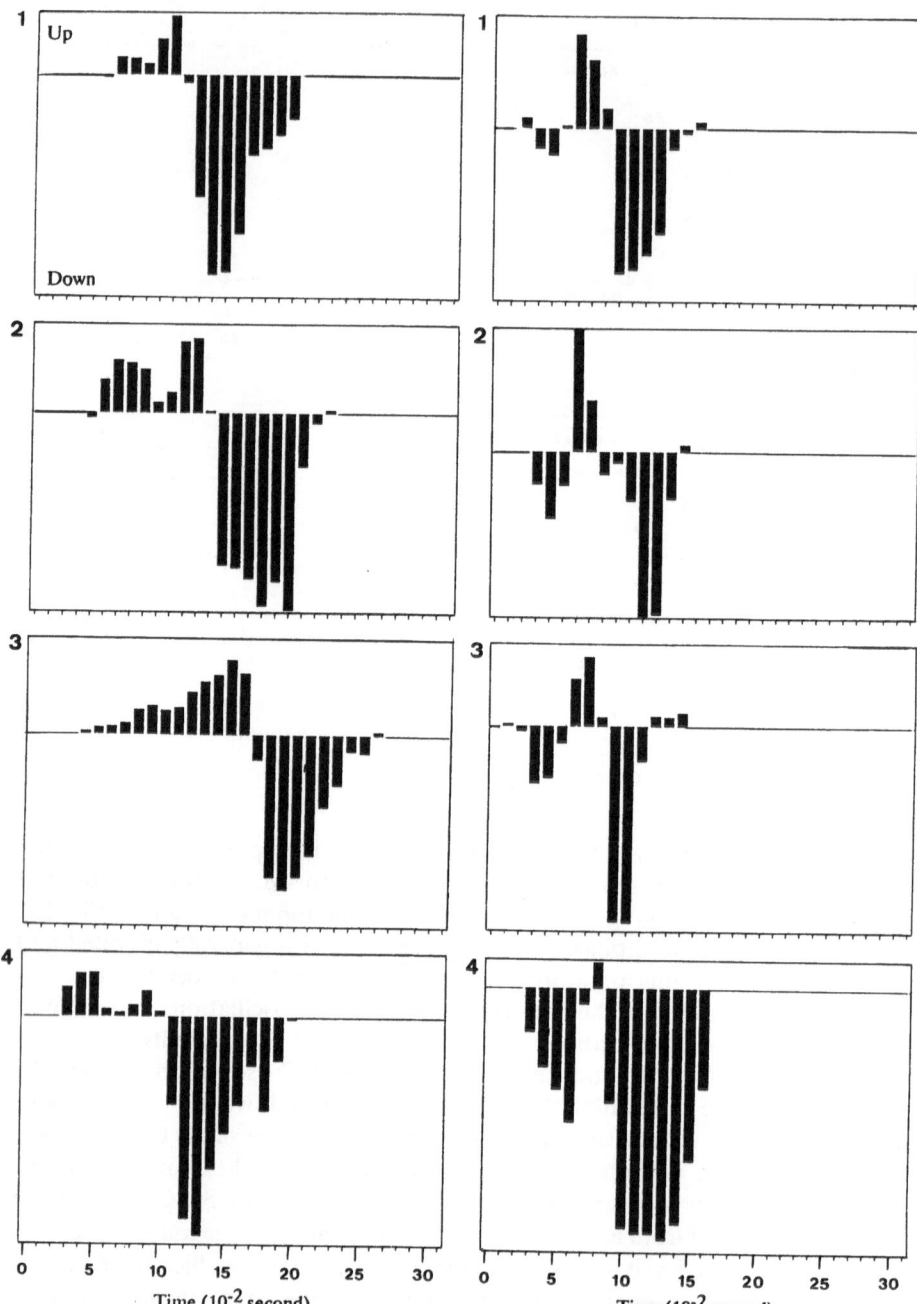

Fig. 14. Variability of tongue displacements during submultiple oscillation in *Lacerta viridis*

flick to the following one, and (2) the tongue protrudes continually during all oscillations in snakes, whereas the opposite occurs in lizards. In lizards and primitive snakes, a tongue flick always begins by an upward movement of the tongue but in advanced snakes, this first movement of the tongue is more variable (Gove 1979). The SMOC in *L. viridis* and *V. griseus* begins by an up *or* down displacement of the tongue. The relationship between horizontal and vertical components of tongue movements is similar to that described for multiple oscillations in snakes.

Smith (1986) describes two bending displacements of the tongue in *Varanus* produced by (1) vertical oscillations during flicking (as in the majority of scleroglossans and in snakes performing SOC, SMOC, and MOC), and (2) wrapping of the tongue laterally around the sides of the mandibles and twisting during "lip lipping". This wrapping tongue displacement is one of the major tongue movements in all the scleroglossans at the end of the feeding sequence ("cleaning" tongue cycles, Bels and Baltus 1989; Delheusy and Bels 1992), and during "eye and mandible lipping" in Gekkonids. These upward movements of the tongue should be produced by the simultaneous bilateral activity of the "circular" muscles and dorsal oblique muscles, and the downward movement by unilateral or dominant activity of one of the two central, longitudinal muscles (M. hyoglossus) (Smith 1986). Twisting should be produced by unilateral contraction of one of the dorsal oblique muscles; the direction of tongue twisting depends on the stiffness of the active layer (Kier and Smith 1985; Smith 1986).

Although the hyolingual system may participate significantly during feeding in lizards, this functional role is lost in snakes. Consequently, in snakes, the hyolingual system is dedicated almost exclusively to vomerolfaction, and it is specialized accordingly. The hypobranchium, often complex in lizards, is simplified in snakes. Only one pair of cornua is present that may be united at the midline and support an unpaired lingual process. The snake hypobranchium is partially ossified in some typhlopids, but it is entirely cartilaginous in all other groups.

On the basis of morphology, four hyoid types can be recognized (Fig. 15; Langbartel 1968), which are characteristic of four groups of snakes. The scoleophidian hyoid is usually V-shaped, often partially ossified, and may bear a prominent lingual process. The anomalepidid hyoid forms a W- or M-shape. The henophidian hyoid is V-shaped but posterior cornua diverge; in some booids and anilioids, cornua are separate, but in others they unite anteromedially. The caenophidian hyoid is of a parallel type; cornua are long, and run parallel to another. They are united anteriomedially where they usually bear a short lingual process.

Simplification of the snake hypobranchium results from loss of the hypohyals, ceratohyals, and ceratobranchials. On the basis of embryological work (Kamal and Hammouda 1965a,b) and a careful comparative anatomical study, Rieppel (1981) identified stages in these reductions. Generally, the primitive hyoid condition is represented in scolecophidian snakes which retain a basihyal and

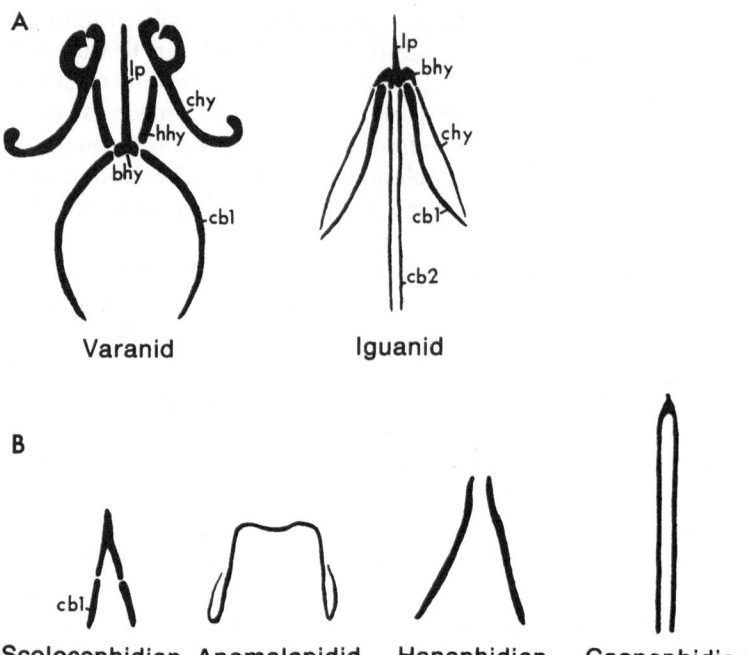

Fig. 15. Hyobranchia. **A** Hyobranchia of lizards, diagrammatic. **B** Hyobranchia of snakes showing the four hyoid types. The scolecophidian is V-shaped, often ossified, with separate ceratobranchials (as shown) or fused to the basihyal; the anomalepidid is M-shaped; the henophidian is V-shaped with divergent cornua and either separated (as shown) or fused medially across the midline; the caenophidian is of a parallel type usually bearing a short lingual process medially. Note the relative simplification of snake hyobranchia compared to lizards. *bhy* Basihyal; *cb1* ceratobranchial I; *cb2* ceratobranchial II; *chy* ceratohyal; *lp* lingual process

usually separate ceratobranchials. Within alethinophidians, a lingual process and diverging cornua represent the retention of primitive character states. However, ceratobranchials have been lost in alethinophidians, replaced by cornua developed by the elongation of posterolateral projections of the corners of the basihyal (McDowell 1972; Rieppel 1981).

Accompanying reduction of the hypobranchium is reduction or consolidation of throat muscles. All snakes lack Mm. geniohyoideus and branchiohyoideus; the M. sternohyoideus is lost in alethinophidians. Through fusions, the M. costomandibularis is present in alethinophidians; the M. neuromandibularis is present in all snakes except *Achrochordus* (Langebartel 1968).

The intrinsic tongue morphology in snakes is rather complex, consisting of a paired M. hyoglossus and bundles of circular and transverse muscles (Smith and Mackay 1990). Unlike some lizards which project their tongues ballistically by exerting force against and sliding along a stable lingual process (Wainwright and Bennett 1992b), the snake tongue acts against no such element of the hypobranchium. Instead, the snake tongue is mechanically a muscular-hydrostatic organ (Kier and Smith 1985). During tongue flicking, the tongue extends lengthwise during protrusion and shortens during retraction; no significant anterior-posterior displacement of the hypobranchium accompanies tongue flicking (Fig. 16). The arrangements of intrinsic muscles – longitudinal, perpendicular, and circular muscles – suggest that the tongue is a constant-volume organ. Elongation is produced by the contraction of circular or perpendicular muscles. The tongue has a high resting length-to-width ratio, so a small decrease in diameter produces a proportionately larger length elongation (Kier 1982; Kier and Smith 1985; Smith 1986; Smith and Kier 1989). Contraction of the hyoglossal muscle, running parallel to the long axis, retracts the tongue, shortening it. Bending of the tongue involves simultaneous interaction between longitudinal, vertical, and horizontal muscle bundles (Smith and Mackay 1990).

Within the snake tongue, regional differentiation of the musculature suggests regional functional specialization. The posterior tongue seems specialized for protraction/retraction of the whole organ. It is composed of the longitudinal hyoglossal muscle, which retracts the tongue, surrounded by perpendicular bundles that reduce the cross section and protrude the tongue. Anteriorly, these relationships are reversed. Two sets of longitudinal muscles are located peripherally, which bend the tongue, and perpendicular musculature tends to be centrally located, which stiffens the tongue (Smith and Mackay 1990).

The protracted tongue sweeps the air in front of the snake, often making contact with the substrate as well (Gove 1979). Chemicals collected during protraction are transported on the tongue back into the mouth. The tongue tines apparently do not directly transport these chemicals into the ducts to the vomeronasal epithelium. Instead, as the tongue returns, the collected chemicals are wiped from it especially by the anterior lingual processes on the floor of the mouth, which are then elevated, lifting transferred chemicals to the vomeronasal ducts (Gillingham and Clark 1981a; Young 1990).

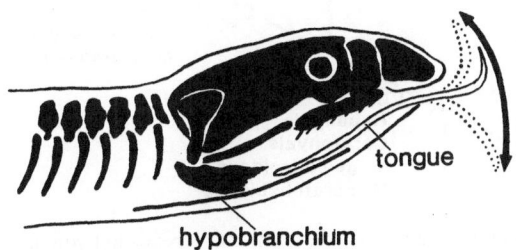

Fig. 16. Tongue flicking in snakes. The hyobranchium, located on the floor of the throat, does not protract during tongue protrusion. Protrusion of the tongue is accomplished primarily by elongation of the tongue itself, based upon a muscular-hydrostatic mechanism. Up to three phases constitute a tongue flick: protrusion, oscillation, withdrawal. During the oscillation phase, the tongue may bend upward and downward in front of the snout sweeping an area from which vomodors are collected

3 Evolutionary Comparison of Hyolingual Biomechanics

Information on feeding, drinking, and displaying motor patterns from a large number of Squamata is sparse. This makes it difficult to provide a functional and evolutionary interpretation of the transformation of the hyolingual apparatus in relation to the behaviors in which it is involved.

From actual data, we suggest the following conclusions (Fig. 17):

1. The tongue and hyoid skeleton evolved independently. Constraints explaining evolutionary changes are different for both elements (Fig. 18). Main constraints in hyoid skeletal evolution result from the displaying function and main constraints acting in tongue transformations result from the chemoreception function. However, modifications of the lingual process are related to tongue transformations.

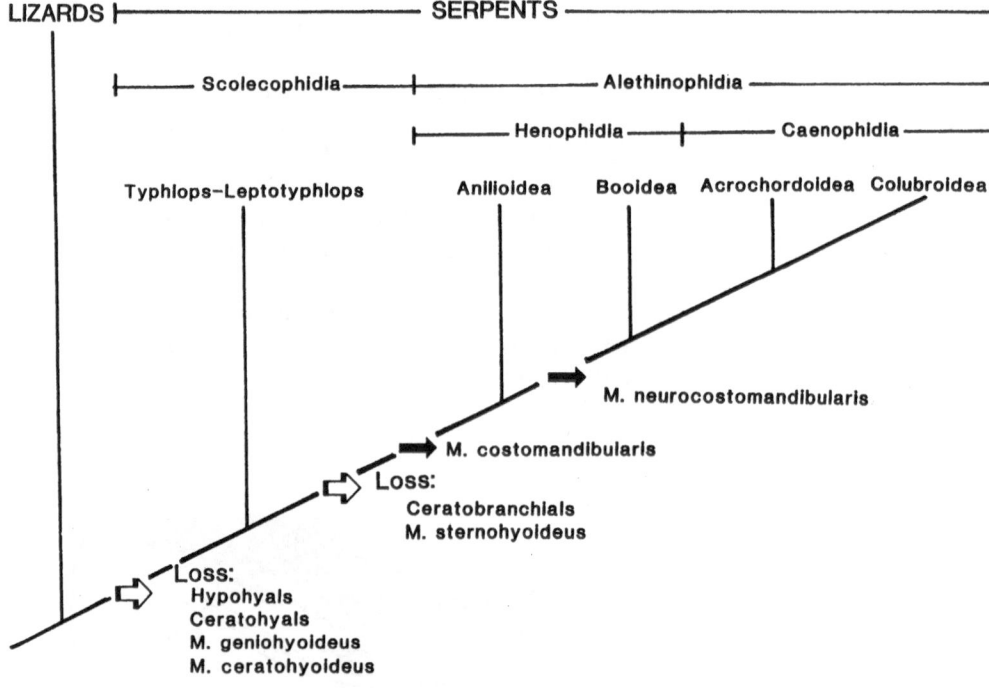

Fig. 17. Summary of evolutionary trends in hyoid and associated musculature in snakes. Lost in snakes are elements of the hyobranchium (hypohyals, ceratohyals) and associated musculature (Mm. geniohyoideus and ceratohyoideus). Ceratobranchials I are present in Typhlops-Leptotyphlops, but these together with M. sternohyoideus are lost in more derived snakes. By consolidation, two new muscles are recognized: M. costomandibularis and M. neurocostomandibularis. Modified from Langebartel (1968), whose muscle terminology has been used, although synonymies are preferred by some (Smith 1986): M. geniohyoideus (M. mandibulohyoideus III), M. ceratohyoideus (M. branchiohyoideus)

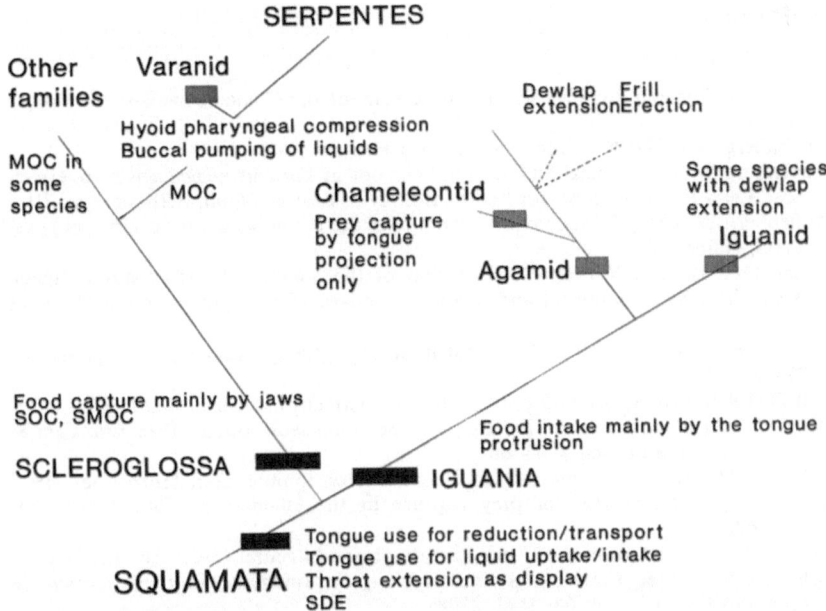

Fig. 18. Summary of the evolutionary trends of the tongue and the hyoid apparatus in the iguanian and scleroglossan lizards. The morpho-functional features of the tongue in Iguania have not greatly changed from those of the Squamata ancestors. The tongue is used in feeding, drinking, and chemoreception functions without any large modifications of the morphological characters, except in the chameleontids. By contrast, transformations of the tongue (i.e., surface, morphology) in the scleroglossan lizards and snakes are very important in relationship to chemoreceptive functions. The evolutionary trends of the hyoid apparatus are completely reversed. In iguanians, the structural and functional modifications of the hyoid elements (i.e., ceratobranchials I in *Pogona*, ceratobranchials II in *Anolis*) are very important in relationship to the displaying function. By contrast, modifications of the hyoid apparatus in scleroglossans are less important. They mainly concern the absence of ceratobranchials II

2. Tongue transformations related to chemoreception in derived squamates make it less suitable for capture and transport functions. This function seems to be consequently performed by more frequent use of the hyoid apparatus and jaws.
3. In most display-action patterns, throat displacements are produced by the same anatomical and mechanical design.
4. Modifications of the tongue and hyoid design do not influence the drinking function, except in varanids and snakes.

Acknowledgments. We are grateful to V. Delheusy, V. Goosse, and J.M. Urbani for their invaluable help with parts of this paper. We thank the Journal of Zoology, Journal of Morphology, and Journal of Experimental Biology for permission to publish several figures originally appearing in these journals (J. Exp. Biol. for Fig.1; J. Zool. for Fig. 6; J. Morphol. for Figs. 9 and 10).

References

Altevogt R, Altevogt R (1954) Studien zur Kinematik der Chamäleonenzunge. Z Vergl Physiol 36: 66–77

Auffenberg W (1972) Komodo dragons. Nat Hist 81: 52–59

Auffenberg W (1978) Social and feeding behavior in *Varanus komodoensis*. In: Greenberg N, Mac Lean PD (eds) Behavior and neurology of lizards. Nimh, Bethesda, pp 301–331

Auffenberg W (1981) The behavioral ecology of the Komodo monitor. University of Florida Press, Gainesville

Avery DF, Tanner WW (1971) Evolution of the iguanine lizards (Sauria, Iguandiae) as determined by osteological and myological characters. Brigham Young Univ Sci Bull 3: 1–71

Avery DF, Tanner WW (1982) Buccal floor of reptiles, a summary. Great Basin Nat 42: 273–349

Bell D (1984) Tongue use and prey capture in chameleons. Am Zool 24: 108A

Bell D (1987) Identification of perikarya in the chameleon tongue. Proc Ord Gen Meet Soc Eur Herpetol Nijmegen 4: 63–66

Bell D (1989) Functional anatomy of the chameleon tongue. Zool Jahrb Anat 119: 313–336

Bell D (1990) Kinematics of prey capture in the chameleon. Zool Jahrb Physiol 94: 247–260

Bell T (1826) Observation sur la structure du gosier du genre *Anolis*. Ann Sci Nat 7: 191–195

Bels VL (1990a) Quantitative analysis of prey-capture kinematics in *Anolis equestris* (Reptilia : Iguanidae). Can J Zool 68: 2192–2198

Bels VL (1990b) The mechanism of dewlap extension in *Anolis carolinensis* (Reptilia; Iguaniade) with histological analysis of the hyoid apparatus. J Morphol 206: 225–244

Bels VL (1992) Functional analysis of the ritualized behaviuoral motor pattern in lizards: evolution of behavior and the concept of ritualization. Zool Jahrb 122: 2141–159

Bels VL, Baltus I (1987) First analysis of the feeding sequence of *Chameleo dilepis*. Proc Ord Gen Meet Soc Eur Herpetol Nijmegen 4: 67–70

Bels VL, Baltus I (1989) First analysis of feeding in *Anolis* lizards. In: Splechtna H, Hilgers H (eds) Fortschritte der Zoologie/Progress in Zoology, Band/vol 35. Trends in vertebrate morphology. Fischer, Stuttgart, pp 141–145

Bels VL, Goosse V (1989) A first report of relative movements within the hyoid apparatus during feeding in *Anolis equestris* (Reptilia: Iguanidae). Experientia 45: 1088–1091

Bels VL, Goosse V (1990) Comparative kinematic analysis of prey capture in *Anolis carolinensis* 1 (Iguania) and *Lacerta viridis* (Scleroglossa). J Exp Zool 255: 120–124

Bels VL, Goosse V, Kardong K (1992) Kinematic analysis of drinking by the lacertid lizard, *Lacerta viridis* (Squamates, Scleroglossa). J Zool Lond 229: 659–682

Brücke E (1872) Über de Zunge des Chameleonen. Sitzungsber Math-Nat Kl Akad Wiss Wien 8: 62–70

Burghardt GM (1970) Chemical perception in reptiles. In: Johnson JW, Moulton DG, Turk A (eds) Advances in chemoreception. Communication by chemical signals, vol 1. Appleton-Century-Cropts, New York, pp 241–308

Burghardt GM (1990) Chemically mediated predation in vertebrates: diversity, ontogeny, and information. In: Macdonald DW, Müller-Schwarze D, Natynczuk S (eds) Chemical signals in vertebrates. Oxford University Press, New York, pp 475–499

Camp CL (1923) Classification of the lizards. Bull Am Mus Nat Hist 48: 289–481

Carpenter CC (1978) Ritualistic social behaviors in lizards. In: Greenberg N, Mac Lean (eds) Behavior and neurology of lizards. Nimh, Rockville, pp 253–267

Carpenter CC, Ferguson GW (1977) Variation and evolution of stereotyped behavior in reptiles. In: Gans C, Tinkle DW (eds) Biology of Reptilia, vol 7. Ecology and behavior A. Academic Press, London, pp 335–554

Chemin A (1899) L'appareil hyoïdien et son fonctionnement chez *Calotes versicolor*. Note pour servir à l'étude de l'anatomie comparée de l'os hyoïde. Bibl Anat 7: 114–123

Chiszar D, Scudder K, Kinght L (1976) Rate of tongue flicking by garter snakes (*Thamnophis radix haydeni*) and rattlesnakes (*Crotalus v. viridis, Sistrurus catenatus tergeminus*, and

Sistrurus catenatus edwardsi) during prolonged exposure to food odors. Behave Biol 18: 273–283

Chiszar D, Radcliffe CW, Scudder KM (1977) Analysis of the behavioral sequence emitted by rattlesnakes during feeding episodes. I. Striking and chemosensory searching. Behav Biol 21: 418–425

Chiszar D, Radcliffe DW, Scudder KM, Duvall D (1983) Strike-induced chemosensory searching by rattlesnakes: the role of envenomation-related chemical cues in the post-strike environment. In: Müller-Schwarze D, Silverstein RM (eds) Chemical signals III. Plenum Press, New York, pp 1–24

Chiszar D, Melcer T, Lee R, Radcliffe CW, Duvale D (1990) Chemical cues used by prairie rattlesnakes (*Crotalus viridis*) as they follow the trail of rodent prey. J Chem Ecol 16: 79–86

Cooper WE (1989) Strike-induced chemosensory searching occurs in lizards. J. Chem Ecol 15(4): 1311–1320

Cooper WE (1990a) Prey odour discrimination by lizards and snakes. In: Macdonald DW, Müller-Schwarze D, Natynczuk SE (eds) Chemical signals in vertebrates. Oxford University Press, New York, pp 533–538

Cooper WE (1990b) Prey odor detection by teiid and lacertid lizards and the relationship of prey odor detection to foraging mode in lizard families. Copeia 1990: 237–242

Cooper WE, Burghardt GM (1990) Vomerolfaction and vomodor. J Chem Écol 16: 103–105

Cundall D (1987) Functional morphology. In: Seigel RA, Collins JT, Novak SS (eds) Snakes-ecology and evolutionary biology. Macmillan, New York, pp 106–142

Cundall D, Gans C (1979) Feeding in water snakes: an electromyographic study. J Exp Zool 209: 189–208

Delheusy V, Bels VL (1992) Kinematics of feeding behaviour in *Oplurus cuvieri* (Reptilia: Iguanidae). J Exp Biol 170: 155–186

Duvall D, Chiszar D (1990) Behavioural and chemical ecology of vernal migration and pre- and post-strike predatory activity in prairie rattlesnakes: field and laboratory experiments. In: Macdonald DW, Müller-Schwarz D, Natynczuk SE (eds) Chemical signals in vertebrates 5. Oxford University Press, New York, pp 539–554

Duvall D, Müller-Schwarze D, Silverstein RM (1986) Chemical signals in vertebrates 4: ecology, evolution and comparative biology. Plenum Press, New York

Duvernoy GL (1836) Sur les mouvements du la langue de chameleon. C R Hebd Séanc Acad Sci Paris 2: 349–351

El-Toubi MR (1947) Some observations on the osteology of *Uromastix aegyptia* (Forskal). Bull Fac Sci Cairo Fouas I Univ 25: 1–10

Estes R, Pregill G (1988) Phylogenetic relationships of the lizard families. Stanford Univ Press, Stanford

Font E, Kramer M (1989) A multivariate clustering approach to display repertoire analysis: head-bobbing in *Anolis equestris* (Sauria, Iguanidae). Amphib-Reptilia 10: 331–344

Font E, Rome LC (1990) Functional morphology of dewlap extension in the lizard *Anolis equestris* (Iguanidae). J Morphol 206(1990): 245–258

Frazzetta TH (1966) Studies on the morphology and function of the skull in the Boidae (Serpentes). Part II. Morphology and function of the jaw apparatus in *Python sebae* and *Python molurus*. J Morphol 118: 217–296

Gandolfi H (1908) Der Zunge der Agamidae und Iguanidae. Zool Anz 32: 56

Gans C (1967) The chameleon. Nat Hist 76: 52–59

Gillingham JC, Clark DL (1981a) Snake tongue-flicking: transfer mechanism to Jacobson's organ. Can J Zool 59: 1651–1657

Gillingham JC, Clark DL (1981b) An analysis of prey-searching behavior in the western diamondback rattlesnake, *Crotalus atrox*. Behav Neural Biol 32: 235–240

Gnanamuthu CP (1930a) The anatomy and mechanism of the tongue of *Chamaeleon cararatus*. Proc Zool Soc Lond Part II: 467–486

Gnanamuthu CP (1930b) The mechanism of the throat-fan in a ground lizards, *Sitana ponticeriana*. Cuv Rec Ind Mus 32: 149–159

Gnanamuthu CP (1937) Comparative study of the hyoid and tongue of some typical genera of reptiles. Proc Zool Soc B: 1–66

Graves BM, Halpern M (1990) Roles of vomeronasal organ chemoreception in tongue flicking exploratory and feeding behaviour of the lizard, *Chalcides ocellatus*. Anim Behav 39: 692–698

Goosse V, Bels VL (1990) Analyse comportementale et fonctionnelle des touchers linguaux lors de l'exploration et de la prise de nourriture chez le lézard vert (*Lacerta viridis* Laurenti 1768). Bull Soc Herp Fr 53: 31–33

Goosse V, Bels VL (1992a) Kinematic and functional analysis of feeding behaviour in *Lacerta viridis* (Reptilia: Lacertidae). Zool Jahrb 122: 187–202

Goosse V, Bels VL (1992b) Tongue movements during chemosensory behavior in the European green lizard *Lacerta viridis*. Can J Zool 70: 1886–1896

Gorniak GC, Rosenberg HI, Gans C (1982) Mastication in the tuatara *Sphenodon punctatus* (Reptilia: Rhynchocephalia): structure and activity of the motor system. J Morphol 171: 321–353

Gorman GC (1968) The relationships of *Anolis* of the *roquet* species group (Sauria: Iguanidae): Comparative study of display behavior. Breviora 284: 1–31

Gove D (1979) A comparative study of snake and lizard tongue-flicking with an evolutionary hypothesis. Z Tierpsychol 51: 58–76

Graves BM, Halpern M (1989) Chemical access to the vomeronasal organs of the lizard *Chalcides ocellatus*. J Exp Zool 249: 150–157

Greenberg N (1977) A neuroethological study of the display behavior in the lizard *Anolis carolinensis* (Sauria Iguanidae). Am Zool 17: 191–201

Greene HW (1982) Dietary and phenotypic diversity in lizards: why are some organisms specialized? In: Mossakowski D, Roth G (eds) Environmental adaptation and evolution. Fischer, New York, pp 107–128

Greene HW (1983) Dietary correlates of the origin and radiation of snakes. Am Zool 23: 431–441

Halpern M (1983) Nasal chemical senses in snakes. In: Ewert JP, Carpina RR, Ingle DJ (eds) Advances in vertebrate neuroethology. Plenum Press, New York, pp 141–176

Halpern M (1987) The organization and function of the vomeronasal system. Annu Rev Neurosci 10: 325–362

Halpern M, Furmin N (1979) Roles of the vomeronasal and olfactory systems in prey attack and feeding in adult garter snakes. Physiol Behav 22: 1183–1189

Halpern M, Kubie JL (1984) The role of the ophidian vomeronasal system in species-typical behavior. Trends Neurosci 7(12): 472–477

Heidweiller J, Zweers GA (1990) Drinking mechanisms in the zebra finch and the Bengalese finch. Condor 92: 1–28

Houston J (1828) On the structure and mechanism of the tongue of the chameleon. Trans R Ir Acad 15: 177–201

Iverson JB (1982) Adaptations to herbivory in iguanine lizards. In: Burghardt GM, Rand AS (eds) Iguanas of the world; their behavior ecology and conservation. Noyes, Park Ridge, pp 60–76

Iwasaki S (1990) Fine structure of the dorsal lingual epithelium of the lizard *Gekko japonicus* (Lacertilia Gekkonidae). Am J Anat 187: 12–20

Kamal AM, Hammouda HG (1965a) The chondrocranium of the snake *Eryx jaculus*. Acta Zool 46: 167–208

Kamal AM, Hammouda HG (1965b) The development of the skull of *Psammophis sibilans*. J Morphol 116: 197–246

Kardong KV (1974) Kinesis of the jaw apparatus during the strike in the cottonmouth snake, *Agkistrodon piscivorous*. Forma Functio 7: 327–354

Kardong KV (1977) Kinesis of the jaw apparatus during swallowing in the cottonmouth snake, *Agkistrodon piscivorous*. Copeia 1977: 338–348

Kardong KV, Dullemeijer P, Fransen JAM (1986) Feeding mechanism in the rattlesnake *Crotalus durissus*. Amphib-Reptilia 7: 271–302

Kardong KV, Haverly J (1993) Drinking by the common boa, *C. Boa constrictor*. Copeia 1993: 808–818

Kathariner L (1894) Anatomie und Mechanismus der Zunge des Vermiliguer, Jena Z Naturwise 29: 247–270.

Kent WS (1895) Observations on the frilled lizard *Chlamydosaurus kingii*. Proc Zool Soc Lond 46: 712–719

Kestevelen HL (1944) The evolution of the skull and cephalic muscles: a comparative study of their development and adult morphology. Part III. The Sauria (Reptilia). Aust Mus Sid Mem VIII 3: 237–269

Kier WM (1982) The functional morphology of the musculature of squid (Loliginidae) arms and tentacles. J Morphol 172: 179–192

Kier WM, Smith KK (1985) Tongues, tentacles and trunks: the biomechanics of movement in muscular-hydrostats. Zool J Linn Soc 83: 307–324

Kraklau DM (1991) Kinematics of prey capture and chewing in the lizard *Agama agama*. J Morphol 210: 195–212

Kubie JL, Halpern M (1979) Chemical senses involved in garter-snake prey trailing. J Comp Physiol Psychol 93(4): 648–667

Kubie JL, Vagvolgyi A, Halpern M (1978) The roles of the vomeronasal and olfactory systems in the courtship behavior of male garter snakes. J Comp Physiol Psychol 92: 627–641

Langebartel DA (1968) The hyoid and its associated muscles in snakes. III Biol Monogr 38: 1–156

List JC (1966) Comparative osteology of the snake families Typhlopidae and Leptopyphlopidae. III Biol Monogr 36: 1–112

Losos J (1985) Male aggressive behavior in a pair of sympatric sibling species. Breviora 484: 1–30

McDowell SB (1972) The evolution of the tongue of snakes and its bearing in snake origins. In: Dobzhansky T, Hecht MK, Steere WC (eds) Evolutionary biology, vol 6. Meredith, New York, pp 192–273

Meredith M, Burghardt GM (1978) Electrophysiological studies of the tongue and accessory olfactory bulb in garter snakes. Physiol Behav 21: 1001–1008

Mivart SG (1870) On the myology of *Chameleo parsonii*. Proc Sci Meet Zool Soc Lond 57: 850–890

Murphy JB, Mitchell LA (1974) Ritualized combat behavior of the pygmy monitor lizard *Varanus gilleni* (Sauria: Varanidae). Herpetologica 30: 90–97

Oelrich TM (1956) The anatomy of the head of *Ctenosaura pectinata*. Misc Publ Mus Zool Univ Mich 94: 1–122

Owasa G (1898) Beiträge zur Anatomie der *Hatteria punctata*. Arch Mikrosk Anat 51: 481–691

Presch W (1974) A survey of the dentition of the macroteiid lizards (Teiidae: Lacertilia). Herpetologica 30: 344–349

Rabinowitz T, Tandler B (1986) Papillary morphology of the tongue of the American chameleon: *Anolis carolinensis*. Anat Rec 216: 483–489

Richter H (1933) Das Zungenbein und seine Muskulatur bei den Lacertilia vera. Jena Z Naturwiss 66: 395–480

Rieppel O (1981) The hyobranchial skeleton in some little known lizards and snakes. J Herpetol 15: 433–440

Rieppel O, Labhardt L (1979) Mandibular mechanics in *Varanus niloticus* (Reptilia: Lacertilia). Herpetologica 35: 158–163

Romer AS (1956) Osteology of the reptiles. University Chicago Press, Chicago

Schwenk K (1982) Lizard tongue morphology: disparate functions and comprehensive designs. Am Zool 22: 923

Schwenk K (1985) Occurrence, distribution and functional significance of taste buds in lizards. Copeia 1985: 91–101

Schwenk K (1986) Morphology of the tongue in the Tuatara *Sphenodon punctatus* (Reptilia: Lepidosauria) with comments on function and phylogeny. J Morphol 188: 129–156

Schwenk K (1987) Evolutionary determinants of cranial form and function in lizards. Am Zool 27: 105A

Schwenk K (1988) Comparative morphology of the Lepidosaur tongue and its relevance to squamate phylogeny. In: Estes R, Pregill G (eds) Phylogenetic relationships of the lizard families, Essays commemorating C.L. Camp. Stanford Univ Press, Stanford, pp 569–598

Schwenk K, Bell DA (1988) A cryptic intermediate in the evolution of chameleon tongue projection. Experientia 44: 697–700

Schwenk K, Greene HW (1987) Water collection and drinking in Phrynocephalus helioscopus: a possible condensation mechanism. J Herpetol 21: 134–139

Schwenk K, Throckmorton GS (1989) Functional and evolutionary morphology of lingual feeding in squamate reptiles: phylogenetics and kinematics. J Zool (Lond) 219: 153–176

Shine R (1990) Function and evolution of the frill of the frillneck lizard Chlamydosaurus kingii (Sauria: Agamidae). Biol J Linn Soc 40: 11–20

Smith KK (1984) The use of the tongue and hyoid apparatus during feeding in lizards (Ctenosaura similis and Tupinambis nigropunctatus). J Zool (Lond) 202: 115–143

Smith KK (1986) Morphology and function of the tongue and hyoid apparatus in Varanus (Varanidae Lacertilia). J Morphol 187: 261–287

Smith KK (1988) Form and function of the tongue in agamid lizards with comments on its phylogenetic significance. J Morphol 196: 157–171

Smith KK, Kier WM (1989) Trunks, tongues and tentacles: moving with skeletons of muscle. Am Sci 77: 28–35

Smith K, Mackay M (1990) The morphology of the intrinsic tongue musculature in snakes (Reptilia Ophidia): functional and phylogenetic implications. J Morphol 205: 307–324

So KK, Wainwright PC, Bennett AF (1992) Kinematics of prey processing in Chamaelo jacksonii: conservation of function with morphological specialization. J Zool Lond 226: 47–64

Sondhi KC (1958) The hyoid and associated structures in some Indian reptiles. Ann Zool 2: 157–227

Throckmorton G, De Bavay SJ, Chaffey W, Merrotsy B, Noske BS, Noske R (1985) The mechanism of frill erection in the bearded dragon Amphibolurus barbatus with comments on the jacky lizard A. muricatus (Agamidae). J Morphol 183: 285–292

Tilak R (1964) The hyoid apparatus of Uromastix hardwickii Gray. Sci Cult 30: 244–246

Ulinski PS (1972) Tongue movements in the common boa (Constrictor constrictor). Anim Behav 20: 373–383

Underwood G (1971) A modern appreciation of Camp's "classification of the lizards". Introduction to reprint by SSAR

Von Geldern CE (1919) Mechanism in the production of throat-fan in the chameleon Anolis carolinensis. Proc Calif Acad Sci 9: 313–329

Wainwright PC, Bennett AF (1992a) The mechanism of tongue projection. I Electromyographic tests of functional hypotheses. J Exp Biol 168: 1–21

Wainwright PC, Bennett AF (1992b) The mechanism of tongue projection. II. Role of shape in muscular hydrostat. J Exp Biol 168: 23–40

Wainwright PC, Kraklau DM, Bennett AF (1991) Kinematics of tongue projection in Chamaeleo oustaleti. J Exp Biol 159: 109–133

Willard WA (1915) The cranial nerves of Anolis carolinensis. Bull Mus Comp Zool 59: 1–134

Young BA (1990) Is there a direct link between the ophidian tongue and Jacobson's organ? Amphib-Reptilia 11: 263–276

Zavattari E (1911) I muscoli ioidei dei sauri in rapporto con i muscoli ioidei degli altri vertebrati. Mem Acad Sci Torino 60: 351–392

Zoond A (1933) The mechanism of projection of the chameleon's tongue. J Exp Biol 10: 174–185

Chapter 8

Behavioral Mechanisms of Avian Feeding

G.A. Zweers[1], *H. Berkhoudt*[1] and *J.C. Vanden Berge*[2]

Contents

[1] Neurobehavioral Morphology, Institute of Evolutionary and Ecological Sciences, Leiden University, P.O. Box 9613, NL 2300 RA Leiden, The Netherlands
[2] Northwest Center Medical Education, Indiana University, School of Medicine, 3400 Broadway, Gary, Indiana 46408, USA

Advances in Comparative and Environmental Physiology, Vol. 18
© Springer-Verlag Berlin Heidelberg 1994

1 Introduction

Feeding comprises three consecutive behavioral acts which often merge into one another: (1) foraging; (2) food acquisition; (3) digestion. Foraging is a matter of food searching in which the whole bird is involved and ends with head fixation just above the food. Vision serves as the primary control system. Optimality theory is used to develop a general foraging theory (e.g. Stephens and Krebs 1986).

Food acquisition comprises downstroke of the head and handling the food in the oropharynx, including ingestion and deglutition. The downstroke is a head depression and merges most often into grasping (see Zweers et al., in press, for a review of head-neck mechanics of the downstroke). The downstroke is primarily under optic control (Bischof 1988; see reviews of Delius 1989; Zeigler 1989). Food handling may comprise mechanisms that collect, test, select, manipulate, prepare, store and transport food for digestion. Mechanics and control of behavioral patterning form the major goal of this review. Bühler (1981), Schilling (1992), Vanden Berge (1979) and Vanden Berge and Zweers (1993) reviewed the underlying muscle-bone apparatus of the jaws, tongue and pharynx; Zusi (1985) presented a detailed functional and evolutionary analysis of avian rhyncho-kinesis; Zeigler et al. (in press) review sensorimotor control of the jaws; Gottschaldt (1985) and Berkhoudt (1985) considered avian touch and taste perception while elucidating some relations with food handling. None of these authors, however, integrates oropharyngeal mechanisms and their control. Optimality models are occasionally used to describe food acquisition (e.g. Kingsolver and Daniel 1983; Kooloos 1986; Kooloos and Zweers 1991), but no principles for vertebrate food acquisition have been developed as yet (Lauder 1989).

Digestion in the alimentary tract is a matter of preparing food by mechanical processing (storage, grinding, transport), chemical processing by mucosa-produced enzymatic and microbial fermentation, and absorption. McLelland (1979) reviews relationships of food types eaten and anatomical features in the digestive tract. Ziswiler (1990) demonstrates specialized form-function relationships in the alimentary tract. For example, frugivorous pigeons (Landolt 1987) and granivorous Estrildidae and Ploceidae (Ziswiler 1990) have histologic specializations at mucosal levels that control negative effects of digesting an extremely unbalanced diet.

The wide trophic diversity is often ordered by listing first what birds eat, and then making a further subdivision on the basis of how birds gather food. For example, Pough et al. (1991) distinguished nine groups based on food types and subdivide each group according to 'guilds' of foraging. Alternatively, trophic diversity can be ordered according to specific functioning. For example, McLelland (1979) reviews adaptations of the tongue for collecting, manipulating and swallowing food, which is followed by a detailed account of 17 distinct adaptations of bill shape and functioning.

We have organized the description of trophic diversity on a guilds basis. All neognathic birds show pecking behavior, regardless of whether they have food acquisition specializations. The pecking design is one key to the success of avian trophic diversification. Once the teeth had been replaced by a keratin cover, the skull kinesis was restricted to pro- and rhyncho-kinetic conditions, and the lingual apparatus was restricted to one set of horns, an amazingly plastic trophic design seemed to have developed. Zweers (1991a, b) stated that pecking mechanisms occur in all modern birds. In addition, Zweers et al. (in prep. a) showed that specialist feeding mechanisms, such as waterfowl-like filter feeding and sandpiper-like substrate probing, can be deduced from a generalized pecking mechanism. Therefore, we use the pecking mechanism as a basis and review in what respect recent data on specialist feeding mechanisms differ from that basic pattern. Ontogenetic development and evolutionary diversification processes are beyond the scope of this review.

Foraging, food acquisition and digestion are closely related systems. We restrict this chapter to food acquisition, including downstroke and grasping food, handling of food in the oropharynx, and the integration with touch and taste control systems. For foraging, head-neck mechanics (review of Zweers et al. in press), integration of feeding and drinking mechanisms (review of Zweers 1992), optic control of the downstroke, and digestion we refer to the mentioned reviews. We will discuss progress in understanding mechanics of food acquisition first for pecking mechanisms, then follow specific food-handling mechanisms, e.g. husking, that prepare food for digestion. Acquisition specialisms are considered next, and then the underlying major sensory control systems and motor patterning are treated. Finally, we will formulate a set of tentative principles on avian food acquisition to systematize present knowledge.

2 General Food Acquisition Mechanism: Pecking

2.1 *Columba*

Most pigeons are granivorous and forage in open fields. Their pecking mechanism is maximized for intake speed. The oropharyngeal anatomy was described by Krause (1922), Lucas and Stettenheim (1972) and Zweers (1982b) (Fig. 1). The slender tongue is highly flexible, protractable and elevatable, and it carries caudally two bilateral, erectable, caudally pointing, lingual wings. The mouth floor is densely packed with a bilateral sequence of seven large Gll. mandibulares rostrales; the palate carries dense fields of salivary glands along the

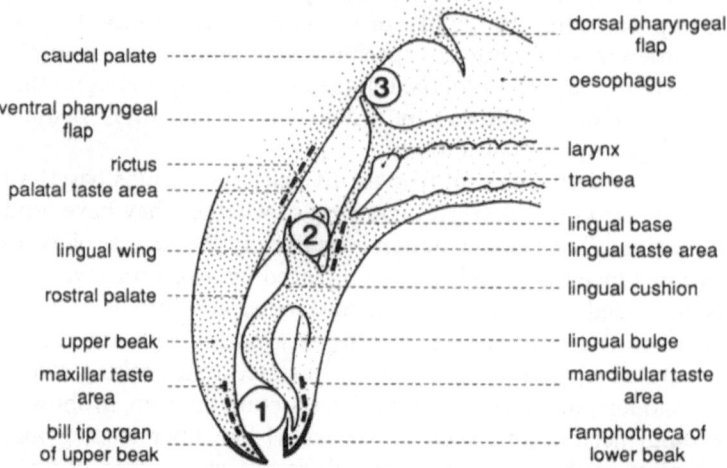

Fig 1. A generalized avian mouth and pharynx illustrating some major anatomical elements. The intraoral transport mechanism by "slide and glue" is shown for seed (*1*). The tongue is protracted and has, while being protracted, scraped saliva from the mouth floor so that the seed will stick to the tongue tip upon its retraction. The intrapharyngeal transport of seed (*2*) is accomplished by retraction of the erected lingual wing and the sticky mucus produced at the lingual base. Intrapharyngeal transport (seed *3*) occurs by retraction of the erected ventral pharyngeal flaps which scrape the seed from the palate. 'Properistalsis' is accomplished by retracting the elevated larynx along the palate. (After Zweers 1985).

secondary choana. Dorsocaudally from the laryngeal apparatus are erectable, pharyngeal flaps pointing caudad.

Pecking mechanics were analyzed by Zeigler et al. (1980; in press), Bermejo et al. (1989) and Zweers (1982a, 1985). In Zweers' 'static and dynamic components' model (Fig. 2), pecking is considered a pattern of alternating static and dynamic phases during which exteroceptive information is gathered during each static phase that controls motor patterning in a subsequent dynamic phase. Pecking starts with a preliminary head fixation during which the head is kept static a few centimeters above the seed, while seed size and position are estimated. A preliminary head approach follows in which the head is depressed slightly and then kept fixated again during a final head fixation. Seed and head-neck positions are retested for composing the final head approach. Then prehension, being a downstroke and a grasp, follows. During the downstroke, the beak tips are directed towards the seed, and the beak gapes are tuned to the seed size; it ends with a grasp of the seed. Now the seed is kept static relative to the beak tips during a short head elevation in which it is tested for edibility and for organizing further transport by estimating size, hardness, taste and position.

The seed is either dropped by opening the beak, thrown away by a head shake and beak opening, or transported through the mouth. The seed may be repositioned in a cyclic 'stationing' phase consisting of small 'throw and catch' motions allowing retesting of the seed and replacing it for better transport

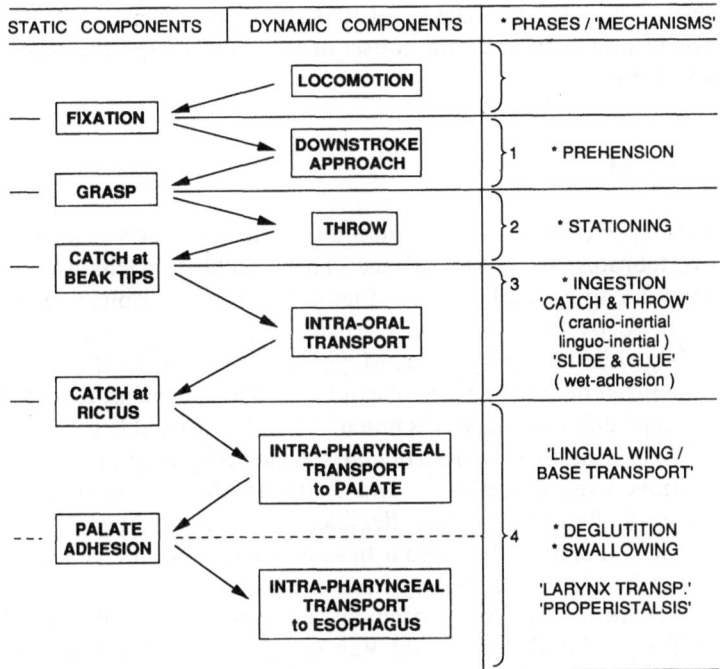

STATIC COMPONENTS	DYNAMIC COMPONENTS	* PHASES / 'MECHANISMS'
— FIXATION	LOCOMOTION	
	DOWNSTROKE APPROACH	1 * PREHENSION
— GRASP		
	THROW	2 * STATIONING
— CATCH at BEAK TIPS		3 * INGESTION 'CATCH & THROW' (cranio-inertial linguo-inertial) 'SLIDE & GLUE' (wet-adhesion)
	INTRA-ORAL TRANSPORT	
— CATCH at RICTUS		
	INTRA-PHARYNGEAL TRANSPORT to PALATE	'LINGUAL WING / BASE TRANSPORT'
--- PALATE ADHESION		4 * DEGLUTITION * SWALLOWING
	INTRA-PHARYNGEAL TRANSPORT to ESOPHAGUS	'LARYNX TRANSP.' 'PROPERISTALSIS'

Fig 2. A generalized and de-specialized scheme of the major behavioral elements of avian pecking according to the 'static-and-dynamic phases' model. Four main phases are shown (*1–4*). See text for explanation

through the mouth. A cyclic 'throw and catch' pattern may occur when large seeds are eaten ('cranio-inertial transport'; cf. Bramble and Wake 1985); it produces a throw of the seed caudad into the oropharynx by a strong head jerk and wide gape. Combinations with a second pattern may occur. Small seeds are transported by a mechanism called 'slide and glue' ('wet adhesion transport'; cf. Bramble and Wake 1985). The protracting tongue slides along the mouth floor pressing the mandibular glands so that mucus is gathered at the lingual tip which is pushed against the seed. Now the head elevates, but does not accelerate and the beak opens while the tongue retracts and the seed sticks to the tongue. The mouth closes and the seed is fixed at the rictus level during which its position and taste is tested. In long-billed pigeons this pattern is repeated until the seed arrives at the rictus level (Van Gennip 1988).

The tongue protracts again and the lingual wings slide underneath the seed. The wings are erected and retracted while the beak opens again so that the seed is carried caudad on the lingual base, while its taste and position are tested again. Lingual retraction pushes the seed against the palate by which action mucus is released so that the seed sticks to the palate. Retesting taste may alternatively result in regurgitation of the seed by strong lingual protraction, wide gaping and head shaking. The seed is fixed along the palate. Now the larynx and ventral pharyngeal flaps depress and protract. Then the flaps are erected and retracted and consequently they scrape the seed caudad into the esophagus.

2.2 *Gallus*

Many species of fowl live in humid climates of forests as foraging omnivores on invertebrates, seeds and foliage. Their pecking mechanism is maximized for forceful gaping and closing. Elements in the oropharynx of chickens were described by Calhoun (1933), Lucas and Stettenheim (1972), and McLelland (1979), while Homberger and Meyers (1989) analyzed the hyoid apparatus. The generalized description above can be used for chickens, except for the following. The chicken's oropharynx is much larger, the mandible carries a large dorsocaudally pointing Proc. retroarticularis, which is lacking in pigeons; the quadrate is positioned more vertically, and a strong Proc. postorbitalis is present. The tongue is fleshier and less flexible. The smaller lingual wings carry strong keratinized spines, the lingual base is extendable, the ventral pharyngeal flaps cannot be erected separately from the larynx.

The chicken's pecking mechanism is described by White (1968), McLelland (1979), Berkhoudt (1985) and Van den Heuvel (1992) and is largely similar to that in pigeons. After a visually controlled head approach, the beaks often bump into the top soil layer. The Proc. retroarticularis serves as an adaptation to forceful gaping against soil resistance. A slight head elevation, head fixation, and again a downstroke occur that ends with a forceful grasp that crunches scaled invertebrates. The Proc. postorbitalis serves as an adaptation to crunching. Catch and throw, and slide and glue mechanisms, as well as lingual inertial transport (cf. Bramblé and Wake 1985) and combinations of both serve food transport to the rictus. The transport through the pharynx occurs by 'pro-peristalsis' in which the larynx plays a major role. The larynx is protracted underneath the food and then retracted while the pharynx floor is elevated so that a constriction moves caudad. Often the food slips back over the larynx and is then picked up in the next cycle of larynx motion, so that as a result, e.g., over ten peas may accumulate at the extended lingual base in the pharynx. The ventral pharyngeal flaps assist the transport.

2.3 *Corvus* and *Nucifraga*

Crows are omnivorous, although they may show adaptations to specialist feeding, e.g. husking shelled seeds (Sect. 3.1). Berkhoudt and Zweers (1990) show that carrion crows (*Corvus corone*) have a flexible pecking mechanism that is basically similar to that in pigeons. The oropharynx of *C. monedula* and *C. corone* was described by Zweers and Berkhoudt (1987). The tongue is fleshy and flexible, and carries a sharp, keratinized and forked tip. The lingual wings hardly protrude, but carry two to three strong, keratinized spines; the lingual base is widely extendable and carries numerous glands. The ventral pharyngeal flaps can be erected in addition to larynx motion by a peculiar intra-laryngeal mechanism.

The crows' pecking mechanism shows the following additional attributes. Soft food is pricked by the lingual tips, then the tongue is retracted, carrying the food caudad; the closing mouth fixes the food. This 'slide and prick' mechanism may be combined with slide and glue and catch and throw applied for eating hard food. Food can be stored at the widely extending lingual base prior to further transport. That transport is a scraping by the ventral pharyngeal flaps in addition to the 'properistalsis' described in chickens.

Storing seeds may also occur at the mouth. Clark's nutcrackers, *Nucifraga columbiana*, harvest, transport and store pine seeds in a communal caching area for use in feeding their young. The critical morphological feature is the presence of a sublingual pouch, a specialization of the mucosal lining of the antelingual oral cavity (Bock et al. 1973). Associated with the sublingual pouch is a modification of the Musc. genioglossus, one of the extrinsic hyolingual muscles, but no significant modification of other hyolingual musculature is observed. Mechanisms for filling the pouch are not known.

2.4 *Anas*

Waterfowl show a pecking behavior that is less flexible than described so far. This is due to the large lingual bulges and lingual cushion which obstruct passage of pecked food (see Sect. 4.3.1 for the anatomy). Mallards peck also by throw and catch but no slide and glue occurs. Kooloos (1986) described that mallards do not adjust their gape size to seed size during the head approach and they do not adapt oropharyngeal motion patterns to seed size during food transport. Instead, Kooloos observed a special pattern maintaining a constant flow of seeds to the pharynx, a throw-and-catch carries the food along the lingual bulges, and a shaker-conveyer belt transports it along the lingual cushion.

Kooloos (1986) and Kooloos and Zweers (1991) developed from experimental analyses a simulation model for the pecking transport mechanism. The model uses four elements. (1) The maximal grasped number of seeds in the rostral mouth cavity. (2) Maximal transport capacity of the rostral mouth tube by throw and catch. (3) Maximal storage capacity of the bilateral caudal mouth cavities. (4) Maximal transport capacity of the caudal mouth tubes by a shaker-conveyer belt mechanism. The simulated performances calculate how many cycles are needed to transport a certain amount of seeds of a specific diameter. Predictions were tested against experimental data. They fit for seed sizes from 2–7 mm. Smaller seeds, however, are handled by a different mechanism that is close to filter feeding. Very large seeds are handled by a new pattern: a lingual cushion depression clears the way to enter the pharynx.

3 Specific Food-Handling Mechanisms: Husking and Storage

Food-handling is by definition an intra-oral behavior. Just prior to food-handling occasionally tools are used, which is by definition the use of an external object as a functional extension of the mouth or beak. For example, Gayou (1982) describes green jays (*Cyanocorax yncas*) probing for insects with twigs, well-known similar behavior is found in woodpecker finches and mangrove finches (*Camarhynchus pallidus* and *C. heliobates*). Marks and Hall (1992) review other examples like Egyptian vultures dropping stones onto ratite eggs. They report cases of tool use by bristle-thighed curlews (*Numenius tahitiensis*) feeding on albatross eggs. As an extension of their common egg-slamming behavior these birds employ a stone thrown at the egg. Very little is known about adaptations to prey killing. Hull (1991) compared differences in the feeding apparatus of 2 species in Falconidae.

3.1 Husking Seeds

3.1.1 Fringillidae, Estrildidae and Ploceidae

Seed-eating oscines show a trophic radiation. Relationships of beak shape and staple diet are well known in finches (Fringillidae) (Morris 1955; Kear 1962). They have strong conical beaks, large jaw muscles, a heavy skull and a strong gizzard, which are all related to eating hard-shelled seeds. Carduelinae show a wide diversity in beak shape fitting to preference for size and shape of seeds, and seedheads of plants, but yet no mechanobehavioral analyses have been studied. For example, goldfinches (*C. carduelis*) probe into thistle seed heads with length-ened tweezer-like bills; hawfinches (*C. coccothraustes*) use a large and powerful beak apparatus to crush hard tree fruits, bullfinches (*P. pyrrhula*) apply their rounded bills to grasp buds, while seeds from closed pine cones are taken by the crossed beaks of crossbills (*Loxia*). Benkman (1987a), in laboratory experiments, measured the efficiency of two species of crossbills in removing seeds from cones of seven species of conifers. He showed how morphological differences in the bill and skull of the two species related to foraging efficiency. Red crossbills (*Loxia curvirostra*) were more efficient in removing seeds from pine cones (*Pinus*), while white-winged crossbills (*L. leucoptera*) were more efficient on spruce cones (*Picea*). Benkman (1988) also clearly separated "handling time" into the time spent extracting seeds from cones and the time spent in husking the obtained seeds. Crossbills have evolved considerably stronger bills for seed extraction, but heavily depend on their specialism. They are less efficient in handling nonconifer seeds compared with other cardueline finches, so they will starve when conifer crops fail. The importance of the crossed mandibles was tested by removing most of the crossed portion. Feeding efficiency was hampered especially in extracting seeds from closed cones (Benkman 1987b). If the diet also comprises insects, beak shapes tend to become slender. Specific, yet unknown adaptations at the muscle-bone level must be present. Unfortunately, only the jaw apparatus of the hawfinch (Sims 1955) and some Geospizinae (Bowman 1961) have been described. Sims (1955) and Bock (1966) calculated the forces jaws must exert on hard seeds. Van den Elzen (1985, 1990) and (Van den Elzen et al. 1987) showed that morphological radiation in African finches, traced by means of morphometrics of skeletal elements, parallels radiation in Eurasian carduelids.

Extensive analyses by Ziswiler and his students (Ziswiler 1965, 1979, 1990; Ziswiler and Trnka 1972; Ziswiler et al. 1972), completed with data from Bock (1978) and Heidweiller and Zweers (1990), have added remarkable features. The keratinized oral palate carries rostrocaudally running ridges and the bilateral maxillary tomium. These ridges end caudally where a transverse protrusion runs (a cavernous body in estrildids; Heidweiller and Zweers 1990), marking the border with the pharynx. The ridges have many different patterns that correspond to the seed types eaten as staple diet. The lower jaw carries a strongly keratinized tomium that may be either blunt or sharp.

Ziswiler (1965, 1979) described two different seed-opening mechanisms: (1) a crushing type in which the shell is opened by pressing it with a blunt mandibular tomium against a maxillary lever ridge. (2) A cutting technique in which the shell is opened by fast pro- and retractions of the sharp mandibular ridge along the fixed seed. In both techniques lingual manipulation is shown to determine the success of all steps of oropharyngeal food handling. The cutting technique is specific for eating spherical dicotyledon seeds and is applied (by Carduelinae and Fringillinae), while crushing is specific for eating elongated monocotyledon seeds (Graminae) (by most Estrildidae, Ploceidae and Emberizidae).

The tongue may be rather slender, carrying a cup of keratinized hairs in Estrildidae; the tongue may also be provided with special stiffening devices to manipulate seeds (Ziswiler 1979) [Emberizidae possess fat tissue and blood sinuses that stiffen the tongue, Ploceidae stiffen the tongue by a bony preglossale (Bock 1978), whereas Estrildidae do this with a large cavernous body that is supported by entoglossal and hypentoglossal bones.] The lingual wings are relatively large, but carry only a few spines. The lingual base is not widely extendable, the rostral ridge of the cricoid is elevatable, and the larynx is of the type that allows erection of the ventral pharyngeal flaps independently of the larynx (Heidweiller and Zweers 1990).

3.1.2 Geospiza

Grant (e.g. 1986), Grant and Grant (1989) and Schlüter (1988), discussed relations of beak morphology and feeding ecology in 14 species of Galápagos ground finches (Geospiza) including work of Lack (1947). Primarily, size and shape modifications related to base-crushing, tip-bitting and probing techniques were compared in the adaptive radiation of six largely granivorous species. Some major aspects are as follows. Seed size varies with body size: the largest species eat the largest and hardest seeds, while having the largest, stoutest and deepest beaks. It was assumed on a performance basis, rather than from mechanical analysis, that the ability to handle large, hard seeds increases with size. Shape, i.e. the curvature of the beak, strongly affects the biting force (Bock 1964), while size, i.e. a lengthened beak, affects in similarly sized species the ability to probe *Opuntia* cactus flowers. The longer beaked *G. scandens* (compared to similarly sized species like *G. fortis*) pierces the fruit and the soft arils around the seeds are eaten. During the wet season, all species forage on the same abundant soft seeds that are easy to handle. Schlüter (1988) stresses that reflection of the diet preferences in the beak shape occurs primarily in the dry season when food is scarce. Intrapopulation variation in diet connected to morphology was mentioned for *G. fortis* in which the smaller individuals consume smaller and softer seeds than larger individuals, since they are less efficient at cracking larger seeds. Grant (1983, 1986) exemplified a case of character displacement in *Geospiza*. *G. fortis* and *G. fuliginosa* are sympatric on Santa Cruz where they have very different beak and body sizes. These species have intermediate sizes on islands where they occur as single species.

3.1.3 Corvidae

Most crows have stout bills and many pound their food (e.g. jackdaw, *C. monedula*; rook, *C. frugilegus*), from which a specialist seed-husking technique may have developed. Zusi (1987) described the chisel-like use of the lower jaw in a technique applied by the Florida scrub jay (*Aphelocoma coerulescens*) in husking nuts. Perching jays were offered fixed peanuts. The head and neck were raised high, the bill was slightly opened and downward blows of the lower jaw were directed tangentially toward the outer edge of the peanut. After some blows, upon penetration of the shell by the lower jaw, the shell was peeled off by using both beaks. Acorns, however, were held against a branch, secured by the inner front toe. Powerful blows of the lower jaw were applied to the acorn until the shell broke, after which pieces of the shell were then grasped by both beaks and torn off by head shaking. The acorn was rotated and the opening enlarged.

Zusi (1987) mentions that Eurasian jay (*Garrulus glandarius*), being twice as large as *Aphelocoma*, husks the thin-shelled acorns of Eurasian oaks, but is unsuccessful in husking thick-shelled acorns of the New World oak. The yellow-billed magpie (*Pica nuttali*) requires hundreds of blows to open the shell, while the scrub jay removes nut pieces after only about 20 blows, and the Mexican jay (*A. ultramarina*) needed only a few strong pecks. The difference is that the latter species has a 'buttress complex' that anchors the lower jaw on the quadrate.

Zusi (1987) carried out a vector-analysis at the moment of impact for the lower mandible and quadrate of a jay lacking the buttress complex and one having it. The complex comprises three elements, in addition to a general corvid jaw apparatus: (1) The lower jaw carries a vertical, unusually steep cotyla rostrally in the articular fossa of the lower jaw at a vertical buttress of the mandibular ramus. (2) The quadrate carries an additional condyle on a rostroventral pedicel which faces the extra cotyla. (3) The quadrato-otic articulation has enlarged meatic and suprameatic processes. Zusi shows that force is conducted most economically when the lower jaw is just opened, and that the presence of the buttress complex reduces torque on the quadrate during pounding.

3.1.4 Psittacidae

Parrots (Psittacidae) forage on a variety of vegetable foods. They have specializations for eating fruit, seeds, buds, nectar, pollen, insects and slime-like fungi. However, all parrots are at least occasionally granivorous and husk seeds or fruit. They have some trophic features in common, which are reviewed in the extensive analyses of Homberger (1980). The lengthened, broad-based upper beak curves strongly downward, and has a strong, sharp-ending rhamphotheca. A few millimeters dorsal to the very tip, a prominent, hard, keratin, transverse ridge, the rostral maxillary ridge, lies along the palate. More caudally lies a hard keratin palate covered by a field of file-like notches. This field is followed by a torus-like, transverse thickening and a soft mucous palate. The trough-like lower jaw is covered with a very hard and large, keratin rhamphotheca, which has a sharp, transverse tomial edge at the tip. Ten or more dermal papillae protrude rostrad in the tip edge of the rhamphotheca, presumably carrying clusters of touch corpuscles (Gottschaldt 1985). The oral cavity is largely filled by the tongue. The lingual mass is supported ventrally by a trough-like hard keratin nail. The rostral lingual mass comprises an inflatable cavernous body and numerous touch corpuscles are present in the dorsal lingual mucosa. Nectar-feeding parrots have less elaborate lingual cavernous bodies, but instead numerous papillae occur dorsally at the lingual tip.

Many authors have described seed husking. Small seeds are grasped with the upper bill tip and lingual tip; large seeds are grasped with the beak tips. The tongue tip positions the seed against the hard transverse maxillary ridge. Now a stationing phase occurs by highly tuned beak and tongue tip motions which finally fix the seed against the rostral maxillary ridge with the seed's groove opposite the tomial edge of the mandibular tip. Then a forceful bite of the lower jaw cracks the seed and the lower jaw is pushed in the crack like a chisel between the scale and kernel to the other end of the seed. Lower jaw motion stops and the lingual tip rotates the seed 180°, while the lower jaw tip edge is kept continuously fixated at the cut edge of the seed. Once the seed is in the reversed position, the lower jaw husks the other side of the seed. Then the husked seed is transported by lingual, inertial action. Large kernels are often bitten to pieces, which is called 'chewing', and each piece is transported to the lingual base where it is stored until a bolus is formed and swallowed. Feeding on fruit and berries is largely similar. The head approach to large fleshy seeds is a blow by which the upper jaw pierces the meat. The elevating lower jaw tip edge cuts a piece off, which is then husked and 'chewed'. Expelling the scales at the tips is always a tuned manipulation of jaws and tongue.

Homberger (1980) explains that the tongue shape is adapted to the functional requirements of feeding. A stiff, rounded, lingual tip is required to touch the food and manipulate it accurately, while a trough-like shape suffices to transport it. Homberger and Brush (1986) relate special chemical characters of lingual keratin elements to the special muscle-bone organization of the tongue and the specialist seed husking mechanism in the African grey parrot (*Psittacus erythacus*). Their results may represent the following adaptations to seed husking. The lingual nail is a stiff, yet pliable, non-extensible, keratin cover of the ventral and lateral tip of the tongue. Lateral and apical cavernous bodies underlie the dorsal edges of the nail, they reach down to the bilateral paraglossals. Homberger and Brush (1986) propose a mechanism of unique tongue muscle action and elastic recoil by the nail that serves a lingual inflation mechanism, which controls the shape of the tongue from flat to spoon-shaped, and is therefore a mechanism to manipulate and transport food.

3.2 Extracting Snails: *Chondrohierax*, *Anastomus* and *Aramus*

Seizing snails with their talons and extracting the body of a snail is a characteristic feeding behavior of the hook-billed kite (*Chondrohierax uncinatus*) (Smith and Temple 1982), the snail kite (*Rostrhamus sociabilis*) and slender-billed kite (*R. hamatus*) (Voous and Van Dijk 1973). After securing the snail with the foot, the hook-billed kite probes in the aperture of the snail and, by leverage action of the upper mandible, it breaks the inner spire by forcing the beak tip towards the apex. The other kite species probe directly to the center of the shell spiral to sever the columellar muscle and thus extract the snail without first probing the shell's aperture.

The peculiar morphology of the bill of the African open-bill stork, *Anastomus lamelligerus*, namely, the permanent "gap" between the upper and lower halves of the beak, is not used for crushing snails. This stork probes the lower jaw into the shell and extracts the right- and left-coiled snails without any significant damage to the shell except for some pecking at the aperture to expose the operculum (Kahl 1971, pers. comm.). Snyder and Snyder (1969) note that the Asian open-bill stork (*Anastomus oscitans*), however, deals with right-coiled snails only. The bill is asymmetric due to a hook-like bending of the lower jaw tip to the bird's right. Snails are extracted by forcing the curved distal tip between the operculum and the columella, while the upper jaw functions as the fulcrum, by which action the operculum and columellar muscles are severed. Snail handling is assisted by 20–30 leathery columnar pads at the tomium of the upper jaw. The operculum is snipped off.

Limpkins (*Aramus guarauna*) feed on apple snails which are captured under visual or tactile control. Captured snails are carried tweezer-like, using a special space between the mandibles. Snails are positioned spire down, aperture up, then the operculum is severed by blows near the columellar muscle. The asymmetric bill, especially the lower bill tip, is bent to the bird's right and is used in a scalpel-like action to cut the columellar muscle (Snyder and Snyder 1969).

3.3 Biting Fruit: *Tachyphonus*, *Didunculus* and *Psittrichas*

Many frugivorous songbirds pluck and swallow fruit whole (e.g. *Turdus*, *Zonotrichia*, *Vireo*) or they take a bite, afterwards they proceed with the common pecking mechanism. Also, a push and bite technique is employed (Foster 1987). If this is the case, a bird e.g. neotropical tanagers like *Euphonia*, may impale a fruit, carry it to a perch and remove the pulp with the assistance of some other structure than the bill. In addition, a special mash technique is shown in E.g. *Tachyphonus cornatus* (Foster 1987). A fruit (*Allophyllus edulis*) is grasped with the tip of the bill which rolls it with the lower mandible and tongue so that the pulp is cut from the seed with the tomium of the upper mandible. A strip of pulp trails below the bill before the tongue pulls the pulp back in the mouth, then the seed is dropped, spit out, shaken off, or wiped onto a branch. Wheelwright (1985) demonstrated that the size of a fruit that can be swallowed is correlated with the width of the bill at the rictus commissure. Also, a correlation between gape width and fruit diameter is present. Foster (1987) concluded from a comparative study that bill dimensions do not always predict the fruit-handling type, and that frugivorous birds do not necessarily handle fruits as efficiently as they are capable of.

Tooth-billed pigeons, *Didunculus strigirostris*, feed primarily on fruits of *Dysoxylum* (Meliaceae), taking the pea-sized seeds only. They hook the tip of the upper mandible into the viscous fruit, while projections at the tip of the lower mandible perform a sawing motion while being pushed in the fruit (Beichle 1987). The seed is then grasped by the beak tips and the common pecking mechanism is set into action. Compared to other Columbidae, *Didunculus* has special anatomical adaptations to perform this fruit handling. The upper mandible is widely elevatable by a hinge-like naso-frontal joint (Zusi 1985), and the upper beak is greatly enlarged and carries a hooked tip. The enlarged lower mandible carries at the rhamphotheca tooth-like extensions. The quadrato-mandibular joint allows rostrocaudal motion of the lower jaw since the quadr-ate condyles can move parallel to the mandibular rami by running along a longitudinal ridge rather than being blocked by the common transversal mandibular ridge in the quadratomandibular joint (Martin 1904). This parallel ridge is also present in parrots and waterfowl, which also show rostrocaudal motion.

Most parrots are granivorous/nucivorous and frugivorous. Whole fruit is grasped and pieces of fruit are bitten and husked as in seed husking (Sect. 3.1.4). Homberger (1980) concluded that, in parrots, very few characters are found that are adapted to fructivory, except that the mandibular salivary glands are elongated. However, in *Psittrichas* which is strictly frugivorous and not nu-civorous/granivorous, Homberger (1980) described birds that grasp a whole fruit and bite a piece, which is husked as described earlier and then transported by a mechanism that is quite comparable to the slide and glue mechanism. Not only adaptations to this mechanism in tongue and palate are mentioned, but also adaptations to granivory that are functionless and contraproductive in frugivory (p. 146).

4 Specific Food Acquisition Mechanisms

4.1 Catching

4.1.1 Striking: Ardeidae, Ciconiidae and Podicipedidae

Ardeidae. In many avian taxa there is no sharp borderline between foraging and food acquisition, since final actions of the locomotory system may be combined with head-neck and jaw motions. This is especially true of feeding behavior of North American herons (see reviews of Kushlan 1976; Hancock and Kushlan 1984, based on many observations of Meyerriecks 1959). A common sequence of phases is found in the grey heron (*Ardea cinerea*) (Creutz 1981). The bird walks slowly or stands still and waits. The head is fixated in an upright position; it may then be somewhat depressed and fixated again. A fast head approach then follows ("darting stroke", Kral 1965) by stretching the cranial portion of the cervical column. The opening of the bill is tuned to the size of the prey which is grasped scissor-like. The head is elevated and small catch and throw actions reposition the fish transversely at the beak tips in a stationing phase. Washing the prey may be followed by head depression plus head shaking as soon as the beak tips are submerged; head elevation follows while the prey is repositioned in another stationing phase, now longitudinally and head first. Meanwhile the head is tipped up. Prey is swallowed by head jerking and properistalsis comprising linguolaryngeal pro- and retractions and caudal running constrictions of the pharynx floor, success of this mechanism being due to gravity. The strike and the powerful tip up action require a special construction of the cervical column (Kral 1965). In addition, a special construction that allows extended lengthening of the lingual base must be present.

One further specialization is observed in the Goliath heron, *Ardea goliath* (Mock and Mock 1980). The strike starts from a final head fixation in a crouched position with full retraction of the neck. The beak tips open 2 cm during the strike and both impale large prey (many slender-billed herons impale fish with closed beaks). While it may take the bird 40 s to land large prey, the fish is disengaged from the bill by head shaking. Additional stabs at the gills are made, then the usual grasping and swallowing actions occur. Boat-billed herons, *Cochlearius cochlearius*, may utilize an additional non-visually controlled method of food acquisition. Mock and Mock (1980) describe a 'touch-hunting' foraging, which may be explained by (1) a snapping bill entrapment; (2) a scooping mechanism; or (3) a differential pressure gradient, generating a vacuming mechanism.

Ciconiidae. A bill-snap mechanism in probing is also described for the wood stork, *Mycteria americana* (Kahl and Peacock 1963). Most storks exhibit a feeding behavior similar to that in herons. However, Guillet (1979) mentioned that the whale-headed stork (shoebill; *Balaeniceps*) may have a motionless head fixation phase that may last hours. Then the bird suddenly collapses into the water, legs folding, keeping the bill axis vertical with no forward component. The bird reemerges with a mouth full of vegetation and a lungfish. Then head shakes occur. Normal posture is regained by pressing head and outstreched neck against the substrate.

Podicipedidae. Grebes usually catch fish while pursuing them. They swallow their prey often while submerged (Berkhoudt 1985). However, the western grebe, *Aecmorphorus occidentalis*, is capable of impaling its prey fish, but the behavior involves diving patterns rather than the terrestrial locomotion described in herons. Observations of Lawrence (1950) suggest that the morphology of the bill is also tuned to the shape and size of the prey, and the strike pattern indicates compensation for these differing parameters. Subsequent stationing of the prey and associated catch and throw actions may occur before the bird emerges from the dive. Additional tipping up of the head may occur. For additional information on feeding mechanisms and behavior in grebes, see Fjeldså (1981).

Some specific form-function relationships of the feeding apparatus in South African cormorants are described by Burger (1978). *Anhinga* and cormorants (*Phalacrocorax* sp.) also feed on fish. The anhinga widens its gape prior to submerging for the dive and the gape remains open during impaling of prey, since two punctures are typically found on the lateral sides of prey fish. Cormorants widen the gape after the dive has commenced, the bird surfaces before stationing, catch and throw transport are supported by gravitational forces since the head is tipped up (Owre 1967). Dullemeijer (1952) described the muscle-bone apparatus of *Phalacrocorax carbo sinensis*.

4.1.2 Aquatic Sweeping: Pelecanidae

An 'aquatic sweep' is postulated for catching prey by the brown pelican, *Pelecanus occidentalis* (Schreiber et al. 1974). Film analyses show that head fixation is maintained during phases of diving, whereby any additional compensation for diving speed or plane of entry into the water is made by a specific change in wing or body profile. In the pelican, the gular pouch, suspended from the mandibular rami, fills with water when the beak opens underwater. A flexible zone in the mandible is reported by Bühler (1981); filling of the gular pouch may occur by active jaw muscle action or passive elastic recoil in the flexible jaw zone. Emptying of the pouch occurs by head elevation, involving a gravitational flow. Catch and throw follows complete emptying of the water from the pouch. However, recent studies of McMahon and Evans (1992) of feeding techniques in the American white pelican (*Pelecanus erythrorhynchos*) suggest that non-visual neurosensorial "probing" for food may be an additional element in the feeding system of pelicans.

4.1.3 Aerial Sweeping: Caprimulgidae

The opened mouth of nightjars has an extremely wide gape, and along the upper jaw rims there are over ten very long keratin hairs. These attributes make this beak an efficient trap to catch insects during flight. Bühler (1981) explains how ligaments in the underlying jaw apparatus allow the development of the wide gaping of the lower jaw. Basically, postorbital and jugomandibular ligaments cause the lower jaw, upon its depression, to rotate laterally around the longitudinal axis of the caudal part of the mandible, in addition, forcing the mandible to hinge laterally, which is allowed by a symphyseal joint. As a result, the slender lower mandible expands widely during jaw depression changing into a spherical shape surrounded by long hairs along the tomial edges. Martin and Stiles (1992) describe for swifts (Apodidae) specialization on Hymenoptera. Stomach analysis shows exploiting of rich highly localized patches of winged ants.

4.2 Probing

4.2.1 *Calidris*

Sandpipers probe air- and water-filled holes as well as dense substrates like sand. These two kinds of probing set different requirements upon the jaw apparatus and oropharynx. For example, probing holes requires slender, curved and elongated beaks, and an accurate inspection capacity by direct touch and taste. The penetration of substrates requires lengthened, flattened, and slenderized, but straight, strong and broad-based beaks. Burton (1974, 1986) describes mouth and jaw muscle anatomy, Zusi (1985) reviews the rhynchokinesis of the jaw apparatus.

Gerritsen (1988), Gerritsen and Heezik (1985) and Gerritsen and Meijboom (1986), Zweers and Berkhoudt (1991), and Zweers et al. (in prep. a) have analyzed probing in sandpipers (*Calidris*). Sandpiper probing differs in several respects from pecking:

1. After the usual final head fixation, a head approach occurs during which the curved beaks are stretched and slightly opened; upon touching the substrate, the optic control of the head approach is taken over by trigeminal touch control as the beaks penetrate the substrate. In contrast to the visually controlled head approach, touch-controlled penetration is interruptable, and the head changes direction during probing.
2. Gerritsen and Meijboom (1986) have calculated from food discrimination experiments in large aviaries that the probability of direct hits upon prey hidden in the substrate was far too small to have directed the selection of the best foraging location. They have calculated that sandpipers have a 'remote touch' capacity of at least 2 cm from the beak tips. Zweers et al. (in prep.a) developed a model for the bill tip touch organ which meets these conditions, by assuming that sandpipers measure the difference in intensity of worm-produced waves that travel along their beak tips.
3. Burrowed prey is grasped by widening the gape through elevation of the upper and depression of the lower mandible tip. Zweers et al. (in prep. b) have shown that for this function a special construction of the jaws enables

'decoupling' of the symphyseal areas from the rest of the jaws. In addition, they deduce what specific muscle-bone adaptations in the jaw apparatus should occur. For example, the horizontal position of the quadrate serves to conduct penetration forces onto the skull, and the greatly enlarged, ventro-caudad pointing Proc. retroarticularis allows gaping of the mandibular tips during penetration.

4. Burrowed prey, like small worms and amphipods, is retracted from the substrate by 'pinching' and forceful pulling by head elevation. Pinching is the reverse action of gaping while beak tips penetrate; the symphyseal areas now enclose prey to that it is firmly held.

4.2.2 Threskiornidae

Most studies on sexual dimorphism and age-related differences address eco-behavioral differences rather than mechanics in the oropharynx. They, however, provide basic information to understand modifiability. The work of Bildstein on juvenile and sex-related feeding performances in ibises exemplifies this. White ibises (*Eudocimus alba*) probe shallow water mud and grass for small animals. Bildstein (1983) studied white ibises, which hunt small fiddler crabs (*Uca pugilator*) by probing crab burrows and pecking at disturbed, moving crabs. He relates foraging efficiency to prey choice, rate of prey capture, frequency of captures, and prey handling time, among other aspects. He finds that juveniles forage at the periphery of hunting flocks and that they capture prey at only 40% of the adult rate. Although a connection to feeding mechanics is not made, Bildstein (1987) estimated the energetic consequences of sexual dimorphism in bill size. Males have disproportionally larger bills than females. They capture prey at similar rates, but males spend 37% more time foraging. Bildstein explains this from an evolutionary perspective – disregarding open mechanical aspects – since the benefits in securing mates and prey by being larger and having larger bills are counteracted by the costs of a longer foraging time to maintain a sufficient energy balance for the larger body. Moreover, Bildstein et al. (1989) report that white ibises prefer eating female and declawed fiddler crabs over male crabs. Since ibises select crabs while pecking as well as while probing, both visual and tactile cues must be present.

4.2.3 *Haematopus*

The probing behavior of oystercatchers has been investigated by a series of workers (e.g. Boates and Goss-Custard 1989; Norton-Griffiths 1967; Heppleston 1970; Hulscher 1976, 1985; Swennen et al. 1983; Hulscher and Ens 1992). Oystercatchers show reversed sexual dimorphism (Jehl and Murray 1986): females have longer bills compared to males. Females capture more deeply buried prey like lugworms and ragworms, while males probe for the shallower buried mussels and cockles. Further, birds with blunt-tipped beaks hammer more, while birds with sharp-tipped beaks stab more. Swennen et al. (1983). Hulscher (1985), and Hulscher and Ens (1992) reported that beak shape differs in relation to dietary differences. Males consume relatively more thick-shelled mussels than females. Thus, the bills of males are better tools to open mussels than the relatively longer, but smaller and thinner bills of females. Durell and Goss-Custard (1984) observed that hammering oystercatchers selected mussels with the thinnest shells, in contrast to stabbing oystercatchers which did not select mussels according to shell thickness. Two types were distinguished. Ventral hammerers selected mussels that were thin on the ventral surface, while dorsal hammerers selected mussels that had eroded, thinner dorsal shells. The authors did not study eventual differences in beak structure.

Eating bivalves is accomplished in two ways: stabbing and hammering (e.g. Norton-Griffiths 1967; Heppleston 1970). Hammering begins with pecking of the mussel, which is then dropped on firm sand which itself is used as an anvil. An average of five blows is applied at the ventral or dorsal margin of a mussel after the final head fixation. A semicircular chip is fractured from the margin, closed beaks are inserted, the adductor muscles are cut by stabbing and biting, then actions are as in stabbing. In stabbing a final head fixation occurs once the beak tips are brought a few centimeters above a gaping mussel. A final head approach follows with a closed beak, which is an accurate blow to the caudal adductor muscle. The beak tips then stab while cutting the muscle, thereafter they push forward, cutting the other adductor again by stabbing and biting. A fast and forceful 90° lateral rotation of the head places the beak tips transversely in the cleft of the mussel, a forceful gaping follows so that the mussel is opened.

4.2.4 *Sturnus* and *Corvus*

Starlings, *Sturnus vulgaris*, search for food in the top soil layers and the overlying vegetation. At the end of the downstroke the bill tips penetrate tufts of vegetation or holes in the soil. Then the mandibles gape forcefully so that a cleft is produced which is optically inspected for food. Lorenz (1949) refers to this action as "Zirkeln", while Tinbergen (1981) calls it "gaping". Lockie (1956) described a similar gaping behavior in rooks, *Corvus frugilegus*, while probing a worm cast or a tuft of grass. In starlings special adaptations to this gaping are mentioned. The eyes are positioned in line with the axis of the mandibular tomial edges, allowing continuous and accurate visual control of the down-stroke. A bilateral, groove-like indentation runs rostrad from the eye ventral to the nostrils, allowing a binocular view in the cleft produced by the gaping beaks. Further, the indentation is covered by short, black, non-glossy feathers eventually absorbing interfering light reflection. In addition, starlings have an enlarged mandibular depressor muscle which is attached to a somewhat enlarged ventro-caudally pointing Proc. angularis posterior at the lower mandible, allowing a forceful gaping.

4.3 Filter Feeding

4.3.1 Anatidae

Waterfowl filter feed by similar mechanisms (Goodman and Fisher 1962), but differences are also found (Kooloos et al. 1989). The mechanism comprises two major parts: a water-pumping and a filtering system. These mechanisms have been analyzed experimentally by Zweers et al. (1977), Kooloos et al. (1989) and Kooloos and Zweers (1991). A rapid, cyclic jaw and tongue motion is released by which, during each cycle, a sequence of mechanisms for pumping, filtering, collecting and transporting develops at 15–20 cycles per second.

The pumping mechanism operates as a suction-pressure pump (Zweers et al. 1977). Four anatomical elements are relevant: (1) The rostral mouth cavity between the beak tips and the lingual tip; (2) The rostral mouth tube, which is the longitudinally running median lingual groove; (3) The bilateral mouth cavity, which is bordered by the lingual lumps area, the maxillary lamellae, the lingual cushion and the lingual combs; (4) The bilateral caudal mouth tubes, which are the lateral, maxillary grooves along the lingual cushion. The mouth is considered a cylinder having an inflow opening at the beak tips, and bilateral outflow openings at the mouth corners. The retracting and elevated lingual bulges suck water into the mouth while the beak opens. The bulges then depress and the tongue protracts, the water flowing by inertia over the lingual bulges. The lingual cushion separates the flow into bilateral streams leading to the bilateral caudal mouth cavities. Then elevating and retracting lingual bulges cause the expulsion of the water mass and a further dose at the beak tips is sucked in.

Two types of mechanisms are involved in filtering. In shovelers (*Anas clypeata*) and mallards (*A. platyrhynchos*) particles are filtered by direct impact whenever they are larger than the mesh width formed by the distance between the maxillary lamellae. A second mechanism operates in mallards and tufted ducks (*Aythya fuligula*) for small seeds, which is called a vortex mechanism (Kooloos et al. 1989). The water flow along the maxillary roof is separated in bilateral streams by the lingual cushion. These streams must funnel down along the maxillary lamellae, forcing the water flow in vortices through the widening interlamellar channels. The vortices centrifuge relatively small seeds by inertial impact deposition. They are held against the lamellar walls until they are brushed up during each cycle by the lateral lingual hairs into the longitudinal maxillary grooves. Then lingual scrapers transport the collected seeds caudad into the rostral pharynx.

Kooloos et al. (1989) have shown that the size of the staple food in mallards and tufted ducks is directly related to the most economic construction and operation of the upper beak, being the major element serving pumping and filtering. They found that filtering of the staple food occurs at the most efficient levels of maxillary rotation and gape size.

4.3.2 *Phoenicopterus* and *Phoeniconaias*

Flamingos filter feed upon a variety of small invertebrates and seeds; in addition, they swallow mud to digest algae (e.g. Allen 1956; Jenkin 1957; Rooth 1965; Kear and Duplaix-Hall 1975). Filter feeding comprises about four cycles per second including three separate mechanisms; pumping water, filtering food and transporting food. The filter mechanism was analyzed by Allen (1956) and Jenkin (1957) via analysis of the tongue and lamellar anatomy and stomach contents.

Two kinds of filter mechanisms were found (Jenkin 1957). First, we consider the type observed in greater and Caribbean flamingos (*Phoenicopterus roseus* and *Ph. ruber*). The bill is bent in the middle and the lower jaw is large, forming a shallow-keeled trough, with a small, but equally wide and lid-like upper jaw. Jenkin (1957) measured the dimensions of lamellar meshes along the closed gape at the mandible margins. Meshes are formed by large, distal, outer lamellae plus smaller, interpositioned, outer lamellae at the maxilla and juxtapositioned, low serrations along the rims of the mandible. The tongue is distally large and fleshy, and carries proximally a bilateral row of some 20 flexible and large spines. Food kernels are 1–10 mm in size (Allen 1956; Jenkin 1957). Zweers et al. (in prep. b) observed that filtered seed sizes range from 1.0–8.5 mm, while the maximum filter capacity peaks sharply at sizes of 2–4 mm. Flamingos also discriminate preferred seed sizes from mixtures of seeds. Performances peak (1) when sorghum seeds (4.3 mm) were *excluded* while grass seeds (2.8 mm) were filtered, and (2) when grass seeds were filtered while broken millet (1.4 mm) was *washed away*.

Allen (1956) suggests that filter feeding occurs by pumping movements of the gular region so that suction occurs in the mouth opening, while streams of water are forced outward when the upper jaw is clamped shut, whereby certain portions are retained within the lingual spines. Jenkin (1957), however, proposes that filtering of particle sizes of 1–10 mm is caused by the stationary, outer lamellar filter when the water suspension is pumped out, while the beak is kept closed. Raising the upper jaw admits food through the narrow gape, acting as an excluder, and closing the bill reforms the filter. Meshes are formed by the maxillar marginals, submarginals and mandibular ridges. The pumping mechanism produces primarily a back and forth flow along the same main route (Zweers et al. in prep. b). Adjustment of gape size is accurately tuned to food sizes preferably gathered from mixtures, in contrast to the concept that gapes are set at just one open and one closed position (contra Jenkin 1957). Lamellae may be applied flexibly as either a filter or as an excluder, and not only as an excluder when mud is gulped (cf. Jenkin). Bilateral series of lingual spines serve the transport of filtered food into the pharynx, rather than filtering food (contra Allen 1956).

Jenkin (1957) shows that greater and lesser flamingos feed in the same lakes without competing for food, because the greater flamingo filters food sizes > 0.5 mm, while the lesser flamingo (*Phoeniconaias minor*) filters food sizes < 0.4 mm. Jenkins explains this trophic separation by a difference in lamellar construction and possibly in water flow. Lesser flamingos have a fringed series of inner lamellae on both jaws of deep-keeled beak leaving meshes in the outflow through closed beaks of 0.01×0.05 mm, while the excluder mesh size formed by the outer lamellae is 1.0×0.4 mm. She assumes that the filter mechanism involves back and forth flow, although flow directions do not run fully in parallel.

4.3.3 *Pachyptila* and *Puffinus*

Morgan and Ritz (1982) analyzed filter feeding of krill in the mutton bird (*Puffinus tenuirostris*) and the fairy prion (*Pachyptila turtur*). Their hydrodynamic studies of fixed heads in a flow tank predict that as heads with an open gape move through the water, the bill architecture produces a lower pressure in the mouth which sucks in water and krill. In muttonbirds the tongue has longitudinal rows of caudally pointing papillae that overlap rows of recurved, rostrally pointing papillae on the palate. If water and krill are squirted into the closed bill, the flow erects the papillae, which trap krill > 2.5 mm, it then leaves the mouth along a permanent gap at the rictus. In fairy prions, however, the lingual papillae are much smaller, the palate is now smooth except along the margins where a series of waterfowl-like lamellae is found. The mouth floor has several longitudinal grooves so that a gular pouch develops if the floor is depressed. If water and krill are squirted through the esophagus into the mouth, water leaves the mouth along the lamellae and krill < 2.5 mm are filtered. The pouch may serve to pump a food-bearing current of water out through the filter of the bill. Further analysis of the filter mechanisms may include the possibility of vortex generation by lamellae so that prey much smaller than the inter-lamellar width may also be filtered (e.g. Kooloos et al. 1989).

Klages and Cooper (1992) analyzed the bill morphology and diet of broad-billed prions (*Pachyptila vittata*), showing that they feed upon an exclusive diet of copepods of 0.7–4.1 mm size which is consistent with palatal lamellae positioned 0.16 mm apart in two rows. They agree with Harper (1987), stating that out of four feeding methods the suction method is the most common, suggesting that suction is created by the tongue to draw in water and food particles; the water is then squeezed out again with the tongue while palatal lamellae filter copepods. A large tongue and a mandibular pouch are the suitable adaptations for this method of feeding.

4.4 Adhesion and Capillarity

4.4.1 Trochilidae

Hummingbirds have straight or slightly curved slender bills designed to probe into tubular corollas. The tongue is slender and strongly elongated (Weymouth et al. 1964). The strongly lengthened epibranchials wind around the skull and terminate on the left side of the culmen; the pro- and retractor muscles are also very long. This construction allows a large lingual protrusion. Microscopic sections of the tongue show that there are bilateral trough-like tubes that open dorsolaterally into the mouth. The lingual tip is forked, largely keratinized, and mostly fimbriated.

Ecophysiological analyses of Hainsworth and coworkers elucidate form-function relationships. Hainsworth and Wolf (1976) investigated nectar preference in choice experiments and demonstrated that birds selected primarily on the basis of sugar concentration, not composition, and secondarily on the rate of intake. They preferred sugar concentrations which were close to that found in nature. Ewald and Williams (1982) observed that hummingbirds tend to forage on corollas which have a similar length as their beak and explained this as an adaptation in maximizing transit speed. They showed that the tongue groove volume was sufficient to hold the total amount of food gained per lick and that the drop adhering to the tongue tip is only a minor part. From film analysis nectar feeding was described as the intake in the lingual grooves by capillarity, while a constriction of the tongue against the palate squeezes the nectar into the pharynx.

Advanced model simulations of Kingsolver and Daniel (1983) elucidate the mechanical determinants of nectar feeding in hummingbirds. They develop a fluid dynamic model which describes the relationships of the mechanics and energetics of capillary feeding, while predicting nectar intake rates and nectar volumes per lick. They show that if the intake rates of energy are maximized, tongue morphology and licking behavior determine the optimal nectar concentration. Moreover, they make clear that there is a critical value for the food canal diameter above which suction feeding is superior to capillary feeding. Comparison of predictions and the actual anatomy shows that most tongues of hummingbirds are above this critical value. The authors proposed that development of suction feeding was constrained by the elastic properties of the flexible tongues, which developed their initial shape by elastic recoil.

4.4.2 Psittacidae

Homberger (1980) describes fruit juice and nectar feeding in Loriinae. These parrots have many papillae dorsal to the lingual tip, forming a brush, and the lingual wings are strongly reduced. Berries are pressed against the hard palate by the edge of the lower jaw and the expelled juice is collected by dips of the brush-like tongue tip. The juice is gathered by adhesion, and the fruit pulp is expelled by head shaking. The tongue is retracted; by protraction the tongue tip is pressed against the bulging, soft part of the palate and the juice is pressed out. Nectar is eaten similarly, except that the tongue is protruded further and the papillae at the tongue tip are actively spread and then dipped into the nectar. Homberger (1980) assumes that seed and nectar feeding mechanisms pose different requirements on the shape of the tongue tip: the inflatable lingual bulge versus the papillary field.

4.4.3 *Calidris* and *Phalaropus*

The feeding behavior of several sandpiper species (*Calidris*) was analysed by Gerritsen (1988). He observed that as the head elevates after probing mud, a drop of water was often left in the mouth. The drop may be expelled by a head shake, but he also observed that the beak was carefully opened and closed. By this motion the drop runs up by capillary action and adhesion, and runs down when the beaks are closed. Zweers (1991a) called this the 'tweezer' effect, and Gerritsen (1988) called it a 'fetch and carry' mechanism.

Jehl (1985) describes several foraging techniques for phalaropes. For example, they probe and peck shrimp and larval flies from the upper millimeters of the water. Most typical is the spinning behavior used to stir food while swimming. Phalaropes are planktivorous most of the year, being at sea on a diet of copepods (Brown and Gaskin 1986). Rubega and Obst (inpress) have analyzed films of red-necked phalaropes (*Phalaropus lobatus*) feeding on shrimp. They observed that the tweezer effect really transports shrimp without use of suction. They call this feature 'surface tension transport'. When a shrimp is lost while a drop runs up, the beak is closed somewhat so that the drop runs down again and picks up the shrimp. Rubega and Obst consider this mechanism the simplest form of filter feeding. Mahoney and Jehl (1985) consider a filter feeding mechanism in *Phalaropus tricolor* from beak and tongue morphology.

4.4.4 Drepanididae

Honeyeaters (Meliphagidae) and Hawaiian honeycreepers (Drepanididae) forage upon a mixture of insects, nectar and fruit. Although 'suction' feeding is most typically described in these taxa according to the mechanism of nectar intake, a mechanism that operates by capillarity is more likely since a comparison of their lingual structures and those in hummingbirds shows that 'tube' parameters match very closely (Bock 1972 mentions a tube of 9 mm and diameters at the tip of 0.5, increasing to 1 mm at the posterior end in *Ciridops anna*). The reader is also referred to the comments on *Lampornis clemenciae* by Wagner (1952), and the review article of Paton and Collins (1989).

The adaptive radiation of the Drepanididae was studied by Richards and Bock (1974) and Bock (1972). They compare the balance between the major food types and tongue and bill form (see also Raikow 1977). For example, *Loxops virens virens* is primarily an insect gleaner and secondarily a nectar feeder. Small insects are brushed up with the tongue and larger ones are grasped with the slender bill. The tongue shape and apparatus show elements that meet nectar intake requirements. The long, slender tongue is slightly decurved and tubular with edge and tip in the form of a fringe. Further, the hyoid horns are relatively long and the lingual pro- and retractor muscles correspond to the lengthened horns, allowing lingual protrusion. *Loxops maculata*, however, feeds on arthropods, not on nectar, so on compromises to nectar feeding requirements are expected. Here, the tongue is stiff, forked and straight, not tubular, the paraglossalia are long, and hyoid horns and branchiomandibular muscles are short.

Carothers (1982) examines the effect of trophic morphology and behavior in three Drepanididae upon foraging rates on flowers of *Vaccinum calycinum*, which have straight corollas. Comparison of the maui creeper, amakihi and i' iwi, indicate five parallel shifts: (1) the intake rate to empty a corolla changes from 3.9 s/flower to 2 s/flower; (2) bills change from short and straight to long and decurved shapes; (3) primacy of feeding changes from insectivorous to nectarivorous; (4) changes from probing for insects to probing for nectar; (5) tongue shape changes from a tongue typical for insect gleaning (stiff, flat and forked, long paraglossal, short horn, like in *L. maculata*) to a tongue typical for collection of nectar (long, slender, tubular, brush-tipped, long hyoid horns, like in *L. viridens*). Carothers (1982) explains that the nectar intake rate and adapted bill shape are inversely related compared to their intake performance data; apparently, differences in the mechanical appearance of the tongue are dominant over that of the bill.

5 Sensory Control and Motor Patterning

5.1 Senses of the Oropharynx

The sensory control of food acquisition depends on a multitude of extra- and intraoral receptors. Two separate sensory systems occur superimposed in the skin of the avian oropharynx: (1) a superficial oral taste system, based on contact chemoreceptors in the form of taste buds of the general vertebrate type, located in the superficial strata of the epidermis (reviewed in Berkhoudt 1985); (2) a deeper located cutaneous somatosensory system, the end-organs of which are located almost exclusively in the dermis (reviewed in Gottschaldt 1985).

Taste. In birds, a rather small amount of taste buds is found, innervated by facial and glossopharyngeal nerve branches, but sensitivity is present (e.g. Berkhoudt 1985). Very few taste buds occur in the epithelium of the dorsum of the tongue. Coupled with the fact that most of the modern studies only checked the tongue for taste buds, this resulted in the still persistent notion that birds are poorly equipped with taste organs. On the contrary, taste buds are located in the non-cornified mucosa of the palate and lower mandible, adjacent to the tip and rims of the tongue. This shift may be connected to the involvement of the beaks in prehension and the prominent role of the bird tongue in manipulation and transport of food.

Behavioral experiments show that screening of food for gustatory information occurs at different levels in many speices. In black-chinned hummingbirds (*Archilochus alexanderi*) Stromberg and Johnsen (1990) found that not physical mechanisms of viscosity but rather chemosensory mechanisms are responsible for the sensory evaluation of sucrose nectars and their subsequent selection.

Berkhoudt (1985) reports rejecting behavior when pigeons and chickens grasped peas dipped in acid solutions. In birds not previously subjected to experiments this rejection occurred in much more caudal areas than in birds previously tested, where it occurred after barely touching the food. The presence of taste buds in these areas was demonstrated in both species. Gerritsen and Heezik (1985) present evidence that many species of sandpipers (*Calidris*) can also use gustatory information at beak tip levels in preference experiments, and Gerritsen and Sevenster (1985) add additional proof by sketching the distribution of taste buds in these species. Berkhoudt (1985) reports from films that grebes (*Podiceps cristatus*) tested moldy fish during stationing behavior somewhat caudally to the beak tips, where taste buds were observed, before they rejected it.

Touch. The cutaneous somatosensory receptors in the rhamphotheca of the bill are innervated by the three main branches of the trigeminal nerve. Receptors for at least four modalities are present in the bill skin: mechanoreceptors, thermoreceptors, nociceptors and chemoreceptors. Generally, four distinct morphological types of receptors are ordered according to size and complexity (ct. Gottschaldt 1985): free nerve endings, Merkel cells (often arranged into Merkel corpuscles), Grandry corpuscles, and lamellated Herbst corpuscles. Only recently, structure-function relationships for all types of end organs have been established (Gottschaldt 1985). All are involved with different submodalities of mechanoreception: the majority is rapidly adapting, while Merkel cells detect amplitude, Grandry corpuscles velocity and Herbst corpuscles acceleration components in mechanical stimuli. The morphology of the different types of mechano-receptors varies largely with species (Malinovsky 1967; Gottschaldt 1985) and their distribution closely matches the functional demands on the tactile system in rhamphotheca and tongue (Krulis 1978; Berkhoudt 1980). Increased sensitivity can be obtained by enlarging the dermal area available for mechanoreceptor positioning. This is clearly demonstrated in waders, where extra space is gained by a dermis invading the bone, in the form of depressions of the premaxilla and mandible (Bolze 1968). In extreme forms, such as snipes, bone structure mimics the shape of a honeycomb. Zweers et al. (in prep. a) used these anatomical findings in combination with behavioral observations on sandpipers to arrive at a remote-touch concept, in contrast to direct touch, which is viewed as a telereceptor to obtain information on distant prey.

In the context of our listing of food-acquiring mechanisms, the occurrence of two structurally related mechanoreceptors deserves special mention: Merkel corpuscles are predominantly found in birds employing the pecking technique (chicken, pigeon and quail; Gottschaldt 1985; Nafstad 1986; Halata and Grim 1993). The highest numbers and most complex formations of Merkel corpuscles are found in the tongues of birds using specific food-handling techniques such as seed husking (Krulis 1978; Toyoshima and Shimamura 1991; Toyoshima et al. 1992). Grandry corpuscles seem to be restricted to waterfowl, but some misnomers exist in the literature where this name is used to indicate typical Merkel corpuscles in chickens and Fringillidae (cf. Gottschaldt 1985; Toyoshima 1993).

It is tempting to connect the presence of Grandry corpuscles to the detection of the velocity of flowing, water-containing food during filter feeding (cf. Zweers et al. 1977; Berkhoudt 1980).

In many birds maximal touch sensitivity occurs in a special sensory structure, called the bill tip organ (Gottschaldt 1985). It consists of rows of dermal papillae stacked with mechanoreceptors, which run in the tubules of the horny beak tips, and is particularly prominent in the nail of Anseriformes (Berkhoudt 1976; Zweers et al. 1977) and Psittaciformes (Homberger 1980). It also occurs in pecking birds. Zweers (1982b) mentioned it in the pigeon and Gentle and Breward (1986) described it in the chicken.

In some storks, under certain conditions, head approach and grasping food merge into oropharyngeal processing. The optic control of head-neck motion is then taken over by the trigeminal touch system. For example, Kahl and Peacock (1963) discuss a control feature in the wood stork (*Mycteria americana*). When partially blinded storks foraged in the company of normal storks, both were equally successful in catching prey. When the open, groping bill made contact with a fish, it was snapped immediately, apparently upon touch and without any visual cue. It seems that a touch-trigeminal takeover from the optic system had occurred. Similar features are discussed by Zweers and Berkhoudt (1991) for filter feeding waterfowl and flamingos, as well as for probing sandpipers. These authors indicate that a major difference between the visually and the touch-controlled head-neck motion is that the optically controlled approach phase, once released, cannot be interrupted, but that touch control steers head-neck motion continuously.

5.2 A Comparator Control Model

Integration of sensory information channeled by the various cranial nerves already occurs at lower brainstem levels (Dubbeldam 1984). For instance, tactile glossopharyngeal information derived from the tongue is processed in the princeps nucleus of the trigeminal nerve. Similarly, gustatory information from facial and glossopharyngeal sources converges to specific subnuclei in the solitary nucleus. Detailed wiring diagrams were developed also for higher order neuronal connections, especially in the trigeminal system, primarily concerned with mechanoreception (Dubbeldam 1984).

Models of control of food acquisition were seldom formulated, investigators primarily focused on the optic control of grasping (e.g. Friedman 1975; Goodale 1983; Bischof 1988; Zeigler 1989). Striking of underwater prey poses a special problem in controlling the strike. Katzir and Intrator (1987) present a model on how a reef heron (*Egretta gularis schistacea*) copes with light refraction at the air/water interface. They found that birds grasping submerged prey show a 'pre-strike' and 'strike' which differ remarkably in mean path angle and mean velocity, while these values are much closer for herons catching unsubmerged prey.

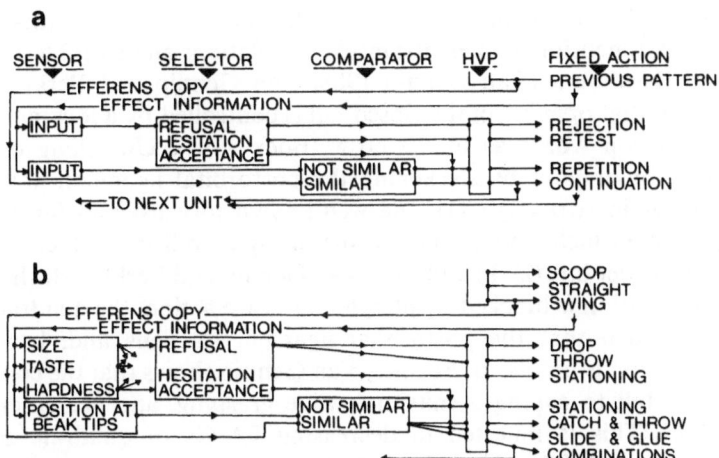

Fig 3. A comparator model for the first static-dynamic unit of pecking (in this case: head fixation-downstroke-grasp). **a** General unit applicable to all four basic units of food acquisition. **b** Specific work from the first control unit of food acquisition. *HVP* Highest value passage unit. See text for explanation. (After Zweers 1985)

Zweers (1982a, b, 1985) proposes a 'comparator' model based on the assumption that pecking can be viewed as a sequence of static and dynamic phases (See. Sect. 2.1). The model conceives a control unit for each of the different pairs of static and dynamic phases, consisting of three subunits: a sensoselector, a comparator and a highest value passage (HVP) unit (Fig. 3). The sensoselector selects an input set from the food. This set is compared to an internal reference value, which may be either innate or acquired by experience. This unit tests edibility and may send information to the HVP unit parallel to the comparator unit, or it may release the output of the comparator. The comparator unit compares the incoming information about the food with an expectancy based upon an efference copy of the pattern in the previous dynamic phase. Basically, the present situation is compared to the intended result of the last dynamic phase. The result of this feedback is passed to the HVP unit. The HVP unit only releases the highest stimulated pattern, which is a modification of three types: rejection, retesting or proceeding intake.

5.3 Motor Patterning

Columba. The muscle-bone apparatuses of jaws and tongue in pigeons have been described by Bhattacharyya (1980, 1991), Van Gennip (1986) and Zweers (1982b). The beak is slender and no large adductor forces are required since the small food grains are gently lifted in grasping and held securely, but not forcefully during subsequent transport. This is reflected in the myology, as shown by Van Gennip (1988) who calculated maximal torques during pecking. Three-dimensional kinematic model simulations and EMG tests by Van Gennip and Berkhoudt (Van Gennip 1988; Van Gennip and Berkhoudt 1992) have shown that opening of the closed bill is controlled by a locking system which in the pigeon works as in the crow (Bock 1964). Unlocking of the quadrato-mandibular joint by slackening the postorbital ligament, however, occurs in pigeons in two ways. (1) The well-known forward and inward swing of the quadrate which is coupled to raising the upper bill and unlocking the depression of the lower bill. (2) In addition, Van Gennip and Berkhoudt show that, if upper bill elevation is blocked by pterygoid muscle action, the quadrate is still able to move such that the postorbital ligament slackens and, hence, lower beak depression is unlocked. Further, Van Gennip shows that the contribution of the upper bill to gaping is maximal while grasping, and decreases in subsequent stages. This is reflected in decreasing EMGs of quadrate protractors and increasing EMGs of jaw depressors.

Bout and Zeigler (in prep.) and Zeigler et al. (in press) add that amplitude of the gape is proportional to the size of the food prior to grasping, during stationing and during the slide-and-glue transport. The relative contribution of jaw opening velocity and rise time (the time needed to arrive at a certain gape) in these pecking phases were studied. In grasping, both contribute equally, in stationing and transport the greater contribution is made by velocity. Each phase also has a unique muscle action pattern. While grasping, gape scaling is accomplished by quadrate protractor and jaw depressor muscles, however, during stationing the antagonist pterygoid muscles modulate gape amplitude, while both modulated during transport.

Deich and Balsam (1993) investigated the components of pecking in adult and squab ring doves by a very precise gape transduction system. They concluded that task-specific experience is needed for the development of the adult food-peck. Deich and Balsam (in press) also suggest that ring dove squabs learn what to peck at through a Pavlovian association between objects and positive ingestional consequences. In addition, they suggest that stereotyped, species-typical behavior need not reflect an inherited central motor mechanism due to its development in a domain that is highly constrained by the anatomy.

Gallus. The anatomy of the jaw apparatus in chickens was reviewed by Vanden Berge (1979). The lingual apparatus was carefully analyzed by Homberger and Meyers (1989). Special attention was given to the hyoid suspension at the skull by a sheath-like fascia. The lingual and jaw apparatuses of the chicken show all the major features that are specific to a pecking system; in addition, some particular features are present. For example, an enlarged Proc. postorbitalis is present, and also a bar- like Proc. retroarticularis points dorsocaudad. While the adductor and depressor complexes are relatively forceful, the paraglossal is small, but relatively broad, supporting a flexible, thickened and enlarged tongue. EMG analyses of Suzuki and Nomura (1975) show that jaw motions are predictable . Van den Heuvel (1992) describes four ways in which the quadrato-mandibular joint is unlocked for lower jaw depression. Unlocked opening is assisted by or even solely caused by protraction of the tongue before a seed is transported by lingual retraction.

Anas. Muscle-bone apparatuses of the jaws and tongue in the mallard were described and reviewed by Goodman and Fisher (1962), Vanden Berge (1979), and Zweers (1974). The following anatomical aspects are typical. Large horizontal Proc. postorbitalis and angularis posterior allow the attachment of large adductor and depressor muscle complexes. The pterygo-quadrate protractor and quadrato-mandibular adductor muscle complexes can change the position of the quadrate condyles along the mandibular facet so that the fulcrum may shift along the mandible. The lingual apparatus has a flattened, lengthened paraglossal bone which supports the large mass of the lingual bulges. Pro- and retractor muscles and also elevator and depressor muscles are forceful complexes.

An electromyographical analysis of all jaw and tongue muscles was made by Zweers (1974) and Zweers et al. (1977). The same locking mechanism as mentioned previously was shown to control gape widening, but unlocking was shown to be primarily a case of shifting the quadrate condyles along the mandible while rotating forward and inward. Gape tuning during filter feeding and, consequently, fine tuning of the filter mesh width (cf. Kooloos et al. 1989) were explained by vector analysis. Thus, guide and power muscle complexes were distinguished. The guide complex (quadrate protractors and quadrato-mandibular adductors) primarily controls fine tuning of the gape (mesh width) by controlling the shift in the position of the fulcrum of the quadrate and mandible along the mandible. The power complex (jaw adductors and depressors) primarily delivers the power for strong motions needed to pump water through the mouth. In addition, these authors explained how the paraglossal is moved by a complex pro- and retractor, and elevator-depressor action pattern so that the large lingual bulges can serve as a pumping piston and also as a working valve in the cylinder-like mouth.

6 Principles of Avian Food Acquisition

The pecking design is for several reasons one key to systematizing the wide diversity of avian trophic designs. All neognath birds show pecking, regardless of whether or not they have food-handling and acquisition specializations. Comparison of the reviewed specialist feeding and pecking mechanisms makes clear that specializations can be considered modifications of specific behavioral and anatomical elements of the pecking design. Moreover, this comparision makes clear that only a restricted set of different qualities can be altered in, or imposed upon the pecking design. Within the initial conditions of the pecking design, modifications seem to rely upon (1) extending or altering physical qualities such as carrying, cutting, penetrating, filtering, suction, pressure and capillarity, and (2) extending sensory capacities like direct touch to remote touch (cf. Gerritsen 1988), or integrating motion control by different qualities like vision and touch (cf. Zweers and Berkhoudt 1991). Further, each modification is in itself very modifiable because of the plasticity of the initial pecking design (see Zweers et al. in prep. a, for a discussion). These features are illustrated below with respect to some major aspects (Fig. 4).

Pecking. Comparison of all analyses of pecking shows that they have a similar basic pattern which was described for chickens and pigeons (Sects. 2.1 and 2.2). Many 'pecking' taxa have attributes superimposed on the basic pattern that add specific food-handling or specialist acquisition capacities (Fig. 4). These attributes do not disrupt the basic static-dynamic phase order, but rather modify elements in these phases. For example, *Columba* has (1) a slender, flexible

ELEMENTS of PECKING		ELEMENTS MODIFIED in SPECIALIST FEEDING				
STATIC	DYNAMIC	FOOD HANDLING Peck ⏐ Husk	CATCH	PROBE	FILTER	ADHESION CAPILLARITY
Prel. Head Fixation		411				441 444
	Prel. Head Approach	411				441 444
Final Head Fix.		411				
	Final Head Approach	313	411 412 413	421 422 423	431 432	441 444
Grasp at Jaw Tips		22 311 32 312 33 313 314	411	421 422 423 424		
	Stationing Head-Shake	24 314	411			
Catch at Jaw Tips		313 314				
	Intra-Oral Transport	21 311 22 312 23 313 24 314	411 412 413		431 432 433	441 442 443 444
Catch at Rictus / Li. Wing						
	Intra-Pharyngeal Transport	21 22				

Fig 4. Scheme describing diversity of food acquisition based upon the static-dynamic phases and comparator models. See text for explanation. The *numbers* refer to sections where the special modifications of particular elements of the basic pecking design are explained.

tongue allowing specialized 'slide and glue' transport, and (2) erectable pharyngeal scrapers that greatly increase swallowing efficiency (Sect. 2.1; See also food-handling column in Fig. 4). *Gallus* has (1) a strong Proc. retroarticularis allowing muscular attachment for a forceful gaping that breaks the top soil at the end of a downstroke, and (2) a strong postorbital process allowing a forceful grasp (Sect. 2.2). *Corvus* has (1) a split tongue tip allowing intraoral transport by pricking soft food, and (2) a widely extendable lingual base allowing storage of food (Sect. 2.3). *Nucifraga* has a lingual pouch in the mouth floor allowing storage of seeds (Sect. 2.3). Pecking in *Anas* illustrates that certain elements of the basic pattern cannot occur as a consequence of filter-feeding modifications, e.g. the slide and glue mechanism (Sect. 2.4).

Specific Food Handling: Husking Seeds. Prior to swallowing, certain mechanisms may prepare food for later digestion: seeds are husked and snails are extracted. Seed husking is performed through grasping, stationing and intraoral transport. Carduelinae, Fringillinae, and Estrildidae have (1) modified beaks that allow the favorable grasping of certain seed types, and (2) modified tongue elements, e.g. a seed cup or an inflatable, cavernous body which allows highly accurate manipulation of seeds into a favorable position between palatal and tomial ridges to either crush or cut the shell and husk the seed (Sects. 3.1.1 and 3.1.2). *Aphelocoma* has a 'buttress' complex at the quadrato-mandibular joint that allows the bird to pound acorns very effectively at the end of the downstroke to husk them (Sect. 3.1.3). *Psittacus* has (1) a chisel-like lower mandible rhamphoteca, (2) a curved upper mandible which carries a transversal ridge, and (3) a tongue tip with an inflatable, cavernous body that allows continuous (re-) positioning and husking of the seed (Sect. 3.1.4).

Extracting Snails. At the end of the downstroke some kites, storks and limpkins extract a snail from its shell by probing the aperture and detaching the flesh. *Chondrohierax* has a hooked bill producing a lever arm, and *Anastomus* and *Aramus* have an asymmetry at the beak tips that produces a hooked lower mandible tip which cuts the columellar muscle (Sect. 3.2).

Biting Fruit. At the end of the downstroke tooth-billed pigeons, *Didunculus*, bite into a fruit and saw a piece of pulp from the fruit prior to grasping the seed by using special lower mandible adaptations. *Tachyphonus* mashes a whole, grasped fruit prior to swallowing it. Parrots like *Psittrichas* bite large pieces from a whole, grasped fig and tongue and palate adaptations transport it by slide and glue (Sect. 3.3).

Catching: Striking. A specific downstroke in combination with a specific grasp forms a modification to catch prey by striking, aerial and aquatic sweeping. Ardeidae, Ciconidae and Podicipaedidae have (1) spear-like, lengthened beaks and (2) special modifications in the cervical column that allow strong downstrokes in combination with a specific locomotory and beak-gaping pattern.

Impaling fish with closed and opened beaks and forceful head shakes are modifications. *Cochlearis* and *Mycteria* apply a touch-hunting component producing a bill-snapping reflex, which means that touch control takes over from optic control during head approach and grasping (Sect. 4.1.1).

Catching: Aquatic and Aerial Sweeping. Downstrokes are integrated with locomotion and grasping is omitted to allow prey catching by sweeping. Pelecanus has (1) a lengthened and flexible lower mandible and (2) a large gular pouch to sweep water and fish while swimming (Sect. 4.1.2). *Caprimulgus* has (1) an extra joint in the lower mandible and (2) a specific quadrato-mandibular joint causing a very wide gape to allow sweeping the air and insects in flight (Sect. 4.1.3).

Probing. A specific downstroke that merges into penetrating a substrate and that is followed by a special grasp of prey forms a modification to finding and gathering prey by probing. *Calidris* has (1) a lengthened, flattened beak, (2) a special bill tip organ for remote touch, and (3) a rhynchokinetic skull and a retroarticular process that produce beak tip motion by moving the symphyseal areas separately from the mandibular rami, to allow grasping of burrowed prey (Sect. 4.2.1). *Haematopus* has (1) lengthened, flattened and very sharp beaks, and (2) a very large ventrocaudad pointing retroarticular process that allows the penetration and opening of mussels (Sect 4.2.3). *Eudocimus* (Sect. 4.2.2), *Calidris* and *Haematopus* have a visually controlled downstroke that is taken over by touch control which steers the penetration. *Sturnus* and *Corvus* penetrate the top soil layer, then they gape and simultanously inspect binocularly the hole made for prey. An enlarged Proc. angularis posterior allows the attachment of a large depressor muscle (Sect. 4.2.4).

Filter Feeding. Grasping is omitted and a specific intraoral mechanism transports a medium from which food is gathered in the filter modification of food acquisition. Anatidae and Phoenicopteridae have (1) a cylinder-like mouth and a large tongue that perform (different) pumping mechanisms, and (2) a lamellar system along the mandibular margins and a lingual papillar organization that produce (different) filtering mechanisms. Postorbital and retroarticular processes, the special quadrato-mandibular joint and the connected muscles have a specific organization that allows tuning of the pumping and filtering systems. Also, the bill tip organ is highly specialized for direct touch which primarily controls gape and mesh size to select preferred food during straining (Sects. 4.3.1 and 4.3.2). Also, *Pachyptila* and *Puffinus* filter feed. Lingual papillae and palatal lamellae filter krill from streams that develop either from inflow from an open beak while swimming, or outflow from an emptying gular pouch (Sect. 4.4.3).

Adhesion and Capillarity. Grasping is omitted and a special intraoral, lingual mechanism transports a fluid medium by capillarity and adhesion. Trochilidae and Meliphachidae have (1) slender, lengthened bills, and (2) a long, slender, groove-like tongue with a fringed tip to allow nectar feeding by capillarity (Sects. 4.4.1 and 4.4.4). Psittacidae have a brush-like lingual tip that allows nectar feeding by dipping the tip in nectar through which adhesion of the fluid occurs (Sect. 4.4.2). *Calidris* and *Phalaropus* have a long and slender beak that allows careful tuning of the transport of a water drop along the rami, including eventual prey that might be in the drop. This is a special intraoral transport allowed by adhesive forces (Sect. 4.3.3).

Conclusion. The reviewed knowledge on trophic diversity may be systematized by some tentative principles:

1. All neognath birds have a pecking mechanism, regardless of the specialist food acquisition mechanism they have.
2. The generalized and de-specialized pecking mechanism represents the basic elements of the initial pecking design.
3. Pecking and all kinds of food acquisition mechanisms can be described by a sequence of static and dynamic components which is controlled by a comparator system.
4. The domain of the avian feeding design is formed by modifications from the pecking design.
5. The domain is defined by (a) altering basic elements, (b) adding or extending physical attributes, within (c) certain initial and boundary conditions.

Re a. For example, changing size, shape, presence, amplitude, and frequency of anatomical and behavioral elements.

Re b. For example, incorporating/extending carrying, cutting, catching, probing (including penetration), pressure, suction, filtering, capillarity and adhesion, specializing sensory modalities.

Re c. For example, fixed order in time and space of the elements in the initial pecking design.

Systematizing a wide trophic diversity is different from elucidating the evolutionary process of trophic diversification. The latter feature is beyond the scope of this review. It is noted that Zweers (1991a, b), assuming that the pecking design is the ancestral avian feeding design, discusses a methodology to connect the domain of trophic diversity to evolutionary diversification processes.

Acknowledgments. We are indebted to Dr. R. Bout, E.M.S.J. Van Gennip, A.F.C. Gerritsen, W.J. Van den Heuvel, J. Hulscher, P. Snelderwaard, H.P. Zeigler, and V. Ziswiler for access to unpublished material.

References

Allen RP (1956) The flamingos: their life history and survival. Res Rep 5. National Audubon Society, New York, 285 pp

Beichle U (1987) Lebensraum, Bestand und Nahrungsaufnahme der Zahntaube, *Didunculus strigirostris*. J Ornithol 128: 75–89

Benkman CW (1987a) Crossbill foraging behavior, bill structure, and patterns of food profitability. Wilson Bull 99: 351–368

Benkman CW (1987b) On the advantages of crossed mandibles: an experimental approach. Ibis 130: 288–293

Benkman CW (1988) Seed handling ability, bill structure, and the cost of specialization for crossbills. Auk 105: 715–719

Berkhoudt H (1976) The epidermal structure of the bill tip organ in ducks. Neth J Zool 26: 561–566

Berkhoudt H (1980) The morphology and distribution of cutaneous mechanoreceptors in the bill and tongue of the mallard (Anas platyrhynchos L.). Neth J Zool 30: 1–34

Berkhoudt H (1985) Structure and function of avian taste receptors. In: King AS, McLelland J (eds) Form and function in birds, vol 3. Academic Press, London, pp 463–496

Berkhoudt H, Zweers GA (1990) Food intake in the carrion crow Corvus corone. Acta XX Cong Int Ornithol, pp 476

Bermejo R, Allan RW, Houben D, Deich JD, Zeigler HP (1989) Prehension in the pigeon. I. Descriptive analysis. Exp Brain Res 75: 569–576

Bhattacharyya BN (1980) The morphology of the jaw and tongue musculature of the common pigeon, Columba livia, in relation to its feeding habit. Proc Zool Soc, Calcutta 31: 95–127

Bhattacharyya BN (1990) The functional morphology of the lingual apparatus of two species of imperial pigeons, Ducula aenea nicobarica and D. badia insignis. Proc Zool Soc, Calcutta 43: 65–93

Bildstein KL (1983) Age-related differences in the flocking and foraging behavior of white ibises in a South Carolina salt marsh. Colon Waterbirds 6: 45–53

Bildstein KL (1987) Energetic consequences of sexual dimorphism in white ibises (Eudocimus albus). Auk 104: 771–775

Bildstein KL, McDowell SG, Brisbin IL (1989) Consequences of sexual dimorphism in sand fiddler crabs, Uca pugilator: differential vulnerability to avian predation. Anim Behav 37: 133–139

Bischof H-J (1988) The visual field and visually guided behavior in the zebra finch (Taeniopygia guttata). J Comp Physiol A 163: 329–337

Boates JS, Goss-Custard JD (1989) Foraging behaviour of oystercatchers Haematopus ostralegus during a diet switch from worms Nereis diversicolor to clams Scrobicularia plana. Can J Zool 67: 2225–2231

Bock WJ (1964) Kinetics of the avian skull. J Morphol 114: 1–42

Bock WJ (1966) An approach to the functional analysis of the bill shape. Auk 83: 10–51

Bock WJ (1972) Morphology of the tongue apparatus of Ciridops anna (Drepanididae). Ibis 114: 61–78

Bock WJ (1978) The preglossale of passer – a skeletal neomorph. J Morphol 155: 99–110

Bock WJ, Balda RP, Vander Wall SB (1973) Morphology of the sublingual pouch and tongue musculature in Clarck's nutcracker. Auk 90: 491–519

Bolze G (1968) Anordnung und Bau der Herbstchen Körperchen in Limicolenschnäbeln in Zusammenhang mit der Nahrungsfindung. Zool Anz 181: 21–355

Bout R Zeigler HP Jaw muscle (EMG) activity and amplitude scaling of jaw movements during eating in pigeon (Columba livia) (in prep).

Bowman RI (1961) Morphological differentiation and adaptation in the Galápagos finches. Univ Calif Publ Zool 58: 1–302

Bramble DM, Wake DB (1985) Feeding mechanisms in lower vertebrates. In: Hildebrand M, Bramble DM, Liem KF, Wake DB (eds) Functional vertebrate morphology. Belknap Press Cambridge, pp 230–261

Brown RGB, Gaskin D (1986) The pelagic ecology of the grey and red-necked phalaropes Phalaropus fulicarus and P. lobatus in the bay of Fundy, eastern Canada. Ibis 130: 234–250

Bühler P (1981) Functional anatomy of the avian jaw apparatus. In: King AS, McLelland J (eds) Form and function in birds, vol 2. Academic Press, London, pp 439–468

Burger AE (1978) Functional anatomy of the feeding apparatus of four South African cormorants. Zool Afr 13: 81–102

Burton PJK (1974) Feeding and feeding apparatus in waders. Br Mus Nat Hist Publ 719, 150pp

Burton PJK (1986) Curlews' *Numenius* bills: some anatomical notes. Bird Study 33: 70

Calhoun ML (1933) The microscopic anatomy of the digestive tract of *Gallus domesticus*. Iowa State Coll J Sci 7, 3: 261–382

Carothers JH (1982) Effects of trophic morphology and behavior on foraging rates of three Hawaiian honeycreepers. Oecologia 55: 157–159

Creutz G (1981) Der Graureiher, *Ardea cinerea*. Neue Brehm Bücherei, Wittenberg

Deich JD, Balsam P (1993) The form of early pecking in the ring dove squab (*Streptopelia risoria*). J Comp Psychol 107: 261–275

Deich JD, Balsam PD Development of the prehensile feeding in ring doves (*Streptopelia risoria*). In: Green P, Davis M (eds) Perception and motor control in birds. Springer, Berlin, Heidelberg, New York (in press b)

Delius JD (1989) The peck of the pigeon: free for all. In: Lowe CF, Richelle M, Blackman DE, Bradshaw CM (eds) Behavioral analysis and contemporary psychology. Erlbaum, Hillsdale, New York, pp 53–81

Dubbeldam JL (1984) Brainstem mechanisms for feeding in birds. Brain Behav Evol 25: 85–98

Dullemeijer P (1952) The correlation between muscle system and skull structure in *Phalacrocorax carbosinensis*, Shaw and Nodder. Proc K Ned Akad Wet C 55: 95–102

Durell S, Goss-Custard JD (1984) Prey selection within a size-class of mussels, *Mytilus edulis*, by oystercatchers, *Haematopus ostralegus*. Anim Behav 32: 1197–1203

Ewald PW, Williams WA (1982) Function of the bill and tongue in nectar uptake by hummingbirds. Auk 99: 573–576

Fjeldså J (1981) Comparative ecology of Peruvian grebes – a study in mechanisms for evolution of ecological isolation. Vidensk Med Dan Naturhist Foren 144: 125–246

Foster MS (1987) Feeding methods and efficiency of selected frugivorous birds. Condor 89: 566–580

Friedman MB (1975) How birds use their eyes. In: Wright P, Caryl PG, Vowles DM (eds) Neural and endocrine aspects of behavior in birds. Elsevier, Amsterdam, pp 181–204

Gayou DC (1982) Tool use by green jays. Wilson Bull 94: 593–594

Gentle MJ (1975) Gustatory behaviour of the chicken and other birds. In: Wright P, Caryl PG, Vowles DM (eds) Neural and endocrine aspects of behaviour in birds. Elsevier, Amsterdam, pp 305–318

Gentle MJ, Breward J (1986) The bill tip organ of the chicken (*Gallus gallus* var. *domesticus*). J Anat 145: 79–85

Gerritsen AFC (1988) Feeding techniques and the anatomy of the bill in sandpipers (*Calidris*). Thesis, University of Leiden

Gerritsen AFC, van Heezik YM (1985) Substrate preference and substrate related foraging behavior in three *Calidris* species. Neth J Zool 35: 671–692

Gerritsen AFC, Meijboom A (1986) The role of touch in prey density estimation by *Calidris alba*. Neth J Zool 36: 530–562

Gerritsen AFC, Sevenster J (1985) Foraging behaviour and bill anatomy in sandpipers. Fortschr Zool 30: 237–240

Goodale MA (1983) Visually guided pecking in the pigeon (*Columba livia*). Brain Behav Evol 22: 22–41

Goodman DC, Fisher HI (1962) Functional anatomy of the feeding apparatus in waterfowl. South Ill Univ. Press, Carbondale

Gottschaldt K-M (1985) Structure and function of avian somatosensory receptors. In: King AS, McLelland J (eds) Form and function in birds, vol 3. Academic Press, London, pp 375–462

Grant PR (1983) Inheritance in size and shape in a population of Darwin's finches, *Geospiza conirostris*. Proc R Soc Lond B 220: 219–236

Grant PR (1986) Ecology and evolution of Darwin's finches. Princeton Univ Press, Princeton

Grant BR, Grant PR (1989) Natural selection in a population of Darwin's finches. Am Nat 133: 377–393

Guillet A (1979) Aspects of foraging behavior of the shoebill. Ostrich 50: 252–255

Hainsworth FR, Wolf LL (1976) Nectar characteristics and food selection by hummingbirds. Oecologia 25: 101–113

Halata, Z, Grim M (1993) Sensory nerve endings in the beak skin of Japanese quail. Anat Embryol 187:131–138

Hancock J, Kushlan J (1984) The herons handbook. Croom Helm, London

Harper PC (1987) Feeding behaviour and other notes on 20 species of Procellariiformes at sea. Notornis 34: 169–192

Heidweiller J, Zweers GA (1990) Drinking mechanisms in the zebrafinch and the Bengalese finch. Condor 92: 1–28

Heppleston PB (1970) Anatomical observations on the bill of the oystercatcher (*Haematopus ostralegus occidentalis*) in relation to feeding behaviour. J Zool (Lond) 161: 519–524

Homberger DG (1980) Funktionell-morphologische Untersuchungen zur Radiation der Ernährungs- und Trinkmethoden der Papageien (Psittaci). Bonn Zool Monogr 13: 1–192

Homberger DG (1986) The lingual apparatus of the African grey parrot, *Psittacus erithacus* Linné (Aves: Psittacidae). Ornithol Monogr 39: 1–233

Homberger DG Brush AH (1986) Functional-morphoological and biochemical correlations of the keratinized structures in the African grey parrot, *Psittacus erithacus* (Aves). Zoomorphology 106: 103–114

Homberger DG, Meyers RA (1989) Morphology of the lingual apparatus of the domestic chicken, *Gallus gallus*, with special attention to the structure of fasciae. Am J Anat 186: 217–257

Hulscher JB (1976) Localisation of cockles (*Cardium edule* L.) by the oystercatcher (*Haematopus ostralegus* L.) in darkness and daylight. Ardea 64: 292–310

Hulscher JB (1985) Growth and abrasion of the oystercatcher bill in relation to dietary switches. Neth J Zool 35: 124–154

Hulscher JB, Ens BJ (1992) Is the bill of the male oystercatcher a better tool for attacking mussels than the bill of the female? Zool Jahrb Anat 122:219–223

Hull C (1991) A comparison of the morphology of the feeding apparatus in the peregrin falcon, *Falco peregrines*, and the brown falcon, *F. berigora* (Falconiformes). Austral J Zool 39: 67–76

Jehl JR (1985) Biology of the eared grebe and Wilson's phalarope in the non-breeding season. Stud Avian Biol 12: 1–74

Jehl JR, Murray BG (1986) The evolution of normal and reverse sexual dimorphism in shorebirds. In: Johnston RF (ed) Current Ornithol 3:1–86

Jenkin PM (1957) The filter feeding and food of flamingoes (Phoenicopteri). Phil Trans Roy Soc Lond B, 240: 401–493

Kahl MP (1971) Food and feeding behavior of open-bill storks. J Ornithol 112: 21–35

Kahl MP, Peacock LJ (1963) The bill-snap reflex: a feeding mechanism in the American wood stork. Nature 4892 : 505–506

Katzir G, Intrator N (1987) Striking of underwater prey by a reef heron, *Egretta gularis schistacea*. J Comp Physiol A 160: 517–523

Kear J (1962) Food selection in finches with special reference to interspecific differences. Proc Zool Soc Lond 138: 163–204

Kear J, Duplaix-Hall N (1975) Flamingos. Poyser, Berkhamstead

Kingsolver JG, Daniel TL (1983) Mechanical determinants of nectar feeding strategy in hummingbirds: energetics, tongue morphology, and licking behavior. Oecologia 60: 214–226

Klages NTW (1992) Bill morphology and the diet of a filter-feeding seabird: the broad-billed prion *Pachyptila vittata* at South Atlantic Gough Island. J Zool (Lond) 227:385–396

Kooloos JGM (1986) A conveyer-belt model for pecking in the mallard (*Anas platyrhynchos* L.). Neth J Zool 36: 47–87

Kooloos JGM, Zweers GA (1991) Different integrations of pecking, drinking and filter feeding mechanisms in waterfowl. Acta Biotheor 39:107–140

Kooloos JGM, Kraaijeveld AR, Langenbach GEJ, Zweers GA (1989) Comparative mechanics of filter feeding in *Anas platyrhynchos*, *Anas clypeata* and *Aythya fuligula* (Aves, Anseriformes). Zoomorphol 108: 269–290

Kral B (1965) Functional adaptations of Ciconiiformes to the darting stroke. Acta Soc Zool Bohemoslov 29: 377–391

Krause R (1992) Mikroskopische Anatomie der Wirbeltiere, II. Vögel und Reptilien. de Gruyter, Berlin

Krulis V (1978) Struktur und Verteilung von Tastrezeptoren im Schnabel-Zungenbereich von Singvögeln, im besonderen der Fringillidae. Rev Suisse Zool 85: 385–447

Kushlan JA (1976) Feeding behavior of North American herons. Auk 93: 86–94

Lack D (1947) Darwin's finches. Cambridge Univ Press, Cambridge

Landolt R (1987) Vergleichend funktionelle Morphologie des Verdauungstraktes der Tauben (Columbidae) mit besonderer Berücksichtigung der adaptiven Radiation der Fruchttauben (Treronidae). Zool Jahrb Anat 116: 169–215, 285–316

Lauder G (1989) How are feeding systems integrated and how have evolutionary innovations been introduced. In: Wake DB, Roth G (eds) Complex organismal functions: integration and evolution in vertebrates. Wiley, Chichester, pp 97–115

Lawrence GE (1950) The diving and feeding activity of the western grebe on the breeding grounds. Condor 52: 3–16

Lockie JD (1956) The food and feeding behaviour of the jackdaw, rook and carrion crow. J. Anim Ecol 25: 421–428

Lorenz KZ (1949) Uber die Beziehungen zwischen Kopfform und Zirkelbewegung bei Sturniden und Ikteriden. In: Mayr E, Schutz E (eds) Ornithologie als biologische Wissenschaft.

Lucas AM, Stettenheim PR (1972) Avian anatomy, integument. Agricultural handbook 362. US Government Printing Office, Washington DC

Mahoney SA, Jehl J (1985) Adaptations of migratory shorebirds to highly saline and alkaline lakes: Wilson's phalarope and American avocet. Condor 87: 520–527

Malinovsky L (1967) Die Nervenendkörperchen in der Haut von Vögeln. Z Mikr Anat Forsch 77: 279–303

Marin M, Stiles FG (1992) On the biology of five species of swifts (Apodidae, Cypseloidinae) in Costa Rica. Proc Western Found Vert Zool 4: 287–351

Marks JS, Hall CS (1992) Tool use by bristle-thighed curlews feeding on albatross eggs. Condor 94: 1032–1034

Martin R (1904) Die vergleichende Osteologie der Columbiformes unter besonderer Berücksichtigung von *Didunculus strigirostris*. Zool Jahrb 20: 167–352

McLelland J (1979) Digestive system. In: King AS, McLelland J (eds) Form and function in birds, vol 3. Academic Press, London, pp 69–181

McMahon BF, Evans RM (1992) Nocturnal foraging in the American white pelican. Condor 94: 101–109

Meyerriecks AJ (1959) Food-stirring feeding behavior in herons. Wilson Bull 71: 153–158

Mock DW, Mock KC (1980) Feeding behavior and ecology of the Goliath heron. Auk 97: 433–448

Morgan WL, Ritz DA (1982) Comparison of the feeding apparatus in the mutton-bird, *Puffinus tenuirostris* (Temminck) and the fairy prion, *Pachyptila turtur* (Kuhl) in relation to the capture of krill, *Nyctiphanes australis*. J Exp Mar Biol Ecol 59: 61–76

Morris D (1955) Seed preferences of certain finches under controlled conditions. Avic Mag 61: 271–287

Nafstad PHJ (1986) On the avian Merkel cells. J Anat 145: 25–33

Norton-Griffiths M (1967) Some ecological aspects of the feeding behaviour of the oystercatcher *Haematopus ostralegus* on the edible mussel *Mytilus edulis*. Ibis 109: 412–424

Owre OT (1967) Adaptations for locomotion and feeding in the anhinga and the double crested cormorant. In: Storer RW (ed.) AOU Ornith Monogr 6, pp 138

Paton DC, Collins BG (1989) Bills and tongues of nectar-feeding birds: review of morphology, function and performance, with intercontinental comparisons. Austral J Ecol 14: 473–506

Pough HF, Heiser JB, McFarland WN (1991) Vertebrate life. Macmillan, New York

Raikow RJ (1977) The origin and evolution of the Hawaiian honeycreepers (Drepanidae). Living Bird 15: 95–117

Richards LP, Bock WJ (1974) Functional anatomy and adaptive evolution of the feeding apparatus in the Hawaiian honeycreeper genus *Loxops* (Drepanididae). Ornithol Monogr 15: 1–173

Rooth J (1965) The flamingos on Bonaire (Netherlands Antilles): habitat, diet and reproduction of *Phoenicopterus ruber ruber*. Natuurws Stud Suriname Ned Ant 141: 1–151

Rubega MA, Obst BS Surface tension feeding in phalaropes: discovery of a novel feeding mechanism. Auk (in press)

Schall U (1989) Sensory control of pecking in the pigeon (*Columba livia*). Thesis, University of Konstanz

Schilling C (1992) Structure and dry mass spectrum of the jaw muscles in birds with different food intake strategies. Zool Jb Anat 122:275–285

Schlüter D (1988) Character displacement and adaptive divergence of finches on the islands and continents. Am Nat 131: 799–824

Schlüter D, Price TD, Grant PR (1985) Ecological character displacement in Darwin's finches. Science 227: 1056–1059

Schreiber RW, Woolfenden GE, Curtsinger WE (1974) Prey capture by the brown pelican. Auk 92: 649–654

Sims RW (1955) The morphology of the head of the hawfinch (*C. coccothraustus*). Bull Br Mus Nat Hist 13: 371–393

Smith TB, Temple SA (1982) Feeding habits and bill polymorphism in hook-billed kites. Auk 99: 197–207

Snyder NFR, Snyder HA (1969) A comparative study of mollusc predation by limpkins, everglade kites, and grackles. Living Bird 8: 177–223

Stephens DW, Krebs JR (1986) Foraging theory. Princeton University Press, Princeton, 247 pp

Stromberg MR, Johnson PB (1990) Hummingbird sweetness preferences: taste or viscosity? Condor 92: 606–612

Suzuki M, Nomura S (1975) Electromyographic studies on the deglutition movement in fowl. Jpn J Vet Sci 37: 289–293

Swennen C, de Bruijn LLM, Duiven P, Leopold MF, Marteijn ECL (1983) Differences in bill form of the oystercatcher (*Haematopus ostralegus*): a dynamic adaptation to specific foraging techniques. Neth J Sea Res 17: 57–83

Tinbergen JM (1981) Foraging decisions in starlings, *Sturnus vulgaris*. Ardea 69: 1–67

Toyoshima K (1993) Are Merkel and Grandry cells two varieties of the same cell in birds? Arch Histol Cytol 56:167–175

Toyoshima K, Seta Y, Shimamura A (1992) Fine structure of the Herbst corpuscles in the lingual mucosa of the finch (*Lonchura striata*). Arch Histol Cytol 55:321–331

Toyoshima K, Shimamura A (1991) Ultrastructure of Merkel corpuscles in the tongue of the finch, *Lonchura striata*. Cell Tissue Res 264:427–436

Vanden Berge JC (1979) Myologia. In: Baumel JJ, King AS, Lucas AM, Breazille JE, Evans HE (eds) Nomina Anatomica Avium. Academic Press, London, pp 175–219

Vanden Berge JC, Zweers GA (1993) Myology. In: Baumel JJ (ed) Handbook of Avian Anatomy. Nutt Ornithol Club, Cambridge, pp 241–277

Van den Elzen R (1985) Systematics and evolution of African canaries and seed eaters (Aves, Carduelidae). In: Schuchmann K-L (ed), Proc Int Symp Afr Vertebr, pp 435–451

Van den Elzen R, Nemeschkal HL (1991) Radiation of African canaries – a comparison of different classificatory approaches. Acta XX Congr Int Ornithol, New Zealand Ornithol Congr Board, Wellington, pp 459–467

Van den Elzen R, Nemeschkal HL, Classen H (1987) Morphological variation of skeletal characters in the bird family Carduelidae. Bonn Zool Beitr 38: 221–239

Van den Heuvel WF (1992) Kinetics of the skull in the chicken (*Gallus gallus domesticus*). Neth J Zool 42:561–582

Van Gennip EMJS (1986) The osteology, arthrology and myology of the jaw apparatus of the pigeon (*Columba livia* L.). Neth J Zool 36: 1–46

Van Gennip EMJS (1988) A functional morphological study of the feeding system in pigeons (*Columba*). Thesis, University of Leiden

Van Gennip EMJS, Berkhoudt H (1992) Skull mechanics in the pigeon, *Columba livia*. A three dimensional kinematic model. J Morphol 213: 197–224

Voous KH, Van Dijk T (1973) How do snail kites extract snails from their shells? Ardea 61: 179–185

Wagner HO (1952) Beitrag zur Biologie des Blaukehlkolibris *Lampornis clemenciae* (Lesson). Veröff Mus Bremen 1952 A, 2: 6–43

Weymouth RD, Lasiewski RC, Berger AJ (1964) Tongue apparatus in hummingbirds. Acta Anat 58: 252–270

Wheelwright NT (1985) Fruit size, gape width, and the diets of fruit-eating birds. Ecology 66: 808–818

White SS (1968) Mechanisms involved in deglutition in *Gallus domesticus*. J Anat 104: 177

Zeigler HP (1989) Neural control of the jaw and ingestive behavior. Ann NY Acad Sci 563: 69–86

Zeigler HP, Levitt PW, Levine R (1980) Eating in the pigeon (*Columba livia*): movement patterns, stereotypy and stimulus control. J Comp Physiol Psychol 94: 783–794

Zeigler HP, Bermejo R, Bout R, Ingestion, prehension and the sensorimotor control of the jaw. In: Green P, Davies M (eds) Perception and motor control in birds. Springer, Berlin Heidelberg NewYork (in press)

Ziswiler V (1965) Zur Kenntnis des Samenöffnens und der Struktur des hörnernen Gaumens bei körnerfressenden Oscines. J Ornithol 106: 1–48

Ziswiler V (1979) Zungenfunktion und Zungenversteifung bei granivoren Singvögeln. Rev Suisse Zool 86: 823–831

Ziswiler V (1990) Specialization in extremely unbalanced food: possibilities and limits of investigation exclusively by functional morphology. Neth J Zool 40: 299–311

Ziswiler V, Trnka V (1972) Tastkörperchen im Schlundbereich der Vögel. Rev Suisse Zool 79: 307–318

Ziswiler V, Güttinger HR, Bregulla H (1972) Monographie der Gattung *Erythura* Swainson, 1873 (Aves, Passeres, Estrildidae). Bonn Zool Monogr 2: 1–158

Zusi RL (1985) A functional and evolutionary analysis of rhynchokinesis in birds. Smithson Contrib Zool 395: 1–40

Zusi RL (1987) A feeding adaptation of the jaw articulation in New World jays (Corvidae). Auk 104: 665–680

Zweers GA (1974) Structure, movement and myography of the feeding apparatus of the mallard (*Anas platyrhynchos* L.). Neth J Zool 24: 323–467

Zweers GA (1982a) Pecking of the pigeon (*Columba livia* L.) Behaviour 81: 173–230

Zweers GA (1982b) The feeding system of the pigeon (*Columba livia* L.). Adv Anat Embryol Cell Biol 73: VII + 108

Zweers GA (1985) Generalism and specialism in the avian mouth and pharynx. Fortschr Zool 30: 189–201

Zweers GA (1991a) Pathways and space for evolution of feeding mechanisms in birds. In: Dudley EC (ed) The unity of evolutionary biology, Proc Int Congr Evol Biol, 1990. Dioscorides Press, Portland, pp 530–547

Zweers GA (1991b) Transformation of avian feeding mechanisms: a deductive approach. Acta Biotheor 39: 15–36

Zweers GA (1992) Behavioral mechanisms in avian drinking. Neth J Zool 42:60–84

Zweers GA, Berkhoudt H (1987) Larynx and pharynx of crows (*Corvus corone* L. and *C. monedula* L., Passeriformes: Corvidae. Neth J Zool 37: 365–393

Zweers GA, Berkhoudt H (1991) Recognition of food in pecking, probing and filter feeding birds. Acta cc Congr Int Ornithol, New Zealand Ornithol Congr Trust Board, Wellington, pp 897–902

Zweers GA, Gerritsen AFC, van Kranenburg-Voogd PJ (1977) Mechanics of feeding of the mallard (*Anas platyrhynchos* L., Aves, Anseriformes). Contrib Vertebr Evol 3: IX + 109

Zweers GA, Bout RG, Heidweiller J, Motor organization of the avian head-neck system. In: Green P, Davies M (eds) Perception and motor control in birds. Springer, Berlin Heidelberg New York (in press)

Zweers GA, Gerritsen AFC, Bock WJ (in prep a) Morphological modifications for probing and filter feeding in the avian pecking mechanism

Zweers GA, Jong F de, Berkhoudt H (in prep. b) Filtermechanismus des Roten Flamingos (*Phoenicopterus ruber*; Aves, Phoenicopteridae)

Chapter 9

Evolutionary Approach of Masticatory Motor Patterns in Mammals

W.A. Weijs

Contents

Department of Functional Anatomy, ACTA, 15 Meibergdreef, 1105 AZ Amsterdam, The Netherlands

Advances in Comparative and Environmental Physiology, Vol. 18
© Springer-Verlag Berlin Heidelberg 1994

1 Introduction and Scope

The diversity of mammalian masticatory systems has attracted considerable attention for a long time. In the first half of this century, numerous descriptive studies appeared, including comparative ones, e.g. Fiedler (1953, Insectivora); Starck (1933, platyrrhine primates; 1935, ursids); Storch (1968, Chiroptera); later, significant studies were published by Gaspard et al. (1976), Schumacher (1961) and Turnbull (1970). Bluntschli (1929) and followers (Müller 1933; Zey 1939) published morphological papers, taking into account postnatal development.

Interest in the actual process of feeding increased as techniques to record jaw movements and muscle activity became available. Dentistry set the trend, registering in humans 3-D jaw movements by X-ray cinematography (Hildebrand 1931) and jaw muscle EMG by surface electrodes (Moyers 1949). As the techniques were further refined jaw movement studies became possible (Ardran et al. 1958) and were followed by EMG studies in unrestrained animals (since Kallen and Gans 1972).

The differences among species in relative mass and orientation of the four jaw closing muscles and the jaw openers and in shape of the jaw articulation invited biomechanical speculation. An explanation to (asymmetric) jaw action was attempted already in 1921 by Gysi, and comparative approaches appeared subsequently (Kühlhorn 1938; Arendsen de Wolf-Exalto 1951; Becht 1953). A widely cited paper by Smith and Savage (1959) considered the mandible as a two-dimensional free body, subjected to muscle, joint and bite forces. These forces must, in a static bite, cancel each other. By assuming full contraction of the muscles, their forces could be predicted and their effect upon bite force and joint force determined. The shape of the temporomandibular joint and the mode of action of the teeth could, to some extent, be explained. Later, refinements of this approach included the use of three-dimensional models (Greaves 1978), the subdivision of the architecturally complex muscles into subunits with separate lines of action and accounting for differential contraction of muscles and muscle units (Weijs and Dantuma 1975). All these approaches considered the jaws as rigid bodies. An important complementary biomechanical approach deals with the way the bones themselves are constructed to withstand functional stresses (Endo 1965). Hylander (1979a, b) measured mandibular strains in vivo. This not only made the determination of the kind of loading the jaws experienced possible, but also the elucidation of the role of mandible and skull as structures resisting external loads.

Understanding occlusion is vital for understanding feeding mechanics. The evolution of a limited tooth replacement system is a mammalian characteristic and makes the development of specific areas for shearing and crushing on opposing teeth possible. The presence of wear facets and striations (Mills 1966) indicates how teeth were actually used and completes the coarser movement information obtained by (X-ray) film or displacement transducers. Jaw move-

ment is controlled jointly by jaw muscles, temporomandibular joint and occlusal pattern. Study of occlusion in fossil mammals is an important aid in the
reconstruction of evolution of chewing in mammals (Crompton and Hiiemae
1970).

The cooperation pattern of jaw closer and jaw opener muscles has been
investigated for two decades now in many different mammalian species. A basic
uniformity of this pattern was claimed already by Hiiemae (1978). This paper
attempts to analyze the now available information, and to hypothesize how
different motor patterns arose from a primitive pattern. However, it should be
realized, first, that motor patterns are flexible and adjust to the mechanical
demands of food and second, that there is not only inter- but also intramuscular
heterogeneity in structure and activity patterns (Herring et al. 1979; Weijs and
Dantuma 1981; Blanksma and van Eijden 1990).

Motor patterns are determined by central programs that can be modified by
peripheral feedback. Periodontal receptors, sensitive to magnitude and direction
of bite force per tooth and muscle spindles, and known to be heterogeneously
distributed, are thought to be important in influencing motor patterns. Recent,
available reviews on motor control (Lund and Enomoto 1988; Rossignol et al.
1988; Taylor 1990) show that only little neurophysiological information concerning task differentiation within and between jaw closers has become known. It is
also too early to try to make comparisons between different animals. Therefore,
this review will concentrate on a comparison of anatomy, biomechanics and
peripheral motor patterns. It will outline the presumed primitive condition of
the mammalian jaw apparatus, as it is now viewed in the light of fossil evidence
and the study of recent primitive mammals. The main adaptive modifications
are then treated. I hope to show that similar motor patterns evolved from a
basic, primitive pattern in different, independent lineages in response to similar
demands on the masticatory system.

2 The Evolutionary Development of the Mammalian Jaw Apparatus

The evolutionary transition from primitive, mammal-like reptiles (pelycosaurs,
of the Upper Carboniferous and Lower Permian) to advanced mammal-like
reptiles (therapsids, of Upper Permian and Lower Triassic, among which the
cynodonts are presumed mammalian ancestors) and to the earliest mammals
(*Eozostrodon*, Late Triassic) is relatively well documented, not only osteological-
ly (Crompton 1963; Kermack et al. 1973, 1981; Romer 1974) but also with
respect to the (possible) arrangement of the external adductor mass [the
homologue of the later temporalis (TEM) and masseter (MAS) muscles]
(Barghusen 1968, 1973) and (possible) development of external and middle ear
structures (Allin 1975).

Within several therapsid lineages the following marked phylogenetic changes occur in the jaw apparatus (Barghusen 1968; Allin 1975; Crompton and Hylander 1986): (1) the lower jaw, in primitive forms, is formed by the dentary, bearing the teeth and a number of postdentary bones, among others the articular, forming the jaw articulation with the quadrate. The postdentary bones and the quadrate become reduced markedly in size and the connection between postdentary bones and dentary, firm in pelycosaurs and extant reptiles, becomes loosened. In the most advanced therapsids an additional joint develops between one of the postdentary bones (in *Probainognathus*) or the dentary (in *Diarthrognathus*) and the squamosal of the skull. (2) A coronoid process of the dentary is formed, rising high above the level of the jaw articulation and occlusal plane. (3) The mass of the external adductor (mammalian TEM) muscle increases and it attains a more posterior direction of pull. Some of the muscle insertion moves forward under the zygomatic arch in order to attach to the lateral side of the jaw (mammalian deep masseter, MASDEEP). In many advanced therapsids a superficial masseter (MASSUP) develops lateral to the MASDEEP; it attains a horizontal, forward-pulling orientation because it originates from the anterior end of the zygomatic arch and inserts into a newly formed process of the dentary, the mandibular angle. (4) Despite a marked increase in jaw muscle mass, the jaw articulation becomes much smaller. Allin (1975) showed that the mammalian middle ear (whose malleus and incus developed, respectively, from the articular and quadrate) did not develop after the dentary had increased in size and taken over the function of jaw articulation but simultaneously with it. He showed that cynodonts must have possessed an eardrum connected to the postdentary bones. The reduction of these bones and their loosening from the dentary were hypothesized to be the result of selection for improved auditory acuity. At the same time, the musculature must be rearranged to reduce the loads imposed on the jaw joint.

The mechanical effects of the drastic changes in jaw system morphology were analyzed by DeMar and Barghusen (1973), who interpreted them as a means to increase the leverage of the musculature, and by Bramble (1978) and Crompton and Hylander (1986), who stressed the increased possibilities of unloading the jaw joint. These studies made clear that raising the coronoid process, shifting the line of action of the TEM from obliquely posterior to purely posterior and adding an anteriorly pulling MASSUP to the system in advanced therapsids lead to a situation in which the jaw joint remains unloaded or is even subjected to tensile loads during biting. The exact loading pattern depends on the bite point; but the muscles are arranged in such a way that compressive loads can always be avoided by their selective use. The most advanced therapsids (*Probainognathus*, tritylodonts) had tiny jaw joints and most likely a bilateral occlusion.

The earliest mammals such as *Eozostrodon* continue the trends of reduction of the postdentary bones, while the quadrato-articular joint was replaced by a dentosquamosal joint. The most striking difference with advanced cynodonts is the appearance of occlusal wear facets on the postcanine teeth, indicating (1) unilateral occlusion and lateral to medial occlusal movement; and (2) the inward rotation of the hemimandible during occlusal movement, taking place about its longitudinal axis. The dentosquamosal joint (called the temporomandibular joint or TMJ in this work) was clearly larger than in *Probainognathus*. Applying the same (three-dimensional) model on this fossil, and assuming unilateral occlusion and participation of muscles of the non-occluding (balancing) side, it was found by Crompton and Hylander (1986) that significant compressive loads must have been exerted to the balancing side of the TMJ and, hence, that a strong joint is a prerequisite for unilateral occlusion.

3 Mammalian Jaw Muscles and Feeding Cycles: Definitions

3.1 Jaw Muscles

Many systems for subdividing the 'external adductor mass' of mammals have been proposed. For the present purpose a very crude system is adopted. The masseter (MAS) originates from the lateral face of the zygomatic arch and can usually be subdivided into a more horizontally oriented superficial (MASSUP) and a more vertically oriented deep (MASDEEP) portion. Medial to the MAS, the zygomaticomandibular (ZYMA) is found, originating from the medial face of the zygomatic arch. It is often possible to discern an anterior and a posterior portion, separated by the masseteric nerve. The posterior ZYMA is closely associated with the TMJ and has a posterior direction of pull. In some rodents, ZYMA has a separate anterior portion, extending through an enlarged infraorbital foramen, the infraorbital (or maxillomandibular) portion. The temporalis muscle (TEM) is routinely, but arbitrarily divided into an anterior (TEMAN) and posterior (TEMPO) portion. Often, a deep portion (TEM-DEEP), inserting down the anterior margin of the ramus, can be found. Medial pterygoid (MPT) and lateral pterygoid (LPT) are the other jaw closers present. The LPT is absent in some carnivores and MPT and LPT are absent in monotremes.

In this chapter relative weight data will be cited, expressing the weight of a jaw muscle as a percentage of MAS + TEM + MPT + LPT + digastric. Although for interorder comparisons muscle weights expressed in relation to body weight or skull size would be preferable, too few data are available (Cachel 1984; Hurov et al. 1988; Langenbach and Weijs 1990) to make such an approach useful.

3.2 Feeding Sequences

A feeding sequence consists of (1) ingestion; (2) food transport to the cheek teeth; (3) chewing plus bolus formation; and (4) swallowing. Ingestion and biting have been little investigated and are extremely species- and food-dependent. Rodents and rabbits have developed rapid, repetitive gnawing, involving open/close and forward/backward jaw movements, reminiscent of chewing cycles. The food is moved to the cheek teeth in so-called transport I movements (Hiiemae and Crompton 1985): the tongue protrudes while pressed against the palate, often provided with transverse rugae. In one to several movement cycles, food items are displaced backward on the tongue dorsum. In rodents, these transport cycles also involve cyclic mandibular depression (See Sect. 8.2).

Chewing or mastication is a cyclic movement of the mandible and tongue/hyoid apparatus, whereby food is reduced between cheek teeth and simultaneously chewed material is formed into a bolus. In most mammals, swallowing occurs without interruption of the masticatory rhythm. Swallowing cycles can sometimes be recognized from their prolonged duration.

Food is usually chewed only or mainly on one side at a time, the active or working side (ws). The other side is the balancing side (bs). Normally, in a power stroke (PS) the ws lower teeth move medially and often forward across the upper teeth. In most mammals, including man, the ws and bs are swapped frequently, sometimes so regularly that an innate reversal mechanism has been proposed (Kallen and Gans 1972). In contrast, many ungulates and rabbits chew for prolonged periods, sometimes for days, on the same side. Few animals (camels, some rodents) chew alternately on their left and right sides. True, bilateral chewing with little or no transverse motion is mainly found in rodents, but always occurs in puncture/crushing cycles (see below).

Chewing cycles can be conveniently divided into jaw opening, fast closing (FC) and power stroke (PS) phases (Hiiemae 1978). From maximal opening, the jaw closes at a high velocity and with little muscle activity, until tooth-food-tooth contacts begin; the FC/PS transition is sometimes arbitrary. The PS involves vertical, medial and forward movement components in various proportions, depending on the animal. In the pig the final closing movement and the PS movement are separated in time: the ws moves laterally first while closure is completed and then medially (Herring and Scapino 1973). During PS, food is usually displaced lingually by the lower cheek teeth. In the first phase of the opening movement the tongue and hyoid apparatus move forward. Food is then probably redeposited on the teeth by upward tongue movement, while sufficiently processed food is displaced backward on the tongue dorsum (Franks et al. 1984; Cortopassi and Muhl 1990). Eventually, this material is collected into a bolus under the soft palate. This is phase II transport. In some mammals (insectivores, carnivores) vigorous head movements assist the transport process. The forward movement of the tongue and hyoid system is reversed abruptly about halfway through the opening movement. In many insectivores and lower primates this reversal is accompanied by a sudden increase in jaw opening speed, so that a slow opening (SO) and fast opening (FO) phase can be discerned. In other species (rodents, ungulates, higher primates) transition is more gradual. In cats (Gorniak and Gans 1980) and rabbits (Weijs et al. 1989a) a marked transverse jaw displacement (back to the midline) occurs during SO. Lund and Enomoto (1988) propose subdividing jaw opening into O_1, O_2 (a pause at the end of O_1) and O_3 (FO), a convention that will be followed here. O_2 corresponds to the resting position of the jaw so that O_3 is the true start of a chewing cycle, at least in the rabbit.

Crompton et al. (1977) first described these tongue/hyoid cycles for the opossum; they have been found in many species (three shrew, Fish and Mendel 1982; fruit-eating bat, deGueldre and de Vree 1984; macaque, Franks et al. 1984; hyrax, German and Franks 1991; cat, Hiiemae et al. 1981; rabbit, Anapol 1988; armadillo, Smith and Redford 1990; sloths, Naples 1985). The forward movement of the tongue always stops at the beginning of O_3, but the rest of the movement cycle is more variable, and perhaps less stereotyped than originally suggested (Mendel et al. 1985). Bramble and Wake (1985) have expanded the idea of a food transport cycle basic to all feeding modes to non-mammalian terrestrially feeding tetrapods. A motor pattern involving adductor, neck, tongue and hyoid musculature would have been conserved in tetrapod evolution.

In many species the first chewing cycles are different from the rest because the jaw fails to reach occlusion as the food mass intervenes between the teeth. These cycles were called 'puncture crushing' cycles, in contrast to 'normal' chewing cycles (Hiiemae and Crompton 1971). The marked horizontal component of the power stroke is often completely absent in such cycles. In most animals (e.g. cat, Gorniak and Gans 1980; rabbit, Lund and Enomoto 1988) further changes in movements and motor patterns can be seen during the rest of the chewing bout.

Workers studying wear facets (e.g. Mills 1967) have discerned two phases in the power stroke movement, a buccal phase (phase I), whereby the active jaw moves medially until maximum intercuspidation is reached and a lingual phase (phase II), whereby the jaw moves out of occlusion and further medially, often in a different direction. Crompton and Hiiemae (1970) have stressed that buccal and lingual phases depend on tooth morphology, not on the position of the jaw relative to the skull. In rodents, the upper teeth are often closer together than the lower ones so that the buccal phase starts with the mandible in a symmetric position. Hylander and Crompton (1986) have shown, however, that during the lingual phase bite forces have already decreased to negligible levels so the importance of the lingual phase for food reduction must be questioned.

The masticatory motor patterns depend on the food being chewed. EMG amplitude increases with food toughness (see, for instance, Horio and Kawamura 1989), while an increase in food particle size induces an earlier onset of jaw closing activity, as the power stroke becomes longer (de Vree and Gans 1976; Gorniak and Gans 1980). In the rabbit, changes in the relative amplitude of LPT and posterior ZYMA muscles determine the relationship between lateral and vertical movement (Weijs and Dantuma 1981). In man, bs MPT activity determines the amount of lateral jaw excursion during FC.

Chewing frequency decreases regularly with the size of the animal, similar to heart and respiratory rate, but is also determined by the hardness and size of food. Thexton et al. (1980) showed that in cats variation in chewing cycle time is mainly due to variation in slow opening time and is related to food type; in rabbits it is due to variation in PS length in early chewing cycles and in O_2 length in late cycles (Lund and Enomoto 1988). Otten (1987) has suggested that the force velocity properties of the jaw muscles are the principal parameters determining the chewing rate. The intrinsic velocity of all myosin types decreases with increased animal size (Rome et al. 1990). Chewing rate is controlled by a central pattern generator located in the brainstem (see Goldberg and Chandler 1990, for review).

4 Feeding in Primitive Mammals

4.1 Models for Primitive Mammals

From Late Triassic to the end of the Mesozoic many fossil mammals have been recovered, belonging to different lineages. They include the primitive morganucodonts (for example, *Eozostrodon*) and their triconodont descendents, the widespread and highly successful herbivorous multituberculates and the kueneotheriids from which both monotremes and therians (marsupials and placentals) are believed to descend (Jenkins 1990). All groups possessed multicusped cheek teeth with wear facets, evidencing precise molar occlusion. Like in their reptilian ancestors, mammalian teeth are divided into incisor, canine and usually molar and premolar series. Tribosphenic cusp patterns of molar teeth, characteristic of all living mammals, appear relatively late, in the Lower Cretaceous, although the line toward tribosphenic dentitions can be followed back into the Jurassic (Clemens 1971). Triassic mammals had different dentitions. Tribosphenic dentitions are thought to be suitable for insectivory and carnivory, but not for herbivory, and act by both crushing and cutting food (Crompton and Hiiemae 1970). The evolution of these teeth has been linked to the increase in the fauna of terrestrial invertebrates concurrently with the rise in angiosperms (Clemens 1971).

As marsupials represent a more primitive pattern of mammalian organization than placentals, a generalized marsupial such as the opossum is a good model for a primitive mammal. It has a tribosphenic dentition only slightly different from primitive, close relatives (*Alphadon*) from the Upper Cretaceous. The most primitive placental mammals are the insectivores and among them the tenrecomorphs are often considered to be closest to the ancestral stock (Eisenberg 1981). In the course of their long evolution on Madagascar these animals have retained a suite of primitive characters. Their originally tribosphenic molars lost the protocone (upper molars) and also much of the talonid (lower molars) and, thus, secondarily obtained a molar morphology, reminiscent of Jurassic therian ancestors.

4.2 Morphology

The Insectivora generally have slender, elongated skulls with tooth rows consisting of sharply pointed and well-differentiated molars, premolars, canines and incisors occupying most of the length of the jaws (about 70% in soricids, Dötsch 1982). The adductor mass is differentiated into an MAS, ZYMA, TEM, MPT and (two-headed) LPT muscles (see Fiedler 1953 for comparative myology). Relative weight data are given by Turnbull (1970) and Dötsch (1982, 1983a). The TEM is the dominant jaw muscle and occupies about 60–70% of the total (adductor plus digastric) weight. Its main line of action is posteriorly directed, due to its insertion to the large coronoid process, reaching far above the level of tooth row and joint. The MAS and ZYMA muscles occupy only 15–25% of the total mass. MASDEEP and ZYMA are weakly developed. The MASSUP is often oriented quite horizontally due to its anterior attachment to the maxilla (the zygomatic arch lacking in some families) and to a long, posteriorly directed angular process. The pterygoid muscles, particularly the LPT, are weakly, the digastrics strongly developed. The TMJ has a flat glenoid fossa, a postglenoid process and either a transverse cylindrical or a double mandibular condyle (in soricids, see Dötsch 1983b); insectivores have a mobile mandibular symphysis.

The biomechanical layout of the jaw closers is quite variable. In opossums and, for instance, *Erinaceus* and *Echinosorex* (Turnbull 1970), MASDEEP and ZYMA muscles are relatively strong and the adductor musculature is a virtually continuous block (Hiiemae and Jenkins 1969) including forward, vertical and backward pulling fibers. Under such conditions TMJ loading tends to become reduced at the ws (Crompton and Hylander 1986). In a number of insectivore families (soricids, tenrecids) the zygomatic arch is incomplete or absent, while the MAS and MPT muscles are very small and horizontally oriented. Most of the TEM is pulling horizontally backward. Simultaneous action of these muscles puts the TMJ under tension. Soricids show this pattern to the extreme and their extra, ventrally facing condylar joint facet might be an adaptation to prevent joint dislocation.

4.3 Motor Patterns

Jaw movements and jaw muscle EMG patterns were recorded in opossums (Crompton and Hiiemae 1970; Hiiemae and Crompton 1971; Crompton et al. 1977), in the tenrec (Oron and Crompton 1985) and some information is available for a shrew (*Suncus marinus*; Dötsch and Dantuma 1989).

Both in opossums and tenrecs jaw opening is accompanied by protrusion, jaw closing by retrusion. During mastication, gape (angle between upper and lower occusal surface) depends on food type and reaches 30°–40° in the opossum and 15°–21° in the tenrec. The mandible can, however, open much further. In both the opossum and tenrec fast opening is produced by simultaneous lifting of the skull and depression of the jaw. The skull movement may produce inertial forces that assist in positioning the food back on the teeth after a power stroke.

Two movements accompany the sagittal jaw movements. First, there is occlusal plane rotation of the entire mandible about a vertical axis located at the level of the condyles, or posterior to them. The movement may be different for both hemimandibles. At either FO, FC or both, the ws hemimandible is moved toward the ws, while the return movement toward the median plane occurs in the late FC and PS. As a consequence, the lower teeth move inward across the upper teeth at the ws. For the opossum the movement is small but no quantitative data are available. For the tenrec, the transverse movement is large: 0.5 – 1.0 cm (jaw length about 9 cm). The amount of movement decreases with harder food. Second, an inward rotation of the ws hemimandible (inversion) takes place about its longitudinal axis. For the opossum, no data are available on the extent of the inversion (but see Hiiemae and Crompton 1985 for an example); for the tenrec it is estimated be quite large (about 10°). In *Suncus* considerable hemimandibular inversion was also observed (Dötsch and Dantuma, pers. comm.). In all cases, the bs hemimandible rotates in the opposite direction, but partly independently from the ws. The inversion movement makes shearing action of the cheek teeth possible without obstruction by the large, interlocking canines (Mills 1966). It necessitates a mobile symphysis to prevent dislocation of the bs condyle and a rounded condyle articulating in a relatively flat genoid fossa. Crompton and Hylander (1986) claim that the Triassic morganucodont *Eozostrodon* has a similar TMJ morphology as the tenrec and might have used the same occlusal movement patterns as recent primitive mammals.

How are these movements generated? Crompton et al. (1977) are not very explicit with regard to differences in timing and amplitude among the MAS, MPT and TEM of either side during mastication in the opossum. They show that MAS and MPT can also be active during slow opening. Crompton and Hylander (1986) give more information, pointing to MPT and TEMPO activity during SO. Furthermore, they indicate that TEMAN and MASSUP are equally active at both sides, while MASDEEP and TEMPO are more active at the ws. In the tenrec (Oron and Crompton 1985) a number of muscles act clearly asymmetrically, either in firing level or in timing. During FC, the jaw is moved to the ws by differential action of the TEM (ws TEM starting to act first) and LPT (bs active, ws silent). During PS, the ws LPT is active and may be instrumental in pulling the mandible back to the midline. The TEM is then active with equal strength at both sides but MASSUP and MPT act asymmetrically, particularly with soft food. In case of hard food action is more symmetrical and the extent of lateromedial movement is smaller. A variable degree of antagonistic coactivation can be observed, depending on food and other conditions. During O_{12}, the MAS muscle can be active bilaterally. During O_3, not only digastric and bs LPT but also MASSUP are often active. Conversely, digastrics are often active in SC and PS phases. The same phenomenon of checked, non-ballistic motion has been reported for *Suncus*, a shrew (Dötsch and

Dantuma 1989). This type of antagonistic activity is much more variable than the agonistic ('normal') activity burst. It is probably related to movement control as it was reported that antagonistic activity increases with harder food.

4.4 Conclusion

Primitive mammals chew unilaterally; the PS consists of an upward and inward movement plus inversion of the ws hemimandible. Compensatory movement takes place at the mobile symphysis. Asymmetrical timing of action of the muscles can be seen and brings about horizontal jaw movement, especially in tenrecs. This action takes the form of differential timing of activity onset in the large TEM muscle or as completely unilateral contraction in the small LPT and tenrec MAS muscles. During the PS, the TEM muscles act mostly symmetrically and pulls both hemimandibles upward and inward while inverting them. The symphysis transmits most of the bs vertical force to the ws bite point and shows special ligamental strengthening against the invoked vertical shearing forces (Beecher 1979). The MAS and MPT muscles are weakly developed, horizontally oriented and at the bs do not contribute consistently to the chewing force.

4.5 Mammalian Diversity

Turnbull (1970) uses the type of dentition, the direction of the jaw movement in the PS and the relative mass of the four jaw closers to subdivide mammals according to masticatory type into five categories:

1. A 'generalized' group, with many multicusped, sharply pointed teeth, showing vertical movement and a predominant TEM muscle; the group includes marsupials, insectivores, bats and primates and is thought to represent the 'primitive', 'unspecialized' condition.
2. A 'carnivore shear' group, with a hinge jaw joint, vertical movements, a dentition specialized for shearing and a predominant TEM muscle.
3. An 'ungulate grinding' group with grinding cheek teeth, mainly transverse jaw movements and a predominant MAS muscle.
4. A 'rodent gnawing' group with specialized incisors for gnawing, anteroposterior jaw movements and a predominant MAS muscle.
5. A 'miscellaneous' group of specialized mammals.

With this system 'typical' representatives of mammalian orders/species (e.g. tiger for carnivores) can be readily classified. Its implicit suggestion that the functional classification coincides with classification according to shared ancestry raises problems in almost every group. Should wombats be included in the rodent group, and kangaroos with the ungulates? What is the position of the edentates? Within the here defined groups, enormous diversity remains, in both dietary habits and the relative size of muscles (in group 3 the TEM muscle of bovids comprises 11% of total muscle mass, of camels 43%). I will treat mammalian diversity along the lines of Turnbull's system and will show that in each major group different masticatory modes, including generalized ones, can be recognized.

5 Bats

5.1 Morphology

Bats have masticatory systems closely similar to those of marsupials and insectivores. The members of this large group show a wide range of dietary specializations (insects, vertebrates, blood, fruit, nectar and pollen) and associated behavioral diversity. Different modes of phonation, for instance, have a profound influence on cranial morphology (Czarnecki and Kallen 1980). The suborders Megachiroptera and Microchiroptera differ in many respects and the differences in masticatory systems can be related to herbivorous specialization in the first group. According to Storch (1968), the Megachiroptera are characterized by a relatively flat glenoid fossa (lacking a retroarticular process) and a well-developed MAS/ZYMA complex. These muscles comprise 32–40% of the total jaw closing mass in the frugivorous members, contrasting sharply with the 10–22% in the Microchiroptera. Within the latter group, frugivorous habits have not led to much increase in MAS mass. Here, the position of the face relative to the cerebral skull is extremely variable and seems to determine the degree of elevation of the condyle above the occlusal plane (Czarnecki and Kallen 1980). Despite this variation, the basic primitive mammalian features, such as TEM predominance, low position of the TMJ and a long tooth row with posteriorly placed muscles, are maintained. Typical for bats is a small, parallel-fibered and horizontally oriented MPT muscle, extending far into the orbit (Storch 1968). In orientation and position, the muscle resembles the LPT of other mammals. The LPT itself is extremely small.

5.2 Motor Patterns

These have been investigated for a generalized Microchiroptera, *Myotis* (Kallen and Gans 1972) and for a Megachiroptera, *Pteropus*, a fruit-eating bat (deGueldre and deVree 1984, 1988, 1990). Czarnecki and Kallen (1980) summarize jaw movement data on a few other species. Both *Myotis* and *Pteropus* have unilateral mastication, involving lateral displacement of the active hemimandible in the open jaw position. However, in *Myotis* there is a considerable medial displacement during the PS (about 1 mm at the canine, involving 6° horizontal rotation). In *Pteropus* large interlocking canines and a fused symphysis prevent transverse occlusal movement. Instead, the jaw moves forward during O_{12}, effecting some shearing action between the molars. Nevertheless, a very small transverse component in O_3 and FC has been retained. For *Myotis*, but not *Pteropus*, a longitudinal rotation of the hemimandible during the power stroke has been reported.

The motor patterns are quite different in the two species. In *Pteropus* muscle activity is more symmetric, both in amplitude and timing. Jaw closing starts with ZYMA and TEMDEEP activity (the ws leading). At the start of PS, these muscles are joined by the rest of the TEM and MAS, so that all jaw closers fire simultaneously and with comparable intensity during PS. In *Myotis*, the bs MPT is the first to act (starting before maximal opening). Here, the timing of all muscles is asymmetric and simultaneous action in the power stroke does not occur. The TEM and ZYMA fire asymmetrically, but in reverse order as in *Pteropus* and most other mammals. The lack of quantitative treatment of EMG in the *Myotis* study makes it necessary to exert caution in interpreting the many subtle differences in the timing described. In both animals the TEM muscle seems capable of differential activity. The ZYMA acts with the TEMDEEP. During O_{12}, the MPT muscles (only the ws MPT in *Myotis*) are active, together with the anterior digastric; jaw opening is produced by the digastrics plus, in *Pteropus*, the LPTs. In *Myotis*, the ws LPT is active in late PS and O_{12}, a pattern also known from other mammals. Digastric asymmetry is another feature of the motor pattern in both animals.

5.3 Conclusions

Myotis and *Pteropus* show that chewing motor patterns can be quite different in animals belonging to the same order, sharing a basic muscular layout. *Myotis* has not departed far from primitive mammals, although motor patterns are more asymmetric. *Pteropus* has a jaw system stabilized in occlusion. Limitation of power stroke movement has simplified the motor pattern to a predominantly symmetric mode. In both animals both LPT and MPT (either unilaterally or bilaterally) are involved in jaw opening. This is found in many other mammals.

6 Primates

6.1 Morphology

Primates differ from primitive mammals by modifications of the skull, due to reduction of olfactory function, emphasis on stereoscopic vision, increase in brain volume and attainment of upright posture, engendered by their diurnal, arboreal lifestyle. Dietary specializations range from pure insectivory to folivory (see Fleagle 1988 for comprehensive data). Food specialization can change musculoskeletal morphology to a considerable extent, as is highlighted by the rodent-like dentition, TMJ and musculature of *Daubentonia* (Biegert 1956). Nevertheless, clear-cut trends can be seen in the primate jaw system, going from prosimians to hominids.

In the higher forms the TMJ attains a more elevated position (Biegert 1956). In prosimians it is hardly above the occlusal plane, in most anthropoids it is moderately elevated (to a degree of about 20–25% of skull base length) and in hominids it is raised high (30–40%) above the occlusal plane. In folivorous species of different lineages (*Alouatta, Megalodapis, Gorilla*) the TMJ is elevated more than in related frugivorous species. The TMJ of *Tupaia*, an intermediate between insectivores and primates, is characterized by a relatively short, transversely oriented fossa allowing only limited transverse jaw excursion. In higher prosimians, the fossa becomes longer; in anthropoids an articular eminence develops in front of the fossa, making possible protrusion and extensive side to side movements (Biegert 1956). While in recent prosimians the mandibular symphysis remains unfused, in larger fossil prosimians and higher primates a synostosis replaces the fibrocartilaginous joint.

Relative to the insectivores, the masticatory muscles show a trend of increased weight and complexity of the MAS relative to the TEM, and much stronger MPT and LPT muscles (Turnbull 1970). The pterygoids comprise 14–23% of jaw muscle mass, twice the value of primitive mammals. Their development is probably related to the increased capacity for translatory condylar movements, made possible by the longer fossa. Although quantitative data are lacking it seems that primitive prosimians (*Tupaia*) hardly differ in muscle arrangement from omnivorous insectivores like hedgehogs (Fiedler 1953). Increase in MAS size has probably taken place in *Perodicticus* (Turnbull 1970) and, judging from anatomical descriptions (Toldt 1905, *Lemur*; Fish 1983, *Tupaia*; Fiedler 1953, *Tarsius*), in other prosimians; in anthropoid primates the MAS and pterygoid muscles have increased their size further at the cost of the TEM. The digastric muscle is less developed than in insectivores.

Cachel (1979) observed an increase in anterior TEM mass relative to MAS mass in primates with large incisors. Hylander (1975) has shown that large incisors are associated with frugivory, as fruits need more incisor preparation. As it is known that both MAS and TEM are strongly involved in chewing and biting (Hylander and Johnson 1985), it is unlikely that there is a direct link between anterior TEM and incisor size. In view of comparisons within other mammalian groups, it is suggested that MAS and pterygoid size are related to food preparation by the cheek teeth.

The dentition of lower primates closely resembles that of primitive mammals. In the majority of primates a dietary trend toward folivory and insectivory is associated with the retention of a fairly high cusped, shearing, crushing and grinding dentition, while a trend toward frugivory can be linked to reduced tooth size and cusp height (Kay 1975; Kay and Hylander 1978). In Old World monkeys changes in molar patterns increase the total crushing/grinding surface of the cheek teeth (Kay 1977).

6.2 Motor Patterns

Fish and Mendel (1982) have registered jaw movements in the tree shrew, *Tupaia*, Kay and Hiiemae (1974) in *Tupaia, Galago* (a prosimian), *Saimiri* and *Ateles* (New World monkeys), Luschei and Goodwin (1974), Byrd et al. (1978) and Byrd and Garthwaite (1981) in macaques and Hildebrand (1931), Gibbs et al. (1971) and many others in man. Generally, mastication is unilateral (but see Stohler 1986, who describes habitual mastication in man as often bilateral). The precise movement pattern varies between cycles, foods and species, but without exception, the jaw is displaced laterally during either FO or FC and approaches maximal intercuspidation in a movement of the ws hemimandible from lateral to medial. Retrusion and protrusion movement is associated with closing and opening, respectively. Moreover, movement of the jaw to the ws during closing involves bs condylar protrusion. The resulting complicated patterns of pro- and retrusion of the condyles seem to be similar in man (Gibbs et al. 1971) and macaque (Hylander et al. 1987).

In *Tupaia*, possessing an unfused mandibular symphysis, changes in both the distance between the rami (Fish and Mendel 1982) and hemimandibular inversion (Kay and Hiiemae 1974) have been reported. In animals with a fused symphysis such movements do not take place to a measurable extent. However, the muscles acting upon the jaw invoke mandibular deformations, reflecting similar loading patterns. Hylander (1979b, 1981, 1984), using in vivo applied strain gauges, showed that the following deformations occur during mastication: (1) movement of the hemimandibles from and toward each other in the horizontal plane ('wishboning'); (2) shearing of the hemimandibles relative to one another in the midsagittal, vertical plane through the symphysis; (3) twisting of the hemimandibles about their long axis; and (4) bending of both hemimandibles, particularly the bs one, in the sagittal plane. In *Galago*, with an unfused symphysis, Hylander (1979a) found the same deformations; he noted that the ratio of ws/bs strain was much higher than in macaques, implicating less transfer of jaw muscle force from bs to ws.

The actual movements of the jaw during the PS are less accessible to experimental analysis and much has been derived from the occlusal cusp patterns and wear facets (Mills 1967; Crompton and Hiiemae 1970; Kay and Hiiemae 1974). From these patterns buccal and lingual phases of PS could be inferred. In animals with a fused symphysis contralateral contacts between the molars might occur in the lingual phase. However, Hylander et al. (1987) have combined cineradiographic, electromyography and bone strain analysis to show in the macaque that in chewing of hard food occlusal forces have normally decreased to zero by the time the lingual phase is reached. They concluded that the wear patterns of the lingual phase facets are most likely produced during buccal phase movement.

Electromyographic data on mastication have been published for two species of macaques (Luschei and Goodwin 1974; McNamara 1974; Byrd and Garthwaite 1981; Miller et al. 1982) and for man (Møller 1966; Hannam et al. 1977;

Hannam and Wood 1981; Stohler 1986; Widmalm et al. 1987). I know of no data on prosimians or other primates.

In man, asymmetric firing is most clearly present in the MPT, much less in the MASSUP and almost absent in the TEMAN and TEMPO. Hannam and Wood (1981) showed by correlation of movement and EMG signals that in man the bs MPT, but not the MAS and TEM, determines the amount of jaw excursion to the ws during closing. On the other hand, the ws MPT becomes active only just before intercuspidation and remains active well after the other jaw closers have stopped firing. This muscle might be instrumental in moving medially and disengaging the teeth at the end of the power stroke.

For macaques, no quantitative descriptions of muscle activity are available except from Hylander et al. (1992) concerning asymmetric firing levels. The descriptions do point to a role of MASSUP, TEM and MPT muscles in effecting the transverse jaw movements, like in humans. Hylander et al. (1987) stress the role of the bs posterior MASDEEP, acting late in the power stroke, presumably together with the ws MPT. They show convincingly from the strain patterns recorded on the mandible that during the PS the mandibular rami are first pulled together, by action of the bs MPT pulling the bs ramus toward the stabilized ws ramus. Then, as the latter muscle becomes silent, action of the bs deep MAS pulls the balancing ramus out, while the teeth start to disengage. This type of action has not been described yet for humans but may be present, given the high level of bs deep MAS activity (Belser and Hannam 1986). The here-mentioned studies clearly demonstrate different roles for superficial and deep masseter in higher primates.

The LPT muscle in macaques can be active during PS, jaw opening, or both phases (Miller et al. 1982) and the activity pattern may be different for the superior and inferior bellies of the muscle (McNamara 1974). For humans the ambiguities in the recordings may stem from uncertainties in electrode location, but certainly also originate from real variability in motor patterns, as demonstrated for the inferior head of this muscle by Wood et al. (1986). Wood (1987) concludes on the basis of the unusual variability in activity during the second half of PS (when most muscles are relaxed) and clenching that this muscle must have a fine tuning role in directing and controlling forces on the condyle.

6.3 Conclusions

It appears from the mostly qualitative macaque data that the motor patterns of macaques and man are well comparable. In unilateral chewing ws and bs muscles act asynchronously. The bs MPT and MAS act prior to their ws counterparts. A reverse, but less apparent difference in timing is present in the ws and bs TEM muscles. These patterns are similar to those of primitive mammals, including bats. There are clear indications that the relatively large MPTs most strongly determine asymmetries in the free movement paths, while asymmetric firing of these muscles and MASDEEP (and perhaps the deep, medially pulling portions of the TEM) effects medial displacement of the ws hemimandible during the occlusal stroke. The morphological constraints for lateral jaw movement have been reduced by the increased size of the TMJ fossa and decreased cheek tooth intercuspidation.

Higher primates differ from recent prosimians by an increased depth of the mandibular corpus. Hylander (1979a) demonstrated that the use of bs muscula-ture in unilateral biting or mastication occurs to a greater degree in higher primates than in prosimians. The bs muscles are used more often if hard food is being chewed. Use of bs muscles makes the exertion of higher chewing pressures possible. The bs hemimandible is subjected to a bending moment, produced by the upward forces of the jaw closers and downward forces at the symphysis and TMJ. Hylander showed that fusion at the symphysis and increased height of the corpus are adaptations to the same function: use of bs muscles to increase chewing forces. Among the higher primates, it could be shown that folivorous species have higher mandibles than frugivorous species.

Diets among primates differ considerably, implicating different functional demands with respect to the amount of incisal preparation relative to chewing, amount of gape necessary, chewing movements (crushing insects and fruit; cutting and grinding leaves; grinding seeds) and chewing pressures. Neverthe-less, only a few clear-cut correlations with morphology have been demonstrated. The dentition adapts readily to diet, but the relative size of the individual jaw muscles stays within particular limits, with few exceptions; the same applies to the robustness of the temporomandibular joint (Smith et al. 1983) and the mandibular dimensions (Smith 1983). Although 'primate' features of the chewing apparatus can in part be attributed to the vegetarian habits of most species, it is suggested that the shortening of the jaws is not related to dietary habits but to the reduction of the nasal cavity and downward displacement of the face relative to the cerebral skull.

7 Carnivores

7.1 Morphology

Several lineages of carnivores developed from insectivores. The direct ancestors of modern carnivores as well as recent canids, mustelids and many viverrids are characterized by the development of a puncture/crushing type of molar series and carnassial teeth, formed by the last upper premolar and first lower molar. The carnassials are vertical shearing blades with a scissor-type action; (type II dentition of Scapino 1981). The TMJ has developed a strong retroarticular and a preglenoid process, bordering a transversely oriented fossa into which the cylindrical condyle fits snugly. The mandibular symphysis can be either fused or fibrocartilaginous. It tends to be a rigid synostosis in larger felids and ursids (Scapino 1981). Increased brain and orbital volumes further separate modern carnivores and their direct ancestors from insectivores (Romer 1974). Secondarily, dentitions have evolved with reduced postcarnassial molars and emphasized carnassials (type I dentition of Scapino, e.g. in the purely carnivorous felids) or reduced carnassials and a low cusped, crushing/grinding cheek tooth battery (type III dentition in omnivorous and herbivorous ursids, procyonids and many viverrids).

Radinsky (1981a, b, 1982) analyzed functionally important skull proportions. Corrected for skull size, ancestral carnivores have relatively wide zygomatic arches accomodating large TEM and also large MAS muscles, long tooth rows with anteriorly placed carnassials and small brains. Most recent carnivores could be classified into four groups showing the following modifications relative to the primitive condition: (1) viverrids, with narrower skulls and somewhat shortened jaws; (2) canids, with more slender skulls, retaining the long jaws; (3) felids, with increased brain and eye size, short and wide skulls, backward-placed carnassials and a relatively important MAS muscle; (4) mustelids, with reduced eye size, shortened jaws and a reduced MAS muscle. Radinsky showed further that many procyonids, ursids and aberrant mustelids cannot be readily classified in this way and occupy much of the rest of the available 'morphospace'. Muscle weight data (Turnbull 1970) show that the organization of the jaw musculature in carnivores is similar to that of insectivores with regard to the relative weight of TEM, MAS and MPT muscles. However, compared to insectivores MPT has a poor mechanical advantage (see Turnbull's 'useful power' formula). The LPT is reduced or completely absent while the digastric is strongly developed, particularly in aquatic species, where it can occupy up to 25% of the jaw closer plus digastric mass (Scapino 1976). Felids and mustelids are the carnivores which emphasize carnassial shear most. They have shortened jaws and a strong bite (Ewer 1973). They differ markedly, however, with respect to the relative sizes of the MAS and TEM muscles. In mustelids, the TEM occupies 70% of the total muscle mass (including digastric), and the MAS is little developed (15%). In felids, the figures are 50% for TEM and 30% for MAS; felids have a correspondingly larger MPT muscle. This suggests that mustelids have less capacity for transverse movement and the generation of transverse forces between the carnassials (see below). This suggestion is supported by the fact that some mustelids have almost pure hinge jaw joints where the jaw cannot be dislocated from the macerated skull. In most other carnivores the relative weight of the TEM is between 50–70% and that of the MAS between 20–30%; no significant trends in the differences can be discerned, neither between the major families, nor between members of different dietary groups. However, Davis (1955, 1964) found a much stronger ZYMA in animals with a crushing/grinding dentition such as herbivorous *Ailuropoda* (giant panda) and *Tremarctos* (spectacled bear). The ZYMA is relatively undeveloped in canids and felids.

Although size-independent cranial shape variation can be quite impressive, no general rules describing a relationship with diet have been found (Radinsky 1981b). Only in a few cases have changes in skull proportions been linked convincingly to changes in function. A good example is the decrease in coronoid size and reorientation of the cerebral skull and nuchal plane in different independent lineages of saber-toothed tigers (Emerson and Radinsky 1980).

7.2 Motor Patterns

In type I dentitions (Scapino 1981) a small medial shift of the ws mandible is predicted as the carnassials shear past one another. In full occlusion, the interlocking canines and incisors make further transverse movements impossible. During closing, the jaw must be displaced to the ws (laterally) to bring the carnassials in occlusion. In type II dentitions additional medial movement is allowed between the molars after the carnassials have slid past one another. In type III dentition a transverse PS movement is well developed and can be seen, for example, in ursids (Starck 1935). As far as known chewing is unilateral in all carnivores (Scapino 1981; and references below).

Jaw movements during feeding were studied in detail in *Procyon* (raccoon; Gorniak 1986) and *Felis* (cat; Gorniak and Gans 1980). Both animals follow the mammalian pattern, exhibiting, in frontal view, orbital jaw movement with a very small transverse component. The canines travel about 1–1.5 mm in a transverse direction in cats and 0.5 mm in raccoons.

Differential activity of jaw muscles during chewing was investigated in the cat (Gorniak and Gans 1980). The LPT muscles act only during closing and thus differ in this from other mammals. Digastric activity was marked during both opening and closing. The closer muscles fire more or less simultaneously, but small differences in the timing of the start and end of activity can be seen. The vertical ZYMA muscles are the first to fire, already before maximal opening. The muscles becoming active next during FC are the bs MPT and ws TEM, forcing the mandible to the ws. The ws MPT and MASDEEP and the TEM of both sides fire until relatively late in the power stroke. Scapino (1981) has argued that considerable horizontal forces are needed to keep the carnassials in contact during the power stroke. Such forces could be provided by cooperation of ws TEM and bs MAS/MPT during the power stroke, but there is no evidence for such activity. In the cat there is also no evidence for an *active* medial move of the ws ramus during PS as left and right TEM and superficial and deep MAS muscles (together 90% of the jaw closers) act simultaneously.

Differences in EMG amplitude between bs and ws muscles appear to be fairly consistent, bs activity in all muscles varying between 50–80% of the ws amplitude. Dessem (1989) compared ws and bs TEM and MAS activity in dogs during carnassial biting (on bones) and chewing and found that both sides were active, the bs muscles firing at about 40–80% of the ws level in both activities. Interestingly, in contrast to the situation in primates (Sect. 6.2) the amount of bs muscle participation does not depend on the level of bite force (high in bone biting, low in chewing).

7.3 Conclusions

Carnivores show a clear range of adaptive features in their dentitions but as far as known their TMJ and arrangements of jaw muscles are rather uniform, and show only weak correlations with diet. The motor and movement patterns have only been investigated in the cat. The results confirm that only a slight transverse movement component is present; muscle activity patterns are closest to an alternate opener-closer pattern of any mammal investigated.

8 Ungulate Grinding Type

8.1 Morphology

The masticatory systems of the ungulate orders Hyracoidea, Perissodactyla and Artiodactyla as well as those of lagomorphs will now be discussed. Despite their incisors the jaw system of lagomorphs is closer to ungulates than to rodents. Ungulates and lagomorphs have a transverse power stroke. They are usually herbivorous but there is a trend toward omnivory. Herbivory (Janis and Fortelius 1988) implies (1) a greater volume of food to be consumed; (2) a more thorough mastication per unit of volume; (3) more abrasion of the tooth surface. To meet the functional demands several tooth forms have evolved from the high cusped tribosphenic teeth of ancestral therians. From bunodont teeth with low cusps and the possibility of a transverse sliding movement, a variety of lopho-dont teeth developed where individual cusps were combined to form extended cutting ridges (see Janis and Fortelius 1988, for a review). In the case of highly abrasive diets, such as grasses or fibrous materials, the cheek teeth become high-crowned and often grow indefinitely. An increase in the grinding surface area is attained when molar size and molarization of premolars increase. Although the extent of the transverse movement in the masticatory power stroke can be small in species with high-cusped dentitions and/or large canines (tapir, peccary, hippopotamus), there is a clear trend toward extension of this movement. This is made possible by the often strongly tilted, relatively flat occlusal surfaces, and by a TMJ with a fossa providing ample space for horizontal rotation of the transversely oriented, but relatively narrow condyle (Schumacher 1961; Savalle et al. 1990 for rabbits). Usually, a retroarticular process braces the condyle in its most caudal position. Its development is associated with the size and degree of posterior pull of the TEM muscle in artiodactyls (Herring 1985). In rabbits, in which most of the temporalis originates from the posterior wall of the orbit and has a fairly vertical pull (Weijs and Dantuma 1981), a retroarticular process is lacking completely. Many selenodont artiodactyls and rabbits have a patent symphysis allowing some intermandibular movement (Greaves 1978). In others, such as horse and swine, the symphysis is fused in the adult.

Radinsky (1985) identified common cranial traits in ungulates and related forms, living and fossil. The clearest size-independent feature separating these animals from carnivores/insectivores is an increased size of the angular process, indicating increased MAS and MPT size. This feature evolved at least eight times independently in ungulate evolutionary history and also occurs in lagomorphs, herbivorous marsupials and primates. In addition, ungulates are characterized by a reduced temporal fossa and coronoid process. Interestingly, development of a large jaw angle does not always coincide with elevation of the TMJ above the occlusal plane. Particularly in primitive ungulates the TMJ is still located at the level of the teeth and the TEM may have been relatively large.

Muscle weight data [(Turnbull 1970); see also Janis (1983) for hyrax and Herring (1985) for peccaries and pigs] document the great increase in MAS and MPT mass. Typically, the MAS comprises 30–50% of the jaw closer plus digastric mass, the MPT 20–30%, the LPT 3–6% and the TEM 20–30%. The digastric is relatively small (5–7%). However, in at least three lineagés (represented by horse, cow and rabbit) the TEM muscle has been even further reduced, to 10–15%. Langenbach and Weijs (1990) showed that in rabbits the reduction in size is caused by negative allometric, postnatal growth. The lines of action of MAS and MPT are quite vertical (Schumacher 1961) and situated behind the cheek teeth.

Parallel with the increase in size, the MAS muscle has increased its complexity by enlargement of the number of parasagittally oriented aponeuroses in sheep, deer and camel (Schumacher 1961) and rabbit (Weijs and Dantuma 1981). Another form of pinnation has been developed in the hyrax (Janis 1983) and some pigs (Herring 1980) showing transversely oriented tendon septa, linked to the parasagittal aponeuroses. In suids, Herring considered this an adaptation to increase muscle strength.

8.2 Motor Patterns

All investigated animals of this group use a medially directed unilateral grinding stroke in mastication. This applies to hyrax (Janis 1979), pig (Herring and Scapino 1973), goat (de Vree and Gans 1976), rabbit (Schwartz et al. 1989; Weijs et al. 1989a) and many selenodont artiodactyls and perissodactyls (see Hendrichs 1965, also for older references). Most species chew for a prolonged period at one side of the jaw, others change sides regularly (pigs) and a few (camel, giant forest hog) do so every stroke. The total amount of transverse movement as well as its timing relative to the vertical movement vary between individuals and species.

After the mandible has reached a symmetrical position in the PS, it continues to move to the bs at least in the rabbit and pig. In the pig, where the transverse distances between maxillary and mandibular teeth are the same (a condition called isognathy, in contrast to anisognathy), it is uncertain if food is being crushed at the bs; this does not happen in the anisognathous rabbit. The movement to the opposite side takes place while the jaw closers are already relaxing or relaxed, making it improbable that it plays a role in food reduction at the ws. However, in anesthetized rabbits chewing in response to electrical forebrain stimulation, this movement becomes more pronounced when material is placed between the molars (Inoue et al. 1989). In the hyrax a comparable 'phase II' movement in an anteromedial direction was found to occur after maximal occlusion (Janis 1979). In pigs at the end of FC the jaw often moves further laterally while hard food is crushed (Herring and Scapino 1973), and then moves back medially during PS. I believe, in contrast to Hiiemae (1978), that this pattern is not a gross modification of the mammalian pattern. A true two-way (medial followed by lateral, or reverse) power stroke was described for armadillos (Smith and Redford 1990).

Little information is available on the precise character of jaw movement in ungulates. Dorsoventral X-ray films showed that in the goat the movement to the ws is a pure rotation about the ws condyle, placed in a retruded position and possibly braced against the postglenoid process. As far as known the open/close rotation of the jaw coincides with anteroposterior translation (hyrax, rabbit). This implies that, like in man, the center of rotation of the mandible is below the condyle. In the rabbit we (Weijs et al. 1989b) found it consistently located at the level of the occlusal plane.

Motor patterns during mastication were investigated in goat (de Vree and Gans 1976), rabbit (Weijs and Dantuma 1981; Schwartz et al. 1989; Weijs et al. 1989a) and pig (Herring and Scapino 1973; Herring et al. 1979). In the three species there is more or less simultaneous action of ws TEM with bs MAS + MPT, swinging the jaw to the ws about the ws condyle. I will call this group triplet I. The jaw is swung back to the midline (or beyond) by action of the mirror image triplet II, i.e. the bs TEM and ws MAS + MPT. There is a large overlap between the two triplets but they start, reach peak amplitude and end with a clear phase difference.

The action of the ZYMA muscle (called deep MAS in our rabbit studies) is even more asymmetric. This muscle, located close to the joint, pulls the mandible backward in the three species. The ws ZYMA (in the pig, both ZYMAs) is the first jaw closer to become active. It starts before maximum opening and activity is confined to FC. The bs ZYMA fires largely in concert with the second triplet but the peak and end of activity are reached even later, in the last portion of the power stroke. In rabbit and pig (but apparently not in the goat) this bs ZYMA activity is accompanied by (mainly) ws LPT activity. The two muscles have very little jaw closing power and seem to effect the lingual phase PS movement taking place while the teeth disengage. In the rabbit the extent of lingual phase movement and the amplitude of bs ZYMA activity change in parallel with food: low in carrot, intermediate in pellet and high in hay mastication. The ws LPT is most active in late PS and O_1 in rabbit, pig and goat. The bs LPT is most active in O_3 in the three animals.

ZYMA and PTL muscles determine jaw movement at times in which few other muscles are active, i.e. (1) around maximal opening (ws ZYMA) or (2) after full intercuspidation is reached in the power stroke (bs ZYMA and ws PTL). Both muscles can influence the patterning of the 'free' transverse movements and have only a feeble capacity to produce bite force, because their lines of action pass close to the jaw joint. In rabbits, unlike the situation in other jaw closers, ZYMA and PTL firing amplitudes do not increase with food hardness. In all species the other main function of the muscles appears to be stabilization of the condyle.

For rabbit and pig differential activity within the structurally complex masseter has been demonstrated. In the rabbit, with masseter compartments arranged from superficial to deep, graded differences in timing also run from superficial to deep. At the bs, the more horizontal, superficial layers fire prior to the more vertical, deep fibers, at the ws the reverse happens (Weijs and Dantuma 1981). In the pig, separate anterior and posterior compartments show antero-posterior differentiation in EMG, whereby the posterior, more horizontal portion shows prolonged activity at the ws. Hence, in both animals MAS activity differentiation leads to differences in protractive forces exerted on the two hemimandibles and this produces the horizontal rotatory movements.

8.3 Conclusions

Parallel developments characterize the masticatory systems of ungulates and lagomorphs. The asymmetric action of MAS, MPT and TEM muscles, already apparent in primitive mammals, is further enhanced and cooperation of TEM muscle of one side with MAS/MPT of the other is clearer. The capacity of this triplet to produce transverse movement is increased by the much increased size of the MPT muscle. Furthermore, in the frontal plane the occlusal surfaces are often placed in such a way (the upper molars facing inward) that an almost vertical MAS/MPT force will produce a net medial displacement of the ws mandible (Becht 1953).

In addition, emphasis is placed upon a secondary mechanism to continue transverse movement while the teeth are disengaging, formed by LPT and posterior ZYMA muscles. Although the LPT is strong in primates, particularly in higher primates, and the ZYMA occurs in insectivores, primates and carnivores, the close cooperation of the two at the start of FC and end of PS may represent a new motor pattern, related to the extended horizontal plane rotation in this group. The ZYMA muscle thereby assumes the role of joint stabilizer, pulling the bs hemimandible back against the retroarticular process.

In modern ungulates a conspicuous feature of cranial morphology is the high jaw joint position. The most simple explanation for it is that this increases the available space for the MAS/MPT muscles. These muscles generate the medial PS movement of the ws hemimandible. Although in theory the bs TEM can produce the same effect, the retainment of a large TEM is mechanically unfavorable for three reasons. (1) Most of the force would have to act over a large distance and across the symphysis, making necessary massive, mandibular reinforcement. (2) Most of the vertical force would then be exerted at the bs and would put the working condyle under tension (Greaves 1978). (3) It is known

(Taylor 1990) that muscle spindles control jaw closer action during shortening, making adjustment of muscle force possible during movement perturbation. This mechanism will work best if occlusal movement is coupled rigidly to muscle movement, in other words, if the muscle is as close as possible to the teeth; this is the case for ws MAS and MPT muscles. It is striking that in many animals MASDEEP, lying very close to the molars, has the highest spindle density of all jaw muscles (Bredman et al. 1991, and references therein).

Greaves (1974) has stressed the importance of differences in distance between upper tooth row and joint, on the one hand, and lower tooth row and joint, on the other: if the distances are equal and the jaw rotates about the joint, all teeth come into occlusion simultaneously, and this would be the case in large herbivores. However, movement studies have clearly shown that the jaw joint is *not* the center of open/close rotation. This implies that in herbivores the position of the jaw joint is irrelevant in regard to the way the tooth rows meet. The translatory capacity of the condyles in their joints allows rotation of the mandible about any transverse axis. In the rabbit (Weijs et al. 1989b) the axis is consistently located at the level of the tooth row, far below the TMJ, so that the lower tooth row approximates the upper one in a vertical path. The TMJ is the site of reaction force generation and its position determines the static efficiency of the system. We (Weijs et al. 1989b) have shown that moving the rotation axis down allows rabbits to reduce the length changes in their large MAS and MPT muscles, while their high mechanical advantage is retained. The lowered position of the rotation axis causes an increase in opening-induced length changes in the small TEM muscle and this is 'countered' by a reduced coronoid process.

The herbivore group has moved far away from the primitive mammalian pattern. An increased capacity for transverse grinding movements was obtained by changes in motor program and morphology.

9 Rodents

9.1 Morphology

Rodents are more specialized in their masticatory systems than any of the other major groups and yet able to exploit almost every available food source (Landry 1970). A key feature is a pair of large ever-growing first incisors, strengthened with an anterior band of enamel and making the cropping of vegetable materials, the killing of small animals by piercing, the cutting of flesh, the handling of small food objects and the removal of bark possible (Landry 1970). Incisor action takes place independently from cheek teeth action as simultaneous occlusion is impossible. The size of the jaw muscles, particularly the MAS, has increased considerably. The length of cheek teeth row (three molars and sometimes one or two molariform premolars) is reduced; the teeth are placed far back in the mouth and the mechanical advantage of the jaw muscles is very high (unity in the rat, Weijs and Dantuma 1975). Wood (1965) recognizes three grades of rodent masticatory adaptation, occurring in parallel in many taxa. The first grade is the acquisition of a pair of ever-growing incisors, while molars are still cuspidate. Although the MAS is well developed, there is still a large TEM muscle. Primitive paramyid and sciurid rodents, but also more advanced species such as the hamster, follow this pattern. According to Butler (1984), these rodents have a horizontally oriented power stroke with a medially directed buccal phase and an obliquely forward-directed lingual phase; they were omnivores. The second grade of development is characterized by lophodont teeth where complicated systems of flat enamel ridges evidence adaptation to a vegetable diet. In a number of independent lineages these teeth become high-crowned or even indefinitely growing and this is the third grade of development. Moreover, a number of myomorph rodents are insectivorous or carnivorous and have high-cusped teeth reminiscent of those of primitive mammals.

The TMJ of rodents is characterized by an anteroposteriorly oriented fossa without pre- and postglenoid processes and usually a similarly elongated condyle. The possibilities for anteroposterior translation and transverse rotations are rather large. The mandibular symphysis can be either patent or fused. The cheek tooth rows vary with respect to their degree of anisognathy; the upper molars are closer together than the lowers. The jaw movements near occlusion can be restricted by the cusp pattern of individual teeth or by the orientation of the occlusal surfaces. In most rodents the upper occlusal surfaces face outward and the lower surfaces inward. Lingual occlusal movement at one side produces separation of the occlusal surfaces at the other side. Furthermore, in many species (for instance, beavers, voles; Kesner 1980) both upper and lower occlusal surfaces are curved in lateral view, forcing the jaw along a fixed anteroposterior movement path. The molar rows can be either parallel or can converge anteriorly. Finally, in low-cusped teeth, e.g. of rats, longitudinal ridges and valleys are formed by the cusps on the upper and lower teeth, fitting reciprocally and guiding jaw motion forward.

The weight of the jaw muscles in rodents relative to skull size or body mass exceeds that of most other mammals. An extensive review of rodent jaw muscles is given by Tullberg (1899). The relative weight data of Turnbull (1970) show the great development of the MAS muscle, occupying 55–70% of the jaw closer plus digastric mass. Hurov et al. (1988) have shown that at least in the rat this mass develops by a strongly positive, allometric, postnatal growth, particularly in the masseter. Rodents share the relatively large mass of the MAS with ungulates. However, in rodents of grade 2 the muscle has developed in a unique way by showing: (1) a forward shift relative to the cheek teeth battery, so that the latter is, in side view, completely or almost completely covered by the MAS. In some species (e.g. mole rats) the anterior margin of the masseter reaches almost as far as the incisors. (2) An increase in inclination of the muscle to a strongly protractive position. In primitive (grade 1) rodents the MAS is neither forwardly displaced nor exceptionally large and the TEM and coronoid process are well developed. In the other rodent groups either the MASSUP and MASDEEP originate far forward from the side of the maxilla underneath and in front of the orbit (sciuromorphs) or the anterior portion of the ZYMA (infra-orbital) does so via a greatly enlarged infraorbital foramen (hystricomorphs) or both developments take place (myomorphs). These developments have occurred independently in different lineages so that sciuromorphs, hystricomorphs and myomorphs are no longer considered as monophyletic groups (Wood 1965). Hystricomorphs, such as *Cavia*, show the greatest development and differentiation of the MAS and reduction of the TEM. In *Pedetes*, the springhare, having uncertain affinities to these major groups, the size of the TEM has been reduced to a sparse 3% of the total jaw closer mass. The MPT and LPT muscles are less strongly developed than in ungulates but still account for approximately 10–20% of the jaw closer plus digastric mass (Turnbull 1970). The MPT does not follow the extreme horizontal orientation of the MAS. The space for this muscle is reduced by the posteriorly placed molars. Because of its vertical orientation the line of action passes closely in front of the jaw joint. The contribution of this muscle to the jaw closing moment is small (see Turnbull's 'useful power' values). In *Cavia* the muscle can even act as a jaw opener as soon as the mandible is protracted (Chen and Herring 1986).

Rodent skull shape is extremely variable and in part determined by life-style (aquatic, fossorial), size of the nasal cavity, middle ear and orbit, and the often very large roots of their incisors. Dietary habits can modify the occlusal pattern of the cheek teeth. The basic layout of the jaw muscles is rather determined by group membership. For example, differences in molar structure, but not in jaw muscle structure, were found in a comparative study of two myomorphs, the insectivorous *Deomys* and omnivorous *Rattus* (Lemire 1966).

9.2 Motor Patterns

Rodents use their incisors in prolonged bites to chip off particles from a larger object. In rats, the mandible can thereby vary in its anteroposterior position (Weijs 1975). There is alternating activity of the digastric and jaw closing muscles, while the LPT fires continuously at high levels (Weijs and Dantuma 1975). In addition, cyclic ingestive movements in which incisor occlusion is often reached are common in guinea pig (Byrd 1981), rat (Hiiemae and Andran, 1968), hamster (Gorniak 1977), springhare (Offermans and de Vree 1990) and mole rat (Meirte 1986). These movements are different for different foods and species and are often faster than masticatory movements. In lateral view the incisor path is orbital. After reaching edge-to-edge occlusion the lower incisors translate backward along the lingual bevel of the upper incisors. In *Cavia*, alternate activity of openers and closers has been reported while LPT activity was high throughout the cycle; in *Pedetes* PTL and digastric activity alternate (Offermans 1990).

After ingestion food is transported across the diastema to the molar region by the action of the tongue against the hard palate. In the rat and mole rat the transport cycles are accompanied by fast repetitive jaw movements. In the rat they are brought about by rhythmic firing of the jaw openers while jaw closers remain silent.

Offermans and de Vree (1990) generalize the existing information on chewing movements as follows. The presumedly primitive pattern is present in the more or less isognathous mountain beaver (*Aplodontia*), hamster (*Mesocricetus*) and squirrel (*Eutamias*). It involves unilateral mastication and is not principally different from the primitive mammalian pattern. Cheek teeth intercuspidation and incisor placement leave no possibilities for a forward power stroke. The movement of the ws hemimandible is from lateral to medial and starts either from a symmetrical position of the mandible (hamster) or from slightly lateral to that position. On the basis of occlusal patterns, Butler (1984) concludes that some vertical, closing movement is involved in the buccal phase, and opening movement in the lingual phase of the power stroke, similar to the situation in insectivores. Unilateral chewing is also known to occur with some foods in the anisognathous *Cavia* (Byrd 1981). In this case a more extensive medial movement, starting when the jaw is in the midline, is accompanied by an anterior movement of about half this magnitude. It is highly likely that medial and forward PS movement is combined in the majority of anisognathous rodents.

A second pattern, occurring in strictly isognathous rats and springhares and some hystricomorphs (Woods and Howland 1979), is bilateral mastication. The lower molars move forward across the uppers over a considerable distance (about 4 mm in springhares and 2 mm in rats). The movement is guided by a pattern of longitudinal ridges and valleys formed by the molars. It has been claimed (Byrd 1988) that rats chew alternately at the left and right sides but their frontal view jaw trackings show less than 0.3 mm of transverse travel of the incisors between up- and downstrokes and even less, if any, transverse travel during the PS. This implies that grinding must take place at both sides simultaneously. Electromyographically, however, alternating contraction patterns can sometimes be seen (Weijs and Dantuma 1975) and alternating left and right jaw excursion can be seen in rats during other oral activities. The springhare has an immobile symphysis but in the rat symphyseal movement occurs during chewing: both rami move inward during the power stroke, adding a slight lingual component to the forward movement.

The third pattern is alternating left- and right-sided mastication. It has been found in hystricomorph rodents such as *Cavia*, *Myocastor* and *Hydrochoerus* (Müller 1933) and in the myomorph mole rat, *Tachyoryctes* (Meirte 1986). In these animals the occlusal planes are relatively flat and the tooth rows converge anteriorly. The symphysis can be either mobile or immobile. Occlusal movement consists of a large transverse and a smaller forward component.

In none of the investigated rodents is there much evidence for a clearly marked slow opening movement, although tongue displacements similar to those of primitive mammals have been described. During opening and closing, there are corresponding pro- and retrusions of the condyle. Phase differences between rotation and these translations lead to differently shaped orbits in lateral view. In animals with a strong protrusive component in their PS, retrusion takes place during FC.

The motor pattern of the unilateral chewing cycles of the hamster has much in common with that of primitive mammals (Gorniak 1977). Jaw closure occurs in the midline. It is brought about by symmetric contraction of all jaw closers. The muscles of triplet II (ws MAS + MPT, bs TEM) dominate PS movement and pull the jaw to the bs. The ws LPT is active throughout PS till early jaw opening. The free movements of the jaw are determined by asymmetric activity of LPTs and digastrics.

In the rat (Weijs and Dantuma 1975) and springhare (Offermans 1990) jaw muscle activity is usually bilaterally symmetric. The TEM, MASDEEP and ZYMA muscles fire during FC. The large protractive MASSUP muscles lag considerably behind in action and supply the main power to the grinding movement. Jaw closing muscle activity is sometimes also found around maximal opening in the infraorbital portion of the ZYMA and reflected portion of the MASSUP (this also occurs in the mole rat of group 3). The MPTs are active together with the LPTs during the power stroke and early jaw opening. In the rat O_{12} activity was found in the geniohyoid and mylohyoid muscles, prior to digastric activity, in accordance with the basic mammalian pattern.

The alternating unilateral chewing motor pattern has been studied in *Cavia* (Byrd 1981) and the (taxonomically unrelated) mole rat (*Tachyoryctes*, Meirte 1986). The movement pattern is complicated because opening paths do not coincide for left and right cycles. The result is horizontal rotation of the jaw with half the frequency of the open-close movement. From the extreme open position at the left, for instance, the jaw starts moving to the right while closing, goes through a left-side power stroke and moves further to the right until maximum opening. The jaw moves back to the left while closing, goes through a right-side power stroke until, at maximum opening, it has completed its horizontal cycle. Movement and motor pattern are surprisingly similar for the two animals and quite different from other mammals. It is possible, however, that other alternately chewing mammals, such as the giant forest hog (see Hiiemae 1978) and camel, have a similar pattern. Motor activity is very asymmetric; the MAS and MPT muscles are most active at the ws, the TEM at the bs. Cooperation of the muscles of triplet I, followed by those of triple II, can be seen in both species. Small differences in timing between the triplets exist, but, more dramatically, the amplitude of the muscles triplet I has fallen to very low levels, in *Cavia* even virtually to zero. This motor pattern is clearly related to the absence of a movement of the jaw toward the ws during closing. Because the upper cheek tooth rows are so close together, midline closure of the jaw already leads to an approach of the lower molars to the lateral side of the upper molars. It should be noted that both animals regularly chew at a single side for a short time. In mole rats these cycles are similar to the ones seen in alternating mastication; in guinea pigs they appear more like 'normal' mammalian cycles, including displacement of the jaw back to the midline during opening. In the latter case, the ws MPT acts during late PS and jaw opening to disengage the teeth and effect jaw opening.

9.3 Conclusions

In primitive rodents (first grade of Wood) jaw movements are characterized by the typical mammalian, medially directed power stroke; the masticatory motor pattern is slightly asymmetric in timing and amplitude and similar to that of primitive mammals.

Two motor patterns evolved from this pattern. In myomorphs (rat, mole rat) and some hystricomorphs a bilateral, predominantly or exclusively forward (propalinal) power stroke developed, with little asymmetry and simultaneous action of both triplets. The molar rows become important in guiding the occlusal movement, produced by early TEM and late MASSUP activity. Free mandibular movements are produced by digastric and LPT muscles, while anterior ZYMA and the pars reflexa portion of MAS play a role in checking these movements. It is likely that intermediate masticatory patterns between the first two exist, characterized by a unilateral, obliquely forward PS.

In other hystricomorphs (*Cavia, Myocaster*) and myomorphs (*Tachyoryctes*) a long, obliquely forward and medial power stroke is used in alternating left- and right-sided mastication. The maxillary occlusal planes slope outward and downward and make simultaneous occlusion at both sides impossible. Power stroke action becomes a pure rotation in the occlusal plane, taking place (in the mole rat, at least) about an oblique axis through a stabilized bs condyle. The rotating moment is produced by triplet II, the other triplet fires at a very low level. The movement pattern integrates two chews in such a way that each jaw muscle fires strongly only once every two cycles.

In both myomorphs and hystricomorphs the protractive character of the musculature is strongly enhanced and serves to move the mandible in a forward translation or a horizontal rotation. The horizontal position of MAS and MPT muscles increases the possibilities to attain large gapes, necessary for incisor function (Herring and Herring 1974). Woods and Howland (1979) compared bilateral and alternately chewing in capromyid hystricomorph species and found that the latter have less distinctly protractive MAS and MPT and less retractive TEM muscles, while the LPTs were better developed and the digastrics deviated considerably more from the sagittal plane.

10 Discussion

10.1 Cranial Muscle Cooperation

Jaw, tongue and hyoid movements in mammals involved in feeding are produced by a relatively uniform motor pattern, including the jaw closer, infra- and suprahyoid and tongue muscles. A food transport cycle, as seen in lapping, eating of soft food and perhaps in suckling, consists of a slow close phase characterized by hyoid and tongue retraction and little jaw closer activity and a slow open phase, with hyoid and tongue protraction and suprahyoid muscle activity (Hiiemae and Crompton 1985). During chewing of solid food, the cycles are modified in two ways. First, maximum gape is increased by addition of extra opening (O_3) and fast closing (FC) phases. O_3 is produced by simultaneous action of infra- and suprahyoid muscles, supported by activity of the neck extensors and often the LPT muscles. Jaw closers, particularly the MPT, can also be active. Activity of digastrics and LPT muscles is often asymmetric. Thexton and Crompton (1989) suggested that solid food triggers the simultaneous action of supra- and infrahyoid muscles in a jaw opening reflex. This would effect clearing of the food bolus from the palate, so that tongue retraction can start. Second, jaw closing muscles are much more active than in transport cycles, and a power stroke movement replaces the slow closing movement. Differential jaw closer activity determines the precise movement.

The motor programs involved are stored in the brainstem; they differ between species. Forebrain stimulation in different areas in a single species elicits different motor patterns, including puncture-crushing cycles, grinding cycles, gnawing cycles and perhaps transport cycles (small amplitude, vertical jaw movements) at least in the rabbit (Lund et al. 1984). In this animal further modification of movement profiles in automatic grinding cycles occurs under the influence of periodontal and muscle spindle receptors (Inoue et al. 1989).

10.2 Jaw Closer Cooperation

Jaw closer asymmetry effects unilateral occlusion and a medially directed power stroke. The primitive mammalian condition can be derived from fossil evidence and is presently found in generalized marsupials, insectivores, insectivorous bats, lower primates, generalized rodents, generalized ungulates and some carnivores (bears, raccoons). To effect an occlusal slide between the cusped molars, the jaw is rotated horizontally toward the ws during closure. As the teeth start to interlock, they are guided lingually by the jaw closers and by the cusp pattern.

I have attempted to group masticatory muscles according to function (Fig. 1) by separating muscles that (1) fire early during FC, in a symmetric fashion (symmetric closers); (2) move the jaw toward the bs during FC (triplet I); and (3) move the jaw toward the ws during PS (triplet II). In primitive mammals (see above) these groups can be discerned and fire in an overlapping but usually distinct sequence. The symmetric closers are the ZYMA (if present) and usually also the TEMAN and TEMDEEP muscles, in other words, the vertically oriented muscles. They fire from the start of FC in tenrec, hamster, fruit bat, cat, goat and pig. In the little brown bat, rabbit and human there is no symmetric closing activity, presumbly because the jaw moves to the ws from maximum opening. It is tempting to hypothesize that these muscles are controlled by jaw muscle spindles, behaving as passive length receptors and firing about maximal opening (Fig. 7.7 in Taylor 1990).

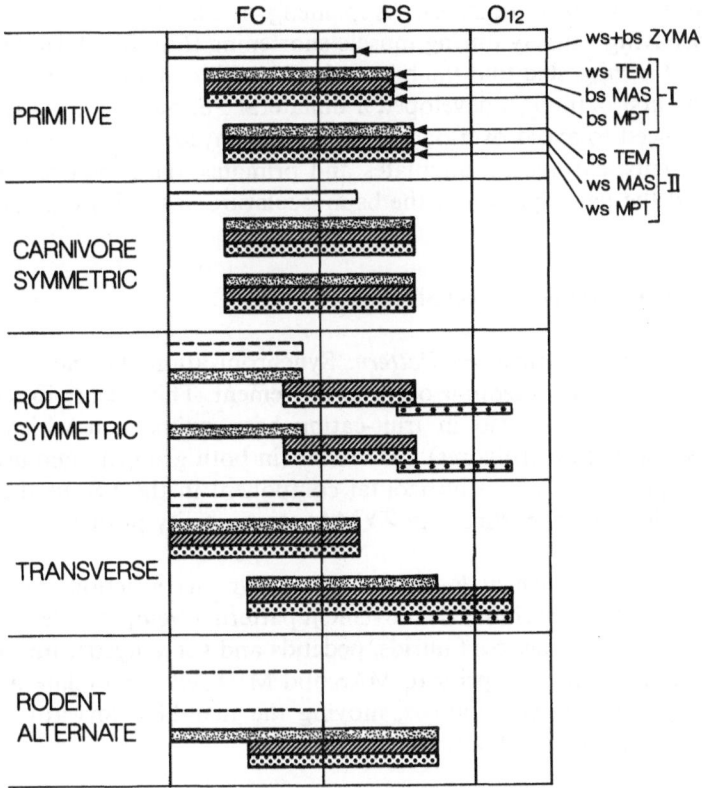

Fig. 1. Diagram of five mammalian jaw closer motor patterns during mastication. Of the masticatory cycle only fast closing (*FC*), power stroke (*PS*) and initial opening (O_{12}) have been represented and activity of jaw closers during O_{12}, being often inconsistent and variable, has been ignored. The muscles have been subdivided into four groups. From *top to bottom* the activity of the symmetric closers (*ZYMA* and other vertically oriented muscles; *open bar*), triplet I and triplet II muscles is indicated. In addition, in rodents and transverse grinders, working-side lateral pterygoid activity is indicated by *asterisk-filled bars. Bars with dashed outlines* represent very low level or inconsistent muscle activity

Triplet I (ws TEM and bs MASSUP + MPT) acts during FC; horizontal rotation back to the midline starts late in FC and continues through PS by action of triplet II (bs TEM and ws MASSUP + MPT). These patterns can be best recognized from differences in onset, peak and end of EMG activity between ws and bs muscles. A bs lead is almost universally present for MASSUP and MPT, a ws lead for TEM. The amount of asynchrony is associated with the degree of transverse movement during PS. It is large in the little brown bat and tenrecs and small in the opossum. The actions of both triplets involve both movement steering and exertion of bite force. They can therefore be controlled by both a positive periodontal receptor feedback, known to be especially strong at the FC/PS transition (Appenteng et al. 1982), and by muscle spindles behaving actively during muscle shortening (Fig. 7.7 in Taylor 1990).

It is suggested that the bilateral contraction of jaw closers was retained when the first mammals developed a unilateral PS, but the firing level at the bs was reduced to avoid overloading of the symphysis. In many groups, such as some rodents, carnivores, ungulates and primates, the symphysis was strengthened and the participation of the bs musculature in the PS was increased.

10.3 Mammalian Specializations

1. Carnivore Symmetric Pattern. Synchronization of triplets I and II results in a reduction of horizontal occlusal movement. This occurs in felid and mustelid carnivores, but also in fruit-eating bats, animals in which the amount of carnassial (in carnivores) and canine (in both groups) intercuspidation reduces the possibilities of a horizontal component in the PS. In these animals symmetric closers (particularly ZYMA) start activity prior to the triplet muscles.

2. Rodent Symmetric Pattern. A purely symmetric action of jaw closers, leading to a bilaterally symmetric movement pattern developed independently in at least three rodent lineages (murids, pedetids and some hystricomorphs). TEM muscles fire during FC prior to MAS and MPT, effecting bilateral retrusion. MAS and PTM activity follows, moving the mandible forward during FC while bilateral grinding takes place.

3. Transverse Pattern. Further separation in time of action of triplet I and II muscles has developed independently in many ungulates, lagomorphs and higher primates. These animals have flattened (lophodont or bunodont) occlusal planes. The tooth rows are usually anisognathic, the upper rows being further separated than the lower ones. Because from maximal opening the jaw starts to swing out to the ws, triplet I acts as a fast closing group as well as a mover to the ws; no symmetric action of the vertical jaw closers can be seen in macaque, man and rabbit. The onset of activity of the second triplet could be determined by the establishment of tooth-food contacts at the beginning of the power stroke. The triplet II muscles often fire more strongly than the triplet I muscles. The ws MPT and LPT muscles are the last muscles to stop firing; their activity often extends into O_1 while the mandible crosses the plane of symmetry and the ws occlusal surfaces disengage. Such disengagement is effected particularly by ws pterygoids (displacing the ws hemimandible medially) and bs deep masseter (displacing bs hemimandible laterally).

4. Alternate Pattern. Reduction of symmetric closer and triplet I activity to very low levels leads to the most extreme form of horizontal mastication, with dominating triplet II activity. This motor pattern has evolved independently in at least two rodent lineages. These animals usually chew alternately at the left and right side, but sometimes mastication at one side occurs during a few cycles. This is a development from chewing with rapidly alternating ws and bs, as seen in primitive mammals and rodents. The functional advantage of this system is the reduction of the number of (strong) contractions of the jaw closers to one per two cycles, of course at the cost of a smaller total chewing force per bite. It would be interesting to know whether camel and giant forest hog, also chewing alternately, do so by completely asymmetrizing the actions of their jaw muscles.

10.4 Evolution and Adaptation

The key feature of mammalian mastication is unilateral occlusal movement consisting of an inward horizontal rotation plus hemimandibular inversion, produced by differential use of protrusive and retrusive forces provided by MAS, MPT and TEM muscles of both sides. This pattern is found in generalized members of most investigated orders. The relative uniformity of the mammalian jaw closer motor pattern has been pointed out before (Hiiemae 1978; Gorniak 1985). Horseradish peroxidase labeling has demonstrated a somatotopic organization of the mammalian trigeminal motor nucleus of a surprisingly uniform character, in guinea pig (Tal 1980; Uemura-Sumi et al. 1982), rat (Rokx et al. 1985 and references therein), rabbit (Matsuda et al. 1978), cat (Mizuno et al. 1975) and macaque (Mizuno et al. 1981). In some species, including rat, rabbit and pig (Chap. 1, this Vol) differentiation of the MAS muscle is reflected in its motoneuron organization.

The diversity in the morphology of jaw systems, on the other hand, is impressive and must be crucial in determining its adaptation to diet. However, only few direct links between morphology and diet have been demonstrated. Herbivory induces flattening of molar cusps and an increase in molar grinding surface. Furthermore, a consistent trend in herbivores belonging to many different mammalian orders is the increase in MAS size, relative to TEM and the increase in elevation of the TMJ above the occlusal plane. A comparison of omnivores such as opossum, raccoon, rat, macaque and pig leaves the definite impression that mammalian order membership exerts a far more profound influence upon the masticatory system layout than alleged uniformity of diet. Other important functions, such as locomotion (head posture), vision, hearing, olfaction and brain volume, characteristically developed in different orders, may heavily constrain the possibilities of masticatory system development. Rodents are the only group where the masticatory system truly dominates the skull, particularly if large incisors have to be accomodated and powered.

A limited number of new motor patterns have developed from the primitive pattern, involving (1) left/right synchronization, in the case of *Pteropus* and *Felis* due to increased vertical tooth overlap and guidance and in the case of some rodents due to the development of bilateral occlusion; (2) enhancement of left/right asynchrony in the case of unilateral mastication involving an extended, occlusal plane grinding movement in many ungulates, lagomorphs and higher primates. The pattern becomes almost 100% asymmetric in the alternate masticatory cycles that developed independently in several rodent and ungulate forms.

Hence, the evolutionary scenario of masticatory specialization is one of parallel development. The primitive tribosphenic teeth, the differentiation into incisors, canines, premolars and molars and the general layout of non-parallel masticatory muscles provided enough flexibility to utilize most food resources. Emphasis on carnivory or herbivory led to parallel changes in motor pattern and morphology in the same direction in the different mammalian orders.

Acknowledgments. I am very grateful to Prof. W.L. Hylander and Prof. F. de Vree for providing me with unpublished manuscripts, to H. Rolleman-ter Heurne for preparing the manuscript and to J. H. van Horssen-Medema for preparing the figure.

References

Allin EF (1975) Evolution of the mammalian middle ear. J Morphol 147: 403–438

Anapol F (1988) Morphological and videofluorographic study of the hyoid apparatus and its function in the rabbit (*Oryctolagus cuniculus*). J Morphol 195: 141–157

Appenteng K, Lund JP, Seguin JJ (1982) Intraoral mechanoreceptor activity during jaw movement in the anaesthetized rabbit. J Neurophysiol 48: 27–37

Ardran GM, Kemp FH, Ride WDL (1958) A radiographic analysis of mastication and swallowing in the domestic rabbit: *Oryctolagus cuniculus* (L). Proc Zool Soc Lond 130: 257–274

Arendsen de Wolf-Exalto E (1951) On differences in the lower jaw of animalivorous and herbivorous mammals. Proc K Ned Akad Wet C 54: 237–246, 405–410

Barghusen HR (1968) The lower jaw of cynodonts (Reptilia, Therapsida) and the evolutionary origin of mammal-like adductor jaw musculature. Postilla 116: 1–49

Barghusen HR (1973) The adductor musculature of *Dimetrodon* (Reptilia, Pelycosauria). J Paleontol 47: 823–834

Becht G (1953) Comparative biologic-anatomical researches on mastication in some mammals. Proc K Ned Akad Wet C 56: 508–527

Beecher RM (1979) Functional significance of the mandibular symphysis. J Morphol 159: 117–130

Belser UC, Hannam AG (1986) The contribution of the deep fibers of the masseter muscle to selected tooth-clenching and chewing tasks. J Prosthet Dent 56: 629–635

Biegert J (1956) Das Kiefergelenk der Primaten. Morphol Jahrb 97: 249–404

Blanksma NG, van Eijden TMGJ (1990) Electromyographic heterogeneity in the human temporalis muscle. J Dent Res 69: 1686–1690

Bluntschli H (1929) Die Kaumuskulatur des Orang-Utan und ihre Bedeutung für die Formung des Schädels. Morphol Jahrb 63: 531–606

Bramble DM (1978) Origin of the mammalian feeding complex: models and mechanisms. Paleobiology 4: 271–301

Bramble DM, Wake DB (1985) Feeding mechanisms of lower tetrapods. In: Hildebrand M, Bramble DM, Liem KF, Wake DB (eds) Functional vertebrate morphology. Harvard Univ Press, Cambridge, pp 230–261

Bredman JJ, Weijs WA, Brugman P (1991) Relationships between spindle density, muscle architecture and fibre type composition in different parts of the rabbit masseter. Eur J Morphol 29: 297–307

Butler PM (1984) Homologies of molar cusps and crests, and their bearing on assessments of rodent phylogeny. In: Luckett WP, Hartenberger JL (eds) Evolutionary relationships among rodents. NATO ASI Series A, vol 92. Plenum Press, New York, pp 381–401

Byrd KE (1981) Mandibular movement and muscle activity during mastication in the guinea pig (*Cavia porcellus*). J Morphol 170: 147–169

Byrd KE (1988) Opto-electronic analysis of masticatory mandibular movements and velocities in the rat. Arch Oral Biol 33: 209–215

Byrd KE, Garthwaite CR (1981) Contour analysis of masticatory jaw movements and muscle activity in *Macaca mulatta*. Am J Phys Anthropol 54: 391–399

Byrd KE, Milberg DJ, Luschei ES (1978) Human and macaque mastication: a quantitative study. J Dent Res 57: 834–843

Cachel SM (1979) A functional analysis of the primate masticatory system and the origin of the anthropoid post-orbital septum. Am J Phys Anthropol 50: 1–18

Cachel S (1984) Growth and allometry in primate masticatory muscles. Arch Oral Biol 29: 287–293

Chen EK, Herring SW (1986) An unusual function for the medial pterygoid muscle in the guinea pig. Arch Oral Biol 31: 781–783

Clemens WA (1971) Mesozoic evolution of mammals with tribosphenic dentitions. In: Dahlberg AA (ed) Dental morphology and evolution. Univ Chicago Press, Chicago, pp 181–192

Cortopassi D, Muhl ZF (1990) Videofluorographic analysis of tongue movement in the rabbit (*Oryctolagus cuniculus*). J Morphol 204: 139–146

Crompton AW (1963) On the lower jaw of *Diarthrognathus* and the origin of the mammalian lower jaw. Proc Zool Soc Lond 140: 697–753

Crompton AW, Hiiemae K (1970) Molar occlusion and mandibular movements during occlusion in the American opossum, *Didelphis marsupialis* L. Zool J Linn Soc 49: 21–47

Crompton AW, Hylander WL (1986) Changes in mandibular function following the acquisition of a dentary-squamosal jaw articulation. In: Hotton N, MacLean PD, Roth JJ, Roth EC (eds) The ecology and biology of mammal-like reptiles. Smiths Inst Press, Washington DC, pp 263–282

Crompton AW, Thexton AJ, Parker P, Hiiemae K (1977) The activity of the jaw and hyoid musculature in the Virginian opossum, *Didelphis virginiana*. In: Gilmore D, Stonehouse B (eds) Biology of the marsupials, vol 2. MacMillan, London, pp 287–305

Czarnecki RT, Kallen FC (1980) Craniofacial, occlusal, and masticatory anatomy in bats. Anat Rec 198: 87–105

Davis DD (1955) Masticatory apparatus in the spectacled bear *Tremarctos ornatus*. Fieldiana Zool 37: 25–45

Davis DD (1964) The giant panda – a morphological study of evolutionary mechanisms. Fieldiana Zool 3: 1–339

de Gueldre G, de Vree F (1984) Movements of the mandibles and tongue during mastication and swallowing in *Pteropus giganteus* (Megachiroptera): a cineradiographical study. J Morphol 179: 95–114

de Gueldre G, de Vree F (1988) Quantitative electromyography of the masticatory muscles of *Pteropus giganteus* (Megachiroptera). J Morphol 196: 73–106

de Gueldre G, de Vree F (1990) Biomechanics of the masticatory apparatus of *Pteropus giganteus* (Megachiroptera). J Zool (Lond) 220: 311–332

DeMar R, Barghusen HR (1973) Mechanics and the evolution of the synapsid jaw. Evolution 26: 622–637

Dessem D (1989) Interactions between jaw-muscle recruitment and jaw-joint forces in *Canis familiaris*. J Anat 164: 101–121

de Vree F, Gans C (1976) Mastication in pygmy goats (*Capra hircus*). Ann Soc R Zool Belg 105: 255–306

Dötsch C (1982) Der Kauapparat der Soricidae (Mammalia, Insectivora). Zool Jahrb Anat 108: 421–484

Dötsch C (1983a) Morphologische Untersuchungen am Kauapparat der Spitzmäuse *Suncus murinus* (L.), *Soriculus nigrescens* (Gray) und *Soriculus caudatus* (Horsfield) (Soricidae). Säugetierkd Mitt 31: 27–46

Dötsch C (1983b) Das Kiefergelenk der Soricidae (Mammalia, Insectivora). Z Säugetierkd 1: 65–77

Dötsch C, Dantuma R (1989) Electromyography and masticatory behavior in shrews (Insectivora). Prog Zool 35: 146–147

Eisenberg JF (1981) The mammalian radiations. Univ Chicago Press, Chicago

Emerson SB, Radinsky L (1980) Functional analysis of sabertooth cranial morphology. Paleobiology 6: 295–312

Endo B (1965) Distribution of stress and strain produced in the human facial skeleton by the masticatory force. J Anthropol Soc Jpn 73: 9–22

Ewer RF (1973) The carnivores. Weidenfeld & Nicolson, London

Fiedler W (1953) Die Kaumuskulatur der Insectivora. Acta Anat 18: 101–175

Fish DR (1983) Aspects of masticatory form and function in common tree shrews, *Tupaia glis*. J Morphol 176: 15–29

Fish DR, Mendel FC (1982) Mandibular movement patterns relative to food types in common tree shrews (*Tupaia glis*). Am J Phys Anthropol 58: 255–269

Fleagle JG (1988) Primate adaptation and evolution. Academic Press, San Diego

Franks HA, Crompton AW, German RZ (1984) Mechanism of intraoral transport in macaques. Am J Phys Anthropol 65: 275–282

Gaspard M, Laison F, Lautrou A (1976) Le plan général d'organisation de la musculature masticatrice chez les mammifères. Acta Odontostomatol (Paris) 113: 65–100

German RZ, Franks HA (1991) Timing in the movement of jaws, tongue, and hyoid during feeding in the hyrax, *Procavia syriacus*. J Exp Zool 257: 34–42

Gibbs CH, Messerman T, Reswick JB, Derda HJ (1971) Functional movements of the mandible. J Prosthet Dent 26: 604–620

Goldberg LJ, Chandler SH (1990) Central mechanisms of rhythmical trigeminal activity. In: Taylor A (ed) Neurophysiology of the jaws and teeth. McMillan Press, Houndmills, pp 268–293

Gorniak GC (1977) Feeding in golden hamsters, *Mesocricetus auratus*. J Morphol 154: 427–458

Gorniak GC (1985) Trends in the actions of mammalian masticatory muscles. Am Zool 25: 331–337

Gorniak GC (1986) Architecture of the masticatory apparatus in eastern raccoons (*Procyon lotor lotor*). Am J Anat 176: 333–351

Gorniak GC, Gans C (1980) Quantitative assay of electromyograms during mastication in domestic cats (*Felis catus*). J Morphol 163: 253–281

Greaves WS (1974) Functional implications of mammalian jaw joint position. Forma Functio 7: 363–376

Greaves WS (1978) The jaw lever system in ungulates: a new model. J Zool (Lond) 184: 271–285

Gysi A (1921) Studies on the leverage problem of the mandible. Dent Digest 27: 74–84, 144–150, 203–208

Hannam AG, Wood WW (1981) Medial pterygoid muscle activity during the close and compressive phases of human mastication. Am J Phys Anthropol 55: 359–367

Hannam AG, De Cou RE, Scott JD, Wood WW (1977) The relationship between dental occlusion, muscle activity and associated jaw movement in man. Arch Oral Biol 22: 25–32

Hendrichs H (1965) Vergleichende Untersuchung des Wiederkauverhaltens. Biol Zentralbl 84: 651–751

Herring SW (1980) Functional design of cranial muscles: comparative and physiological studies in pigs. Am Zool 20: 283–293

Herring SW (1985) Morphological correlates of masticatory patterns in peccaries and pigs. J Mammal 66: 603–617

Herring SW, Herring SE (1974) The superficial masseter and gape in mammals. Am Nat 108: 561–575

Herring SW, Scapino RP (1973) Physiology of feeding in miniature pigs. J Morphol 141: 427–460

Herring SW, Grimm AF, Grimm BR (1979) Functional heterogeneity in a multipinnate muscle. Am J Anat 154: 563–576

Hiiemae KM (1978) Mammalian mastication: a review of the activity of the jaw muscles and the movements they produce in chewing. In: Butler PM, Joysey KA (eds) Development, function and evolution of teeth. Academic Press, London, pp 359–398

Hiiemae KM, Ardran GM (1968) A cinefluorographic study of mandibular movement during feeding in the rat (*Rattus norvegicus*). J Zool (Lond) 154: 139–154

Hiiemae K, Crompton AW (1971) A cinefluorographic study of feeding in the American opossum, *Didelphis marsupialis*. In: Dahlberg AA (ed) Dental morphology and evolution. Univ Chicago Press, Chicago, pp 299–334

Hiiemae KM, Crompton AW (1985) Mastication, food transport, and swallowing. In: Hildebrand M, Bramble DM, Liem KF, Wake DB (eds) Functional vertebrate morphology. Harvard Univ Press, Cambridge, pp 262–290

Hiiemae K, Jenkins FA (1969) The anatomy and internal architecture of the muscles of mastication in *Didelphis marsupialis*. Postilla 140: 1–49

Hiiemae KM, Thexton AJ, McGarrick J, Crompton AW (1981) The movement of the cat hyoid during feeding. Arch Oral Biol 26: 65–81

Hildebrand GY (1931) Studies in the masticatory movements in the human lower jaw. Scand Arch Phys Suppl 61

Horio T, Kawamura Y (1989) Effects of texture of food on chewing patterns in the human subject. J Oral Rehabil 16: 177–183

Hurov J, Henry-Ward W, Phillips L, German R (1988) Growth allometry of craniomandibular muscles, tendons, and bones in the laboratory rat (*Rattus norvegicus*): relationships to oromotor maturation and biomechanics of feeding. Am J Anat 182: 381–394

Hylander WL (1975) Incisor size and diet in anthropoids with special reference to Cercopithecidae. Science 189: 1095–1098

Hylander WL (1979a) The functional significance of primate mandibular form. J Morphol 160: 223–240

Hylander WL (1979b) Mandibular function in *Galago crassicaudatus* and *Macaca fascicularis*: an in vivo approach to stress analysis of the mandible. J Morphol 159: 253–296

Hylander WL (1981) Patterns of stress and strain in the macaque mandible. Craniofac Biol 10: 1–35

Hylander WL (1984) Stress and strain in the mandibular symphysis of primates: a test of competing hypotheses. Am J Phys Anthropol 64: 1–46

Hylander WL, Crompton AW (1986) Jaw movements and patterns of mandibular bone strain during mastication in the monkey *Macaca fascicularis*. Arch Oral Biol. 31: 841–848.

Hylander WL, Johnson KR (1985) Temporalis and masseter muscle function during incision in macaques and humans. Int J Primatol 6: 289–322

Hylander WL, Johnson KR, Crompton AW (1987) Loading patterns and jaw movements during mastication in *Macaca fascicularis*: a bone-strain, electromyographic, and cineradiographic analysis. Am J Phys Anthropol 72: 287–314

Hylander WL, Johnson KR, Crompton AW (1992) Muscle force recruitment and biomechanical modelling: an analysis of masseter muscle function during mastication in *Macaca fascicularis*. Am J Phys Anthropol 88: 365–387

Inoue T, Kato T, Masuda Y, Nakamura T, Kawamura Y, Morimoto T (1989) Modifications of masticatory behavior after trigeminal deafferentation in the rabbit. Exp Brain Res 74: 579–591

Janis CM (1979) Mastication in the hyrax and its relevance to ungulate dental evolution. Paleobiology 5: 50–59

Janis CM (1983) Muscles of the masticatory apparatus in two genera of hyraces (*Procavia* and *Heterohyrax*). J Morphol 176: 61–87

Janis CM, Fortelius M (1988) On the means whereby mammals achieve increased functional durability of their dentitions, with special reference to limiting factors. Biol Rev 63: 197–230

Jenkins FA (1990) Monotremes and the biology of Mesozoic mammals. Neth J Zool 40: 5–31

Kallen FC, Gans C (1972) Mastication in the little brown bat, *Myotis lucifugus*. J Morphol 136: 385–420

Kay RF (1975) The functional adaptations of primate molar teeth. Am J Phys Anthropol 43: 195–216

Kay RF (1977) The evolution of molar occlusion in the Cercopithecidae and early catarrhines. Am J Phys Anthropol 46: 327–352

Kay RF, Hiiemae KM (1974) Jaw movement and tooth use in recent and fossil primates. Am J Phys Anthropol 40: 227–256

Kay RF, Hylander WL (1978) The dental structure of mammalian folivores with special reference to Primates and Phalangeroidea (Marsupialia). In: Montgomery GG (ed) The ecology of arboreal folivores. Smiths Inst, Washington DC, pp 173–191

Kermack KA, Musset F, Rigney HW (1973) The lower jaw of Morganucodon. Zool J Linn Soc 53: 87–175

Kermack KA, Musset F, Rigney HW (1981) The skull of Morganucodon. Zool J Linn Soc 71: 1–158

Kesner MH (1980) Functional morphology of the masticatory musculature of the rodent subfamily Microtinae. J Morphol 165: 205–222

Kühlhorn F (1938) Anpassungserscheinungen am Kauapparat bei ernährungsbiologisch verschiedenen Säugetieren. Zool Anz 121: 1–17

Landry SO (1970) The Rodentia as omnivores. Q Rev Biol 45: 351–372

Langenbach GEJ, Weijs WA (1990) Growth patterns of rabbit masticatory muscles. J Dent Res 69: 20–25

Lemire F (1966) Particularites de l'appareil masticateur d'un rongeur insectivore Deomys ferrugineus (Cricetidae, Dendromurinae). Mammalia 30: 454–494

Lund JP, Enomoto S (1988) The generation of mastication by the mammalian central nervous system. In: Cohen AH, Rossignol S (eds) Neural control of rhythmic movements in vertebrates. Wiley, New York, pp 201–283

Lund JP, Sasamoto K, Murakami T, Olsson KA (1984) Analysis of rhythmical jaw movements produced by electrical stimulation of motor-sensory cortex of rabbits. J Neurophysiol 52: 1014–1029

Luschei ES, Goodwin GM (1974) Patterns of mandibular movement and jaw muscle activity during mastication in the monkey. J Neurophysiol 37: 954–966

Matsuda K, Uemura M, Kume M, Matsushima R, Mizuno N (1978) Topographical representation of masticatory muscles in the motor trigeminal nucleus in the rabbit. A HRP study. Neurosci Lett 8: 1–4

McNamara JA (1974) An electromyographic study of mastication in the rhesus monkey (Macaca mulatta). Arch Oral Biol 19: 821–823

Meirte D (1986) Functioneel morfologische studie van het kauwen bij de Afrikaanse molrat Tachyoryctes splendens. Thesis, Univ of Antwerp

Mendel F, Hicks W, Kallen F, Fish D (1985) Jaw/hyoid movements in two species of bats: is there a basic eutherian pattern? J Mammal 66: 774–777

Miller AJ, Vargervik K, Chierici G (1982) Electromyographic analysis of the functional components of the lateral pterygoid muscle in the rhesus monkey (Mucaca mulatta). Arch Oral Biol 27: 475–480

Mills JRE (1966) The functional occlusion of the teeth of the Insectivora. J Linn Soc Zool 47: 1–25

Mills JRE (1967) A comparison of lateral jaw movements in some mammals from wear facets on the teeth. Arch Oral Biol 12: 645–661

Mizuno N, Konishi A, Sato M (1975) Localization of masticatory motoneurons in the cat and rat by means of retrograde transport of horseradish peroxidase. J Comp Neurol 164: 105–116

Mizuno N, Matsuda K, Iwahori N, Uemura-Sumi M, Kume M, Matsushima R (1981) Representation of the masticatory muscles in the motor trigeminal nucleus of the macaque monkey. Neurosci Lett 21: 19–22

Møller E (1966) The chewing apparatus. Acta Physiol Scand 69 (Suppl) 280: 1–229

Moyers RE (1949) Temporomandibular muscle contraction patterns in Angle class II division 1 malocclusions: an electromyographic analysis. Am J Orthod 35: 837–857

Müller A (1933) Die Kaumuskulatur des Hydrochoerus capybara und ihre Bedeutung für die Formgestaltung des Schädels. Morphol Jahrb 72: 1–59

Naples VL (1985) Form and function of the masticatory musculature in the tree sloths, Bradypus and Choloepus. J Morphol 183: 25–50

Offermans M (1990) De springhaas, *Pedetes capensis*: een functioneel-morfologische studie. Voedselopname en voortbeweging. Thesis, Univ Antwerp

Offermans M, de Vree F (1990) Mastication in springhares, *Pedetes capensis*: a cineradiographic study. J Morphol 205: 353–367

Oron U, Crompton AW (1985) A cineradiographic and electromyographic study of mastication in *Tenrec ecaudatus*. J Morphol 185: 155–182

Otten E (1987) A myocybernetic model of the jaw system of the rat. J Neurosci Methods 21: 287–302

Radinsky LB (1981a) Evolution of skull shape in carnivores 1. Representative modern carnivores. Biol J Linn Soc 15: 369–388

Radinsky LB (1981b) Evolution of skull shape in carnivores 2. Additional modern carnivores. Biol J Linn Soc 16: 337–355

Radinsky LB (1982) Evolution of skull shape in carnivores 3. The origin and early radiation of modern carnivores. Paleobiology 8: 177–195

Radinsky LB (1985) Patterns in the evolution of ungulate jaw shape. Am Zool 25: 303–314

Rokx JTM, Jüch PJW, van Willigen JD (1985) On the bilateral innervation of masticatory muscles: a study with retrograde tracers. J Anat 140: 237–243

Rome LC, Sosnicki AA, Goble DO (1990) Maximum velocity of shortening of three fibre types from horse soleus muscle: implications for scaling with body size. J Physiol 431: 173–185

Romer AS (1974) Vertebrate paleontology, 3rd edn. Univ Chicago Press, Chicago

Rossignol S, Lund JP, Drew T (1988) The role of sensory inputs in regulating patterns of rhythmical movements in higher vertebrates. In: Cohen AH, Rossignol S (eds) Neural control of rhythmic movements in vertebrates. Wiley, New York, pp 201–283

Savalle WPM, Weijs WA, James J, Everts V (1990) Elastic and collagenous fibers in the temporomandibular joint capsule of the rabbit and their functional relevance. Anat Rec 227: 159–166

Scapino RP (1976) Function of the digastric muscle in carnivores. J Morphol 150: 843–860

Scapino R (1981) Morphological investigation into functions of the jaw symphysis in carnivorans. J Morphol 167: 339–375

Schumacher GH (1961) Funktionelle Morphologie der Kaumuskulatur. Fischer, Jena

Schwartz G, Enomoto S, Valiquette C, Lund JP (1989) Mastication in the rabbit: a description of movement and muscle activity. J Neurphysiol 62: 273–287

Smith JM, Savage RJG (1959) The mechanics of mammalian jaws. School Sci Rev 40: 289–301

Smith KK, Redford KH (1990) The anatomy and function of the feeding apparatus in two armadillos (Dasypoda): anatomy is not destiny. J Zool (Lond) 222: 27–47

Smith RJ (1983) The mandibular corpus of female primates: taxonomic, dietary and allometric correlates of interspecific variations in size and shape. Am J Phys Anthropol 61: 315–330

Smith RJ, Peterson CE, Gipe DP (1983) Size and shape of the mandibular condyle in primates. J Morphol 177: 59–68

Starck D (1933) Die Kaumuskulatur der Platyrrhinen. Morphol Jahrb 72: 212–286

Starck D (1935) Kaumuskulatur und Kiefergelenk der Ursiden. Morphol Jahrb 76: 104–147

Stohler CS (1986) A comparative electromyographic and kinesiographic study of deliberate and habitual mastication in man. Arch Oral Biol 31: 669–678

Storch G (1968) Funktionsmorphologische Untersuchungen an der Kaumuskulatur und an korrelierten Schädelstrukturen der Chiropteren. Abh Senckenb Naturforsch Ges 517: 1–92

Tal M (1980) Representation of some masticatory muscles in the trigeminal motor nucleus of the guinea pig: horseradish peroxidase study. Exp Neurol 70: 726–730

Taylor A (1990) Proprioceptive control of jaw movement. In: Taylor A (ed) Neurophysiology of the jaws and teeth. McMillan Press, Houndmills, pp 237–267

Thexton AJ, Crompton AW (1989) Effect of sensory input from the tongue on jaw movement in normal feeding in the opossum. J Exp Zool 250: 233–243

Thexton AJ, Hiiemae KM, Crompton AW (1980) Food consistency and bite size as regulators of jaw movement during feeding in the cat. J Neurophysiol 44: 456–474

Toldt C (1905) Der Winkelfortsatz des Unterkiefers beim Menschen und bei den Säugetieren und die Beziehungen der Kaumuskeln zu demselben. Sitzungsber Kais Akad Wiss Wien Math Naturwiss Kl 114: 315–476

Tullberg T (1899) Ueber das System der Nagethiere, eine phylogenetische Studie. Nova Acta R
    Soc Sci Uppsal 18: 1–514
Turnbull WD (1970) Mammalian masticatory apparatus. Fieldiana Geol 18: 149–356
Uemura-Sumi M, Takahashi O, Matsushima R, Takata M, Yasui Y, Mizuno N (1982)
    Localization of masticatory motoneurons in the trigeminal motor nucleus of the guinea pig.
    Neurosci Lett 29: 219–224
Weijs WA (1975) Mandibular movements of the albino rat during feeding. J Morphol 145:
    107–124
Weijs WA, Dantuma R (1975) Electromyography and mechanics of mastication in the albino
    rat. J Morphol 146: 1–34
Weijs WA, Dantuma R (1981) Functional anatomy of the masticatory apparatus in the rabbit
    (Oryctolagus cuniculus L.). Neth J Zool 31: 99–147
Weijs WA, Brugman P, Grimbergen CA (1989a) Jaw movements and muscle activity during
    mastication in growing rabbits. Anat Rec 224: 407–416
Weijs WA, Korfage JAM, Langenbach GJ (1989b) The functional significance of the position
    of the centre of rotation for jaw opening and closing in the rabbit. J Anat 162: 133–148
Widmalm SE, Lillie JH, Ash MM (1987) Anatomical and electromyographic studies of the
    lateral pterygoid muscle. J Oral Rehab 14: 429–446
Wood AE (1965) Grades and clades among rodents. Evolution 19: 115–130
Wood WW (1987) A review of masticatory muscle function. J Prosthet Dent 57: 222–232
Wood WW, Takada K, Hannam AG (1986) The electromyographic activity of the inferior part
    of the human lateral pterygoid muscle during clenching and chewing. Arch Oral Biol 31:
    245–253
Woods CA, Howland EB (1979) Adaptive radiation of capromyid rodents: anatomy of the
    masticatory apparatus. J Mammal 60: 95–116
Zey A (1939) Funktion des Kauapparates und Schädelgestaltung bei den Wiederkäuern. Med
    Inaug Diss, Frankfurt

Chapter 10

Differential Wear of Enamel: A Mechanism for Maintaining Sharp Cutting Edges

A.W. Crompton[1], *C.B. Wood*[2] and *D.N. Stern*[3]

Contents

1 Introduction

Mammals are characterized by unilateral dental occlusion, complex molar patterns and highly organized enamel. The variety of molar types that evolved during the long history of mammals represents adaptations to different diets. In this chapter we wish to discuss the adaptive features of the tribosphenic molars of several insectivorous mammals.

In recent years our knowledge of the early history of mammals has increased considerably. Mammals arose during the Late Triassic ($\pm$ 200 million years ago, Odin et al. 1982) and rapidly attained a worldwide distribution. Recent discoveries have shown that the diversity of early mammals far exceeds that suggested by fossil finds made during the nineteenth century and first few decades of this century (Lilligraven et al. 1979; Jenkins 1990; Hahn et al. 1991; Crompton and Luo 1993, for reviews of recent literature). The most abundant remains of early mammals consist of isolated teeth that include a wide variety of molar types, most of which appear suitable for breaking down small invertebrates. Many different patterns of occlusion are found in the early mammals, but in terms of evolutionary potential, the most successful pattern has

[1] Department of Organismic and Evolutionary Biology, Harvard University, Cambridge, Massachusetts, USA
[2] Natural Science Program Providence College, Providence, Rhode Island, USA
[3] Department of Orthopedic Research , Harvard Medical School and Forsyth Dental Centre, Boston, Massachusetts, USA

Advances in Comparative and Environmental Physiology, Vol. 18
© Springer-Verlag Berlin Heidelberg 1994

been that first encountered in the early mammals classified as therians. We do not wish to imply that this was the prime reason for the adaptive radiation of therian mammals. Therians include several extinct groups, marsupials and placentals, and according to some workers (Archer et al. 1985; Kielan-Jaworowska et al. 1987), monotremes. If this is correct, all living mammals can be derived from early therian mammals that possessed a unique molar pattern. In all early therian mammals the principal cusps of the molars were arranged to form reversed triangles. In the upper molars the cusps formed a trigon, and in the lower molars a trigonid (Fig. 1A). When these molars were in occlusion (Fig. 1B), the near-vertical surfaces on the mesial and distal aspects of the trigon and trigonid sheared past one another. The cutting or leading edges of the shearing surfaces were formed by crests that, in the upper molars, ran from the paracone to both the parastylar and metastylar regions, and in the lower molars, from the protoconid to both the paraconid and metaconid. The molars fitted into the embrasures formed between the molars in the opposite jaw. Tooth to tooth contact between occluding molars was limited to the shearing surfaces. This type of occlusion is often referred to as embrasure shearing . Among early therian taxa, considerable variation was present in the orientation of the leading edges of the shearing surfaces relative to the longitudinal axis of the jaw. In some theria the crests were nearly parallel to the longitudinal axis of the jaw, whereas in others, they were practically transverse. Molars of the latter type are

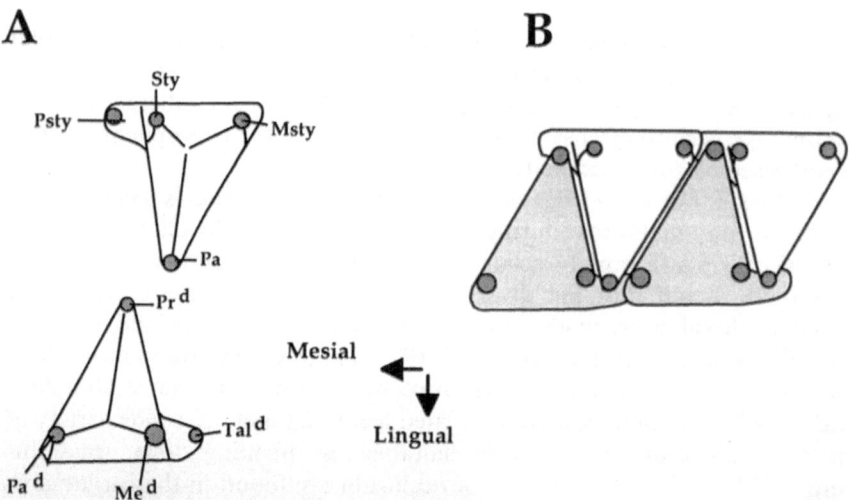

Fig. 1. A Diagrammatic representation of the crown view of an upper and lower molar of a generalized primitive therian. **B** Occlusal relationships. Shearing occurs between the vertical faces on the mesial aspects of occluding molars. Abbreviations in this and in the remaining figures: *ap* aprismatic matrix: *cing* cingulum: *d* dentine; *DEJ* dentine-enamel junction; *En*^{d} entoconid; *Hy*^{d} hypoconid; *ip* interprismatic matrix; *Me* metacone; *Me*^{d} metaconid; *Msty* metastyle; *OES* outer enamel surface; *p* prism; *Pa* paracone; *Pa*^{d} paraconid; *Pr* protocone; *Pr*^{d} protoconid; *Psty* parastyle; *Sty* stylocone; *t* tubule-like opening; *Tal*^{d} talonid

characteristic of insectivorous dryolestid pantotheres, the most common therians of the mid-to-Late Jurassic. Dryolestid pantotheres are known to have survived into the basal Cretaceous of Europe (Clemens and Lees 1971). It was generally accepted that dryolestid pantotheres became extinct during the Early Cretaceous, and were succeeded by mammals with a more complex molar pattern. Bonaparte (1990), however, has shown that dryolestid pantotheres survived in southern South America until the Late Cretaceous. In Europe, North America and Asia dryolestid pantotheres were succeeded by therian mammals with tribosphenic molars (Fig. 2) (Crompton and Kielan-Jaworowska 1978). This type of molar is present in nearly all the Cretaceous therians and retained in many living therians. Tribosphenic molars retain embrasure shearing between vertical shearing surfaces on the mesial and distal surfaces of the molars, but have added shearing surfaces and surfaces between which food can be crushed (a protocone biting into a talonid basin, Crompton 1971). Mammals with tribosphenic molars do not appear to have invaded the southern part of South America until the very end of the Cretaceous. In the absence of these advanced mammals, the more primitive Jurassic dryolestids either survived relatively unchanged or, in terms of molar morphology, underwent an interesting adaptive radiation (Bonaparte 1990; Sigogneau-Russell et al. 1991).

Mammals with embrasure shearing were the dominant mammalian insectivores from the middle of the Jurassic until Late Cretaceous times, a period of over 100 million years. Many groups of mammals that survived, diversified and flourished during the Tertiary retained tribosphenic molars, as do many today. Why was embrasure shearing so well adapted to an insectivorous diet? Part of the answer may be the ultrastructure of prismatic enamel.

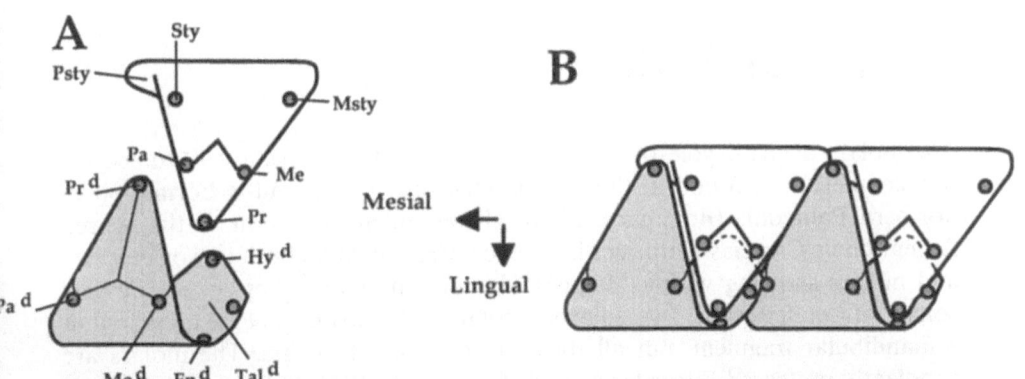

Fig. 2. Diagrammatic representation of an upper and lower tribosphenic molar. In contrast to primitive therian molars, food can be crushed between a protocone and a talonid basin. Additional shearing surfaces have been added between the hypoconid and the embrasure between the metacone and paracone and between the walls of the talonid basin and the protocone

The New World opossums have retained tribosphenic molars that can retain sharp cutting edges to the shearing surfaces despite considerable wear of the apical surface of the tooth crown. Stern et al. (1989), in a study of *Didelphis virginiana*, have suggested that differential wear of the shearing surfaces' cutting edges keep them sharp. The near-vertical shearing surfaces wear much more slowly than the apical surfaces. They correlated this pattern of wear with the relative orientations of groups of enamel crystallites in prisms and interprismatic material. They demonstrated that "at near-vertical shearing surfaces, the prisms approach the outer surface obliquely and are surrounded with IP (interprismatic) crystallites which are perpendicular to the vertical surface." Stern et al. (1992) have recently suggested that the prism tubules are also a factor in differential wear. They suggest that ". . . tubular areas exposed on the outer surface of the tooth represent discontinuities responsible for more rapid erosion. . . ." Because the Virginia opossum is omnivorous, it is not the ideal animal with which to study molar adaptations for insectivory. In addition, some molars of individual opossums do not retain sharp cutting edges. This may be due to the fact that this opossum's diet can include hard substances such as bone. Wood (1992) has recorded considerable diversity of enamel pattern among the opossums, and more work is needed to correlate pattern to functional surface in these animals. Nevertheless, the general conclusions of this report are supported. Smaller members of the group display sharp cutting edges more consistently than *D. virginiana*."

The object of this chapter is, first, to describe the mechanism of embrasure shearing in a group of early therians (pantotheres) and, secondly, to supply additional evidence to support the claim that the arrangement of crystallites within prismatic enamel determines the pattern of wear, so that sharp cutting edges are maintained.

2 Material and Methods

The molars of the dryolestids, *Groebertherium* and *Mesungulatum*, studied for this chapter, were found in the Late Cretaceous Los Alamitos Formation of northern Patagonia (Bonaparte 1990). The complete collection, in the Museo Argentino de Ciencias Naturales in Buenos Aires, includes both unworn molars and molars showing various degrees of wear. The remains of *Groebertherium* consist of isolated teeth. Two adjacent molars of *Mesungulatum* are preserved in a mandibular fragment, but all the rest are isolated molars. The molars are excellently preserved. Wear facets and striations on them indicate tooth movement during occlusion. No upper and lower molars were preserved in occlusion, but occlusal details could be reconstructed from the structure of worn and unworn teeth, and with reference to the more complete dentitions of North American pantotheres.

The teeth described in this study were drawn in several views using a camera lucida attached to a Olympus stereo-binocular microscope. In preparation for scanning electron microscopy, several of the teeth were etched whole with 1% H_3PO_4 for 50 to 90 s. The teeth were cleaned by ultrasonification in tap water before and after the etch. The teeth were then dehydrated in 95% ethanol, air dried, attached to aluminum stubs with silver cement and coated with 25 to 30 nm of palladium-gold. Photomicrographs were taken on an AMR 100 SEM. A single slightly worn molar of *Groebertherium* and large fragments of the crowns of an upper and lower molar of *Mesungulatum* were embedded in Spurr's resin and sectioned, to expose several views through the enamel. Sectioned surfaces were polished with 600 grit paper and powdered alumina. The exposed sections were etched with 1% H_3PO_4 for 6 to 12 s.

3 Results

3.1 *Groebertherium* Molars, Occlusion and Wear

Groebertherium (Figs. 3 and 10) molars are similar to those of the dryolestid pantothere, *Melanodon*, from the Late Jurassic of North America and Europe (Clemens and Lees 1971; Krebs 1971; Prothero 1981). Simple shearing surfaces are present on the mesial and distal surfaces of the molars. The large stylocone of the upper molars (Fig. 3B) is not involved in the shearing facets as it is situated in the center of the labial aspect of the crown. Its function is to puncture. The metaconid of the lower molar is twinned and, when slightly worn, its apex forms a transverse ridge. The facet on the distal surface of the trigonid shears past a facet on the mesial surface of the trigon (Fig. 3A), and a facet on the mesial surface of the lower molars shears past a facet on the distal surface of the upper molars (Fig. 3C). The leading edges of the shearing surfaces, when viewed either from the distal or mesial aspect, are concave. Food is trapped between these crescents and sheared when the lower teeth move dorsomedially, relative to the upper molars, during occlusion (Fig. 3D). Tooth to tooth contact can occur between the shearing surfaces, the tip of the paracone against the cingulum running labially from the talonid and the protoconid against the cingulum running lingually from the parastyle. Occlusal details are best understood if the crowns are viewed parallel to the trajectory of the lower tooth as it moves into occlusion (Fig. 3E). During mastication, the remainder of the crown contacts food only.

For shearing to be effective, it is essential that the leading edges of the shearing surfaces remain in contact during occlusion and are sharp. However, in mammals, with few exceptions, there is a component of medial movement as the teeth come into occlusion. Because the shearing surfaces form reversed triangles, with the triangular teeth of one set biting into the embrasures formed between the teeth of the other set, one might expect that mediolateral movement of the lower

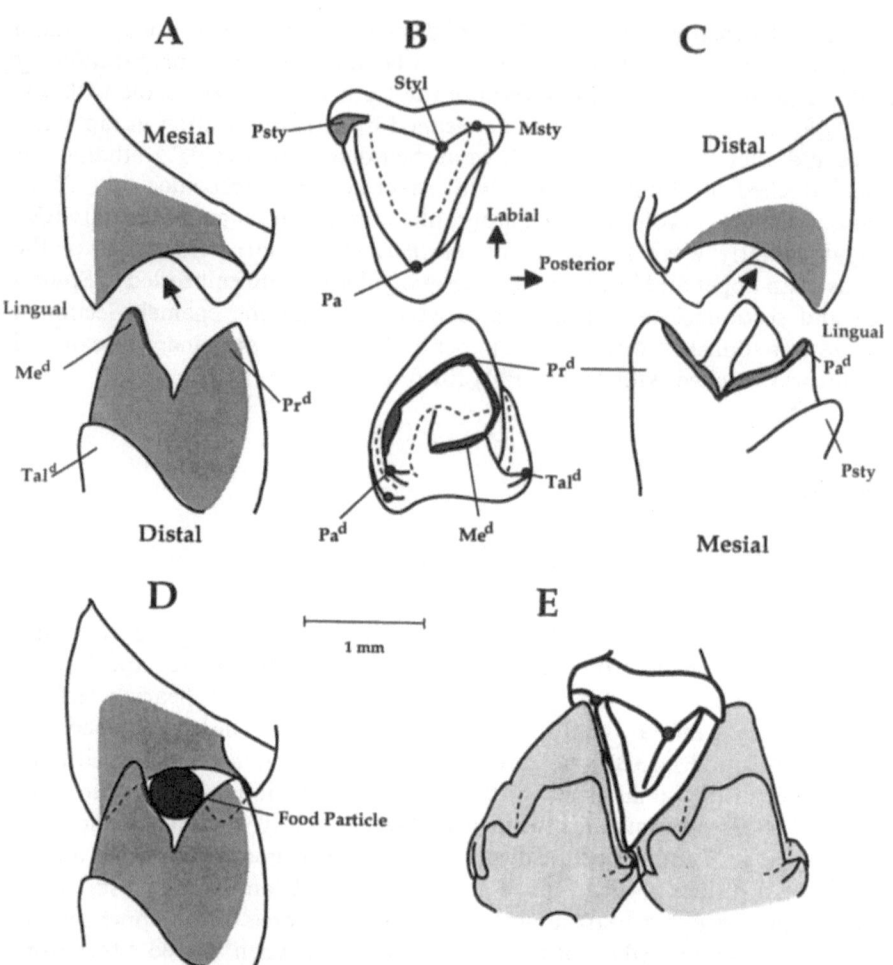

Fig. 3. Molar structure and occlusion in *Groebertherium.* **A, C** The mesial surface of an upper molar shears past the distal surface of a lower molar, and the distal surface of an upper molar shears past the mesial surface of a lower molar. **B** Crown view of an upper and lower molar. **D** The leading edges of the shearing surfaces are concave. Food particles are trapped and held during shearing. **E** Occlusal view parallel to the path of movement of the lower teeth as they move into occlusion. Note that except for the shearing surfaces and the ridges extending from the talonid and parastyle there is no contact between the apical surfaces of the crowns

jaw during occlusion would result in a separation of the shearing surfaces. However, because of the orientation of the shearing surfaces, contact between the surfaces is maintained. Two simple models help to visualize how this is achieved. Imagine a triangular wedge cut from a block (Fig. 4A). If the wedge is placed back into the embrasure of the block so that the sides of the wedge are opposed to the sides of the embrasure, and moved up or down, the surfaces will

A   B

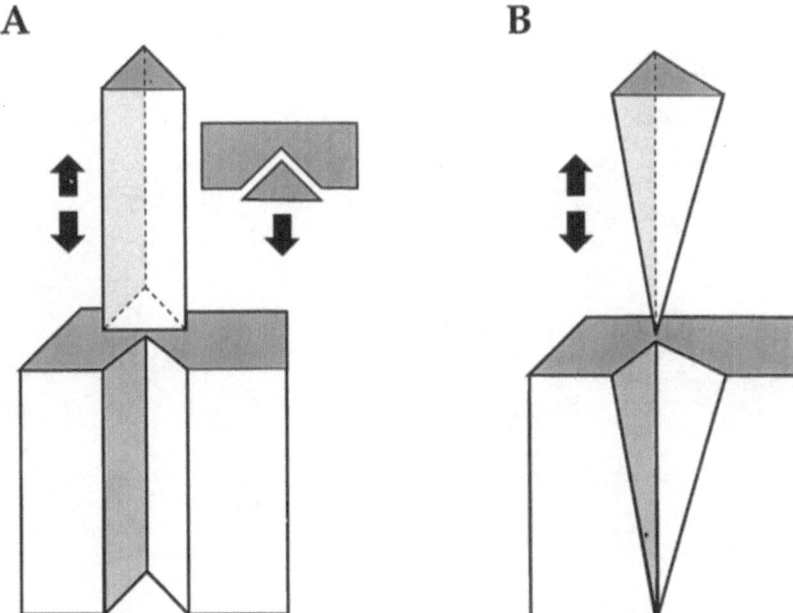

Fig. 4. A A wedge has been cut from a block. If the wedge is replaced within the embrasure in the block, its surfaces will remain in contact with those of the embrasure when the wedge is moved in the vertical direction. A component of lateral movement would separate the surfaces. **B** A wedge that forms a three-sided pyramid is cut from a block. If the wedge is placed within the embrasure and the adjoining surfaces held in contact while the block is moved relative to a stationary wedge, the block will move both upwards and transversely

remain in contact, provided that movement is restricted to the vertical plane. A medial component of movement would separate the surfaces. Now imagine that a wedge that forms a three-sided pyramid has been cut from a block (Fig. 4B). The embrasure at the top of the block is deeper than at the bottom. If the wedge is held in a stationary position and the block moved upwards so that the cut surfaces remain in contact, the block will move upwards and transversely (away from the viewer). The line formed by the intersection of the two sides of the wedge is parallel to the path of the movement of the block. In Fig. 5A, two *Groebertherium* molars are viewed from the labial aspect. The shearing surfaces on the distal and mesial sides of two adjoining lower molars are shaded. Note that the facet on the distal surface of the lower molar is deep, whereas that on the mesial surface of the adjoining molar forms a narrow crescent. The upper molar that occludes into the embrasure is shown above the lower molars. The matching shearing surfaces of this tooth are on the lingual surface of the upper tooth and are not visible in this view. If the planes of the shearing surfaces of the lower teeth are continued in space (Fig. 5B), they form two sides of an inverted, three-sided pyramid. These are similar to the sides of the embrasure cut in the block in Fig. 4B. If the planes of the shearing surfaces on the occluding upper

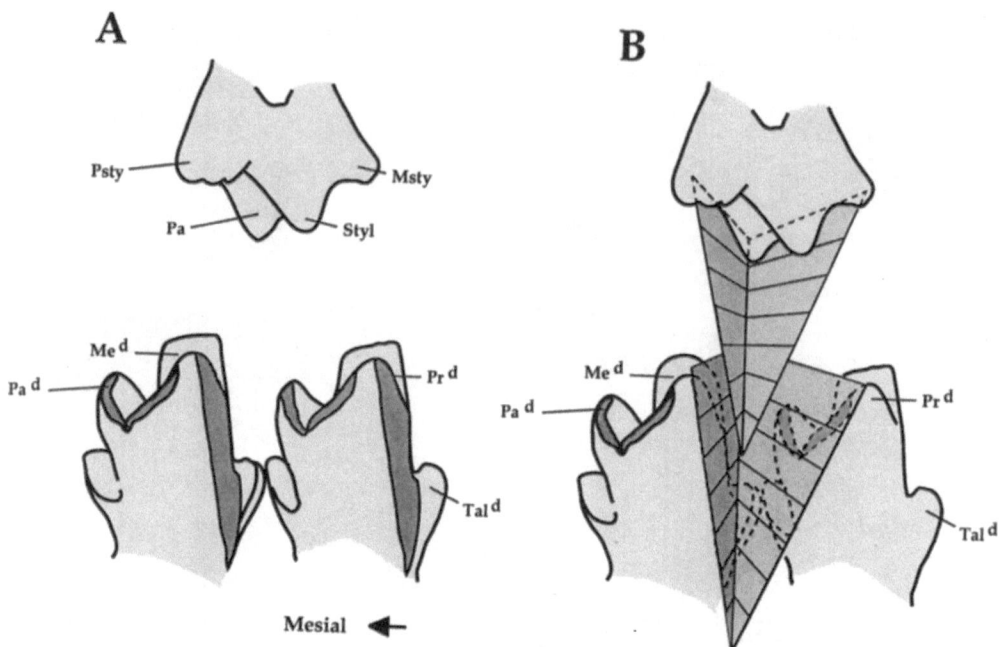

Fig. 5. A Labial view of two lower molars of *Groebertherium* and an occluding upper molar. The shearing surfaces of two adjacent molars, which form a labially directed embrasure, are *shaded*. The shearing surfaces on the lingual aspect of the upper molar are not indicated. **B** The planes of the shearing surfaces of both the upper and lower molars are projected in space. Those of the upper molar form two sides of a three-sided pyramid that fits precisely into the inner surfaces of a three-sided pyramid formed between the two lower molars

molar are continued in space, they will also form two sides of a three-sided pyramid that fits into the embrasure between the lower molars. The orientation of the shearing surfaces makes it possible for the teeth to remain in contact during occlusion, despite dorsomedial movement of the lower jaw. Differential activity of the adductor muscles controls the movement of the lower jaw so that the molars remain within the embrasures in the opposite jaw. Dorsomedial movement during occlusion is important because food sheared by the molars is forced into the oral cavity rather than being trapped between the teeth, as would be the case if jaw movements were strictly vertical. There is another advantage to embrasure shearing: wear results in widening of the embrasure between two adjacent lower molars, and reduction of the mesiodistal width of the molar fitting into the embrasure. If the path of movement of the jaw, primarily controlled by adductor muscles, remained unaltered, the leading edges of the shearing surfaces would cease to contact one another. This situation is avoided by shifting the path of the lower jaw during occlusion sufficiently far laterally to reestablish contact between the occluding teeth, thereby compensating for wear.

The Los Alamitos collection of *Groebertherium* molars contains specimens showing several degrees of wear. A marked feature of the wear pattern is that the

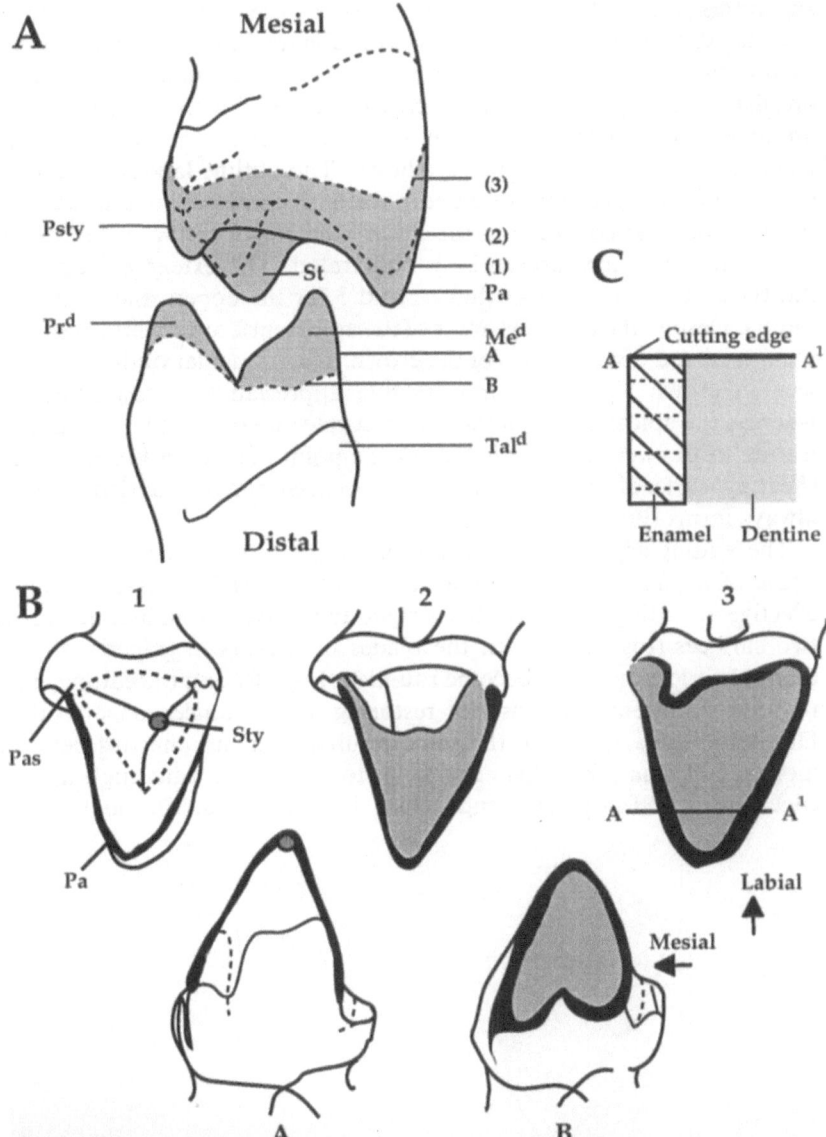

Fig. 6. Wear of *Groebertherium* molars. **A** Outline of the mesial view of an upper molar, and distal views of a lower molar. The *heavy outline* labeled (*1*) in the upper molar and *A* in the lower is that of unworn teeth. The *dotted lines* labeled (*2*) and (*3*) in the upper molar and *B* in the lower illustrate how the crown outline changes as a result of wear. **B** Crown views of upper molars showing three stages of wear, and lower molars showing two stages of wear. The *numbers* above the upper molars and *letters* below the lower molars correspond to the labeled outlines in **A**. **C** Medial view of a transverse section along the line *A-A*^{1} of B_3. A sharp right-angled cutting edge is formed between the vertical and horizontal surfaces of the enamel. *Heavy lines* in the enamel represent the orientation of prisms, and *dotted lines* the orientation of the crystallites within the interprismatic material

shearing surfaces show only limited wear in a mesiodistal direction, whereas wear is extensive in the vertical direction. Figure 6 illustrates varying degrees of wear of three upper and two lower molars. In unworn molars the principal cusps are sharp and high and the entire crown is covered with enamel. The outline of an upper and lower tooth, as seen from behind in the case of the lower, and in front, in the case of the upper, is shown. The outline labeled 1 (upper) and A (lower) shows the contour of unworn teeth. In extensively worn teeth (outline 3 in the upper and outline B in the lower), the crown surface is reduced to a flat plane and the cusps are entirely obliterated. The extent of vertical wear is illustrated by comparing outlines 1 and 3 for the uppers and A and B for the lowers. Despite the extensive wear of the apical surface very little wear occurs on the mesial and distal aspects of the crown. The triangular outline of the teeth, as seen in crown view, does not become appreciably narrower mesiodistally, whereas the apical aspect of the crown surface wears extensively. An important feature of the wear pattern is that at the point of contact between the vertical shearing surfaces and the horizontal occlusal surfaces a right angle almost always forms (Fig. 6C).

The leading edges of the shearing surfaces act in a similar way to the cutting edges of a pair of scissors or a paper cutter (Fig. 7). These instruments are effective at cutting, provided that a right-angled edge is maintained between the two surfaces (Fig. 7A). When the blades of scissors or paper cutters become blunt, their leading edges become rounded (Fig. 7B), and the cutting efficiency is reduced. Sharpening consists of restoring a right-angled edge to the blades. Dryolestid molars achieve the same result by having one surface wear much more rapidly than the other. This helps to maintain a right angle at the cutting edge and an effective shearing action is maintained. The high cusps of an

A

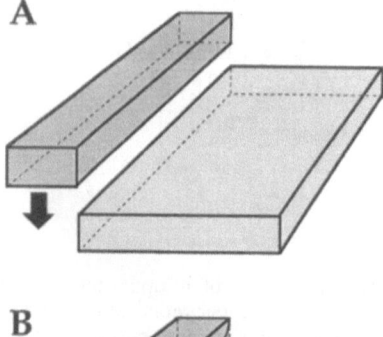

B

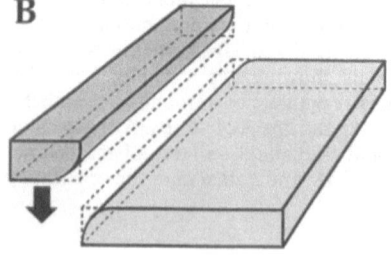

Fig. 7. Illustration of the cutting action of a pair of scissors or a paper cutter. A The cutting edges form a right angle when they are sharp. During shearing, the closing force is concentrated at this edge. B If the edges become blunt, the force is distributed over a wider area and the cutting efficiency reduced

unworn crown were presumably effective as puncturing devices. However, as wear rapidly reduces their effectiveness, it may be concluded that shearing has more importance than puncturing.

3.2 *Mesungulatum* Molars, Occlusion and Wear

Mesungulatum molars combine crushing and shearing. The crowns are bulbous, but the trigon and trigonid, with near-vertical shearing surfaces, remain dominant features of the crown morphology (Fig. 8). As Fig. 8 illustrates, the

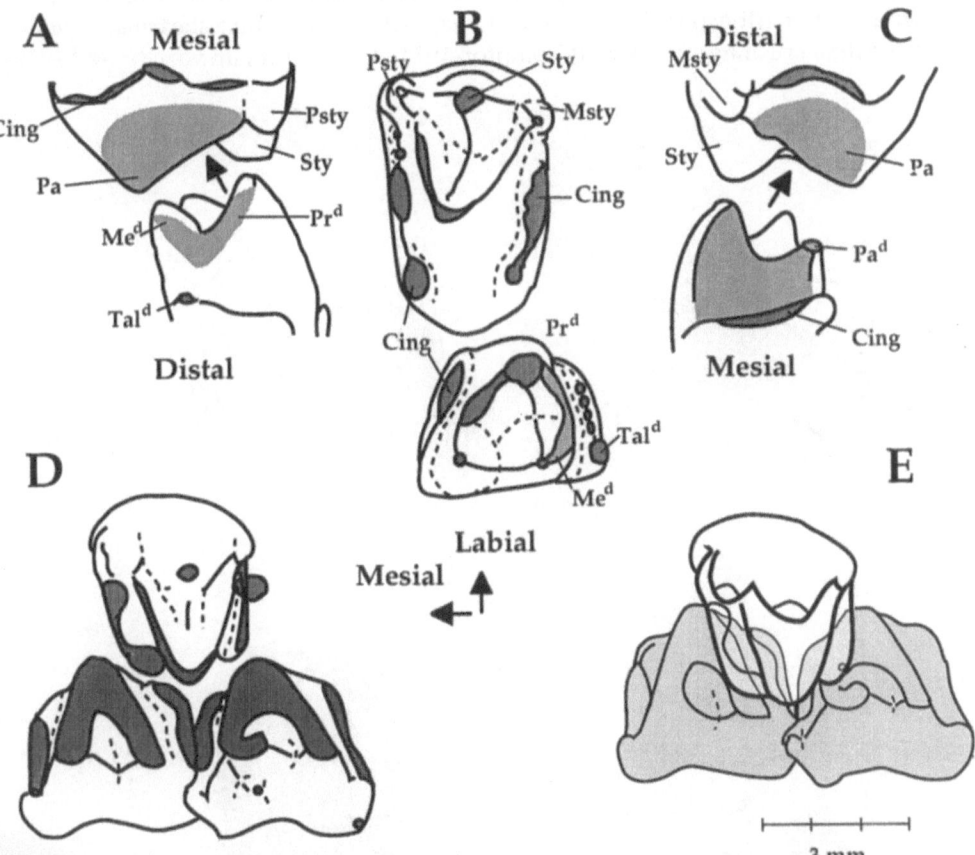

Fig. 8. Molar structure and occlusion in *Mesungulatum*. **A** The leading edge of the facet on the mesial surface of an upper molar shears past the distal surface of a lower molar. **B** Crown view of an upper and lower molar. **C** The distal surface of an upper molar shears past the mesial surface of a lower. When the teeth are fully occluded, the paracone bites into the basin formed between two adjacent lower molars. The apical surface of the mesial and distal edges of a lower molar bite against the cingula of two adjacent upper molars. **D** Occlusal view of upper and lower molars as seen from a view parallel to the trajectory of the lower molar as it moves into occlusion. The molars are separated so structures can be identified. **E** Molars in occlusion. In contrast to *Groebertherium*, there is substantial tooth-tooth contact

structure and orientation of the principal shearing surfaces on the mesial and
distal aspects of the crown resemble those of *Groebertherium*, although the cusps
in unworn molars are, relative to crown size, not as high as those of *Groebertherium*. The characteristic feature of *Mesungulatum* molars are the large and
prominent cingula at the base of the shearing facets. In occlusal view (Fig. 8D), it
will be seen that the cingula of two adjacent lower molars form a basin. The
paracone of an upper molar bites into this basin . A comparable basin, which
receives the apex of the protoconid, is formed between adjoining upper molars.
During occlusion, food is first sheared and then crushed. The basins are formed
by cingula function in the same way as the talonid in tribosphenic molars.
Cingula at the base of shearing surfaces are also present in many Cretaceous
therians with tribosphenic molars, but they act as additional shearing edges
rather than crushing structures (Crompton 1971). In Fig. 9, an unworn (A and B)

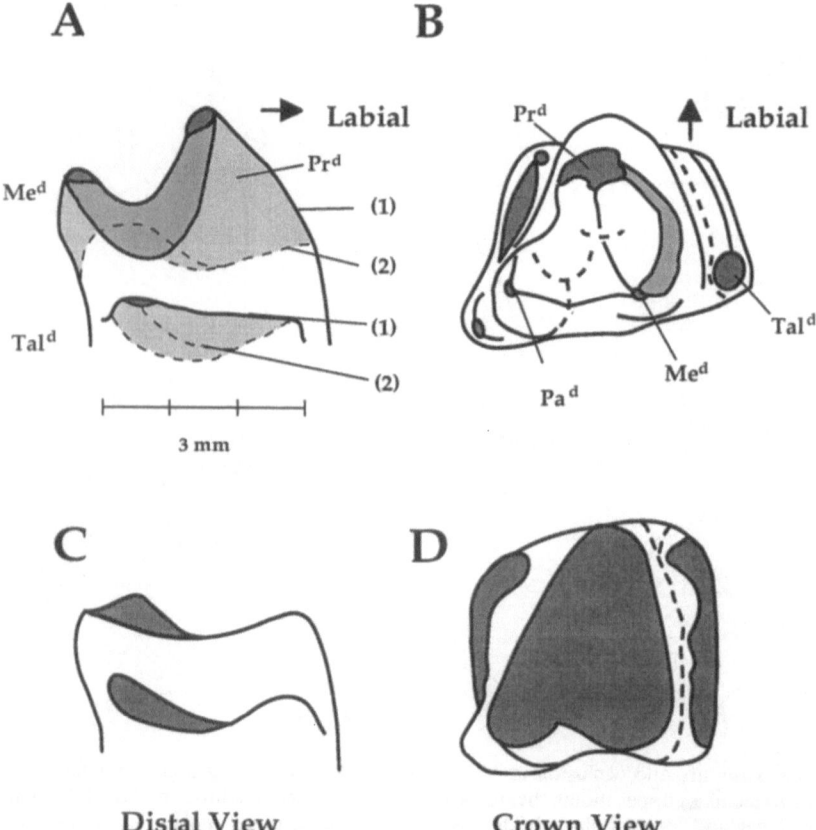

Fig. 9. Wear of a lower molar of *Mesungulatum*. **A** Distal view; **B** crown view of an unworn
molar. **C** Distal view; **D** crown view of a worn molar. The *shaded area* between the outlines
labeled (*1*) and (*2*) in **A** indicates the amount of the apical and talonid ridge surfaces removed
by wear

and a worn (C and D) lower molar are shown in posterior and crown view. The crown of a lower molar is worn away to a featureless horizontal plane. In Fig. 9A, the outline labeled (1) is that of an unworn tooth and (2), the outline of a worn tooth. The shaded area on the posterior view of the unworn lower molar indicates the amount of the crown that has been lost as a result of wear. In the worn tooth, the basin formed by adjoining cingula is deepened. Despite this extensive wear a right-angled (sharp) cutting edge is still present at the leading edges of the shearing surfaces.

Why does one surface wear more rapidly than the other? The answer may lie in the organization of crystallites within the enamel, and the presence of enamel tubules.

3.3 Enamel Structure of *Groebertherium* and *Mesungulatum*

Figure 10 is a stereo SEM photomicrograph of the crown view of a right lower molar of *Groebertherium*. The apical surface of the exposed enamel forms a right angle with the vertical shearing surface. An enlarged micrograph of the horizontal surface of the exposed enamel on the anterior aspect of the tooth is shown in Fig. 11 (the lower edge of the photomicrograph faces the dentine enamel junction, DEJ).

A clear distinction can be drawn between prismatic and interprismatic enamel. Horseshoe-shaped prism sheaths with open ends facing the outer enamel surface (OES) separate the crystallites of the prisms from the interprismatic material. The crystallites within the prisms are directed apically, but

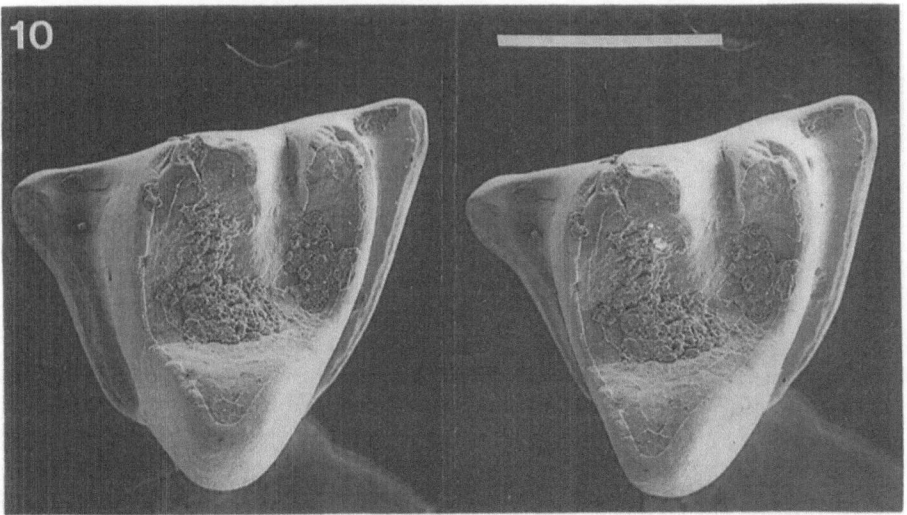

Fig. 10. *Groebertherium.* Stereo-photograph of a right lower molar. *Bar* = 1 mm

tilted slightly towards the OES. The crystallites of the interprismatic material have a horizontal orientation. The interprismatic material forms keyhole-shaped extensions surrounding large tubule-like openings on the external surface of the prisms. Several sections orthogonal to one another were made through the tooth shown in Fig. 10. A horizontal section (Fig. 12, DEJ towards the bottom of the micrograph) was cut below the surface shown in Fig. 11. The organization is similar to that visible in Fig. 11, except that tubule-like openings are not clearly visible. Towards the external surface the prisms fade out and interprismatic enamel is the dominant component. A vertical sagittal section was cut through the metaconid of the tooth shown in Fig. 10. In a photo-micrograph (Fig. 13, DEJ visible on the left) of the exposed enamel that forms the shearing surface on the posterior surface of the tooth, the prisms are directed apically but tilted towards the OES. As a result the section passes through several prisms. The prisms are surrounded by interprismatic material and a thin band of aprismatic enamel is present near the OES. In a sagittal section (Fig. 14, DEJ to the right) cut at right angles to the horizontal section shown in Fig. 12, the plane of section is parallel to that of the prisms. In the inner two-thirds of the section the prisms are directed mesially and apically. The crystallites (Fig. 15) within each prism are directed slightly more apically than the prism itself. Interprismatic material is not well preserved (removed by the etching medium) in the inner two-thirds of the section. Some is visible (Fig. 14) between the upper three prisms. The crystallites of the interprismatic enamel have a horizontal orientation that is normal to the DEJ. The outer third of the enamel consists

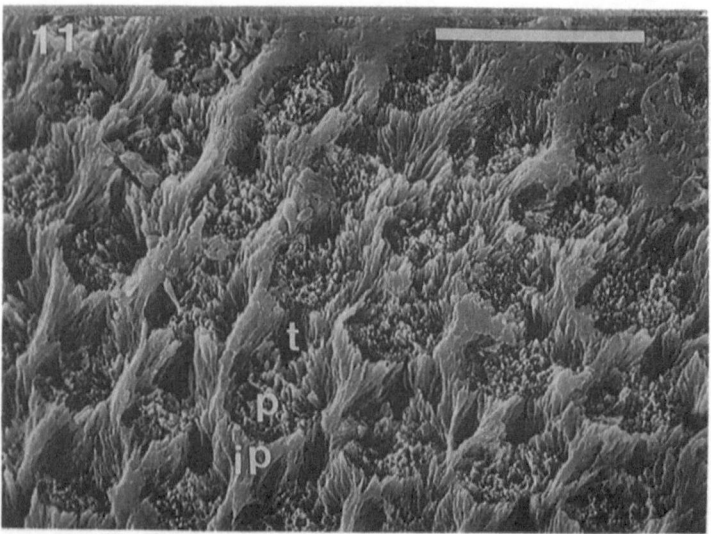

Fig. 11. *Groebertherium.* Naturally worn apical (horizontal) surface of enamel on the mesial side of the lower molar shown in Fig. 1. The ventral edge of the photomicrograph is closest to the DEJ. *Bar* = 10 μm

only of aprismatic material in which the crystallites have a horizontal orientation. This outer aprismatic material is a continuation of the interprismatic. It appears to surround regions in which the enamel substance has been completely etched away. The inner enamel of *Groebertherium* meets the definition for radial

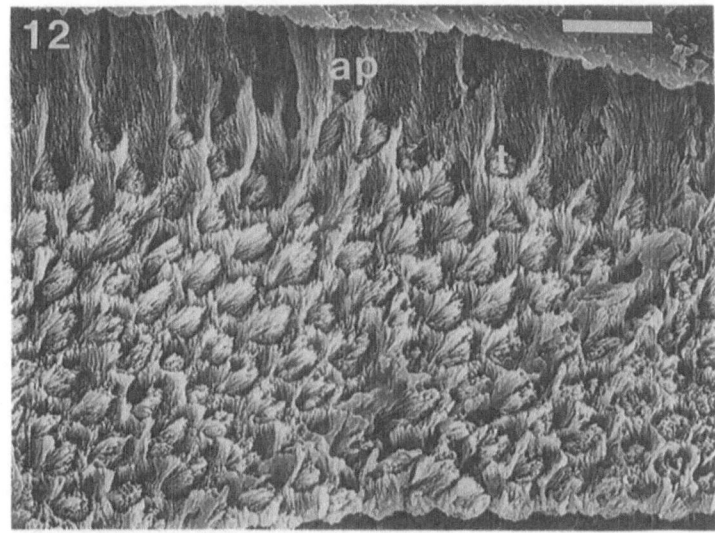

Fig. 12. *Groebertherium.* Horizontal section through the enamel on the mesial side of the lower molar shown in Fig. 10. The DEJ is visible on the lower edge of the figure. *Bar* = 10 μm

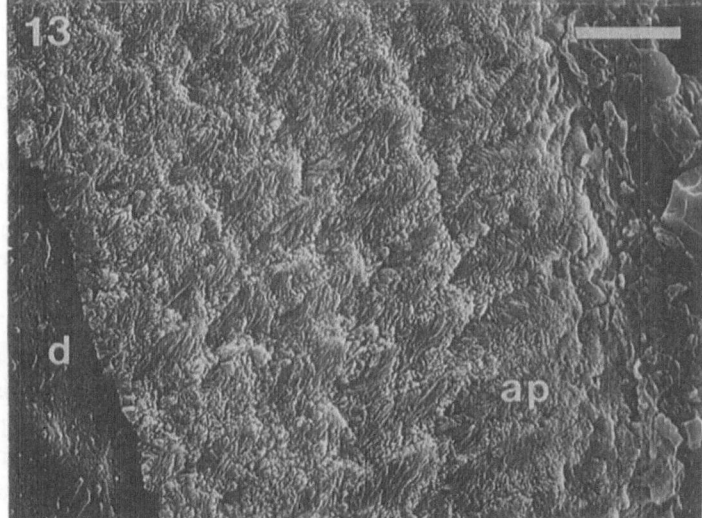

Fig. 13. *Groebertherium.* Vertical section through the metaconid of the tooth shown in Fig. 10. The DEJ is visible on the left. *Bar* = 10 μm

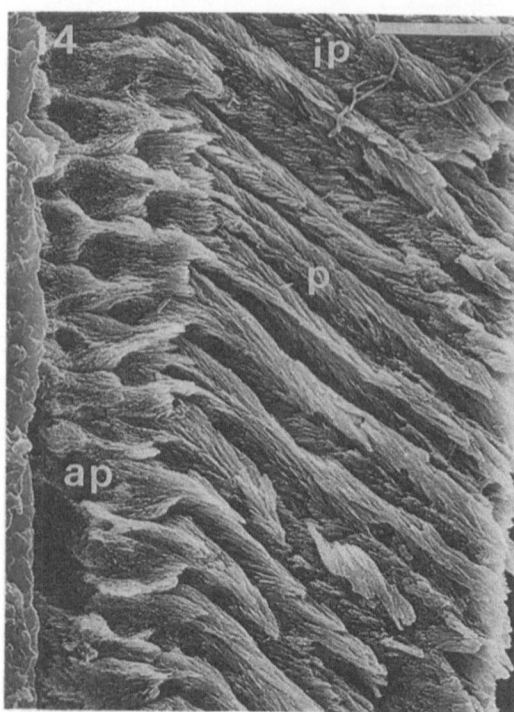

Fig. 14. *Groebertherium.* Sagittal (vertical) section cut at right angles to the vertical section shown in Fig. 12. The DEJ is visible on the right. *Bar* = 10 μm

enamel (Koenigswald and Clemens 1992): ". . .the long axes of the prisms are oriented radially from the DEJ, as seen in a horizontal plane, and rise occlusally towards the surface of the enamel, as seen in a vertical plane." In *Groebertherium*, a layer of aprismatic enamel lies external to the radial enamel. Koenigswald and Clemens (1992) use the term "schmelzmuster" to describe patterns made up of layers of more than one type of enamel.

The enamel of *Mesungulatum* differs from that of *Groebertherium* in that it appears that the ratio of interprismatic to prismatic material is higher in the former. Figure 16 (DEJ below) is a photomicrograph of a ground and polished horizontal section through the anterior surface of the protoconid. The plane of the section is tilted slightly from the horizontal so that the side nearest the DEJ is closer to the apical surface than the side closer to the OES. This section was prepared to ensure that the prisms were perpendicular to the plane of section. Crescent-shaped sheaths are present on the DEJ side of the prisms. Tubule-like openings are sometimes present on the external surfaces of the prisms, but some of the openings lie within the interprismatic matrix. The crystallites of the interprismatic material are parallel to the plane of section and converge towards one another in front of the prisms. A seam can occasionally be seen at the point of convergence of the interprismatic crystallites (Lester 1989; Lester and Koenigswald 1989). The outer region of the enamel, external to that shown in Fig. 16, is

Fig. 15. *Groebertherium.* Enlargement of portion of Fig. 14 to show the transition from radial to aprismatic enamel (DEJ towards the right). *Bar* = 10 μm

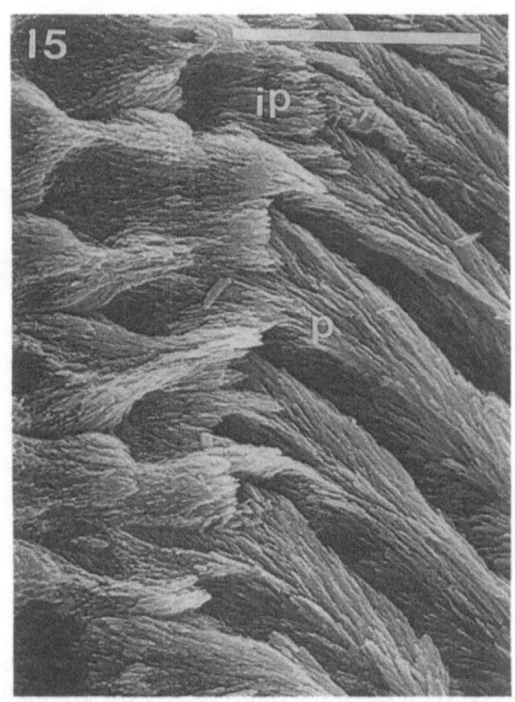

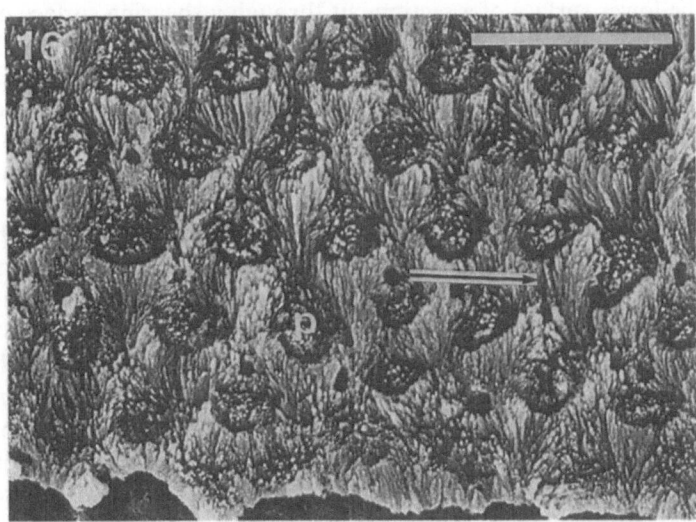

Fig. 16. *Mesungulatum.* Slightly tilted horizontal section (see text) through the protoconid (DEJ below). *Bar* = 10 μm

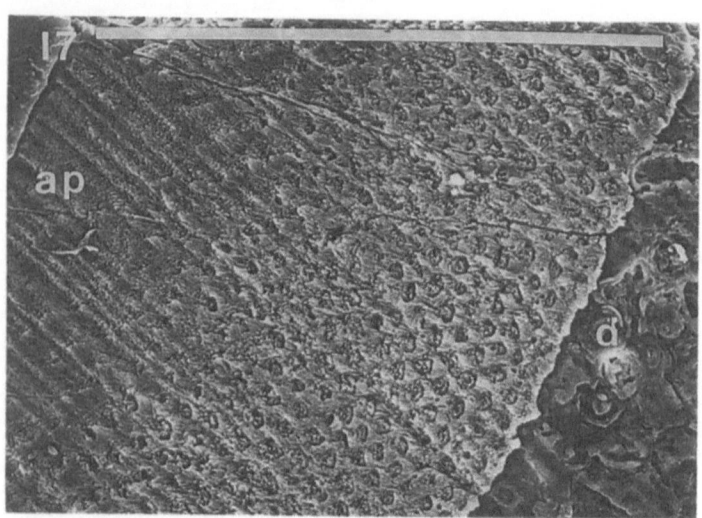

Fig. 17. *Mesungulatum.* Total thickness of enamel from the same region as Fig. 16 (DEJ to the right) *Bar* = 100 μm

aprismatic (Fig. 17, DEJ on the right). In the transition between the two zones (Fig. 18, DEJ to the right) the prisms gradually disappear. If the plane of section through a shearing surface is horizontal relative to the vertical axis of the tooth rather than tilted, so as to obtain a transverse section of the prisms, as in the previous two figures (Figs. 16 and 17), the seams within the interprismatic material, external to the prisms, are clearly visible. These can be seen in a view of the lower surface of a section cut through a shearing surface of an upper molar (Fig. 19, DEJ to the left). A seam is marked with an arrow. In this figure the crystallites of the prism structure has been removed by the etching medium. In this view, cylindrical spaces that housed the prisms are directed externally and away from the plane of section. The tubule-like structures cannot be seen because the prisms that formed an internal border have been lost.

Radial enamel that includes prisms, defined wholly or partially by a major discontinuity (sheath), and regions where crystallites of the interprismatic material converge (seams), have been described by Lester and Koenigswald (1989) in both living and extinct genera, including a Jurassic pantothere from Portugal. Enamel in this form is characterized by a high ratio of interprismatic to prismatic enamel, short incomplete horseshoe-shaped sheaths and well-developed seams. The enamel is similar to that of *Mesungulatum. Groebertherium* enamel is more advanced in that it has a higher ratio of prismatic material and lacks well-defined seams. The Portuguese pantothere molars (Krebs 1971) have a similar morphology to those of *Groebertherium*. It would not be expected that *Mesungulatum*, with its more derived molars, would have a less derived enamel ultrastructure. It is an interesting example of mosaicism in the evolution of dental characters. However, although the enamel pattern of *Mesungulatum*

Fig. 18 *Mesungulatum*. Enlargement of the transition zone between radial and aprismatic enamel shown in Fig. 17 (DEJ beyond the lower border). *Bar* = 10 μm

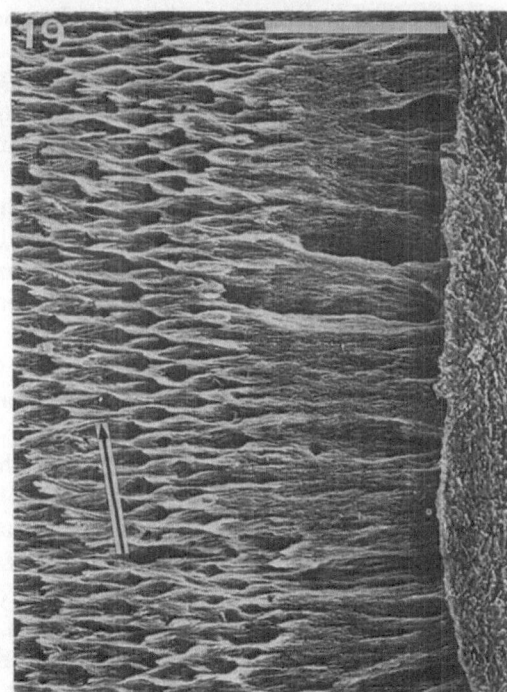

Fig. 19. *Mesungulatum*. Horizontal section through the enamel forming a shearing surface of an upper molar. Seams marked with *arrow*. Etching has removed the prism crystallites (DEJ to the left). *Bar* = 10 μm

differs from that of *Groebertherium*, the orientation of the crystallites relative to the shearing and apical surfaces is the same. The crystallites of the interprismatic are perpendicular to the shearing surface whereas the prisms approach this surface obliquely (Fig. 6).

4 Discussion

It is generally accepted (Fortelius 1985) that enamel, as the hardest tissue in the body, is designed to resist wear. Reptiles that have aprismatic enamel do not possess complex dental occlusion. They lack precise matching shearing surfaces on occluding upper and lower teeth, and a component of horizontal movement of the lower jaw during closure. The crystallites of reptile enamel are perpendicular to the DEJ and the OES. Within nonmammalian cynodonts several different patterns of occlusion developed between the postcanine teeth. In traversodont cynodonts and tritylodontids, for example (Crompton 1972), postcanine occlusion was bilateral and the lower jaw was drawn backwards as the teeth came into occlusion; whereas in the sister group of mammals, the trithelodontids (*Pachygenelus*) and in the Rhaeto-Liassic mammals (*Kuehneotherium*, atherian, and *Morganucodon* or *Dinnetherium*, triconodonts), occlusion was unilateral (Crompton and Jenkins 1968; Crompton 1974). The active side jaw moved dorsomedially as the teeth on the ipsilateral side came into occlusion. Complex occlusion was accompanied by the appearance of enamel that was made up of repetitive domains (Grine et al. 1979; Grine and Vrba 1980; Sigogneau-Russell et al. 1985; Frank et al. 1988; Lester and Koenigswald 1989; Stern 1989). Frank et al. (1988) describe the enamel of early mammals as "preprismatic" and state that "such an enamel consists of a juxtaposition of columns, each of which being constituted by hydroxyapatite crystals in a pinnate disposition, their c-axes forming a divergent angle with the main longitudinal axis of the columns. . ." Lester (1989) and Lester and Koenigswald (1989) refer to this type of enamel as "preprismatic" and relate the orientation of the crystallites to the presence of a conical or pointed Tomes' processes on the ameloblast cells. The reason they used the term "preprismatic" was that the structure they identified did not correspond to the domain of a single ameloblast. Crystallites in taxa of derived mammals apparently line up perpendicular to the secretory surfaces of the Tomes' process. Therefore, the amount of convergence of the crystallites is determined by the angle of the tip of the process relative to the longitudinal axis of the ameloblast. Crystallites formed below that part of the ameloblast cell that is parallel to the DEJ will be approximately perpendicular to the DEJ. Consequently, the cellular domain of this type of enamel consists of a core of converging crystallites and an outer rim of crystallites either perpendicular to the DEJ, or tilted slightly away from the longitudinal axis. If the latter is true, the domain formed by one ameloblast is clearly recognizable. Pseudoprisms that meet this definition have been reported in the trithelodont, *Pachygenelus* (Grine et al. 1979; Stern 1989) and the tritylodontid, *Oligokyphus* (Lester and Koenigswald 1989). Highly organized prismatic enamel is also found in the lizard, *Uromastyx* (Cooper and Poole 1973). This is one of the very few lizards which have developed a shearing type of dental occlusion. Enamel with domains of differing crystallite orientation always occurs in animals that have developed complex occlusion that involves shearing. The independent development of this type of enamel in the distantly related subclasses of vertebrates, the synapsids (mammal-like reptiles and mammals) and diapsids (including lizards), suggests a functional correlation between a more organized enamel structure and complex occlusion.

When a molar erupted in an early mammal such as *Kuehneotherium* or *Morganucodon*, the morphology was such that it did not "fit" the teeth against which it occluded, as does an erupting tribosphenic molar. In these early mammals a considerable amount of the crown had to wear away rapidly before functional matching shearing facets were developed (Crompton and Jenkins 1968). These facets were present on the vertical external surface of the lowers and internal surface of the uppers. The crests joining the principal cusps did not form a cutting edge to the shearing facet. The molars of these early mammals were adapted for puncturing, with high pointed cusps, and rather inefficient shearing on the vertical faces of occluding molars. The pseudoprisms of *Megazostrodon* and *Pachygenelus* are directed apically, and form an angle of about 45° with the DEJ.

The triangular molars of dryolestids and the more complex tribosphenic molars possess a more efficient shearing mechanism. These molars are all characterized by the presence of sharp cutting edges to the shearing surfaces. In these forms different types of radial prismatic enamel are encountered for the first time in mammalian evolution (Lester and Koenigswald 1989). In all of these types of radial prismatic enamel, crystallites of the prisms have a different orientation than those of the interprismatic matrix, and there is a major discontinuity (sheath) between the crystallites of the prisms and those of the interprismatic material. This may result from one or more of the surfaces of the Tomes' processes becoming nonsecretory (Stern, pers. comm., 1989). In general, the interprismatic matrix crystallites have a horizontal orientation, whereas those of the prisms are directed apically and form an angle of 45° or less with the DEJ. A layer of aprismatic enamel often lies external to the radial enamel. Extensive tangential enamel, as defined by Koenigswald and Clemens (1992), and Hunter-Schreger bands have not been observed in dryolestid or tribosphenic molars.

Several suggestions have been made as to the value of prismatic enamel when compared with aprismatic enamel. Lester and Koenigswald (1989) state, "differentiation of enamel to show a convergence line, a seam or even a primitive prism should be of selective value. It is rather difficult at this stage to argue about the specific functional significance of the orientation of the crystallites but, in general it seems to be more advantageous to possess crystallites oriented in different directions that oriented in parallel." They are in agreement with

Koenigswald (1988) and Koenigswald and Pfretzschner (1987) that prisms are"... a significant factor in enhancing the stability of the enamel and prevent cracking." Rensberger and Koenigswald (1980) and Pfretzschner (1986) have made a convincing case for the view that Hunter-Schreger bands, in which the direction of the prisms in one band is different or opposite to that of adjacent bands, represent an effective mechanism for preventing cracking in a plane perpendicular to the individual bands, especially in large teeth where occlusal forces are high.

Relatively little has been published on the relationship between crystallite orientation and wear patterns. Boyde (1984) studied the effect of erosion due to soft-particle abrasion (air-polishing) on different faces of enamel. He concluded that whole prisms that are parallel to an exposed surface are not highly resistant to abrasion because they can be cleaved from the surface. He suggested that enamel can also be preferentially removed at the interfaces of transversely cut prisms and the surrounding interprismatic enamel as well at the boundaries of enamel tubules. The long axes of crystallites are exposed at these locations and can be cleaved from the underlying structure.

Wear patterns can reflect underlying enamel structures. Rensberger and Koenigswald (1980) described ridges on the wear facets of rhinoceros molars. These are parallel to the direction of movement of the lower molar and are formed by differential wear of the exposed Hunter-Schreger bands. When the prisms of these bands are aligned parallel or nearly parallel to an abrasion vector, the prisms exhibit maximum resistance; but when the prisms are perpendicular to the abrasion vector, resistance is minimal. Fortelius (1985) states that both structure and hardness of enamel types should be considered when evaluating the structure-function relations of enamel. In rhinoceros molars, for example, the inner enamel wears more rapidly than the outer enamel. The outer prism layer of nondecussating prisms is more tightly packed and more wear-resistant than the inner, loosely packed but strongly decussating prisms. He suggests that this combination of enamel types is a compromise; decussating prisms prevent cracks and non-decussating prisms reduce wear. Fortelius (1985) also claims that transversely intercepted prisms are more resistant to an attrition vector that is perpendicular to their axes, than prisms that are parallel to the attrition vector. The wear of the molars of *Pedetes* (Koenigswald and Clemens 1992) confirms this observation. *Pedetes* enamel consists of an inner layer of radial enamel in which the prisms are directed apically and an outer layer of horizontally oriented Hunter-Schreger bands in which the prisms are parallel to the occlusal surface. Differential wear is present and the outer layer wears more rapidly than the inner.

In the two dryolestids described in this chapter, the apical surface of the enamel, which only contacts food and forms the leading edges of the shearing surfaces, wears much more rapidly than the vertical surfaces. The differential wear is the mechanism for maintaining a sharp cutting edge to the shearing surfaces. Differential wear was also observed in the molars of the opossum (Stern et al. 1989) and the musk shrew (Crompton and Yang, in prep.). The molars of the latter cannot be strictly defined as tribosphenic, because they have added a hypocone that bites into the trigon basin. However, they have retained embrasure shearing and vertically oriented shearing surfaces with sharp right-angled cutting edges. The radial prismatic enamel of the musk shrew consists of elliptical (i.e., with long axes oriented radially to the DEJ), apically directed prisms, surrounded by interprismatic material with crystallites oriented parallel to the apical surface. As a result of wear, in molars taken from natural populations, the vertical height of the crown is worn at 30 times the rate of the mesiodistal length of the molars. In all stages of wear, the leading edges to the shearing surfaces are sharp.

The enamel of dryolestids, opossums and musk shrews consists of an outer layer of aprismatic enamel and an inner layer of radial enamel. On the apical surface the long axes of crystallites of both enamel types and the transversely cut prisms of the radial enamel are exposed to vertically oriented forces generated by food trapped between the molars. Rapid wear of crystallites parallel to the apical surface confirms Boyde's (1984) experimental results. On the vertical face of the shearing surface in these three taxa, crystallites of the aprismatic and interprismatic matrix are perpendicular to, and those of the prisms, oblique to the attrition vector. Resistance to wear of this face is in agreement with Fortelius' (1985) observation regarding the attrition rates of upper and lower molars of trilophodonts. In the lower molars, the prisms are perpendicular to, and in the upper molars, parallel to the attrition vector. As a result, the upper molars wear more rapidly than the lower molars.

It is concluded that different combinations of crystallite orientations may account for the differential wear of the vertical and apical surfaces of dryolestid and tribosphenic molars. It is possible that the largest forces during shearing are generated at the point of intersection of the apical and vertical surfaces. If the vertical surfaces were worn as readily as the apical, the enamel would rapidly be reduced in thickness and the sharp cutting edge lost. It is not immediately obvious why there is such a marked difference in the hardness of the two faces. A set of experiments designed to determine the distribution of forces on the apical and shearing surfaces and the relative hardness of these surfaces is required before differential wear can be fully explained.

In the transition from nonmammalian cynodonts to primitive mammals and from these to dryolestids and mammals with tribosphenic molars, both enamel structure and occlusal patterns became more complex. Early mammals developed shearing facets, but lacked well-defined cutting edges. Pseudoprisms probably increased the resistance of enamel to wear, and evolved in parallel with the advent of unilateral occlusion. Dryolestids and mammals with tribosphenic molars developed embrasure shearing and sharp cutting edges to the shearing surfaces. Radial prismatic enamel probably helped to prevent cracks. In addition, because it wore differentially, it made possible the evolution and maintenance of sharp cutting edges despite major attrition of the crown surface during life. The success of molars with these attributes in masticating small invertebrates may account for the preponderance of insectivores with tribosphenic molars during the Cretaceous and Tertiary.

Krebs (1971) has suggested that "thegosis" was the mechanism for maintaining sharp edges to the shearing surfaces of dyrolestid molars. He points out that in these forms, "through mastication of food, the cusps and crests of the teeth become worn and dull. . ... Through thegosis, even sharply bordered facets came into existence along the cutting edges." The term, "thegosis" was proposed by Every (1970) and discussed further in Every and Kühne (1971). They claim that teeth are sharpened by tooth-tooth wear in the absence of food, during which the movement of the jaw is opposite to that of normal mastication. Krebs (1971) claims that in dryolestid molars, the grooves in the cingula running lingually and labially from the talonid and parastyle, respectively, and the sharp external and internal edges to these grooves could not have been formed during the closing stroke. But the concept of thegosis, as applied to pantotheres or tribosphenic molars, underestimates the precision with which the musculature can control jaw movement. Further, it relies on mesiodistal wear of the enamel to form sharp edges. We acknowledge that mesiodistal wear occurs, but its rate is only a fraction of that of vertical wear. We conclude that differential wear during mastication sharpens the cutting edges of molars with embrasure shearing.

Acknowledgments. We owe special thanks to Professor J. Bonaparte for allowing one of us (A.W. C.) to study his unique collection of Argentinian Cretaceous dryolestid molars and for permission to section three molars in order to determine the ultrastructure of the enamel. We are deeply indebted to Ms. Catherine Musinsky for preparing the illustrations and for editing the manuscript, to Ms. R. Pinto, for producing the SEM photomicrographs and to Mr. A. Coleman for printing them. This work was carried out with the aid of a grant from the National Science Foundation (8818098).

References

Archer M, Flannery TF, Ritchie A, Molnar RE (1985) The first Mesozoic mammal from Australia – an Early Cretaceous monotreme. Nature 318: 363–366
Bonaparte JF (1990) New late Cretaceous mammals from the Los Alamitos formation, northern Patagonia. Nat Geogr Res 6: 63–93
Boyde A (1984) Dependence of rate of physical erosion on orientation and density in mineralized tissues. Anat Embryol 170: 57–62
Clemens WA Jr, Lees PM (1971) A review of English early Cretaceous mammals. Suppl 1, Zool J Linn Soc 50: 117–130
Cooper JS, Poole DFG (1973) The dentition and dental tissues of the agamid lizard *Uromastyx.* J Zool Soc Lond 169: 85–100
Crompton AW (1971) The origin of the tribosphenic molar. Zool J Linn Soc Suppl 1, 50: 65–88
Crompton AW (1972) Postcanine occlusion in cynodonts and tritylodontids. Bull Br Mus Nat Hist Geol 21: 27–71
Crompton AW (1974) The dentitions and relationships of the southern African mammals,

Erythrotherium parringtoni and *Megazostrodon rudnerae.* Bull Br Mus Nat Hist Geol 24: 397–437

Crompton AW, Jenkins FA Jr (1968) Molar occlusion in Late Triassic mammals Biol Rev 43: 427–458

Crompton AW, Kielan-Jaworowska Z (1978) Molar structure and occlusion in Cretaceous therian mammals. In: Butler PM Joysey KA (eds) *Development, function and evolution of teeth.* Academic Press, London pp. 249–288

Crompton AW, Luo Z (1993) Relationships of the Liassic mammals, *Sinoconodon, Morganucodon oehleri* and *Dinnetherium.* In: Szalay FS, McKenna MC, Novacek MJ (eds) *Mammal phylogeny.* Springer, Berlin Heidelberg New York

Crompton AW, Yang S, A functional interpretation of enamel ultrastructure in the musk shrew, *Suncus murinus.* (in prep.)

Every RG (1970) Sharpness of teeth in man and other primates Postilla 143: 1–30

Every RG, Kühne WG (1971) Bimodel wear of mammalian teeth. Zool J Linn Soc, Suppl 1, 50: 23–28

Fortelius M (1985) Ungulate cheek teeth: developmental, functional and evolutionary interrelations. Acta Zool Fenn 180: 1–76

Frank RM, Sigogneau-Russell D, Hemmerlé J (1988) Ultrastructural study of Triconodont (Prototheria, Mammalia) teeth from the Rhaeto-Liassic. In: Russell DE, Santoro JP, Sigogneau Russell-D (eds) *Teeth revisited, Proceedings of the VIIth International Symposium on Dental morphology.* Mém Mus Natl Hist Nat, Paris 53: 101–108

Hahn G, Sigogneau-Russell, Godefroit P (1991) New data on *Brachyzostrodon* (Mammalia; Upper Triassic). Geol Palaeontol 25: 237–249

Grine FE, Vrba ES (1980) Prismatic enamel: a preadaption for mammalian diphyodonty? S Afr J Sci 76: 139–141

Grine FE, Gow CE, Kitching JW (1979) Enamel structure in the cynodonts *Pachygenelus* and *Tritylodon.* Proc Elect Microsc Soc S Afr 9: 99–100

Jenkins FA Jr (1990) Monotremes and the biology of Mesozoic mammals. Neth J Zool 40: 5–31

Kielan-Jaworowska Z, Crompton AW, Jenkins FA Jr (1987) The origin of egg-laying mammals. Nature 326: 871–873

Koenigswald Wv (1988) Enamel modification in enlarged front teeth among mammals and the various possible reinforcements of the enamel. In: Russell DE, Santoro JP, Sigogneau-Russell D (eds) *Teeth revisited. Proceedings of the VIIth International Symposium on Dental morphology.* Mém Mus Natl Hist Nat, Paris 53: 148–165

Koenigswald Wv, Clemens WA (1992) Levels of complexity in the microstructure of mammalian enamel and their application in studies of systematics. Scanning Microsc 6: 195–218

Koenigswald Wv, Pfretzschner H-U (1987) Hunter-Schreger-Bander im Zahnschmelz von Säugetieren: Anordnung und Prismenverlauf Zoomorphology 106: 329–338

Krebs B (1971) Evolution of the mandible and lower dentition in dryolestids (Pantotheria, Mammalia). Zool J Linn Soc Suppl 1, 50: 89–102

Lester KS (1989) *Procerbus* enamel: a missing link. Scanning Microsc 3: 639–634

Lester KS, Koenigswald W (1989) Crystallite orientation discontinuites and the evolution of mammalian enamel – or, when is a prism? Scanning Microsc 3: 645–663

Lilligraven JA, Kielan-Jaworowska Z, Clemens WA (eds) (1979) *Mesozoic mammals: the first two-thirds of mammalian history.* University of California Press, Berkeley

Odin GS, Curry D, Gale NH, Kennedy WJ (1982) The Phanerozoic time scale in 1981. In: Odin GS (ed) *Numerical dating in stratigraphy.* Wiley, Chichester, p 957–960

Pfretzschner H-U (1986) Structural reinforcement and crack propagation in enamel In: Russell DE, Santoro JP, Sigogneau-Russell D (eds) *Teeth revisited. Proceedings of the VIIth International Symposium on Dental morphology.* Mém Mus Natl Hist Nat, Paris 53: 133–143

Prothero DR (1981) New Jurassic mammals from Como Bluff, Wyoming, and the interrelationships of non-tribosphenic theria. Bull Am Mus Nat. Hist 167: 281–325

Rensburger JM, Koenigswald Wv (1980) Functional and phylogenetic interpretation of enamel microstructure in rhinoceroses. Palaeobiology 6: 477–495

Sahni A, Lester KS (1988) The nature and significance of enamel tubules in therapsids and mammals. In: Russell DE, Santoro J-P, Sigogneau-Russell D (eds) *Teeth revisited Proceedings of the VIIth International Symposium on Dental morphology.* Mém Mus Natl Hist Nat, Paris 53: 85–89

Sigogneau-Russell D, Frank RM, Hemmerlé J (1985) Enamel and dentine ultrastructure in the Early Jurassic therian *Kuehneotherium* Zool J Linn Soc 82: 207–215

Sigogneau-Russell D, Bonaparte JF, Frank RM, Escribano V (1991) Ultrastructure of dental hard tissues of *Gondwanathrium* and *Kuehneotherium.* Zool J LinnSoc 82: 207–215

Stern DN (1989) Structure, function and development of primitive mammalian enamel. PhD Thesis, Harvard University

Stern DN, Crompton AW, Skobe Z (1989) Enamel ultrastructure and masticatory function in molars of the American opossum. *Didelphis virginiana.* Zool J Linn Soc 95: 331–334

Stern DN, Song MJ, Landis WJ (1992) Tubule formation and elemental detection in developing opossum enamel. Anat Rec 234: 34–48

Wood CB (1992) Comparative studies of enamel and functional morphology in selected mammals with tribosphenic molar teeth: Phylogenetic applications. PhD Thesis, Harvard University

Conclusion: A General Theory for Feeding Mechanics?

P. Dullemeijer

Feeding is a necessary activity of the organism to obtain energy for its survival and, as a species member, for its reproduction. In the preceding chapters a great deal has been described of the astonishing diversity of ways in which organisms cope with the problem of obtaining food. Some authors complain that because of this diversity a general theory for the feeding mechanism is still lacking, either for the group which they are studying or for vertebrates in general, notwithstanding the vast amount of knowledge about the structure and function of the feeding system. Others characterize the situation as the outcome of opportunism, a feature quite common in various processes in living structures. Both opinions warrant further discussion as to whether there is a general principle or theory which will explain the feeding process and its diversity.

The title of this book, *Biomechanics of Feeding in Vertebrates*, implies that the explanation is restricted to the mechanical aspects of the feeding process. Thus, if function is deduced from structure with the aid of the theory of mechanics, we might call this theory the general explaining principle which, of course, does not mean that we know how the system works precisely in each particular case (see chapters on the feeding system in amphibians, reptiles, birds and mammals). This statement is obvious, but still unsatisfactory, although much research has been done over the last few decades, the results of which are nicely summarized in the various contributions. The dissatisfaction arises because we want to know which mechanical rules can and must be applied in particular cases and in the theory of mechanics this aspect is not considered. As yet there is not enough information available in particular groups to fill this gap in our knowledge. For some groups, only very general statements, sometimes only suggestions, have been made; for other groups, much detailed information is available on precise quantitative models, supported by experimental evidence. However, even these models are not always sufficient due to the unavoidable, but necessarily simplifying assumptions regarding the properties of the building materials. The theory of mechanics is appropriate for a general superficial description but insufficient when biological materials like bone, connective tissue and muscle are involved, and if there is a theory, it soon becomes too complicated to apply.

Thus, if the theory of mechanics is to be considered a general principle, much more experimental work and theorization must be done to improve the testing

Zoölogisch Laboratorium, Rijksuniversiteit Leiden, Leiden, The Netherlands

Advances in Comparative and Environmental Physiology, Vol. 18
© Springer-Verlag Berlin Heidelberg 1994

of various operational models and to expand the theory of mechanics so that it can cope with the peculiar dynamic visco-elastic properties of biomaterials. However, even then, the theory of mechanics is not satisfactory because it does not say when and where specific rules must be applied. This biological question goes beyond mechanics; the answer depends on various circumstances such as the given bone structure and the environmental conditions under which this structure is used, and all other functional requirements and constraints for the organism as a whole. Obviously, such a complicated interrelation of factors requires very detailed knowledge not only of the environmental factors involved but also of all the structures assumed to participate in the feeding process. Many of these have been discussed, implicitly mentioned or touched upon in the preceding chapters. For research, however, due to the many multiple inter-actions involved in the feeding process, most authors have restricted their studies to the intake of food into the oral cavity, the processing of the food and the act of swallowing, only a few have focused on the neurocybernetic control of these processes. Thus, emphasis was therefore placed on the jaws and the lingual and pharyngeal region, with only occasional extensions to a broader area in which locomotory systems and whole body mechanisms were taken into account.

Therefore, one should bear in mind that not only the head, but also many other regions of the animal body, cooperate in the feeding mechanism. Besides bone and muscle, other structures in the head region, such as sense organs, nerves, blood vessels and epithelial structures must also cooperate in a normal feeding act, particularly if 'searching for food' is considered the very first step of feeding.

General Design

Returning now to the question of a general theory: if the mechanical theory alone is insufficient, can a general, basic structure and function in all the varieties be discovered? In order to answer this question we have to make an abstract generalization regarding all feeding mechanisms, known in some detail, and presume a simple behaviour of the organism. In the following I shall propose a simple model that consists of various stages of development and that can accommodate subsequently the different ways of feeding. This model was developed by abstract, inductive generalization from the information contained in the foregoing chapters. Despite the great variety in feeding mechanisms, environmental conditions and types of food, the general design of the feeding system is indeed very similar in all vertebrates. Being the anteriormost part of the gut, the mouth is basically a widened tube, the walls of which are supported by bony struts and covered with an epithelial lining. The roof of the tube is generally a stable, rigid structure, being the floor of the ethmoidal, the orbital region and the braincase (in short, 'the skull'). This means that the movable parts are always found in the walls and the bottom of the tube. These can be moved as a whole or as separate, loosely connected parts. It is very likely that we are dealing here with a regidification of the peristaltic movement, as can be deduced from fish respiration and feeding (Anker et al. 1966; Sibbing 1982). As the tube is divided into rigid parts or regions, a great variety of constructions originate by different connections of these parts. Activity is constrained by the ultimate possible stretch, which depends on the size and number of rigid elements and their connecting bands. Number, size and shape of these parts are therefore the major parameters determining the various types of feeding.

In most vertebrates feeding involves an act to approach the food, followed by an act to bring the food into the mouth and then an act of transportation to the gut. Variations on this theme depend primarily on environmental factors and the constraining factors in the construction of the tube walls and, to a lesser degree, on the kind of food (Weijs, Chap. 9, this Vol.). Also, in various groups, other functions of the mouth have an effect on the ultimate form and action of the mouth (Vandewalle et al., Chap. 3, this Vol.; Liem 1989). This very general and rather abstract model is the basic design from which the enormous diversity of feeding mechanisms can be derived when confronted with different environmental demands.

To understand the results of this confrontation, it is convenient to classify the general environmental factors. The greatest difference in these general environmental factors is of course that between aquatic and terrestrial ones (see particularly de Vree and Gans, Chap. 4, this Vol.; Lauder and Reilly, Chap. 6, this Vol.; Bels et al., Chap. 7, this Vol.; Liem 1990) and the division of the latter into terrestrial s.s., underground and aerial factors. The reaction of the organisms to these factors is feeding behaviour, easily divided into the acts of searching, approach, intake, reduction, transportation and swallowing. Not all phases are always present. Combining the occurrence of these phases or acts with the environmental factors produces an immense variety and each combination can involve many ways of feeding, depending on the degree of use of the various structures (see examples in de Vree and Gans, Chap. 4, this Vol.; Bels et al., Chap. 7, this Vol.; Zweers et al., Chap. 8, this Vol.).

Passive Feeding

We can visualize a simple structure, as mentioned above, and the animal waiting for food. This wait-and-sit behaviour is well known from sessile animals, filter-feeders which strain their food from the waterstream and the airstream which are caused by the environment and/or respiration (food and oxygen are obtained through the same mechanism). The sit-and-wait strategy is very useful, especially in running water with abundant food, then the animal expends little energy. Although this strategy is known in many terrestrial animals as well (Bels et al., Chap. 7, this Vol.), the outcome is probably less effective. Thus, one can imagine that, as in many invertebrates, attributes were developed for sensing, sieving and selecting food particles.

The chance of obtaining food can be increased by various mechanical and behavioural improvements.

1. The flow of water can be directed by sucking, for which a pumping mechanism is developed from the above-mentioned active respiration (e.g. in long-headed, eel-like fishes, probably also the fossil fish, *Asthenocormus titanius*; Lambers 1992).
2. Devices can be developed to reduce the chance of losing the potential food items, such as tentacles, sieves, valves and dentitions.
3. Special structures and behaviour patterns can be developed to attract prey animals to the mouth, e.g. the angle in the anglerfish and mimic colouration with the corresponding behaviour. To this category belong also the shape and colour signalling of young animals, e.g. birds, who stimulate the parents to bring food and drop it in their mouths.

A further enhancement is combining the sit-and-wait strategy with snapping, a quick closing of the mouth after a motionless open position, or a simultaneous jump and snap. This strategy is known in many aquatic and terrestrial animals.

In the two groups, further improvement follows a different, almost opposite course. In fishes, e.g., the apparatus for respiration with gills can also be used to strain particles for feeding. However, when the feeding changes into active snapping, considerable effort has to be exerted by a complicated mechanism in which the entire head is lifted and the splanchnocranial system is expanded by downward and posterior movement in relatively dense surroundings. This involves all the depressor and cervical muscles. A special triggering mechanism invokes the quick opening and closure in water, necessitating a strong and sudden force, which could not otherwise be achieved by the available muscles (see Muller 1987).

Generally, in terrestrial animals using the sit-and-wait strategy, the upper jaw has to be released and hardly any muscle activity is needed. In crocodiles, the heavy weight of the upper jaw is released by a sudden relaxation of the cervical muscles and by removing a small piece of cartilage between upper and lower jaw, which serves as a lock to keep the upper jaw elevated (van Drongelen and Dullemeijer 1980).

In short, there are many varieties of the sit-and-wait strategy. Having made the step from the sit-and-wait strategy to jump-and-snap, the possibility to obtain more and bigger food items is increased and, gradually, the way to a more active strategy is opened.

Active Feeding

The chance of successful feeding is considerably increased when the organism moves towards the food. This can be done in the simplest way by moving (especially swimming or flying) with an open mouth in an area where abundant food items are freely available, and then sieving and collecting the items.

This pattern is basic, but it has the idealistic simplicity of paradise, just open your mouth and all nice things drop into it and are yours. Actual practice, however, is different, due to many constraints and demands which ultimately determine what can be realized. Constraints are found in the head structure and, consequently, in the function of the head. An open mouth position for long periods seems to interfere with other functions. Many animals therefore follow another strategy, e.g. opening the mouth close to the prey and sucking it in. However, this is not the main reason to employ suction feeding. This strategy becomes increasingly important when there is a greater food demand, the availability of food is reduced or the chance that food items will escape is greater.

In an aquatic environment, simple sieving and collecting require a strong locomotion apparatus. This is well known from fishes having a salmon-like body shape. A similar principle is found in birds which strain the water with a wide-opened mouth. They generally have a special mechanism (i.e. the spreading of a folded sheet) to widen the mouth bottom. An analogous situation is found among birds collecting insects in the air, although the actual catching is slightly more complicated (nightjars, swallows and swifts; see Zweers et al., Chap. 8, this Vol.).

In order to hunt and catch prey in water by pursuit the mechanism has to overcome the feature that the water with the prey is pushed away from the predator due to the great density of the water. This is done by suction feeding. The hydrodynamics of suction feeding has been analyzed mathematically and experimentally in great detail by Osse et al. (1985) and Drost et al. (1988). This mechanism has to cope with water resistance. It consists of a quick opening of the mouth, which effects a tenfold volume increase in the buccal cavity, and the opening of the opercular slit in the posterior part of the mouth, to let the water flow out past the body. A smooth cylindrical or cone-shaped mouth with a circular opening and gill slits overcomes the water resistance.

Mechanically, this process is most efficient when the pressure difference is minimal, which is indeed the case in suction feeding. The measurement of pressure difference (Lauder and Reilly, Chap. 6, this Vol.) is therefore of little importance in understanding this mechanism (van Leeuwen and Muller 1983). More important than the pressure difference is the relative impulse of the water, caused by the water itself or by the forward movement of the fish.

Also, in suction feeding we can imagine a large variety of types, depending on food particle size, food type (quick moving or sessile organism), kind of water (stagnant or flowing) and the way of approaching the prey (ambush versus pursuit). An anatomical account of these various types in cichlid fishes can be found in Barel (1983, 1985). It can be concluded that body shape, muscle size, and the proportion of the various bony bars are relevant parameters for the feeding type. Moreover, these parameters constrain each other mutually. Each combination of these parameters seems optimally adapted to a *specific* type of feeding (Otten 1982, 1983; Barel 1985, 1987; Muller 1987).

For the next step in the development of the model it is important to consider how sucking can be improved and how suction feeding can be transformed. One example may illustrate the problem. Labroid fishes approach the prey and suck it in through a narrow mouth opening producing a strong current just in front of the fish. These fishes feed on light arthropods which are suddenly pulled to the mouth and immediately caught between the front teeth. They are swallowed after manipulation. Although this is certainly not the main development, it shows nevertheless that the dentition can play a crucial role in improvement by presenting a barrier to escaping prey and, in a more active role, by catching the food. Biting is a reasonable strategy to succeed the sucking method. The more so, because only very slight changes in the proportions of the bars of the construction, and consequently of the muscles, seem to determine whether a fish is a biter or a sucker (Otten 1983; Barel 1985, 1987). A large spectrum of intermediate mechanisms can be expected to occur.

Quite different is the situation in terrestrial animals (including those foraging in the air) (see De Vree and Gans, Chap. 4, this Vol.). They do not have to cope with high densities. Moreover, De Vree and Gans correctly point to a number of non-feeding adaptations which are of utmost importance to understand feeding. The major ones are the change in body support by legs, locomotion and the change in respiration.

The support by legs enables the head region to become relatively independent of the rest of the body (which is exemplified nicely in the smaller size of the equilibrium organ in terrestrial animals compared to that in fishes). This relative independence of the head through the presence of a neck means that the cervical muscles are almost the only muscles of the body engaged in the first feeding act, i.e. opening the mouth, and swallowing, and enable the lower jaw to move downward without lifting the entire skull. Traces of this lifting movement can be found in almost all terrestrial vertebrates (Weijs, Chap. 9, this Vol.).

However, most important is that all these movements occur in low density surroundings with a fixed and stable support for the body. For details, the reader is referred to De Vree and Gans (Chap. 4, this Vol.). In most terrestrial animals respiration is not involved in feeding and hardly interferes with it, whereas in aquatic animals, to a great extent, the same head structures are used in feeding and respiration, and in some cases have a large overlap in function. Respiration and feeding structure have a mutually constraining relationship.

Constraints are always affected by the combination of functional components and are observed when the structural configurations acting in different environments are compared. The difference in the relationship between feeding and respiration affected the divergence of food processing between aquatic and terrestrial animals.

In the former, the entire oral area is mainly involved in catching food and serves rarely or only occasionally in the processing which is carried out by the pharyngeal system (Vandewalle et al., Chap. 3, this Vol.; Sibbing 1982). In the latter, the processing takes place only in the oral region, where differentiation of structures occurs and new ones, such as the tongue, are added, in the meantime leaving the pharyngeal region free for other, new functions. Interesting problems arise in secondary aquatic animals, who cannot return to the respiration mechanism found in fishes.

Given these configurations in the aquatic as well as the terrestrial groups, a further development can be visualized in which the feeding systems in both groups become of the biting type. In fact, in terrestrial animals the biting type is almost the only practical possibility although secondary changes may approximate sucking (some whales and some waterfowl, Zweers et al., Chap. 8, this Vol.). Depending on the kind of food to be bitten, additional structural features and functions are developed, first with respect to quantitative changes and then new structures surrounding or included in the biting machinery (many examples in birds, mammals, reptiles).

So far, the general theory explains a good deal of the diversity. To explain the details of this diversity we have to consider the different types of food which gave rise to many types of biting. This involves diversification of the dentitions, the joints, and the muscular systems (Herring, Chap. 1, this Vol.) to a level where also cells and other equally fine structures are adapted to their specific jobs (Vandewalle et al., Chap. 3, this Vol.). It is important to note that it is the physical properties of food, such as size, weight and hardness, rather than the nutritional value or the ecological origin which determine the feeding method (Weijs, Chap. 9, this Vol.). Many of these variations are mainly effected by differences in the quantity and positions of the elements of the constructions and their related functions. These diversifications of the structures are paralleled by diversification in movements (Weijs, Chap. 9, this Vol.; Zweers et al., Chap. 8, this Vol.; Bels et al., Chap. 7, this Vol.).

General Considerations

The rules of mechanics, the basic design, the influence of the environment and types of biting are generally applicable in the majority of vertebrate constructions and superimposed on this general configuration are the specific adaptations. Thus, there is indeed a general theory, demonstrated as a general theme, but with many variations.

In the course of developing the model we met many variations at *each* major step. This was particularly evident in the basic structure. In its pure theoretical form this structure does not exist, as it is always modulated by finer adaptations. On the other hand, it is typical that each variation of the model and examples from nature show features of the basic design. At each step of differentiation, of combining and of adding structures and functions, a large number of variations, mainly in terms of differences in sequence and measure, are superimposed. Most of these are adaptations to specific kinds of food and capturing strategies. Consequently, this results, on the one hand, in many parallel mechanisms and, on the other, in different solutions for similar functions, depending on how much is represented by the general, basic structure, the various stages of the presented model and the specific adaptations.

This picture emerged from the contributions in this volume, where many examples can be found.

With respect to the development of a new structure many possible new paths can be followed. The best example is probably the tongue which, after it was formed, enabled a wide range of possible new ways of feeding (Roth and Wake 1985; Wake and Larson 1987; Bels et al., Chap. 7, this Vol.) (many examples in reptiles, birds and mammals). With this new way of feeding the organisms could enter new niches. Simultaneously, the number of different acts and phases of the feeding mechanism increased (Zweers et al., Chap. 8, this Vol.).

Again and again we see that animals can and do take advantage of the various possibilities provided by the structures that are present or, in contrast, due to the flexibility of their behaviour, they probably gradually improve the structures themselves. Thus, in a way they can be called opportunists. In short, it seems possible that structure and function show a development through repeated, mutual influence (MID, mutually iterative development) which enables the feeding process to extend and diversify.

Thus far, this picture is idealistic and theoretical. It is a model showing that a general theory exists, but it is rather blurred by the enormous diversity and complexity of individual variations. By definition, a general theory does not explain the details regarding a group of variations in the individual members, but it presents only the general, underlying course. For actual research this is a difficult situation. As practical investigations always start with individual cases, at best in a comparative way, the general features and superimposed individual adaptations must be recognized and distinguished at an early stage in research. Classification, implying an ordering in time, would help (Lauder and Reilly, Chap. 6, this Vol.), but unfortunately this cannot be done before this distinction is made.

Can an idealistic, general theory be transformed into a realistic one? To my knowledge this has never been successful, but in practice we can approximate it by agreeing on a number of criteria.

1. Accepting generalizations from a few cases of real observations.
2. Testing predictive models of activity in actual organisms by experimentation.
3. Accepting, in the first instance, the continuous and the most likely trans- formations unless there is clear evidence for jumps or sudden changes. "Most likely" refers to the most feasible small changes, mechanical and/or biological.

However, the conclusions based on a general theory will never be as good as deduction by testing of a model, due to the fact that it remains a generalization from inductive data and due to the rather imprecise formulations of the general changes in structure and function in the successive steps in evolution.

For practical research we can follow the same strategy used by animals during evolution, namely, we can simulate the MID and test the simulation with examples found in nature or obtained by experimental interference.

This seems to be a very difficult task and even when successfully carried out, it gives only a poor approximation in the sense that it reveals that such a model is possible but not (yet) necessary.

The Cause of Development

The general scheme, or hypothesis, rests on the mechanical theory, which answers the question 'how does it work'. It suggests the direction evolution might have taken. However, questions remain: why did it develop in this way or what caused the development to take this direction? If one answer to this question were possible, we could also speak of a general theory, in the sense of a universal theory. Referring simply to general evolution theory is insufficient when no further specific arguments can be provided to show that in particular steps natural selection had an effect on the feeding system. In the presentation of the model of a probable evolution it was assumed that at some time there was an abundance of available food. In such a case, there is then little need for the organisms to change their strategy or structure, which does not imply that change could not occur. A change could have been caused by the dynamics inherent in the structure itself and/or by intraspecific, natural selection among the variants. When there is insufficient food, then there are two possibilities: many organisms try to find new areas, i.e. new niches to occupy, or competition and followed by strong selection occurs.

As every construction can carry out tasks for which it is not specifically adapted and as many structures are rather plastic, entering new areas is always possible, although probably not optimal for those particular animals. Sometimes the variability is so large that some individuals can enter new areas or niches quite easily. Therefore, it is very likely that in evolution feeding *behaviour* took the lead and that, due to individual plasticity, the structure developed later. There is ample evidence from research on cichlid fishes and other groups with closely related species and subgroups that many changes can occur (Hoogerhoud 1986a, b; Witte 1987; Witte et al. 1990).

Considering that overpopulation induces many individual organisms to move to less favourable surroundings, it becomes clear that by extending the area where the species live, new circumstances are introduced. Once new structures have been genetically[1] established, these in turn make new functions possible.

This iterative process (MID) does not exclude a Darwinian evolutionary mechanism, nor does it exclude other mechanisms. Both alternatives seem to be possible, however, no clues are given as to the cause of the precise direction. To determine this we need to know the exact signal that stimulated the organisms to occupy new areas. General research practice seems to be based on the assumption that this information can be deduced from the already present adaptations.

---

[1] It is beyond the scope of this chapter to discuss genetic establishment.

A better argument may be obtained when we can prove that certain pathways remain closed unless the complete reconstruction of almost the entire feeding apparatus takes place. But even then we shall see that this is not sufficient to determine the precise direction of an evolutionary development.

If transformation models are constructed and those models selected which produce the maximum effect on the basis of the least change, we may find a more solid basis for our theory. When these models are compared to real structure and function, the most likely direction can be approximately determined and impossibilities can be shown (see Zweers et al. Chap. 8, this Vol.; Barel 1985).

The next problem is to determine the polarity of the transformation series. In the majority of cases the only criterion for polarity is that already suggested in taxonomic classifications. The polarity of the transformation series in the feeding mechanism can be given a sounder basis when the necessary inter-relations and interactions of the members of the construction are known and when it can be shown where the limitations of the constructions due to the interactions lie. Studies of salamanders (Roth and Wake 1985; Wake and Larson 1987), birds (Zweers et al., Chap. 8, this Vol.) and fishes (Barel et al. 1977) show that in most series both directions are possible, however, most authors tend toward the development of generalist to specialist, or the common or normal to the rare and extreme. Again, this is a subjective choice made by most authors. Moreover, it does not explain the processes underlying evolution which most certainly are connected to the efforts of the animals to find as much food as possible with a minimum of energy. Those animals that are most successful have the best chance of survival as a species. There is actually very little known about these energy balances and factors which may have played a crucial role in evolution. Optimalization theory has provided a few examples of intraspecific behaviour which points to optimal foraging of fishes (Galis and de Jong 1988; Galis 1990). On the other hand, there are examples where optimal foraging is not necessary, the organisms take what is available. In such a case we can actually speak of opportunism. However, opportunism does not exclude optimal foraging. It is important to consider optimal foraging in relation to the capacity of the structure of the organism and the food supply (cf. Hoogerhoud 1986a, b, 1989).

Clearly, quantitative evidence is needed regarding energy consumption vs energy uptake in the feeding process in various, comparable circumstances of closely related species or forms.

General Remarks

It seems possible to describe the various feeding mechanisms of vertebrates in a general dynamic scheme where at the points of major changes many specific adaptations branch off, giving rise to many parallel and convergent developments.

The direction and polarity could not be accurately determined, nor was it easy to determine a common cause. It is apparently difficult to transform this narrative, idealistic scheme into a realistic, objectivistic one.

Acknowledgments. I sincerely thank Dr. C.D.N. Barel and Dr. J.W.M. Osse for their valuable critical remarks and support and Hanna Schut for typing the manuscript.

References

Anker GC, Simons J, Dullemeijer P (1966) An apparatus for direct X-ray cinematography exemplified by analysis of some respiratory movements in *Gasterosteus aculeatus*. Experientia 23: 74–77

Barel CDN (1983) Towards a constructional morphology of cichlid fishes. Neth J Zool 33: 357–424

Barel CDN (1985) A matter of space. Constructional morphology of cichlid fishes. Thesis, Leiden University

Barel CDN (1987) Constructional morphology and phylogenetic reconstruction of relationship within Cichlidae. In: Hovenkamp P et al. (eds) Systematic and evolution, a matter of diversity. Utrecht University, pp 231–232

Barel CDN, van Oijen MJP, Witte F, Witte-Maas ELM (1977) An introduction to the taxonomy and morphology of the haplochromine Cichlidae from Lake Victoria. Neth J Zool 27: 333–389

Drost MR, Osse JWM, Muller M (1988) Prey capture by fish larvae, waterflow pattern and the effect of escape movements of prey. Neth J Zool 38: 23–45

Galis F (1990) The ecology and morphology of ontogenetic changes in food selection by *Haplochromis piceatus*. In: Huges RN (ed) Behavioural mechanisms in food selection. NATO Advanced Study Institute. Springer, Berlin Heidelberg New York, pp 281–302

Galis F, Barel CDN (1980) Comparative functional morphology of the gills of African lacustrine Cichlidae (Pisces, Teleostei). Neth J Zool 30: 392–430

Galis F, de Jong PW (1988) Optimal foraging and ontogeny: food selection by *Haplochromis piceatus*. Oecologia (Berl) 75: 175–184

Hoogerhoud RJC (1986a) Taxonomic and ecological aspects of morphological plasticity in molluscivorous haplochromines. Ann Mus R Afr Cent Sci Zool 251: 131–134

Hoogerhoud RJC (1986b) Ecological morphology of some cichlid fishes. Thesis, Leiden University

Hoogerhoud RJC (1989) Prey processing and predator morphology in molluscivorous cichlid fishes. In: Splechtna H, Hilgers H (eds) Proc 2nd Int Symp Vertebrate morphology. Fortschr Zool 35: 19–21

Lambers P (1992) On the ichthyofauna of the Solnhofen lithographic limestone (Upper Jurassic, Germany). Thesis, Groningen University, 336 pp

Liem KF (1989) Functional morphology and phylogenetic testing within the framework of symeco-morphosis. Acta Morphol Neerl Scand 27: 119–131

Liem KF (1990) Aquatic versus terrestrial feeding modes: possible impacts in the trophic ecology of vertebrates. Am Zool 30: 209–221

Muller M (1987) Optimization principles applied to the mechanism of neurocranium levation and mouth bottom depression in bony fishes (Halecostomi). J Theor Biol 26: 343–368

Osse JWM, Muller M, van Leeuwen JL (1985) The analysis of suction feeding in fish. Fortschr Zool 30: 217–221

Otten E (1982) The development of a mouth-opening mechanism in a generalized *Haplochromis* species: *H. elegans* Trewavas 1933 (Pisces, Cichlidae). Neth J Zool 32: 31–48

Otten E (1983) The jaw mechanism during growth of a generalized *Haplochromis* species: *H. elegans* Trewavas 1933 (Pisces, Cichlidae). Neth J Zool 33: 55–98

Roth G, Wake DB (1985) Trends in the functional morphology and sensorimotor control of feeding behavior in salamanders; an example of the role of internal dynamics in evolution. Acta Biotheor 34: 175–192

Sibbing FA (1982) Pharyngeal mastication and food transport in the carp (*Cyprinus carpio* L.), a cineradiographic and electromyographic study. J Morphol 172: 223–258

van Drongelen W, Dullemeijer P (1980) The feeding apparatus of *Caiman crocodylus*; a functional morphological study. Anat Anz 151: 337–366

van Leeuwen JL, Muller M (1983) The recording and interpretation of pressures in prey-sucking fish. Neth J Zool 33: 425–474

Wake DB, Larson A (1987) Multidimensional analysis of an evolving lineage. Science 238: 42–47

Witte F (1987) From form to fishery. An ecological and taxonomical contribution to morphology and fishery of Lake Victoria cichlids. Thesis, Leiden University

Witte F, Barel CDN, Hoogerhoud RJC (1990) Phenotypic plasticity of anatomical structures and its ecomorphological significance. Neth J Zool 40: 278–298

Zweers GA (1991) Transformation of avian feeding mechanisms: a deductive approach. Acta Biotheor 39: 15–36

Subject Index

Springer-Verlag
and the Environment

We at Springer-Verlag firmly believe that an international science publisher has a special obligation to the environment, and our corporate policies consistently reflect this conviction.

We also expect our business partners – paper mills, printers, packaging manufacturers, etc. – to commit themselves to using environmentally friendly materials and production processes.

The paper in this book is made from low- or no-chlorine pulp and is acid free, in conformance with international standards for paper permanency.